Volunteered Geographic Information and the Future of Geospatial Data

Cláudio Elízio Calazans Campelo
Federal University of Campina Grande, Brazil

Michela Bertolotto
University College Dublin, Ireland

Padraig Corcoran
Cardiff University, UK

A volume in the Advances in Geospatial Technologies (AGT) Book Series

IGI GLOBAL
DISSEMINATOR OF KNOWLEDGE

www.igi-global.com

Published in the United States of America by
IGI Global
Information Science Reference (an imprint of IGI Global)
701 E. Chocolate Avenue
Hershey PA, USA 17033
Tel: 717-533-8845
Fax: 717-533-8661
E-mail: cust@igi-global.com
Web site: http://www.igi-global.com

Library of Congress Cataloging-in-Publication Data

Names: Campelo, Claudio Elizio Calazans.
Title: Volunteered geographic information and the future of geospatial data / Claudio Elizio Calazans Campelo, Michela Bertolotto, and Padraig Corcoran, Editors.
Description: Hershey PA : Information Science Reference, 2017.
Identifiers: LCCN 2017002931| ISBN 9781522524465 (hardcover) | ISBN 9781522524472 (ebook)
Subjects: LCSH: Geographic information systems. | User-generated content. | Geospatial data.
Classification: LCC G70.212 .V65 2017 | DDC 910.285--dc23 LC record available at https://lccn.loc.gov/2017002931

This book is published in the IGI Global book series Advances in Geospatial Technologies (AGT) (ISSN: 2327-5715; eISSN: 2327-5723)

British Cataloguing in Publication Data
A Cataloguing in Publication record for this book is available from the British Library.

For electronic access to this publication, please contact: eresources@igi-global.com.

Advances in Geospatial Technologies (AGT) Book Series

Nilanjan Dey
Techno India College of Technology, India

ISSN:2327-5715
EISSN:2327-5723

MISSION

The geospatial technology field is a fast-paced, high growth industry that is involved in a variety of fields, including military planning, public health, land use, environmental protection, and Google Earth mapping. With such a diverse body of applications, the research in geospatial technologies is always evolving and new theories, methodologies, tools, and applications are being developed. **Advances in Geospatial Technologies (AGT) Book Series** is a reference source and outlet for research that discusses all aspects of geographic information, including areas such as geomatics, geodesy, GIS, cartography, remote sensing, and other areas. Because geospatial technologies are so pervasive in such a variety of areas, AGT also includes books that address interdisciplinary applications of the technologies.

COVERAGE

- Remote Systems
- Military Security and Geospatial Technologies
- Geovisualization
- Environmental Monitoring
- Geospatial Analysis
- Geomatics
- Public Sector Use of Geospatial Technologies
- Disaster Informatics
- Spatial Reference Systems
- Land Surveying

IGI Global is currently accepting manuscripts for publication within this series. To submit a proposal for a volume in this series, please contact our Acquisition Editors at Acquisitions@igi-global.com or visit: http://www.igi-global.com/publish/.

The Advances in Geospatial Technologies (AGT) Book Series (ISSN 2327-5715) is published by IGI Global, 701 E. Chocolate Avenue, Hershey, PA 17033-1240, USA, www.igi-global.com. This series is composed of titles available for purchase individually; each title is edited to be contextually exclusive from any other title within the series. For pricing and ordering information please visit http://www.igi-global.com/book-series/advances-geospatial-technologies/73686. Postmaster: Send all address changes to above address.

Titles in this Series

For a list of additional titles in this series, please visit: www.igi-global.com

Remote Sensing Techniques and GIS Applications in Earth and Environmental Studies
Abhisek Santra (Haldia Institute of Technology, India) and Shreyashi Santra Mitra (Haldia Institute of Technology, India)
Information Science Reference • copyright 2017 • 335pp • H/C (ISBN: 9781522518143) • US $185.00 (our price)

Handbook of Research on Geographic Information Systems Applications and Advancements
Sami Faiz (LTSIRS Laboratory, University of Tunis El Manar, Tunisia) and Khaoula Mahmoudi (LTSIRS Laboratory, University of Tunis El Manar, Tunisia)
Information Science Reference • copyright 2017 • 650pp • H/C (ISBN: 9781522509370) • US $315.00 (our price)

Geo-Intelligence and Visualization through Big Data Trends
Burçin Bozkaya (Sabanci University School of Management, Turkey) and Vivek Kumar Singh (Rutgers, The State University of New Jersey, USA & Massachusetts Institute of Technology, USA)
Information Science Reference • copyright 2015 • 348pp • H/C (ISBN: 9781466684652) • US $225.00 (our price)

Handbook of Research on Emerging Digital Tools for Architectural Surveying, Modeling, and Representation
Stefano Brusaporci (University of L'Aquila, Italy)
Engineering Science Reference • copyright 2015 • 829pp • H/C (ISBN: 9781466683792) • US $525.00 (our price)

Geographic Information Analysis for Sustainable Development and Economic Planning New Technologies
Giuseppe Borruso (University of Trieste, Italy) Stefania Bertazzon (University of Calgary, Canada) Andrea Favretto (University of Trieste, Italy) Beniamino Murgante (University of Basilicata, Italy) and Carmelo Maria Torre (Polytechnic of Bari, Italy)
Information Science Reference • copyright 2013 • 434pp • H/C (ISBN: 9781466619241) • US $180.00 (our price)

Qualitative Spatio-Temporal Representation and Reasoning Trends and Future Directions
Shyamanta M. Hazarika (Tezpur University, India)
Information Science Reference • copyright 2012 • 381pp • H/C (ISBN: 9781616928681) • US $195.00 (our price)

Universal Ontology of Geographic Space Semantic Enrichment for Spatial Data
Tomaž Podobnikar (Scientific Research Centre of the Slovenian Academy of Sciences and Arts, Slovenia & University of Ljubljana, Slovenia) and Marjan Čeh (University of Ljubljana, Slovenia)
Information Science Reference • copyright 2012 • 304pp • H/C (ISBN: 9781466603271) • US $180.00 (our price)

Geospatial Web Services Advances in Information Interoperability
Peisheng Zhao (George Mason University, USA) and Liping Di (George Mason University, USA)
Information Science Reference • copyright 2011 • 552pp • H/C (ISBN: 9781609601928) • US $180.00 (our price)

Editorial Advisory Board

Table of Contents

Detailed Table of Contents

Nowadays an ever-increasing number of applications require complete and up-to-date spatial data, in particular maps. However, mapping is an expensive process and the vastness and dynamics of our world usually render centralized and authoritative maps outdated and incomplete. In this context crowd-sourced maps have the potential to provide a complete, up-to-date, and free representation of our world. However, the proliferation of such maps largely remains limited due to concerns about their data quality. While most of the current data quality assessment mechanisms for such maps require referencing to authoritative maps, we argue that such referencing of a crowd-sourced spatial database is ineffective. Instead we focus on the use of machine learning techniques that we believe have the potential to not only allow the assessment but also to recommend the improvement of the quality of crowd-sourced maps without referencing to external databases. This chapter gives an overview of these approaches.

The advent of user-generated content, crowdsourcing and other forms of lay data generation have led to opposing arguments about the quality and reliability of data in the geoweb,. The main focus of this chapter is an 'experiment' to test the quality, validity and lay monitoring of volunteered geographic information (VGI) data. Given the growing importance of VGI, in particular its very different sources and potential uses, it is important that we also consider how this movement affects the ways in which we re-envision the pedagogy of geographic education. Accordingly, a sub-theme of this paper focuses on the manner in which the VGI experiment is undertaken: the experiment is run with students as a means of complementing their otherwise technical GIS training with primary research that exposes them to the wider social issues and debates relating to geographic data. We discuss the implications of this research project both for observers of the development of VGI and the pedagogy of GIS teaching and learning.

Volunteered Geographic Information is the term used to describe the process of collecting spatial data using a network of volunteers. The approach collects spatial data to build maps which are often freely accessible. The maps and the underlying data can be used by the public, companies and government agencies for a variety of tasks such as route finding. Given that untrained volunteers may collect the spatial data, questions regarding the quality of VGI have been raised. Several studies have emerged to assess the quality (positional, semantic and thematic accuracy) of VGI by comparing the data to ground truth. This approach fails to capture the quality of VGI for domain specific tasks. In this chapter we examine the quality of VGI for an educational planning task in Islamabad, Pakistan, and show that while the data may be suitable for route finding tasks, they are insufficient for educational planning alone.

Crowdsroucing of emotional information can take many forms, from social networks data mining to large-scale surveys. The author presents the case-study of emotional mapping in Ostrava´s district Ostrava-Poruba, Czech Republic. Together with the local administration, the author crowdsourced the emotional perceptions of the location from almost 400 citizens, who created 4,051 spatial features. Additional to the spatial data there were 1,244 comments and suggestions for improvements in the district. Furthermore, the author is looking for patterns and hot-spots within the city and if there are any relevant linkages between certain emotions and spatial locations within the city.

OpenStreetMap (OSM) is a bottom up community-driven initiative to create a global map of the world. Yet the application of OSM to land use and land cover (LULC) mapping is still largely unexploited due to problems with inconsistencies in the data and harmonization of LULC nomenclatures with OSM. This chapter outlines an automated methodology for creating LULC maps using the nomenclature of two European LULC products: the Urban Atlas (UA) and CORINE Land Cover (CLC). The method is applied to two regions in London and Paris. The results show that LULC maps with a level of detail similar to UA can be obtained for the urban regions, but that OSM has limitations for conversion into the more detailed non-urban classes of the CLC nomenclature. Future work will concentrate on developing additional rules to improve the accuracy of the transformation and building an online system for processing the data.

A forest patterns map over a large extent at high spatial resolution is a heavily computation task but is critical to most regions. There are two major difficulties in generating the classification maps at regional scale: large training points sets and expensive computation cost in classifier modelling. As one of the most well-known Volunteered Geographic Information (VGI) initiatives, OpenstreetMap contributes

Foreword

Volunteered Geographic Information (VGI) is spatial data that has been contributed or collected for free by volunteers. These data sources are provided by communities of volunteers, rather than for-profit businesses or government entities. After working as a researcher in VGI for well over a decade it gives me great satisfaction to know that today VGI hardly needs any additional introduction within a book such as this. VGI has been a positive disruptive technology with some authors calling it a 'surprising about turn for GIScience'. It can be fairly argued that VGI has become a research domain in its own right spanning Geography, Computer Science and Information Technologies. VGI has placed citizens at the centre of the production of vast volumes of dynamic, rich and accurate datasets of spatial data. The users and consumers of spatial data of yesterday are now the producers and custodians of spatial data today. New opportunities for research have arisen which would not have been thought possible at the dawn of VGI.

Ardent proponents of VGI indicate that this source of geospatial data and information provides us with a treasure trove of undiscovered knowledge and new data. However this enthusiasm must be tempered and evaluated. Every VGI dataset must be evaluated in terms of how it was collected and the potential use-cases and purposes to which it will be applied. It is important to remember that there are multiple types of VGI. The OpenStreetMap project is certainly dominant in the VGI landscape but it is foolish to consider that it is the only source.

This book carefully evaluates the current use and applications of VGI and its influence on the future of geospatial data. In performing this task it is necessary to consider and evaluate VGI from a range of different viewpoints. One of the remarkable aspects of VGI comes when we integrate it with other forms and sources of spatial data. These include classical geospatial data in the form of Land Cover Maps and Remote Sensing imagery and newer sources representing survey responses of personal emotional perceptions. Reliability of data, quality assessment and evaluation of VGI for a wide spectrum of use cases are crucial towards its increased adoption. Broad sweeping quality assessments are not suitable for VGI. Each dataset or data source will have its own nuances and characteristics which influence and affect its quality assessment.

Unlike some traditional sources of spatial data, VGI is very often collected for one purpose or reason but is potentially usable for many more scenarios. The flexibility of VGI provides opportunities for navigational accessibility, gathering road safety information, creation of models for indoor navigation, development of quality of urban life indicators, and disaster management and resilience. It is easy to see that these types of opportunities are tightly coupled with issues of importance to people and society. The chapters within this book emphasise that VGI cannot be considered as just easily accessible citizen-generated data without consideration of the societal implications in its collection and use. Effective participation from volunteer groups is a complicated and resource-intensive process. The process requires trust building to initiate and sustain contributions. Without these efforts the future of VGI within the landscape of spatial data would be in jeopardy.

The progress and results outlined in this book reaffirm my conviction, and that of many other scholars and practitioners in the domain, that VGI has a major role to play in charting the future of Geospatial Data. This book provides a valuable reference point to measure the progress of research in this domain while also delivering a compelling overview of the state-of-the-art of knowledge which it has inspired. It should be rightfully considered an important source of history, knowledge and inspiration to students, researchers and professionals in the geospatial data industry. As a scholarly work, it makes an important contribution to this exciting and growing area of geographic information and knowledge.

Peter Mooney
Maynooth University, Ireland

Preface

In recent years, an increasing number of volunteers have engaged in the creation of geographic information, a role that for a long time had been predominantly reserved to official agencies and to private companies. This phenomenon has given grounds to the hope that postmodern Geographic Information Systems (GIS) could rely on valuable alternative sources of geographic data.

The term Volunteered Geographic Information (VGI) may refer to maps created through the efforts of volunteers who submit geographic data, such as in the case of the OpenStreetMap project; however, it also includes other forms of geospatial data published on the Web, such as the geotagged photographs of Flickr and geotagged entries in Wikipedia. VGI can be used for a wide range of applications and may have several advantages over official geographic data. These include lower costs to generate and use data and a potentially larger set of up-to-date samples. On the other hand, there are many aspects of VGI that are still challenging for researchers, such as the definition of appropriate methods of assessing accuracy and quality, synthesizing, verifying, integrating, and distributing such information.

The trend in crowdsourcing is not exclusive to the spatial domain. This phenomenon has permeated all different application areas and has contributed to the exacerbation of the problem of information overload. Big data is ubiquitous and has sparked a large volume of research trying to develop new methods to analyse and make sense of the data that is available; find the most relevant and reliable information; exploit aspects of data we did not even think of a decade ago. Machine learning has been identified as a field in which developed techniques have much potential in helping analysing big data and identifying interesting patterns in it. These techniques are now starting to be applied also in the context of VGI data analysis.

While with the advent of crowdsourcing new opportunities have arisen, including exploiting knowledge and expertise of people (for example related to their local environment), and reducing production costs, several other issues are aggravated by the nature of crowdsourced data that were minimised by the level of control applied over authoritative data. So, for example, while Wikipedia has become a very popular source of information, it has been shown to not always be completely accurate. VGI data suffers from the same quality and reliability limitations. However, the advantages it presents in terms of usefulness for critical applications including emergency response planning, improvement to community resilience, gathering of up-to-date road safety critical information and improving accessibility (to name but a few), have shown to outweigh the drawbacks of using volunteered data over authoritative data. Other advantages include the free availability of this data which makes it accessible to all. For example, the use of authoritative spatial data (particularly high resolution maps) for educational and research purposes had in the past been often impeded by its cost and copyright issues. Presenting and divulgating VGI projects to the general public will help the use of this data, for example, to teach GIS in schools and improve geography literacy.

One of the other big revolutions of the last decade relates to the mass adoption of social networks. Given the vast amount of data they constantly record, many people have become interested in analysing this data and discovering ways of exploiting it. An important aspect for this analysis is the ability to know where the data originates and any other spatial information that can be linked to it. Georeferencing social network text is one of the most studied problems in this context and many are looking at exploiting VGI repositories for this.

Notwithstanding the innumerable uses and applications of VGI data, there are still a number of open questions related to its collection, storage and diffusion which need to be further investigated and discussed, including:

- How such an assertive method of collecting geospatial information can be complementary to authoritative methods employed by government agencies and private industry;
- How to deal with VGI preservation, that is, how volunteers can be kept motivated to contribute;
- What are the limitations imposed on VGI by different communication technologies;
- How to assess accuracy and quality of VGI data;
- What are the issues linked with reputation, trust and confidentiality in VGI;
- How to address reliability, accessibility, and sustainability of VGI;
- How can VGI be framed within the larger domain of Big Data;
- What are the practical applications of VGI in research and operation;
- What are the most promising future uses of VGI.

This book set out to examine many of these issues within the topic. All of the above mentioned issues show how VGI is at the centre not only of problems arising in the spatial domain but also in many other areas of research and real practical applications, therefore demonstrating how this book should be of interest to a wide audience.

It is hoped that the examination of the issues presented in this book will highlight open research questions on the topic of VGI and in turn provide a platform for future work both by existing and new researchers in this space.

AIM AND SCOPE

Volunteered Geographic Information and the Future of Geospatial Data is a strong knowledge resource for people interested in learning more about this new paradigm for dealing with geographic data. The idea was to develop a book that was rich on concepts, as well as with timely and up-to-date application examples and case studies.

This book aims to present fundamentals and applications of VGI, its recent developments and future trends. The book intends to help readers reason about the question of whether it is reasonable to count on volunteered efforts to develop high quality geographic data sources in the future, or is VGI a short term phenomenon that will fade away.

VGI is a growing area of information gathering and has attracted the attention of many researchers worldwide. However, the topic has been primarily addressed in research papers in journals or in conference proceedings. The objective of this book is to provide a publication covering a concise yet useful overview of fundamentals, applications and current challenges of VGI, to act as reference material for professionals, researchers and students of several related disciplines. VGI has emerged as a promising resource which can revolutionise GIS, in which individuals are able to assert their own views of their surroundings and play a part in local decision making. In particular VGI can be useful in the context of participatory GIS, citizen government and citizen science.

Given the wide breadth of topics covered, we feel this book can act as an introduction to the topic of VGI for researchers interested in working in this space by providing them with an overview of the research landscape. We hope that this book will prove interesting, useful, and illustrative of these new trends in acquiring, processing, managing and analyzing geographic data, and that it will spark reflections in and provoke questions from readers, therefore contributing to the understanding and progress of this promising field.

TARGET AUDIENCE

We aimed to provide a book that might be read by VGI expert and non-expert readers, by undergraduate and graduate students in related fields, and by practicing managers of sectors that could benefit from the use of VGI-based applications.

The book is intended for a wide audience. It is targeted to professionals and researchers working with Volunteered Geographic Information in various GIScience-related disciplines, such as spatial computing, geography, cartography, information sciences, to name but a few. The target audience is people who want to gain expertise in building, managing and using Geographic Information Systems based on crowdsourced data. The target audience includes students majoring in GIScience related fields who want to begin working with VGI systems.

Other potential readers include those new to VGI, who may not have background knowledge in GIScience. This group comprises researches that are familiar with other crowdsourcing applications and who may be interested in gaining foundational knowledge and skills in VGI. It also includes database researches and data analysts who are seeking different application areas for their investigations. The book is also targeted to managers interested in understanding the challenges and benefits of considering VGI approaches to build their decision support systems.

Our perspective in this book is not only scientific, but also technical and managerial. In writing, selecting and reviewing the chapters, our overriding concern has been to help readers gain knowledge and skills that they can apply to design, develop and manage modern geographic information systems and related technologies.

After completing *Volunteered Geographic Information and the Future of Geospatial Data*, readers should:

- Understand fundamental concepts associated with VGI;
- Be familiar with examples and case studies on VGI;
- Have a broader view of VGI technology resources;
- Know more about the main issues and challenges for dealing with crowdsourced data;
- Have a more sophisticated understanding of how VGI can help companies and governments meet their objectives and improve decision-making;
- Be able to become better developers of end-user VGI applications;
- Be more aware of the impact of VGI on developing solutions.

ORGANISATION OF THE BOOK

Volunteered Geographic Information and the Future of Geospatial Data comprises 15 chapters.

Chapter 1, titled "Engaging with the Participatory Geoweb: Experiential Learning from Practice", examines the issues that arise for engaging people with the participatory processes of VGI contributions. By comparing and contrasting three case studies, the chapter discusses individual and organizational dynamics of engagement, how participants are contributing their information, what their motivations and incentives are, and in what ways users interact with available technologies.

Chapter 2, titled "Quality Evaluation of Volunteered Geographic Information: The Case of OpenStreetMap" presents a systematic review on the quality evaluation process of OpenStreetMap (OSM), a popular VGI platform that allows users to create or edit maps. The chapter describes a case study of London (Canada) for the assessment of completeness, positional accuracy and attribute accuracy.

Still in the field of VGI quality, Chapter 3 gives an overview of the approaches to the use of machine learning techniques to assess the quality of crowdsourced maps. These approaches are in contrast to traditional mechanisms for assessing quality, which require referencing to authoritative maps. This chapter is titled "Traditional versus Machine Learning Techniques for OSM Quality Assessment".

An experiment to test the quality, validity and lay monitoring of VGI is presented in Chapter 4, titled "VGI in the Geoweb: An Experiment to Test Data Reliability". It discusses the implications of a research project on data reliability both for observers of the development of VGI and for the pedagogy of GIS teaching and learning.

Chapter 5, titled "Quality Assessment of Volunteered Geographic Information for Educational Planning", also examines the quality of OSM data; however, it focuses on the data quality for tasks related to educational planning. The city of Islamabad (Pakistan) is adopted as a case study. In this chapter, the classic approach to assessing data quality is taken, in which spatial, semantic and thematic data are compared with authoritative data. However, the approach described is in contrast to other studies assessing the correctness of VGI, which are typically domain independent and focus on assessing the overall quality of OSM without drawing on specific spatial analysis tasks.

In Chapter 6, titled "Emotional and Subjective Volunteered Geographical Information", the problem of emotional mapping is addressed. The chapter presents the methodology and deployment of the web-based application for collecting VGI emotional data. It describes a case-study in Poruba, a district of Ostrava in the Czech Republic. Through the crowdsourcing tool, the emotional perceptions of the location were crowdsourced from almost 400 citizens who created 4,051 spatial features.

Chapter 7 outlines an automated methodology for creating land use and land cover (LULC) maps using the OpenStreetMap platform, a bottom up community-driven initiative to create a global map of the world. The application of OSM to LULC mapping is still largely unexploited due to problems with inconsistencies in the data and harmonization of LULC nomenclatures with OSM. The method described is applied to two regions in London and Paris and uses the nomenclature of two European LULC products: the Urban Atlas (UA) and CORINE Land Cover (CLC). The chapter is titled "Using OpenStreetMap to Create Land Use and Land Cover Maps: Development of an Application".

Chapter 8, titled "Mapping Regional Landscape by Using OpenstreetMap (OSM): A Case Study to Understand Forest Patterns in Maya Zone, Mexico", also discusses the potential of OpenStreetMap to produce land use and land cover maps. The aim is to generate forest cover maps and quantifying road-caused forest fragmentations by using OpenstreetMap in conjunction with remote sensing datasets from

Google Earth Engine. Generating classification maps at regional scale is a difficult task, due to the large training point sets and expensive computation cost for modelling classifiers.

Chapter 9 reviews existing research on the use of VGI as an information source for improving the resilience of cities and communities in the face of natural hazards and extreme weather events. It offers an interdisciplinary perspective of the challenges to be overcome and presents a Geospatial Open collaboRative Architecture, named AGORA, designed to build resilience against disasters and extreme events. The chapter also describes some lessons learned in the context of flood risk management and sets out an interdisciplinary research agenda for leveraging VGI to improve disaster resilience. The chapter is titled "Leveraging Volunteered Geographic Information to Improve Disaster Resilience: Lessons learned From AGORA and Future Research Directions".

Chapter 10, titled "Quality of Urban Life Index From Location-Based Social Networks Data: A Case Study in Belo Horizonte, Brazil" describes how data obtained from contributions of users of Location-Based Social Networks (LBSN) can be used to calculate urban indicators, and hence, how they could be used as an alternative data source for estimating quality of urban life to support urban planning policies.

Chapters 11 to 14 present applications of VGI in different domains. Chapter 11 discusses how VGI can help mitigate accessibility problems and reviews existing research and applications in the field. One of the greatest challenges for physically disabled and visually impaired people is to find paths and areas adapted to their limitations while performing their daily activities, since not all environments they explore have these characteristics. However, traditional mapping technologies have appeared to be inefficient in keeping complete and up-to-date information for this domain. The chapter is titled "Improving Accessibility Through VGI and Crowdsourcing".

Chapter 12, titled "Gathering Road Safety Critical Information from Users", presents an application of VGI to road safety research. It proposes an open platform to report and access information on traffic accidents. Chapter 13 describes an approach to the "Generation of a Data Model for Indoor Navigation Based on Volunteered Geospatial Information (VGI)". Chapter 14 presents a method for the "Adaptive Acquisition of VGI to Fill Out Gaps in Biological Observation Metadata".

The last chapter, titled "Geospatial Digital Rights Management: Challenge to Global Spatial Data Infrastructure", examines a topic that should be of interest to anyone who deals with geospatial data. It discusses the need for general Digital Rights Management (DRM) and its spatial counterpart, Geospatial DRM (GeoDRM). A major motivation for the development of GeoDRM initiatives is the need to manage obstacles to data sharing that are related to data ownership issues in spatial data infrastructure scenarios.

Each chapter also include key terms and definitions to ensure clarity and consistency, improving the overall readability and supporting student and non-expert learning.

FUTURE RESEARCH DIRECTIONS

This book provides an overview of the breadth of applications of VGI data. It also highlights many of the challenges it presents. In particular, although several chapters in the book relate to accuracy and quality of VGI data, this issue will continue to be a topic of debate as automatically controlling it remains an open problem. This is also linked to the issue of curating VGI data, given the voluntary nature of the paradigm. Other research issues that are highlighted in the book include maintaining volunteers' engagement, as well as integration with different datasets and international standards.

In summary, many of the aspects that require further study in the future relate to:

- Geospatial data and interoperability standards;
- Integration of VGI in SDI (Spatial Data Infrastructure);
- Integrating VGI data from multiples sources;
- Integrating VGI and Linked Geo Data;
- Integrating VGI and social network data;
- How to keep the contributors motivated;
- Application of VGI in GIS education;
- Exploring semantics in VGI for improved GIR (Geographic Information Retrieval);
- Reputation, trust and confidentiality issues in VGI;
- Uncertainty and vagueness in VGI.

Finally, very recently several new trends and opportunities for exciting research into the potential of crowdsourced spatial data have been identified that will allow extracting additional interesting information from this rich user-generated data. These include, among many others, studies in sentiment analysis whereby understanding users' emotions could reveal much about the subjectivity of the contributed data. We hope this book will encourage many researchers in undertaking some of these exciting new directions.

Claudio Campelo
Federal University of Campina Grande, Brazil

Michela Bertolotto
University College Dublin, Ireland

Padraig Corcoran
Cardiff University, UK

Acknowledgment

No book of this nature can be produced without the help of others. The editors would like to thank all the people involved in this project. We would like to thank the members of the Editorial Advisory Board for their suggestions and assistance at several stages in the preparation of the book; the support team of IGI Global who assisted us in the publication development; the authors who contributed their valuable time, knowledge and experiences to this book; and the reviewers who shared their expertise with us by critically reviewing the chapters of this book.

Chapter 1
Engaging With the Participatory Geoweb:
Experiential Learning From Practice

Jon M. Corbett
University of British Columbia, Canada

Logan Cochrane
University of British Columbia, Canada

ABSTRACT

Maps were historically used as tools of the elite to maintain and expand power and control. The development of participatory mapmaking and the geoweb have opened new avenues for broader citizen engagement and therefore challenge traditional power dynamics. This chapter analyzes three examples and presents experiential learning around participatory processes and VGI contributions. Specifically we explore who is contributing their information, what are their motivations and incentives, in what ways do users interact with available technologies, and how is this contributing to change? We conclude by discussing the roles of motivations, the type of contribution, organizational capacity and leadership, and objectives. In comparing and contrasting these case studies we examine the individual and organizational dynamics of engagement, and how this can better inform the discourse about VGI.

INTRODUCTION

Over time, maps have been used by society's powerful as tools of control, as means for securing land and resources, and as mechanisms to establish and assert control. In recent decades, participatory mapmaking, and more recently the emergence of the participatory geospatial web (geoweb), has swept "like a pandemic" and acted to purposefully challenge these power dynamics (Chambers, 2006, p. 1). It has been claimed that participatory processes, wherein information is contributed by the 'crowd', can be empowering for those often excluded from the agency of mapmaking (Cochrane, Corbett & Keller, 2014). In the contemporary context of the geoweb, this citizen contribution of spatial information is frequently framed as volunteered geographic information (VGI) (Goodchild, 2007b). As new web-based mapping

DOI: 10.4018/978-1-5225-2446-5.ch001

technologies and mapmaking platforms emerge, there is a need to better understand VGI contributions: who is contributing their information using the participatory geoweb, what are their motivations and incentives, in what ways do users interact with available technologies, and how is this contributing to change? In this chapter we draw upon applied research to approach answering these questions. We analyze the complex, contested and diverse nature of VGI, drawing upon three examples that use the participatory geoweb tool Geolive. In comparing and contrasting these examples we examine the individual and organizational dynamics of engagement, and how this can better inform the discourse about VGI.

BACKGROUND: CONTEXTUALIZING VGI

The contributions made by individuals in participatory mapping projects occur within a broader trend wherein citizens around the world want to voice their opinions and have their concerns acted upon (e.g. McGee, Edwards, Minkley, Pegus & Brock, 2015; World Bank, 2014; 2016). In some instances, these are activities undertaken alongside government initiatives, such as community members informing development planning and natural resource management (Brown, 2006; Brown & Reed, 2009; 2012; Brown & Weber, 2013a; 2013b). However, the needs and priorities of community members who provide VGI and government officials receiving this input may not align (Brandeis & Nyerges, 2015), resulting in a lack of interest and limited impact (Brown, 2012). At the other end of the spectrum, VGI can be a part of the process that challenges and contests government, including re-mapping resources to challenge ownership (Peluso, 1995) and territorial control (Quiquivix, 2014). The choices embedded within maps have multiple impacts creating layers of complexity and dilemmas related to whose worldviews and priorities are conveyed in the process (Hodgson & Schroeder, 2002; Wainwright & Bryan, 2009). It is within this challenging change space that VGI is best engaged with. Participatory mapping and VGI offer opportunities for enhanced citizen engagement as well as a means to influence (or compel) transparency and accountability of decisions that affect the public. At the same time, the limitations and the potential for using these tools to co-opt public opinion or be mechanisms of exclusion must also be considered.

In the context of participatory mapping, VGI is closely associated with crowdsourcing whereby individuals contribute geo-located information as a means to share and engage with geographic issues, processes and problem solving (Brabham, 2008; Estelles-Arolas & Gonzalen-Ladron-de-Guevara, 2012; Howe, 2006). Generally, the act of contributing is voluntary and the viewing of data is open, however in certain instances it is restricted. For many, the 'crowd' is imagined as being representative of citizens for whom the issue at hand is relevant; broad, diverse and "radically distributed beyond the boundaries of professionalism" (Brabham, 2008, p. 75). Platforms, such as Geolive, provide an online space for collective place-based knowledge, experience and wisdom to be captured, shared, exchanged, contested and negotiated. As a group, it is argued, the collective wisdom is greater than the sum of its individual members (Surowiecki, 2004). Yet, the data points and products of participatory mapping projects are not a reflection of all members of society; barriers of accessibility, identity, language, status, skill, political affiliation, location, age and user priorities may alter the composition of the 'crowd' and thus the processes as well as the products (Beischer, Cochrane, Corbett, Evans, Gill & Millard, 2015). While crowdsourcing offers unique opportunities for enhanced citizen engagement, it can also reflect existing inequalities and be a means through which prevailing conflicts may be re-created. It is not possible to make broad generalizations about when crowdsourcing is representative and when it is not; rather, studies of its use can provide insight into the diverse manifestations of the 'crowd' and how analyses of the (lack of) representativeness can enable critical reflections on processes and results.

The contribution of an individual's VGI are intentional and voluntary. Goodchild argues that VGI is "one part of a fundamental transition as society redefines its vision of the role of public information" (2008, p. 239). Emerging from the new forms of participatory, crowdsourced mapping are examples of its power, such as its deployment in response to emergencies, and in improving existing and imperfect field data (Componovo & Freundschuh, 2014; Harvey, 2012; Shekhar, Yang, Gunturi, Manikonda, Olive, Zhou & Lu, 2012). Yet, there are challenges of credibility, accuracy, representativeness, motivations and biases (Flanagin & Metzger, 2008; Goodchild, 2007b; Goodchild & Li, 2012). VGI is not neutral or apolitical (Elwood, 2008; Quiquivix, 2014). Further complicating analyses of these process is that much of the available literature insufficiently outlines the methodologies employed (Sui & DeLyser, 2012) and the impacts that result (Corbett, Cochrane & Gill, 2016). Much more research is needed to bridge the theoretical analyses and critiques with practice-based experiences.

METHODS

An analysis of participatory mapping projects enables the theoretical questions and literature trends to be assessed in specific and concrete ways. In order to systematize the analysis, we have purposefully selected three Geolive projects for comparing and contrasting the manifestations and impacts of VGI in different scenarios. Since Geolive began in 2008 it has been used in approximately forty unique projects. The examples chosen for this chapter represent diverse forms of VGI, requiring different levels of approval and/or user ownership of data and its entry, as well as varying degrees of participant engagement. Since each of the projects started at different times, we analyzed the first year after the project launch, which includes a range of user, visitor and input data. The examples are contextualized in terms of the motivation behind the desire to contribute, the types of contributions, the capacity and leadership offered by the delivering organization, and how these factors align with the overall project objectives. We use these factors to compare and contrast the organizational elements that influence VGI contribution, use and impact.

We recognize that these three examples are not sufficiently broad to represent the entire nature of VGI, furthermore they draw from examples that use the Geolive platform, which uses an invitation-only mechanism to solicit and manage user activity and contributions. In a number of large spatial crowdsourcing projects (such as OpenStreetMap) the contributing volunteers are imagined as broadly as possible. They do not need to be associated with the area being mapped, nor does their involvement in the project require invitation. However, these larger projects are harder to define as examples of participatory mapping, while the uses of Geolive explored in this chapter do serve to provide some experiential insight into the relationship between participatory mapping and VGI.

In each of the three projects outlined below, collaborators have utilized Geolive, a web-based participatory mapping platform. Similar to other web-based spatial crowdsourcing tools such as Maptionnaire[1] and ArcGIS online[2], Geolive allows registered users to create and share their own spatially located information and experiences using a dynamic internet-based map interface. Using Javascript and PHP programming languages, the application is built on top of the Google Maps API and linked to a MySQL database. Information is tied to a specific point and can include embedded text, time, image, tags and video data. Geolive was created in the Spatial Information for Community Engagement (SpICE) lab at the University of British Columbia Okanagan. The platform was initially designed and developed to support the active participation of users in the contribution of location-based content. It has been programmed to be user-friendly to non-technical users (for example data is added to the map using a very structured set of form fillable steps), visually engaging and robust to use. It therefore closely aligns with the intent of participatory mapping practice described in the introduction section. Geolive supports user contributed markers; spatial discussions; interactive layers; variable access control (varying levels for different users); search functions; tags and data filtering; a timeline function to understand how the spread and location of markers change over time; and social media integration.

We have selected three project examples for comparative analyses, these include: (1) British Columbia Marine Trails Network Association, a closed system where elite members crowdsource data about camping and kayaking opportunities, and make that information available to other network members; (2) Inclusive Employment Mapping Tool, an open system about experiences of positive and comprehensive employment for those with intellectual disabilities; and (3) Okanagan Refugee Community Resources, a restricted system connecting community members in the Okanagan with the skills and material possessions needed by newly arriving refugee families. All participants were directly involved with the development and deployment of their specific projects. These examples do not represent the full scope of the Geolive application, but they do provide a means through which some of the complexities of VGI can begin to be analyzed and understood. In addition, they cover a range of types of mapping and VGI experiences, having open, closed and restricted systems.

The examples articulated below draw on: (1) a descriptive narrative based on the experiences of the authors; (2) interview results; and (3) quantitative data analysis. The authors have been intrinsically involved with the design, development, deployment and ongoing maintenance of all three projects and are therefore well positioned to comment on the intent and outcomes of each of the projects. Interviews were conducted with at least two participants from the partnering organizations, further material was derived from email correspondence and material shared through the map. The quantitative data was drawn from database SQL queries. Screenshots of the three examples are presented below, along with descriptive data on usage. We had intended to graph data usage overtime for comparative purposes, however inconsistent data availability did not allow for a representative presentation in a graphed format.

PROJECT EXAMPLES

British Columbia Marine Trails: A Closed System Where Executive Members Contribute Data

The British Columbia Marine Trails Network Association (BCMTNA) is a British Columbia (BC), Canada, based registered non-profit society. The principal aim of the network is to create, in collaboration with governmental and other stakeholders, a marine network of campsites and access points along the coastline of British Columbia. The network is comprised of 10 paddling clubs from Vancouver Island, Salt Spring Island and the mainland of British Columbia. Each club assigns designated representatives to liaise and volunteer with the BCMTNA. Since 2007, when the network was officially established, volunteers have performed site assessments, inventories, cleanups and upgrades along the expanding network of trails (BC Marine Trails Network Association, 2016). The network acts as a focal point for tourists and tourism-related activities and seeks to connect coastal communities and their economies while promoting an ethic of stewardship.

In order to strengthen the objectives of the network, reach out to new members, communicate the extent of the campsites and access points, as well as provide services to the membership base, BCMTNA collaborated with the SpICE Lab to create a crowdsourced and interactive web map. The map was launched in 2010. It originally presented the location of campsites, access points and network sponsors. BCMTNA already had a partial database of this information in a spreadsheet format, however, the data was not consistently geocoded and the dataset contained a large number of incomplete items and fields.

The site was built around a web based interactive map, which was designed to allow users to directly contribute point-based information to the database though a Google Maps interface. The map included: layers that correspond to the data classifications identified above (notably campsites, access points and network sponsors), user registration processes, login data contribution and management tools, as well as search abilities. As the project progressed, BCMTNA requested additional functionality for the map. This included developing an interactive paddling areas function, data export and download tools, as well as the ability for users to create a route and measure the distance and direction from location to location on the map. As the project matured, BCMTNA staff de-emphasized the ability for its broad membership to contribute data to the map and began to limit the ability of anyone but select members (most of them in the network's executive board) to add information to the map, or manipulate existing data through the map's interface. This was partially because of the risks associated with paddling and the need to ensure that the data contributed to the map was accurate.

The de-emphasis on open contributions was also linked to the considerable (and initially unanticipated) effort required by the executive members to clean and reformat the original database. There was a fear that if the ability to add and edit the data was open to the public there was no guarantee that the contributions would be complete and/or accurate, thus diminishing the overall value of the dataset. In this example, VGI was limited by the objectives of the map and therefore may be best understood as a closed input system, having only seven marker creators. However, information is accessed by a large number of members who are part of a pre-existing network. In the initial year there were 263 users, 2,958 markers and 40,619 map displays. Map usage was highly seasonal, as the activities mapped are largely used on a seasonal basis.

The objective of the organization was to create a resource, connect communities and educate users of the coastal areas. In order to do this, a broad network was created of organizations and clubs with similar objectives, in collaboration with government and First Nations communities. The change space for the first objective, or the immediate sphere of influence, was creating useful tools for users, which has been demonstrated as highly effective based on the number of markers and visitors. Due to a lack of baseline information and limitations of on-going monitoring data, we are less able to speak to the impact of the network and map on the second and third objectives: to better connect coastal communities and their economies, and to promote an ethic and practice of stewardship while interacting with the coastal environmental areas. With regard to VGI type, this map demonstrates how different approaches were taken in order to meet the objectives, as well as navigating concerns about safety and risk. The result was a shift away from open, voluntary contributions to a managed and limited entry process, with open access to the data. The success of the map was supported by a pre-existing organization, for which a map was determined to be a key resource that would further its organizational objectives and membership needs, as well as working toward their broader objectives of connecting communities and promoting stewardship beyond the membership.

Figure 1. Screenshot of BC marine trails map

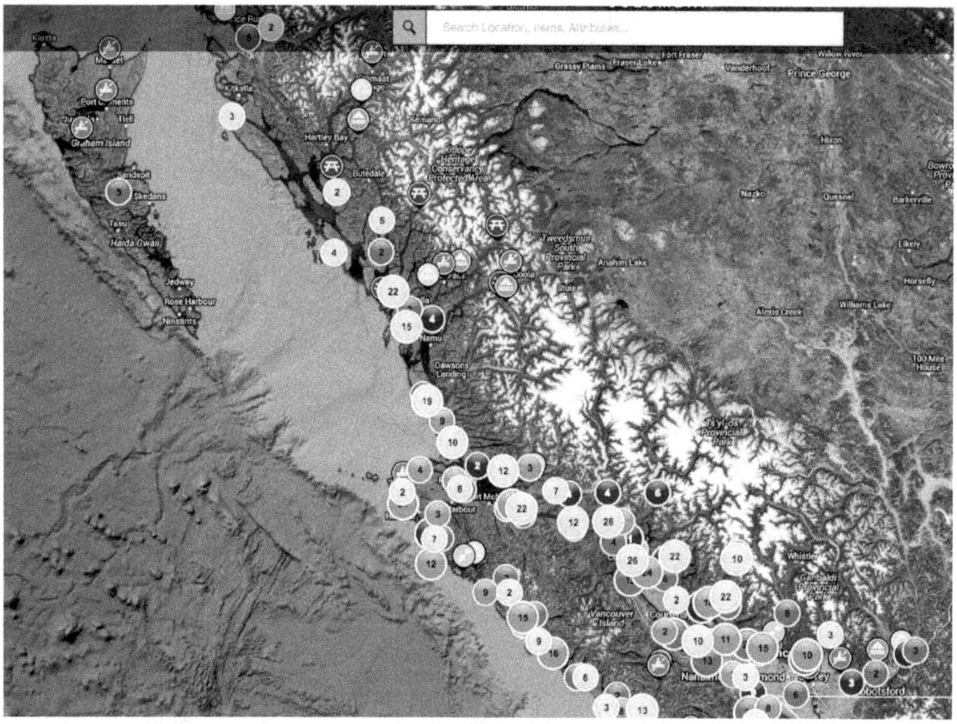

Inclusive Employment Mapping Tool: An Open System Documenting the Experiences of Those With Intellectual Disabilities

The Canadian National Participation and Activity Limitation Survey (Statistics Canada, 2006) identified that 43.7% of people with disabilities were employed compared to 78.4% of persons who do not have disabilities. Despite efforts to develop labor market strategies for persons with disabilities at both the federal and provincial/territorial levels, employment for individuals with disabilities has remained consistently low over the past decade, and the employment situation for people with intellectual disabilities (ID) is even bleaker (CACL, 2006). Because of the disturbingly low employment rates evident in British Columbia, stakeholders throughout the province are calling for action. There has been substantial activity in this area; for example, in 2013 Inclusion BC launched its "Ready, Willing & Able" campaign targeting increased employment for individuals with ID in BC.

To reach these goals, members of the public and employers need to understand the breadth of stakeholder experiences related to the issue. Using the principles of community-based participatory research (CBPR) (Cornwall & Jewekes, 1995) members of the Centre for Inclusive Citizenship, the BC Centre for Employment Excellence and the SpICE lab undertook a collaborative project aimed at helping to transform employment practices for individuals with ID in BC through the design and implementation of a web-based and interactive BC Employment Mapping Tool (EMT).

The BC EMT was co-designed with self-advocates[3] to support individuals with ID to document and share their experiences in the employment sector. Holding true to the principles of CBPR, authentic collaboration between all project partners has been integral throughout each step of the project. We have drawn on an established network of stakeholders (including self-advocates, service providers and government) who collaborate throughout the province to address employment practices for people with ID. Researchers built on existing synergies and relationships to inform and design the implementation of a collaboration strategy in the development of the mapping tool. This directly involved working with individuals and groups of self-advocates. The EMT was foreseen to be a resource for government, service providers, employers, individuals with ID, and family members to learn and to share examples of innovation in the area of employment. In addition, a core objective of this project was to explore the potential of scaling-up this project for applicability with other populations (e.g. youth, immigrants, and Aboriginal populations). As such, the participatory development of the EMT represents an innovative approach to research, knowledge exchange and knowledge translation in the area of employment practices in BC.

For members of the public, the EMT is a web-based application that functions as an accessible and searchable database that documents best practices and positive experiences related to employment for people with ID and embeds the data within map markers that are displayed on an interactive map. As such, the tool is designed to be of value to numerous stakeholders having different needs and interests. Users can search the database for examples of best practices classified by theme. These themes include 'My first job', 'Getting a job', 'Keeping a job', 'Training and supports', 'Transitions/changes in work', 'Benefits of work' and 'Getting to work'. The map markers also contain links to other information, which at the time of finalizing this chapter are primarily video recordings of self-advocates sharing their employment-related experiences. In turn, these materials act as authentic examples, provide clear strategies of how to best support other individuals with ID across the province in preparing for employment, and provide real-life examples of how employers can address barriers to support for individuals with ID. Specifically, the mapping functionality allows stakeholders to search geographic areas for information or other resources of relevance to employment for individuals with ID in their regions.

We encountered a number of challenges related to the design and initial implementation of the EMT. These challenges required us to rethink the development of the Geolive mapping system, and in particular VGI contribution processes, to make it both usable and robust for individuals with ID, as well as other stakeholders. Because of the range of uses and stakeholders, we were required to recognize the diverse familiarity with using online interactive tools and especially social media, and the varying access to the computing tools required to contribute to the map.

In order to address the challenges related to the range of familiarity with using online interactive tools, we focused on creating an accessible and 'friendly' user graphical interface. The development of the interface involved working with focus groups comprised of self-advocates in participatory settings to design icons and workflows for adding information to the map. It became apparent that assumptions around how we facilitated community users involved in past Geolive projects to add VGI needed to be rethought. Adding information to the EMT became a primary bottleneck to the success of the project. Complex, text focused input from self-advocates could not be expected, but nor could it be precluded. Yet, at the same time we had to ensure that other user groups had the ability to add text intensive information. As a result we had to design a data input wizard to share information on the map in as straightforward a way as possible, but one where different users would add different types of information in different ways. Specifically, for self-advocates, we developed a four-step wizard with icon driven input from the user. At no point was a user required to add text using their keyboard. This interface also lent itself well to developing a mobile-friendly version of the tool.

Another important point to note with regard to familiarity with online tools relates to the existing use of social media applications (Facebook and Twitter in particular). A number of the self-advocate partners were active users of social media. We therefore had to develop a system that was not 'oppositional' to existing social media practice. Thus we created the ability for users to login using existing social media credentials and then to share, or repurpose, the information that they add to the EMT through the networks that they already use (Facebook, Twitter and Google+). Not only does this encourage users to share their information through the map (because they do not have to duplicate their efforts), but also it serves to direct more user traffic back to the EMT site.

During focus group meetings it became apparent that almost all self-advocate web activity was conducted using Android-based smart phones and (usually small sized) tablets. We recognized that if we did not purposefully design the EMT to have a mobile-friendly interface we would greatly limit its uptake and use by self-advocates. As a result, we developed the tool to work seamlessly across varying screen sizes (from a 4.5-inch smart phone screen to a full sized desktop computer monitor). As the site recognizes the hardware screen size it automatically adjusts icons, links, fonts and legends to match.

During the first year after launching publicly, there were 42,911 map views, 241 users and 129 markers. The content was added by 14 marker creators. The process of enabling VGI contribution for this map demonstrates how typical mapmaking can exclude voices and decrease participation. The multiple phases of redesign of the map in order to be applicable to diverse needs and audiences highlights how objectives and feedback mechanisms can purposefully support adaptive management of systems to ensure the objectives are being met. This mapmaking experience shows how marginalization and exclusion can be replicated in 'participatory' tools, and how crowdsourcing may not, in fact, represent the entirety of a population due to the design, and therefore limitations, of who is able to contribute.

Okanagan Refugee Community Resources Map: A Restricted System Providing Resources for Recently Arriving Refugees Into the Okanagan Region

A range of VGI-driven crowdsourcing activities have emerged in response to humanitarian emergencies (e.g. Componovo & Freundschuh, 2014; Harvey, 2012; Shekhar et al., 2012). In relation to the 'refugee crisis' a range of mapping related activities have taken place internationally, although not all are reflected in the academic literature (Baranoff & Gonzales, 2015; Xu, Maitland & Tomaszewski, 2015). One of

Figure 2. Screenshot of inclusive employment map

the spaces within which participatory mapping activities has been able to engage is through mobilizing resources and services to support refugees when they have arrived in new host communities. In order to address this need, a resource map was created through a partnership between Kelowna Community Resources (KCR), the University of British Columbia Okanagan's Institute of Community Engaged Research and the SpICE lab. The map is a restricted system that connects community members with the needs of service providers and newly arriving refugees. The content is moderated and access to some information is limited to the public.

Since civil war began in Syria in 2011, over 4 million people have fled the violence. As the war has progressed the actors involved in the conflict have increased in number and the network of relationships between these actors are at best opaque. The civil unrest and division amongst opposition groups has made it unlikely that the hostilities will cease anytime soon. The situation in Syria is a humanitarian crisis with millions of displaced refugees, demanding action from world leaders and states (Cartier, 2016). The Government of Canada, working with Canadians, including private sponsors, non-governmental organizations, and provincial, territorial, and municipal governments, committed to resettle 25,000 Syrian refugees to Canada by the end of February 2016 (Government of Canada, 2016b). Of the 26,166 Syrian refugees that arrived in Canada by that date, 8,950 were privately sponsored (Government of Canada, 2016c). The Private Sponsorship of Refugees Program allows individuals, or a group, to sponsor refugees from abroad who qualify to come to Canada. Sponsors provide financial and emotional support for the refugees for the duration of the sponsorship. This includes help for housing, clothing and food. Most sponsorships last for one year, but some refugees may be eligible for assistance from their sponsors for up to three years (Government of Canada, 2016a).

In the lead up to the arrival of Syrian refugee families in February and March 2016, a large number of local individuals and groups wanted to sponsor, contribute and volunteer. A number of websites and Facebook pages sprung up in anticipation of the needs required by the incoming refugees. However, there was little coordination and groups remained largely disparate and unconnected with each other.

After extensive discussion between the project collaborators, SpICE lab members created a mapping hub that allows people to share information about household items, employment opportunities, services, housing, and volunteering that might be used by incoming Syrian refugees. Information is contributed through the map and is made directly accessible to sponsors of refugees and others who are working with refugees in their transition to Canada. The information shared on the site is also 'pushed' to the websites and Facebook pages that were developing rapidly and independently by groups throughout the region. When developing the hub, we did not intend to create yet another parallel, and competing system, rather to create a hub that would share important and relevant information to existing groups using their own existing communication channels, thus minimizing redundancy and repetition.

In order to contribute information to the map, users must register and log in to the map, drag and drop the marker identifying the appropriate information to the correct location, and fill in a number of required information fields. Users can choose to make the information available to the public, or only to themselves and coordinators from the various groups sponsoring refugees in the area. The map covers the entire Okanagan region, so if other groups in the Okanagan want to have their own Facebook pages automatically included and updated with information added to the map, they simply subscribe to the map and the associated services. Groups working with facilitating the transition of refugees who want to access the private information on the site, needed to contact site administrators before this information was made available.

Figure 3. Screenshot of Okanagan Refugee community resources map

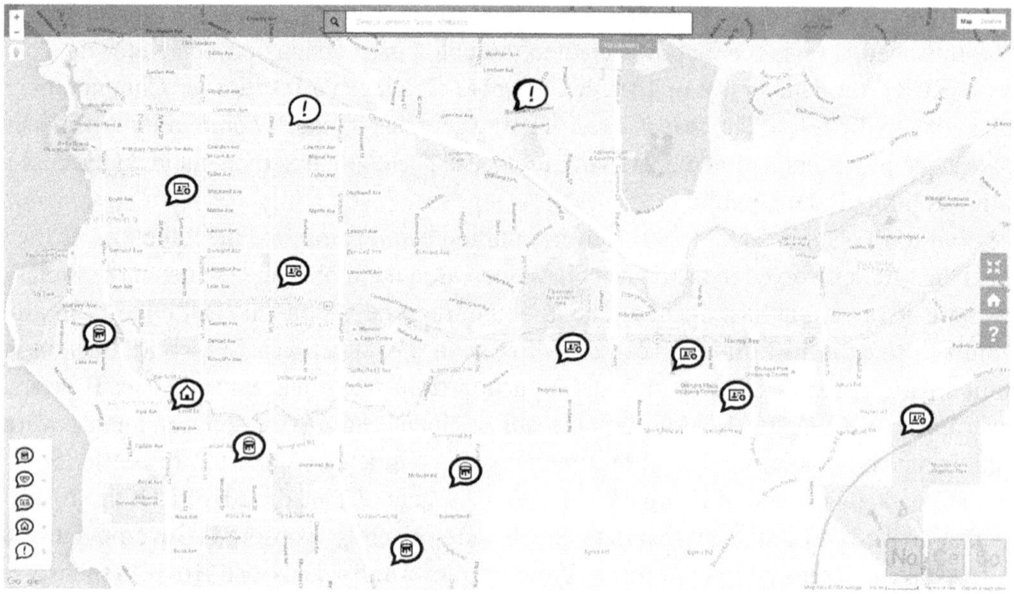

The map was launched in February of 2016, and at the time of writing only four months of data was available (in comparison to the first year after launch used for the other two examples). During that time, 11,810 map views occurred, with 74 users, 28 marker creators and a total of 113 markers. Due to a range of privacy and safety concerns associated with the types of information being shared (especially the location of housing for vulnerable populations), VGI for this map was regulated and access restricted. Although this poses some limitations, as a public service it is designed to allow individuals to volunteer geographic information (such as an address and contact details), which is shared with service providers. In addition, the public facing site enables the public to view the types of goods and services offered and needed, and therefore allows hosts and organizations to make informed choices regarding the most needed contributions. The creation of the map emerged out of a need, as individuals and groups were communicating, but not connecting. The geoweb map sought to centralize this information, as well as operate as a hub that would connect these diverse groups and initiatives. In this example, access to VGI was restricted due to concerns of privacy, whereas in the first example access was open while contributions were restricted due to associated risk.

DISCUSSION

This chapter focuses on experiential learning with regard to the participatory geoweb and VGI, and what these projects can contribute to the theoretical understanding and framing of participatory mapping, the participatory geoweb and VGI. In this section we outline some of the broad findings of the Geolive examples, and discuss the implications for considering and opening a more critical discourse on the way in which VGI-driven participatory projects are researched. The examples were contextualized in terms of the motivation behind the desire to contribute, the type of contribution, the capacity and leadership offered by the delivering organization, and how these factors align with the overall project objectives. We will use these factors to compare and contrast the organizational factors that influence VGI contribution, use and impact. These findings demonstrate how the dynamics of VGI cannot be viewed in isolation, but as being influenced by broader organizational and contextual factors. They also contribute to the literature about participation, VGI and crowdsourcing.

Motivation

Unlike the characterizations of VGI contributors identified by Coleman, Georgiadou and Labonte (2009), who define a spectrum of expertise in user knowledge, all three examples show how users were invited to volunteer their information because they are already invested in the issue that the geoweb tool is being used to shed light on. For example, members of the BCMTNA are avid and engaged kayakers and they are motivated to increase the resources and services available to pursuing this activity. Contributors to the EMT have a direct association to intellectual disabilities and are motivated to improve access to a positive work environment for themselves and other constituents. People motivated to help Syrian refugees share information and resources that they have available in order to facilitate the arrival of the families into the Okanagan region. In other words, in each of the case studies, contributors are drawn by their pre-existing engagement with the issue, and because they are embedded in the matter, they also have a level of experiential understanding that does not need to be classified as expert or 'neophyte' (Coleman, Georgiadou & Labonte, 2009). At the same time, it becomes clear that these maps are not creations of radically diverse groups of individuals, as some literature may portray them. In particular, the EMT map demonstrates that some individuals are excluded by the design of some mapmaking processes.

The motivations for individuals to volunteer their information through the project geoweb interface vary greatly. There is a need to recognize that in each example, the principal agent in the design and delivery of the tool is an organization (which clearly has its own agenda). Unlike the motivation of individuals to contribute to mainstream social media platforms (such as Facebook, Instagram or Twitter), where users have their own dedicated group of followers, being requested to add this data by an organization might mean that contributors do not clearly understand the purpose for doing do, or else they might feel that the value of their contributions might be only realized by the organization. Or, they may have their own motivations and objectives that do not align with those of the organization. Individuals might also feel that they are being encouraged to share information with others who share the same interest in an issue, but perhaps they do not have a personal connection or relationship, or have different expectations of the outcomes of their contributions.

It is also important to note that the motivation to contribute is not necessarily long lasting. Contributors appear to get involved when they first hear about the initiative, but a sustained and long term commitment to continue to add information to the map is not guaranteed. This is why the BCMTNA is perhaps one of the most lasting and sustainable projects. By delegating the responsibility to add, manage and maintain data the organization assumes a role of custodian and therefore may be more likely to ensure sustained activity. Whereas, topic-specific issues, which may rise and fall with interest over time, such as environmental causes and humanitarian responses, do not, in our experience with Geolive, support sustained involvement, with fewer users continuing to contribute over time.

The Type of Contribution

Each of the case studies requires a very different type of contribution and level of interaction. For example, the BCMTNA took a tiered approach. In addition, different types of users contribute and interact with the database in different ways. Administrators have the ability to add map markers, edit existing markers, export data and determine which data is visible to other users (a large portion of the database points are not publically available). Registered and paying BCMTNA members have the ability to use map-related tools such as route planning, kayaking regions and the ability to identify and save a subset of markers, however, they do not have the ability to add new data or manipulate existing data. Non registered users have the ability to view all visible data, they can also undertake some limited manipulation of the data through turning layers on and off, however, this is the extent of their ability to interact. This approach to working with the data ensures that committed members of the network are taking on the most complex and time consuming tasks required to update and maintain the integrity of the database, while also limiting participation due to concerns about accuracy and responsibility regarding safety.

In the case of the EMT, the volunteered information focused on image and videos, and much less on text. Although this form of multimedia information requires more editing, our focus groups indicated that it is far easier to access and understand for people with ID. It is also important to note that the self-advocates and individuals with ID involved in our consultations were in some cases very confident with the technology, and already intensively engaging on social media platforms such as Twitter and Facebook. That said, we anticipated potential usability challenges with the EMT and learned quickly that these were not unique to self-advocates: generational differences in familiarity with online technologies across participants (e.g., family members and service providers) also served as a challenge to contributing information to the EMT. Reflecting on these broader usability and accessibility constraints suggests that mapmaking is representative of some voices, but certainly not all voices, and on-going research needs to better disaggregate VGI input in order to contextualize the design, processes and results and specify the limitations of usability and accessibility more explicitly.

Furthermore, there remain challenges of spatial literacy[4] and map reading within the context of successfully sharing information through the EMT project. Undoubtedly, the map is an important component of the 'friendly' interface and provides a mechanism to examine the relationship of 'place' in relation to access to experiences and services available. Through initial evaluations, a number of users have talked about the importance of the visual aspect of the map. Yet, we are concerned that we have built this on a number of assumptions about the map. These are, firstly that the map is both readable and understandable to users (for self-advocates in particular and members of the public more broadly), and secondly, that the map provides an important framework to structure the volunteered information in a way that makes it more meaningful, or at the very least more accessible. Our user testing has shown that self-advocate users are more drawn to the street view capability built into the EMT and less to the map itself. This remains an ongoing aspect of the research project, and indicates an area for future research in other participatory geoweb initiatives.

The issue of spatial literacy remained important in the Okanagan Refugee Community Resources project. Although the map acted as a means to quickly add and visualize the information, it was not necessarily communicating the information required by the refugee sponsors. A non-spatial database might have been equally useful. The true value of the contributions was the ability for the map to act as a clearing house and mediating tool for the contributed data, as it allowed users to contribute their data using a simple standardized and streamlined interface that was then redirected and pushed to the appropriate non-spatial databases already being used by the relevant organizations. In fact, this became such an issue that we created a completely non-map interface for users to add their information; it still required users to add their physical address, but otherwise the process of adding had no reference to the map.

Organizational Capacity and Leadership

In our experience, organizational capacity is perhaps one of the most important aspects that influence the success of VGI projects. If there is no long-term commitment to market, maintain, prompt the user base and contribute as an administrator, the project is more likely to fail – especially in the long term. In other words, to simply establish and launch a site to collect VGI is not enough, success requires substantial and ongoing support and commitment. It also requires the delivering organization to be clear about the timelines associated with the project goals.

The BCMTNA initiative is the longest running project that uses Geolive. The project began in 2010 and was still running at the time of writing this chapter (2016). In our experience with Geolive, we believe this is partially due to the ongoing relevance of the information contributed and displayed through the map, and is also a testament to the vision and maintenance of the organization and the manner in which they have delegated agency and responsibility. By ensuring a core of dedicated administrators, motivated by both their commitment to kayaking as well as their interest in the technical side of the site, BCMTNA has created a sustainable and continually contemporary system. However, the pressure is placed on a few individuals and not spread through the broader user base, this in turn has its drawbacks, such as the limitation of the breadth of knowledge to draw on and the precariousness of concentrating the requisite skills and knowledge in just a few individuals.

The EMT project, despite having a large number of project partners, did not have a single collaborator prepared to assume the lead role in marketing, maintaining, and prompting the user base after the initial funding was complete. As a result, after an initial period of high public interest (including interviews on CBC radio, the national broadcasting corporation, and coverage in a number of provincial newspapers), during which the project had a funded position for project outreach, the contributions began to dwindle. Although it can be argued that a core base of information regarding employment strategies for individuals with ID was already included on the map, the volunteering of new information does not add to the ongoing usability and contemporaneousness of the site.

However, we do understand that for a range of different reasons organizations do not necessarily want projects to run indefinitely. For example, the Syrian Refugee resource map is only relevant within the immediate context of the humanitarian crisis and the arrival of impacted families in the Okanagan region. Outside of that specific spatial and chronological context, the site and the need for VGI has little value. Clearly there is a paradox that as many of these projects progress we begin to associate value and success with longevity – often times this is neither relevant nor desirable.

Alignment With Objectives

The combination of user motivations and incentives, the types of information contributed and their relevance, as well as the organizational capacity ultimately frame and are framed by the objectives of the project. Any VGI-based project requires a clear set of objectives at the outset. Too often in our experience with implementing Geolive, this not the case. Objectives that are not clearly articulated can greatly complicate the clarity of the messaging within volunteered information. Furthermore, the objectives can shift through the lifespan of a project. If organizations do not have experience with developing and deploying VGI projects (or web-based initiatives more broadly), their preliminary understanding of the technology will determine the initial objectives. However, once the project takes shape and an organization has a better understanding of the range and potential application of the technology, it is easy to broaden the scope of the project and conflate the starting objectives with other organizational goals. Inevitably this leads to confusion in marketing the project objectives to the broader user base and potential contributors.

This conflation of objectives can also occur when there are multiple collaborators involved in a VGI project. Geolive was initially developed as a web based participatory mapping tool. The practice of participatory mapping is in its very nature a collaborative practice and undertaking, so requires the input of all project partners. In the SpICE lab, we therefore aspire to include the ideas and intent of every party involved. However, unaligned objectives can lead to confusion and which in turn can complicate project marketing and outreach, user motivation and the clarity of their contribution.

From our experience, successful projects are built around clear objectives. These projects cannot be determined without their alignment within broader organizational goals. Objectives need to be straightforward to communicate and the technical usability of the platform limited. More objectives do not contribute to greater success; in fact, the opposite tends to occur, the broader the range of objectives, the less clear the contribution and thus the more likely the project to falter. In sum, projects that require members of the public to volunteer their geographic information are more likely to succeed in the long term if their overall objectives are kept simple.

CONCLUSION

Drawing upon diverse examples, this chapter has brought experiential learning as a contribution and to complement developments in the theoretical literature regarding participatory mapping and VGI. We found that motivations tend to center around pre-existing interest and association, as opposed to a broad user base. The three case studies demonstrate how different types of VGI contribution, by design, can offer opportunities and limitations, which offer insight for both developing participatory geoweb projects as well as researching their impact. Organizational capacity emerged as a key determining factor of sustained activity, although we also find that more mapping projects need to actively engage in decision making about the life-span of projects, as not all ought to have a goal of becoming permanent entities. Project objectives, although often clear at the outset, may change or may not be communicated clearly with all users, which can negatively affect motivation to contribute as well create confusion, and potentially frustration when diverse expectations are not met. Whereas works based on a single case study may identify case-specific issues that are not applicable to other projects, the use of comparative examples, as analyzed in this chapter, demonstrate the contribution that comparative studies offer and how experience-based contributions can broaden the discourse, and issues raised, regarding the dynamics of VGI.

REFERENCES

Baranoff, J., & Gonzales, R. I. (2015). Lantern: Empowering Refugees Through Community-Generated Guidance Using Near Field Communication. CHI'15 Extended Abstracts, 7–12.

BC Marine Trails Network Association. (2016). *About the BC Marine Trails - BC Marine Trails Network: Kayak tours, adventure, kayaking maps*. Retrieved August 11, 2016, from https://www.bcmarinetrails.org/about-us/about-the-bc-marine-trails-network

Beischer, A., Cochrane, L., Corbett, J., Evans, M., Gill, M., & Millard, E. (2015). Mapping Experiences of Injustice: Developing a Crowdsourced Mapping Tool for Documenting Good Practices for Overcoming Social Exclusion. *Spatial Knowledge and Information Conference Proceedings*.

Brabham, D. C. (2008). Crowdsourcing as a Model for Problem Solving. *Convergence (London)*, *14*(1), 75–90. doi:10.1177/1354856507084420

Brandeis, M. W., & Nyerges, T. (2015). Assessing Resistance to Volunteered Geographic Information Reporting with Local Government. *Transactions in GIS*, *20*(2), 203–220. doi:10.1111/tgis.12168

Brown, G. (2006). Mapping landscape values and development preferences: A method for tourism and residential development planning. *International Journal of Tourism Research*, *8*(2), 101–113. doi:10.1002/jtr.562

Brown, G. (2012). Public Participation GIS (PPGIS) for Regional and Environmental Planning: Reflections on a Decade of Empirical Research. *Journal of Urban and Regional Information Systems Association*, *25*(2), 7–18.

Brown, G., Kelly, M., & Whitall, D. (2014). Which public? Sampling effects in public participation GIS (PPGIS) and volunteered geographic information (VGI) systems for public lands management. *Journal of Environmental Planning and Management*, *57*(2), 190–214. doi:10.1080/09640568.2012.741045

Brown, G., & Reed, P. (2009). Public Participation GIS: A New Method for use in National Forest Planning. *Forest Science*, *55*(2), 166–182.

Brown, G., & Reed, P. (2012). Values Compatibility Analysis: Using Public Participation Geographic Information Systems (PPGIS) for Decision Support in National Forest Management. *Applied Spatial Analysis and Policy*, *5*(4), 317–332. doi:10.1007/s12061-011-9072-x

Brown, G., & Weber, D. (2013a). A Place-based Approach to Conservation Management using Public Participation GIS (PPGIS). *Journal of Environmental Planning and Management*, *56*(4), 455–473. doi:10.1080/09640568.2012.685628

Brown, G., & Weber, D. (2013b). Using Public Participation GIS (PPGIS) on the Geoweb to Monitor Tourism Development Preferences. *Journal of Sustainable Tourism*, *21*(2), 192–211. doi:10.1080/09669582.2012.693501

CACL. (2006). *The employment and employability of Canadians with intellectual disabilities: Submission by the Canadian Association for Community Living to the House of Commons Standing Committee on Human Resources, Social Development and the Status of Persons with Disabilities*. Toronto: Canadian Association for Community Living.

Camponovo, M. E., & Freundschuh, S. M. (2014). Assessing Uncertainty in VGI for Emergency Response. *Cartography and Geographic Information Science*, *41*(5), 440–455. doi:10.1080/15230406.2014.950332

Cartier, M. (2016). Explaining Differences in the Canadian and American Response to the Syrian Refugee Crisis. *Virginia Policy Review*, 56-82.

Chambers, R. (2006). Participatory mapping and geographic information systems: Whose map? Who is empowered and who disempowered? Who gains and who loses? *The Electronic Journal of Information Systems in Developing Countries*, *25*(2), 1–11.

Cochrane, L., Corbett, J., Evans, M., & Gill, M. (2016). Searching for Social Justice in Crowdsourced Mapping. *Cartography and Geographic Information Science*. doi:10.1080/15230406.2016.1212673

Cochrane, L., Corbett, J., & Keller, P. (2014). *Impact of Community-based and Participatory Mapping. Institute for Studies and Innovation in Community-University Engagement*. University of Victoria.

Coleman, D. J., Georgiadou, Y., & Labonte, J. (2009). Volunteered geographic information: The nature and motivation of produsers. *International Journal of Spatial Data Infrastructures Research*, *4*(1), 332–358.

Corbett, J. (2003). *Empowering technologies? Introducing participatory geographic information and multimedia systems in two Indonesian communities* (Doctoral Dissertation). University of Victoria.

Corbett, J., Cochrane, L., & Gill, M. (2016). Powering Up: Revisiting Participatory GIS and Empowerment. *The Cartographic Journal*, *53*(4), 335–340. doi:10.1080/00087041.2016.1209624

Corbett, J., & Mann, R. (2010). Tlowitsis re-imagined: The use of digital media to build nation and overcome disconnection in a displaced aboriginal community. *International Journal of Information Communication Technologies and Human Development*, *2*(3), 33–54. doi:10.4018/jicthd.2010070103

Corbett, J. M., & Keller, C. P. (2005). An analytical framework to examine empowerment associated with participatory geographic information systems (PGIS). *Cartographica: The International Journal for Geographic Information and Geovisualization*, *40*(4), 91–102. doi:10.3138/J590-6354-P38V-4269

Corbett, J. M., & Keller, C. P. M. (2004). Empowerment and participatory geographic information and multimedia systems: Observations from two communities in Indonesia. *Information Technologies and International Development*, *2*(2), 25–44. doi:10.1162/1544752044193425

Cornwall, A., & Jewkes, R. (1995). What is participatory research? *Social Science & Medicine*, *41*(12), 1667–1676. doi:10.1016/0277-9536(95)00127-S PMID:8746866

ESRI. (2011). *Spatial Literacy Program at the University of Redlands*. Accessed November 14 2015 from http://www.redlands.edu/prospective-students/6304.aspx

Estelles-Arolas, E., & Gonzalen-Ladron-de-Guevara, F. (2012). Towards an Integrated Crowdsourcing Definition. *Journal of Information Science*, *38*(2), 189–200. doi:10.1177/0165551512437638

Flanagin, A. J., & Metzger, M. J. (2008). The Credibility of Volunteered Geographic Information. *GeoJournal, 72*(3-4), 137–148. doi:10.1007/s10708-008-9188-y

Goodchild, M. F. (2007a). Citizens as Sensors: Spatial data infrastructure in the world of web 2.0. *International Journal of Spatial Data Infrastructures Research, 2*, 24–32.

Goodchild, M. F. (2007b). Citizens and Sensors: The World of Volunteered Geography. *GeoJournal, 69*(4), 211–221. doi:10.1007/s10708-007-9111-y

Goodchild, M. F. (2008). Commentary: Whither VGI? *GeoJournal, 72*(3-4), 239–244. doi:10.1007/s10708-008-9190-4

Goodchild, M. F., & Li, L. (2012). Assuring the Quality of Volunteered Geographic Information. *Spatial Statistics, 1*, 110–120. doi:10.1016/j.spasta.2012.03.002

Government of Canada. (2007a, March 31). *Sponsor a refugee*. Retrieved August 11, 2016, from http://www.cic.gc.ca/english/refugees/sponsor/index.asp

Government of Canada. (2015b, November). *Canada's response to the conflict in Syria*. Retrieved August 11, 2016, from http://www.international.gc.ca/development-developpement/humanitarian_response-situations_crises/syria-syrie.aspx?lang=eng

Government of Canada. (2016c, June 21). *Syria – by the numbers*. Retrieved August 11, 2016, from http://www.cic.gc.ca/english/refugees/welcome/commitment.asp

Grira, J., & Bédard, Y. (2009). Spatial data uncertainty in the VGI world: Going from consumer to producer. *Geomatica, 64*(1), 61–71.

Harris, T., & Weiner, D. (1998). Empowerment, marginalization, and community integrated GIS. *Cartography and Geographic Information Systems, 25*(2), 67–76. doi:10.1559/152304098782594580

Harvey, F. (2012). Code/space: Software and Everyday Life by Rob Kitchen and Martin Dodge. *International Journal of Geographical Information Science, 26*(10), 1999–2001. doi:10.1080/13658816.2012.661059

Hodgson, D., & Schroeder, R. (2002). Dilemmas of Counter-mapping Community Resources in Tanzania. *Development and Change, 33*(1), 79–100. doi:10.1111/1467-7660.00241

Howe, J. (2006). The Rise of Crowdsourcing. *Wired, 14*(6), 1–5.

Kuhn, W. (2007). *Volunteered geographic information and GIScience*. NCGIA and Vespucci Workshop on VGI, Santa Barbara, CA.

McGee, R., Edwards, D., Minkley, G., Pegus, C.-M., & Brock, K. (2015). *Making All Voices Count: Research and Evidence Strategy*. Brighton, UK: Institute of Development Studies.

Peluso, N. L. (1995). Whose Woods are These? Counter-mapping Forest Territories in Kalimantan, Indonesia. *Antipode, 27*(4), 383–406. doi:10.1111/j.1467-8330.1995.tb00286.x

Quiquivix, L. (2014). Art of War, Art of Resistance: Palestinian Counter-cartography on Google Earth. *Annals of the Association of American Geographers, 104*(3), 444–459. doi:10.1080/00045608.2014.892328

Shekhar, S., Yang, K., Gunturi, V. M., Manikonda, L., Olive, D., Zhou, X., & Lu, Q. (2012). Experiences with Evacuation Route Planning Algorithms. *International Journal of Geographical Information Science, 26*(12), 2253–2265. doi:10.1080/13658816.2012.719624

Statistics Canada. (2006). *Participation and Activity Limitation Survey 2006: Analytical Report*. Retrieved August 14, 2016, from http://www.statcan.gc.ca/pub/89-628-x/89-628-x2007002-eng.htm

Sui, D., & DeLyser, D. (2012). Crossing the Qualitative-Quantitative Chasm I: Hybrid Geographies, the Spatial Turn, and Volunteered Geographic Information (VGI). *Progress in Human Geography, 36*(1), 111–124. doi:10.1177/0309132510392164

Surowiecki, J. (2004). *The Wisdom of Crowds: Why the Many are Smarter than the Few and How Collective Wisdom Shapes Business, Economies, Societies, and Nations*. New York: Doubleday.

Wainwright, J., & Bryan, J. (2009). Cartography, Territory, Property: Postcolonial Reflections on Indigenous Counter-mapping in Nicaragua and Belize. *Cultural Geographies, 16*(2), 153–178. doi:10.1177/1474474008101515

World Bank. (2014). *Strategic Framework for Mainstreaming Citizen Engagement in World Bank Group Operations: Engaging with Citizens for Improved Results*. Washington, DC: World Bank Group.

World Bank. (2016). *Evaluating Digital Citizen Engagement: A Practical Guide*. Washington, DC: World Bank Group.

Xu, Y., Maitland, C., & Tomaszewski, B. (2015). Promoting Participatory Community Building in Refugee Camps with Mapping Technology.*Proceedings of the Eighth International Conference on Information and Communication Technologies and Development - ICTD '15*. doi:10.1145/2737856.2737883

ENDNOTES

1 https://maptionnaire.com/_

2 https://www.arcgis.com/home/index.html_

3 The self-advocacy movement for individuals with ID seeks to overcome the isolation of people with disabilities and give them the tools and experience to take greater control over their own lives. A self-advocate is an individual with ID who is equipped to speak out for themselves, articulate their rights and capable of making decisions related to matters that affect their lives.

4 Spatial literacy refers to the ability to both read and use maps. Furthermore it relates to the capacity "to visualize and interpret location, distance, direction, relationships, movement and change through space" (ESRI, 2011).

Chapter 2
Quality Evaluation of Volunteered Geographic Information:
The Case of OpenStreetMap

Hongyu Zhang
Western University, Canada

Jacek Malczewski
Western University, Canada

ABSTRACT

A large amount of crowd-sourced geospatial data have been created in recent years due to the interactivity of Web 2.0 and the availability of Global Positioning System (GPS). This geo-information is typically referred to as volunteered geographic information (VGI). OpenStreetMap (OSM) is a popular VGI platform that allows users to create or edit maps using GPS-enabled devices or aerial imageries. The issue of quality of geo-information generated by OSM has become a trending research topic because of the large size of the dataset and the inapplicability of Linus' Law in a geospatial context. This chapter systematically reviews the quality evaluation process of OSM, and demonstrates a case study of London, Canada for the assessment of completeness, positional accuracy and attribute accuracy. The findings of the quality evaluation can potentially serve as a guide of cartographic product selection and provide a better understanding of the development of OSM quality over geographic space and time.

INTRODUCTION

Although a large amount of geospatial data and wide range of applications have made GIS very popular, the users are often unaware of the data quality. New elements were added to the discussion of geospatial data quality in the 21st century through the development of Web 2.0 and the availability of Global Positioning System (GPS). The interactivity of the new web technology helped create a large amount of user-generated content (UGC). UGC with location information is referred to as user-generated geospatial

DOI: 10.4018/978-1-5225-2446-5.ch002

content (Coleman, Georgiadou, & Labonte, 2009), crowd-sourced geodata (Barron, Neis, & Zipf, 2014) or volunteered geographic information (VGI) (Goodchild, 2007). More specifically, using location-based services (LBS), GPS-enabled devices and/or satellite images, VGI users actively upload and share data following an opt-in provision, and VGI can be direct or indirect depending on whether users have local knowledge (Haklay, 2013). The activities of contributing VGI have been termed in different ways as well, including collaborative mapping (Jokar Arsanjani & Vaz, 2015), participatory GIS (Elwood, 2006) and public participation GIS (PPGIS) (Lin, 2013).

Researchers are interested in VGI because of its values. The conventional apprehension about commercial or governmental cartographical products is authoritative, comprehensive and accurate. However, Coleman (2013) and Dobson (2013) concluded that these databases are often out-of-date, incomplete, of inconsistent quality, and costly to maintain. Therefore, VGI is studied as a crowd-sourced alternative to "authoritative" datasets. OpenStreetMap (OSM) is one of the VGI applications that allow users to create and edit maps using satellite images. As of July 2016, the total number of registered users on OSM has passed 2.8 million, creating more than 3.4 billion nodes (data points) accumulatively ("OSMstats - Statistics of the free wiki world map," 2016). This chapter systematically summarizes the quality evaluation process of OSM through literature review and a case study in London, Canada, with focuses on the comparisons of different assessment methods and findings.

BACKGROUND

The term volunteered geographic information (VGI) was suggested by Goodchild (2007) to represent geospatial data contributed by individuals voluntarily. Since VGI is often the most cost-effective solution, the crowd-sourced geodata have been applied in many fields such as participatory planning and spatial decision making. Moreover, VGI is the only source of geodata in some regions because of security or financial concerns. The area of humanitarian relief and crisis management is the most prominent application of VGI. Ushahidi and the Humanitarian OpenStreetMap Team (HOT) are two platforms that have had strong presence on disaster management since 2008 and 2009 respectively. Table 1 compares some VGI applications with OSM. Although OSM is not the project with the longest history, it is the

Table 1. Comparison of volunteered geographic information (VGI) applications

Attributes	OpenStreet Map	Wikimapia	Waze	Moovit	GasBuddy
Founding year	2004	2006	2008	2012	2000
Specialization	Mapping	Mapping	Navigation	Public transit	Fuel prices
Number of users or registered members (in million)	2.8 (in 2016)	1.9 (in 2013)	50 (in 2013)	20 (in 2014)	35 (n.d.)
Coverage in 2016	World	World	World	600+ cities	United States and Canada
License	ODbL	CC BY-SA	Proprietary	Proprietary	Proprietary
Data downloadable	Yes	Yes	No	No	No

Note. ODbL, Open Data Commons Open Database License; CC BY-SA, Creative Commons license Attribution-ShareAlike; data for OpenStreetMap from "OSMstats - Statistics of the free wiki world map" (2016), for Wikimapia from Neis & Zielstra (2014), for Waze from CBC News (2013), for Moovit from "Moovit Company Overview" (2014), and for GasBuddy from "Advertise with us - Gasbuddy Gas Prices" (n.d.).

oldest mapping project in which the geo-information can be applied in more than one field. The number of "registered members" of OSM is relatively small comparing to other specialized applications, but the number of "users" could be a bloated figure and does not represent "active contributors". Jokar Arsanjani and Bakillah (2014), Mooney and Corcoran (2012a) and Yang, Fan, and Jing (2016) provide more insights into the OSM contribution patterns and user behavior. Like Wikimapia and Waze, OSM has a worldwide coverage. The difference is that OSM allows users to freely alter and redistribute its data, which is accessible through multiple servers in different formats. In contrast, Wikimapia only offers its data by a web application programming interface (API) (Neis & Zielstra, 2014), and Waze does not release data from its platform. Therefore, OSM was chosen to be the focus of this chapter. The following subsections starts with the discussion of quality concerns in VGI, introduces OSM in details and ends with a list of the spatial data quality metrics.

Quality Issues of Volunteered Geographic Information (VGI)

Community-based systems, like the review system on Amazon or Airbnb, could be useful to evaluate the relative and latent value of VGI (Feick & Roche, 2013). Data quality assessment is a more explicit way of determining the value of VGI. Quality issues of VGI are typically centered around inconsistency in terms of coverage and accuracy. For instance, remote areas are usually under-mapped (Coleman, 2013). If volunteers are unfamiliar with the remote areas they map, accuracy might be sacrificed because of volunteers' deficiency of local knowledge (Dobson, 2013). In addition to geometrical objects, VGI's metadata is also incomprehensive and inaccurate (Hashemi & Ali Abbaspour, 2015), which creates difficulties for researchers to verify the semantic accuracy of VGI. The demographic composition of VGI contributors has some patterns as well; for example, Haklay (2013) suggested that VGI contributors were mainly male with high income, and Neis and Zipf (2012) found that 72% of OSM users was located in Europe. It is still unclear how this composition influenced VGI quality. Although the International Organization for Standardization (ISO) has published quality principles for geographic information (ISO, 2002), a new quality assurance schema specifically tailored for VGI is needed because of the limitations mentioned above (Van Exel, Dias, & Fruijtier, 2010).

The simplified expression of Linus' Law – "Given enough eyeballs, all bugs are shallow" (Raymond 2001, p. 13) – is often quoted as an underlying theory for discussing the issues of data quality (e.g., Haklay et al. 2010; Miller and Goodchild 2015; Goodchild and Li 2012; Goodchild 2013). However, Linus' Law may not work well in a spatial context (Elwood, Goodchild, & Sui, 2013), and this quotation often misleads readers to conclude that most quality issues will be solved if there are enough testers. The full expression of Linus' Law is that "Given a large enough betatester and codeveloper base, almost every problem will be characterized quickly and the fix obvious to someone" (Raymond 2001, p. 13). This expression specifies that the "eyeballs" must include those from co-developers, who are professionally trained to debug the Linux operating system in the context of the Raymond's article. However, VGI contributors are mainly citizen scientists but not professional cartographers. Moreover, the software "bugs" can be identified during the process of using the software. However, errors on maps cannot be recognized or avoided if map scale is too small, contributors do not have local knowledge, or accuracy is sufficient for certain map applications (i.e., navigation requires less accuracy than road constructions). Furthermore, the contribution pattern of VGI users signifies the necessity of spatial redundancy (Dobson, 2013). For example, 38% of registered OSM members edited at least once, and only 5% of all actively contributed to the project (Neis & Zipf, 2012). Spatial heterogeneity also prevents the existence of consistent spatial errors across the globe that may be corrected all at once. Thus, Linus' Law does not apply to VGI, which means a large number of volunteers may not be enough to ensure the quality of VGI.

OpenStreetMap (OSM)

OSM is a crowdsourced online mapping platform, which aims to provide free and editable digital mapping products under a new copyright license (Haklay & Weber, 2008). Since its initiation in 2004, OSM has been applied in routing and navigation, cartography improvement, Location Based Services (LBS), and 3D city models (Jokar Arsanjani, Zipf, Mooney, & Helbich, 2015). In 2014, high densities of OSM nodes were found in North America, Europe, Russia, Australia and Brazil, while Africa and Greenland were least mapped (Jokar Arsanjani, Zipf, et al., 2015). According to Jokar Arsanjani, Zipf, et al. (2015), Mooney & Corcoran (2014), Neis & Zielstra (2014), Stein, Kremer, & Schlieder (2015) and Vandecasteele & Devillers (2015), OSM can be described by the following key features:

- **Near Real-Time Updates:** Unlike Google Map Maker, which has a review system for submitted edits, OSM publishes modifications just "a few minutes" after contributors save changes;
- **Data Import From Multiple Sources:** OSM supports data generated from Global Positioning System (GPS), smartphones, and other mapping hardware. In the early years of the project, GPS-enabled devices were the most popular data generators. This situation was changed because Yahoo! (from 2007 to 2011) and Microsoft Bing (since 2010) agreed to provide their aerial imageries for OSM enthusiasts to trace data. Some countries such as the United States and Canada also had volunteers to import authoritative datasets into OSM;
- **Data Export in Multiple Formats:** OSM data can be downloaded at different scale (e.g., continental, regional or metro) in different formats (e.g., OSM Extensible Markup Language (XML), Protocol Buffer Binary Format (PBF) or shape file) from several servers (e.g., Planet OSM, Geofabrik or Mapzen);
- **Different Flavours of Editors:** The web-based iD editor has a simple user interface for beginners to immerse into geodata contributions. Besides, Potlatch or JOSM (Java OpenStreetMap Editor) are favoured by advanced mappers. Other editors are available across operating systems and platforms as well;
- **Full Edit History:** OSM keeps all historical edits in its full history dump site ("Index of /planet/full-history," 2016), but only the latest object versions in other forms of extracts;
- **Three Different Object Types:** A "node" represents a point, while a "way" consists of lines or polygons (closed line features). A "relation" connects related nodes and ways with each other;
- **Tags as Metadata:** Attributes of objects are expressed as "key = value" pairs;
- **Spatial Heterogeneity:** Patterns of contributions differ from one place to another;
- **Manifold Collaboration Channels:** The official OSM wiki provides the knowledge base of the project. Other communication methods include Internet relay chats (IRCs) ("IRC - OpenStreetMap Wiki," 2015) and mailing lists ("Mailing lists - OpenStreetMap Wiki," 2016). Community events such as "mapping parties" are organized both online and offline, with the yearly "State of the Map" conference attracting most attendees.

Spatial Data Quality

Spatial data quality can be evaluated internally or externally (Jokar Arsanjani, Mooney, Zipf, & Schauss, 2015). While external quality assesses the fitness of data for a particular purpose, internal quality describes how well data meet specifications, such as:

- **Completeness (C):** Measures the comprehensiveness of a dataset. This criterion not only reports how much data is missing, but also the amount of data that should be excluded;
- **Positional Accuracy (PA):** Measures the relative and absolute accuracy of coordinate values;
- **Attribute Accuracy (AA):** Measures the correctness of attributes associated with geometrical shapes, which is also known as thematic accuracy (ISO, 2002);
- **Logical Consistency (LC):** Measures the internal consistency of a dataset, such as topological correctness and relations of objects;
- **Semantic Accuracy (SA):** Measures whether data objects and their meanings are interpreted correctly;
- **Temporal Quality (TQ):** Measures the validity of changes and the rate of updates in a dataset;
- **Lineage (L):** Measures the history of a dataset from collection to evolution (Van Oort, 2006).

The focus of this chapter is the internal quality of VGI data. However, it has been recognized that the above criteria only assess absolute data quality, while the actual quality is relative to its fitness-of-use (Feick & Roche, 2013; Van Oort, 2006).

A REVIEW OF OPENSTREETMAP QUALITY ASSESSMENT

A systematic survey of literature (as of July 2016) found 60 articles relevant to quality evaluation of OSM (see Appendix). Four databases were used in this process including Web of Science, Scopus, Engineering Village (Geobase) and Proquest (dissertations & theses). 334 articles were found initially using keywords "OpenStreetMap AND (quality OR accuracy)" with the option of anywhere except full text, and the number of relevant articles went down to 202 after removing duplicates. A full-text review of the 202 articles identified 39 articles listed in the Appendix. In addition, 21 relevant articles were found based on an examination of the 39 articles' reference sections. Only studies written in English were retained. It is worth to mention that some excluded articles are not totally irrelevant, but they focus more on method assessment instead of quality of specific areas (Basiri et al., 2016; Brovelli, Minghini, Molinari, & Mooney, 2016; Fan, Yang, Zipf, & Rousell, 2015; Graser, Straub, & Dragaschnig, 2014; Gröchenig, Brunauer, & Rehrl, 2014; Jokar Arsanjani, Mooney, Helbich, & Zipf, 2015; Zhang & Ai, 2015). In Table 10 of the Appendix, time represents the actual time the OSM data was downloaded, which is more accurate than the year of publication. Only years were recorded because of various time precision. Data were retrieved from 2007 to 2014, indicating the discussion of OSM quality assessment started around 2007 and continued as a trending topic until recent times. A limited number of studies were implemented using national data, signifying current exploration stage of OSM quality analysis. Most studies had European regions as their study areas, which was not surprising considering the massive number of European OSM users. Furthermore, most studies used a reference dataset to evaluate the extrinsic quality of OSM data, which include a mix of governmental and commercial databases. For articles that do not have a reference dataset, some constructed frameworks, some analyzed user behavior or data trust, and the rest studied intrinsic quality using data history.

The frequency of examined data quality criteria is shown in Figure 1. Data completeness dominates the quality analysis of OSM, with positional accuracy and attribute accuracy the second and the third most popular criterion. The common evaluation methods of all criteria are explained in the following paragraphs.

Generally, there are two types of methods to measure data completeness: unit-based and object-based (Table 2). The concept behind unit-based methods is to compare total length, area, or number of objects in OSM with those in a reference dataset. Many studies have used this method because of its easiness of implementation. Hochmair et al. (2015) specially considered street network density and visually compared bike lanes with Google street view to avoid potential mistakes. On the other hand, (automated) feature matching is involved in object-based methods using attributes or geometric properties. For example, street segments have orientation and length, and building footprints can be matched by their centroids or overlap ratio between OSM data and a reference. It is worth to mention that the completeness of land use may be calculated without a reference, since a 100% result means everywhere is covered by a land use feature (Jokar Arsanjani, Mooney, Zipf, et al., 2015).

The methods of measuring positional accuracy are categorized by data types (Table 3). A common method for points of interest is Euclidean distance, while buffer analysis is popular for line segments. A buffer of width "x" is created around a road segment from an authoritative dataset, and the percentage of the corresponding OSM road segment that falls within the buffer is calculated (Goodchild & Hunter, 1997). The buffer size differs from one study to another, indicating that there is no theory behind this method, and empirical analysis is the key to determine the buffer size. In terms of polygon features, centroids, corner points and surface are considered for distance measurements.

The methods of measuring attribute accuracy have four types of usages (Table 4). First, presence of OSM tags (e.g., oneway flags of street segments) can be looked up through examining each geometric object. Second, similarities of strings can be calculated by different algorithms. For example, the Lev-

Figure 1. Summary statistics of examined data quality criteria in Appendix

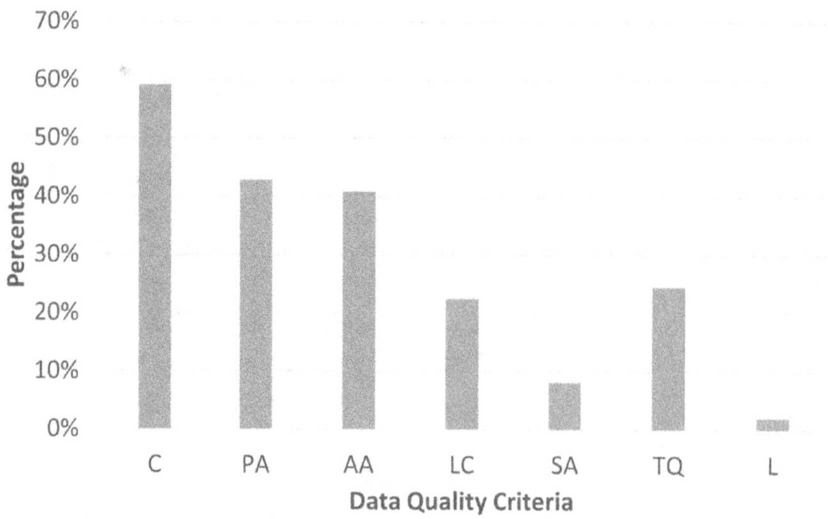

Table 2. Methods of measuring completeness

Types	Criteria	Examples
Unit-based	Number of objects (e.g., attributes, POIs or buildings)	Barron, Neis, & Zipf (2014), Fan, Zipf, Fu, & Neis (2014), Girres & Touya (2010), Haklay (2010), Hecht, Kunze, & Hahmann (2013), Hochmair, Zielstra, & Neis (2015), Jackson et al., (2013), Jokar Arsanjani, Barron, Bakillah, & Helbich (2013), Jokar Arsanjani, Mooney, Zipf, & Schauss (2015), Jokar Arsanjani & Vaz (2015), Mashhadi, Quattrone, & Capra (2015), Neis, Zielstra, & Zipf (2011), Zielstra & Zipf (2010)
	Total length or area	
	Density	Hochmair et al. (2015)
	Visual comparison	
Object-based	Centroids	Hecht et al. (2013)
	Overlap ratio	
	Attribute match (e.g., name)	Jackson et al. (2013), Kalantari & La (2015), Koukoletsos, Haklay, & Ellul (2012), Ludwig, Voss, & Krause-traudes (2011)
	Geometric match (e.g., distance, orientation, length)	

Table 3. Methods of measuring positional accuracy

Data Types	Methods	Examples
Point	Euclidean distance	Girres & Touya (2010)
		Amelunxen (2010)
		Jackson et al. (2013)
Line	Compare actual road conjunction with previous locations	Barron, Neis, & Zipf (2014)
	Hausdorff distance	Girres & Touya (2010)
	Average distance (McMaster, 1986)	
	Buffer analysis (Goodchild & Hunter, 1997; Hunter, 1999)	Haklay, (2010), Jokar Arsanjani, Barron, Bakillah, & Helbich (2013), Ludwig, Voss, & Krause-traudes (2011)
	Bidimentional regression (Friedman & Kohler, 2003; Tobler, 1994)	Helbich, Amelunxen, & Neis (2012)
	G*-statistics (Getis & Ord, 1992)	
Polygon	Surface distance (Vauglin, 1997)	Girres & Touya (2010)
	Average distance of corresponding (corner) points	Fan, Zipf, Fu, & Neis (2014)
	Distance between centroids	Kalantari & La (2015)

enshtein distance is the number of deletions, insertions, or reversals required to transform one string to another. The algorithm was originally developed to tackle the issue of binary information transmission (Levenshtein, 1966). The larger the Levenshtein distance, the greater the differences between strings. Third, numbers can be subtracted, and the absolute values of the results can reflect the differences between them. Finally, thematic accuracy (e.g., for land use accuracy assessment) can be measured by confusion matrix and kappa index.

A framework was constructed exclusively for logical consistency (Hashemi & Ali Abbaspour, 2015). Spatial scenes – sets of spatial objects with spatial relations – are compared in this framework. Topology, distance and direction are some examples of useful spatial relations (Hashemi & Ali Abbaspour, 2015). Here, topology is "the study of qualitative properties that are invariant under distortion of geometric space" (e.g., the London underground map) (Jiang, 2013, p. 128). For instance, two articles from Appendix studied logical consistency of street networks considering topological errors (e.g., connectivity of roads and structure of network), turn restrictions and inter-theme consistency (Girres & Touya, 2010; Neis et al., 2011). Another two articles examined logical consistency of polygons, both using shape similarity ratio in additional to other methods such as turning function distance, number of vertices, mean vertex spacing distance, and feature areas (Fan et al., 2014; Kalantari & La, 2015). Although OSM has a dedicated webpage to record known data errors ("Quality assurance - OpenStreetMap Wiki," 2016), Girres & Touya (2010) mentioned that integrity constraints are not enforced to ensure logical consistency in OSM.

Methods of other data quality criteria are summarized below. Only four out of the 60 articles analyzed semantic accuracy, and two of them compared attributes for the assessment (Girres & Touya, 2010; Jokar Arsanjani, Barron, et al., 2013). Fan et al. (2014) did something special to identify the n:m relations of building footprints between OSM data and a reference dataset. Temporal quality was generally evaluated as a spatial-temporal analysis with the rate and accuracy of changes over time. Level of Details (LOD) assessment can be divided into five schemas including conceptual schema, geometric resolution, semantic resolution, geometric precision and granularity (the size of the minimal features) (Touya & Reimer, 2015). Finally, a number of collected studies analyzed relations between user behaviors or data trust to user information and/or edit history.

Mixed results were found across different locations, times, data types and criteria. Some urban areas with high population density had similar or even better quality than some reference datasets. However, rural areas received less attentions and had scarce coverage. Overall, the findings of collected articles

Table 4. Methods of measuring attribute accuracy

Usages	Criteria	Examples
Measures attribute completeness	Tag presence	Girres & Touya, (2010), Ludwig et al. (2011)
Compares strings (text)	Levenshtein distance (Levenshtein, 1966)	Girres & Touya (2010)
	Similarity ratio (calculated by difflib in Python)	Kalantari & La (2015)
Compares numbers	Difference in speed limits	Ludwig et al. (2011)
Measures thematic accuracy	Classification accuracy by confusion matrix	Estima & Painho (2013), Jokar Arsanjani, Helbich, Bakillah, Hagenauer, & Zipf (2013), Jokar Arsanjani, Mooney, et al. (2015), Jokar Arsanjani & Vaz (2015)
	Kappa index	

follow the two classical geographic theories: Tobler's (1970) first law of geography – near things are more related than others – and the second law of geography – geographic phenomena vary across the globe (spatial heterogeneity) (Goodchild, 2009).

CASE STUDY

According to the Appendix, only a small number of articles evaluated quality of Canadian OSM data (e.g., Meier, 2015; Tenney, 2014). Although Tenney (2014) performed a national study, the results were still preliminary. Thus, there is a need to further evaluate the Canadian OSM quality. The study area here is the Census Metropolitan Area (CMA) of London, Ontario, Canada (Figure 2). London is the eleventh largest CMA in Canada with more than 474,000 inhabitants, including two cities (London and St. Thomas), two municipalities (Thames Centre and Central Elgin) and four townships (Strathroy-Caradoc, Middlesex Centre, Southwold and Adelaide-Metcalfe) (Statistics Canada, 2012). The rate of economic growth in the region was moderate in recent years because of an improved manufacturing sector and a stronger housing market. Two datasets, the source and the reference data, are required for this evaluation of OSM accuracy. The source data are the 2016 OSM metro extracts of London, Ontario from Mapzen[1] in the imposm format[2]. The reference data are the 2015 DMTI road networks from Scholars Geoportal[3], which has a positional accuracy ranging from 0.6 (urban) to 30 m (rural) (DMTI Spatial Inc., 2015). It is therefore hypothesized that urban roads have higher positional accuracy in OSM as well. The 2015 National Road Network (NRN) data collected by Natural Resources Canada is not chosen as the reference dataset because a commercial dataset is preferred when available (Haklay, 2010). The positional accuracy of the NRN data is not specified either (only indicated "in meters") (Natural Resources Canada, 2015).

Methods

The OSM quality, specifically completeness, positional accuracy and attribute accuracy, was assessed using the following techniques and ArcGIS tools (Figure 3 and 4). The attributes were first processed and matched based on Table 5. Evaluation results were classified according to the new road ranks in Table 6. Geometric feature matching was also performed before evaluating the positional and attribute accuracy. The unmatched road segments were identified using the "Detect feature changes" tool in ArcGIS with a search distance of 30 m (the maximum positional offset of the DMTI data) and removed afterwards (Figure 4). The length and density of roads were calculated to analyze the data completeness. This unit-based method was chosen because it is easy to implement and has been used in many previous studies (Table 2). Next, the buffer analysis was used to assess the positional accuracy (Figure 4). This method was validated in the first OSM quality assessment (Haklay, 2010) and other studies (Table 2). Using a self-developed python script and the arcpy library, buffers with widths of 1 to 10 m were created around the DMTI street networks, and the matched OSM road segments that fell within the buffers were clipped for calculating their proportions to the total OSM road length (Figure 5). Finally, the attribute accuracy was evaluated by tag presence, number difference and Levenshtein distance. Tag presence measured whether an OSM road attribute was present if a DMTI road attribute was provided (Figure 4). The absolute difference between two numeric fields were calculated as follows: $d = \left|x - y\right|$.
Levenshtein distance (see the review section for definition) of two text fields was computed using a dynamic programming python script (Levenshtein, 1966).

Figure 2. Study area

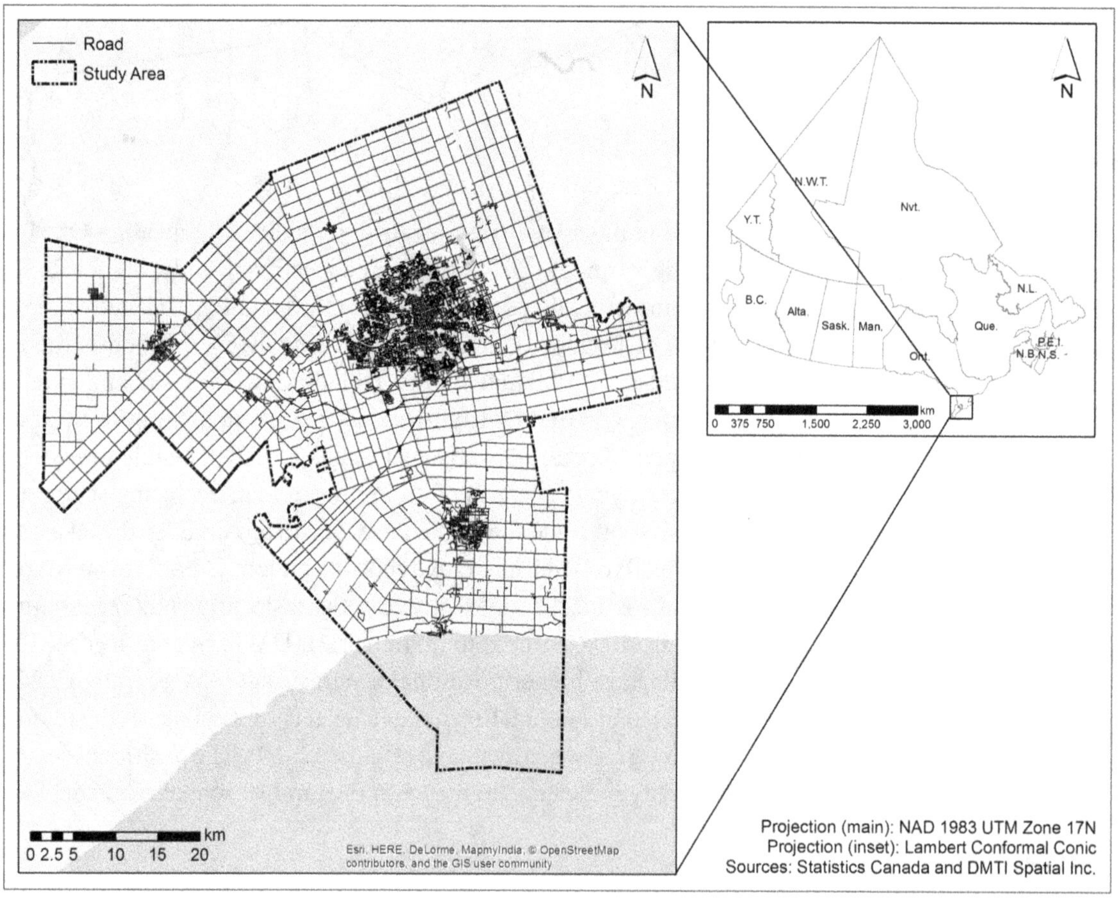

Results and Discussion

Completeness

Figure 6 shows the road lengths by ranks. Many of the ranks have close lengths except rank 0, 5 and 6. Visual examination confirmed that most unclassified (rank 0) road segments of OSM are local roads (rank 5) in suburban areas. Thus, the length difference of rank 5 is actually minimal if the length of rank 0 is added. The difference of rank 6 is large enough to influence the total road lengths because of the large number of footways in the OSM data. This is also the case of the United States (as of 2012) (Zielstra, Hochmair, & Neis, 2013) and Germany (as of 2011) (Neis et al., 2011). If rank 6 is excluded, the difference is significantly reduced. However, OSM has a longer total length than DMTI with or without rank 6, which is different from previous studies since the total length of OSM motorways was still shorter than reference datasets (Neis et al., 2011; Zielstra et al., 2013). The better data completeness potentially benefits from data imports and the increased number of contributors over the years.

The road density of the two datasets is displayed in Figure 7. In general, urban areas especially the City of London and the City of St. Thomas have higher road density, which potentially helps to generate shorter and better routes in navigation applications (Mondzech & Sester, 2011). The location of dense

Figure 3. An overview of the methodology

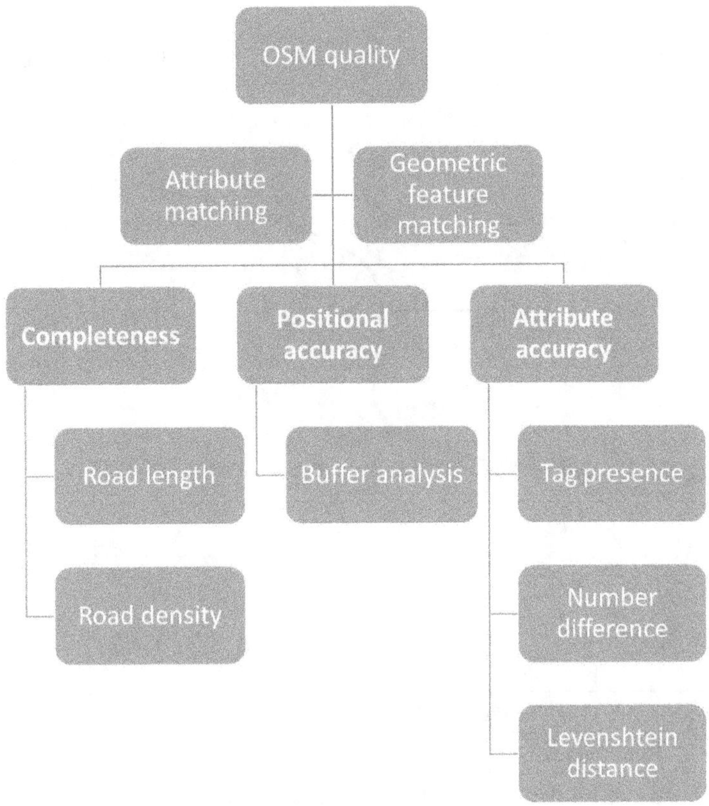

Figure 4. ArcGIS tools used for the evaluation

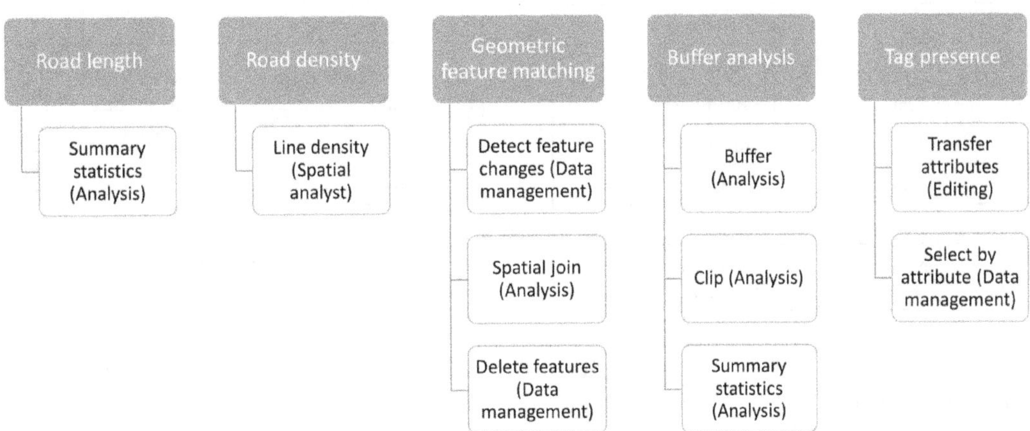

areas verifies that areas with denser population tend to have higher contributions (Jokar Arsanjani & Bakillah, 2014). The maximum density of DMTI is about 12 m/km^2, which is significantly less than the 30 m/km^2 of OSM. The difference is reflected in urban areas where the OSM map has a much darker color. The significant disparity of rank 6 should have great influence on the road density as well.

Figure 5. Example of the buffer analysis
(adapted from Goodchild & Hunter, 1997)

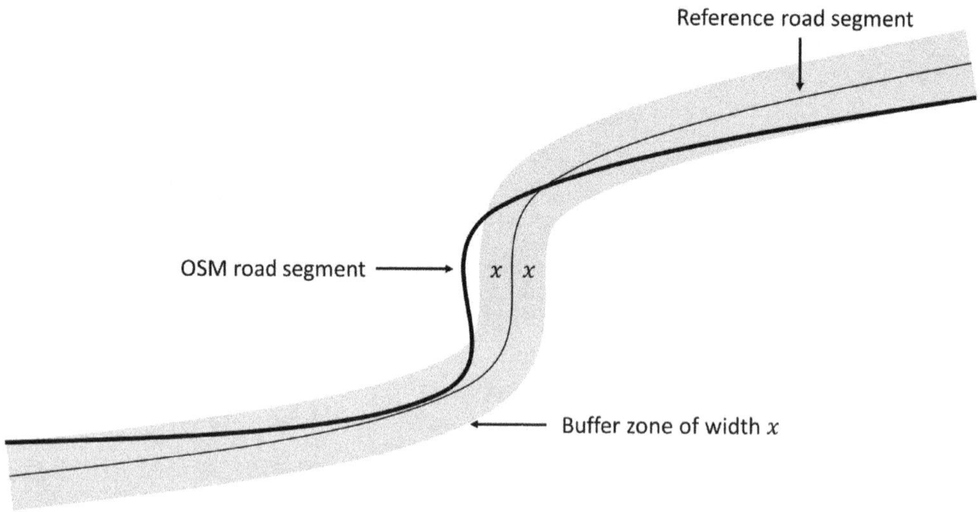

Table 5. Matches of attributes

Field Name	Field Type	Field Description
name	Text	Full street name
length	Number	Length of the road segment
rank	Number	New road classifications
UID	Number	Unique ID
preDir	Text	Prefix direction
preType	Text	Prefix street type
stName	Text	Street name component
sufType	Text	Suffix street type
sufDir	Text	Suffix direction
tunnel	Number	1 = tunnel; 0 = not tunnel
bridge	Number	1 = bridge; 0 = not bridge
oneway	Number	1 = oneway; 0 = two ways; -1 = incorrect input

Table 6. Matches of road classifications

New Rank	DMTI Road Types	OSM Road Types	
0	N.A.	Unclassified	
1	Expressways	Motorway	Motorway_Link
2	Primary Highways	Trunk	Trunk_Link
3	Secondary Highways	Primary	Primary_Link
4	Major Roads	Secondary	Secondary_Link
5	Local Roads	Tertiary	Tertiary_Link
		Residential	Service
6	Trails	Footway	Steps
	Proposed Roads	Path	Track
		Raceway	Cycleway

Positional Accuracy

To improve the results of the geometric feature matching, rank 6 is excluded from the following analysis. Figure 8 shows the proportions of OSM road segments that fall within the buffers of DMTI road segments with a range from 1 to 10 m. Approximately all ranks of roads have a logarithmic increase of their positional accuracy. The average positional offset is 2.3 m, which is significantly better than the results in London, UK and England in 2007 (5.8 m) (Haklay, 2010) and 2009 (7.9 m) (Antoniou, 2011).

Figure 6. Classified road lengths in London, Canada

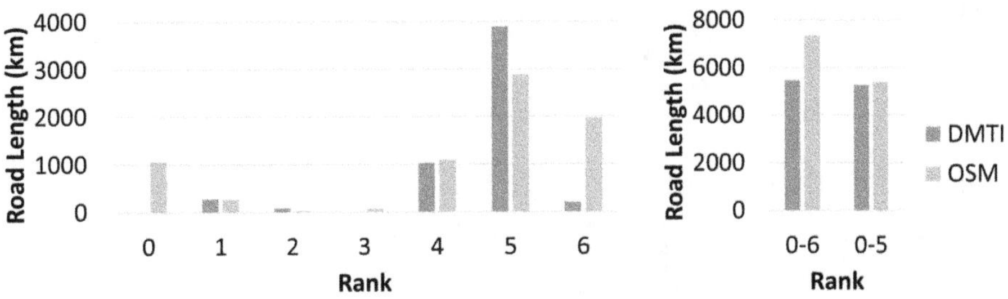

Figure 7. Road density (m/km²) in London, Canada

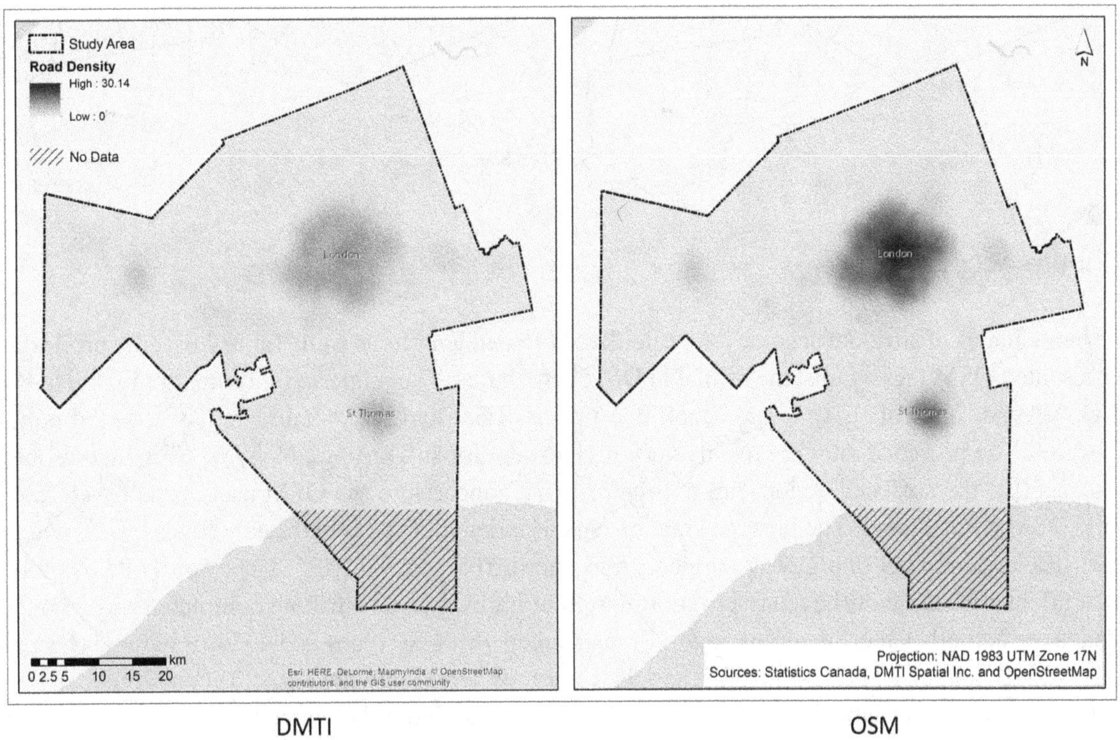

At buffer size of 1 m, the positional accuracy ranges from 14.9 to 59.6%. The accuracy increases at a relatively fast rate until 6 m. After that, the accuracy starts to only increase gently. Over 86% of road segments have positional errors within 5 m, which is also better than 73% of road segments in Germany in 2009(Ludwig et al., 2011). At buffer size of 10 m, most ranks have over 91% of positional accuracy except rank 2 and 3. However, the lengths of roads in these two ranks of exception are relatively short (Table 4), which means their results may not be representative. The most accurate rank at the 10-m buffer is rank 0 (local roads in suburban areas).

Figure 8. Trends of the OSM positional accuracy in London, Canada

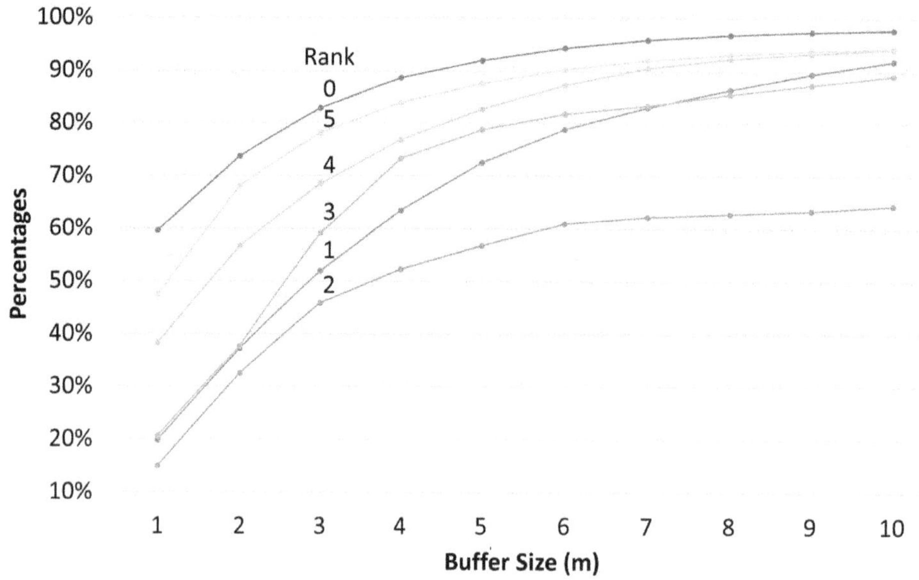

Attribute Accuracy

The percentages of attribute accuracy are calculated by road lengths as well. Table 7 lists the proportions of presented OSM tags against the available DMTI attributes. The numeric fields are not included since all OSM road segments have a rank (rank 0 = unclassified) and the remainders have limited number of entries. The presence rates are mostly very high except for sufDir (e.g., N, S, W, E), which probably indicates that the suffix directions are not the primary concerns to the OSM users or not well-known to the OSM contributors. The presence rate of rank 1 under sufType is extremely low as well, which is because a large number of highway segments miss the suffix type "RAMP". The overall rate of sufType is not affected because of the relatively short length of highway. The attribute completeness of London, Canada are actually superior comparing to French streets (85% for types and 43% for names) (Girres & Touya, 2010) and German streets (82.5% to 94.4% for names) (Ludwig et al., 2011) in 2009.

Table 8 presents the absolute difference of the numeric attributes between the OSM and DMTI data. Only 70.6% of the OSM road segments have matched road classifications, which is largely due to the unclassified local roads in suburban area (the 21.1% that have a difference of 5). The rest of the fields have almost perfect accuracy; however, the results need to be interpreted with caution because of the short total length of tunnels, bridges and oneway roads. Still, the nearly 98% of oneway flag accuracy in London, Canada is better than the 16% completeness in France in 2009 (Girres & Touya, 2010).

Table 9 lists the Levenshtein distance of the text fields. Overall, the longer the field content, the larger the Levenshtein distance. Therefore, preDir and sufDir have excellent accuracy since the length of these fields is one letter. Another reason of the nearly perfect accuracy of preDir is due to its small number of entries, and so does preType. The accuracy results of stName and sufType are lower than the others, but

Table 7. Tag presence of the text fields

OSM Rank	Percent	OSM Rank	Percent
preDir		sufType	
4	100.0%	0	99.5%
5	91.6%	1	27.9%
Overall	93.2%	2	92.8%
preType		3	79.0%
1	99.7%	4	98.5%
3	94.1%	5	97.2%
Overall	99.5%	Overall	96.8%
stName		sufDir	
0	99.6%	0	0.0%
1	97.7%	1	42.0%
2	100.0%	4	69.3%
3	100.0%	5	46.8%
4	99.5%	-	-
5	97.4%	-	-
Overall	98.4%	Overall	62.1%

Table 8. Number difference of the numeric fields

Difference	Percent	Difference	Percent
rank		bridge	
0	70.6%	0	99.5%
1	7.1%	1	0.5%
2	0.9%	oneway	
3	0.0%	0	97.9%
4	0.2%	1	2.1%
5	21.1%	2	0.0%
tunnel		-	-
0	100.0%	-	-
1	0.0%	-	-

Table 9. Levenshtein distance (LD) of the text fields

LD	Percent	LD	Percent
preDir		sufDir	
0	99.7%	0	97.3%
1	0.3%	1	2.7%
preType		sufType	
0	99.4%	0	89.2%
3	0.6%	1	0.0%
stName		2	4.6%
0	86.1%	3	2.2%
1 to 3	3.1%	4	3.8%
> 3	10.9%	5	0.2%

still above 85%. A Levenshtein distance of 1 to 3 usually represents spelling mistakes (Girres & Touya, 2010). However, a small portion of stName and sufType have large Levenshtein distance that is greater than 3. The large Levenshtein distances do not affect the overall accuracy as the average Levenshtein distance of stName is only 0.8, which is significantly smaller than the same variable (4.96) of lake names in France in 2009 (Girres & Touya, 2010).

CONCLUSION

Although OSM has better data completeness and overall good positional and attribute accuracy comparing to DMTI, it still has some quality issues. For example, the majority of local roads in rural areas remain unclassified. Misspelling of street names and suffix types still exists, and a large number of suffix directions are missing as well. Still, the general OSM quality of London, Canada in 2016 has greatly improved comparing to previous studies of US and European regions. An interesting finding is that the local roads in suburban areas (rank 0) actually have the highest level of positional accuracy, which violates the assumption brought up at the beginning of the case study section. This high accuracy of local roads in rural areas is perhaps due to the data import from an old version of NRN starting in 2008 ("Canada Import Status - OpenStreetMap Wiki," 2015) and the limited user-editing afterwards. Hence, it is worth to explore the OSM quality at a larger scale. For instance, there are no reference roads classified as secondary highways (rank 3) in the London CMA, which will not be a problem once the study area is expanded to the national level. In addition, an exploration is still needed for evaluating the trail data (rank 6) if a reference dataset is available. Other future research questions pertaining to OSM and VGI are as follows:

- Which data source, the commercial organization, the governmental data bureau or VGI, should be used under which circumstances?
- Are there better and more efficient methods to evaluate the extrinsic (when a reference dataset is available) and intrinsic (e.g., data history analysis) OSM quality?
- How can one accurately automate the quality assessment process?
- How can one improve OSM quality in general?

REFERENCES

Advertise with us - Gasbuddy Gas Prices. (n.d.). Retrieved July 3, 2016, from https://blog.gasbuddy.com/advertising.aspx

Amelunxen, C. (2010). On the suitability of Volunteered Geographic Information for the purpose of geocoding. *GI_Forum*, 1–8.

Antoniou, V. (2011). *User Generated Spatial Content: An Analysis of the Phenomenon and its Challenges for Mapping Agencies*. Retrieved from http://discovery.ucl.ac.uk/1318053/

Ballatore, A., & Zipf, A. (2015). A conceptual quality framework for volunteered geographic information. *12th International Conference on Spatial Information Theory, COSIT 2015*. Springer Verlag. doi:10.1007/978-3-319-23374-1_5

Barron, C., Neis, P., & Zipf, A. (2014). A Comprehensive Framework for Intrinsic OpenStreetMap Quality Analysis. *Transactions in GIS*, *18*(6), 877–895. doi:10.1111/tgis.12073

Basiri, A., Jackson, M., Amirian, P., Pourabdollah, A., Sester, M., Winstanley, A., … Zhang, L. (2016). Quality assessment of OpenStreetMap data using trajectory mining. *Geo-Spatial Information Science*, *19*(1), 56–68. 10.1080/10095020.2016.1151213

Brovelli, M. A., Minghini, M., Molinari, M., & Mooney, P. (2016). *Towards an Automated Comparison of OpenStreetMap with Authoritative Road Datasets. Transactions in GIS.* doi:10.1111/tgis.12182

Buehler, R., & Pucher, J. (2012). Cycling to work in 90 large American cities: New evidence on the role of bike paths and lanes. *Transportation, 39*(2), 409–432. doi:10.1007/s11116-011-9355-8

Camboim, S. P., Meza Bravo, J. V, & Sluter, C. R. (2015). An investigation into the completeness of, and the updates to, OpenStreetMap data in a heterogeneous area in Brazil. *ISPRS International Journal of Geo-Information, 4*(3), 1366–1388. 10.3390/ijgi4031366

Canada Import Status - OpenStreetMap Wiki. (2015). Retrieved July 3, 2016, from http://wiki.openstreetmap.org/wiki/Canada_Import_Status

Canavosio-Zuzelski, R., Agouris, P., & Doucette, P. (2013). *A Photogrammetric Approach for Assessing Positional Accuracy of OpenStreetMap© Roads. International Journal of Geo-Information, 2(2).*

CBC News. (2013). *CBC News - Google buys Waze mapping app.* Retrieved July 3, 2016, from http://www.cbc.ca/m/touch/business/story/1.1306058

Cipełuch, B., Jacob, R., Mooney, P., & Winstanley, A. C. (2010). Comparison of the accuracy of OpenStreetMap for Ireland with Google Maps and Bing Maps. In *Proceedings of the Ninth International Symposium on Spatial Accuracy Assessment in Natural Resources and Environmental Sciences* (p. 337). CONF, University of Leicester.

Coleman, D. J. (2013). Potential Contributions and Challenges of VGI for Conventional Topographic Base-Mapping Programs. In D. Sui, S. Elwood, & M. Goodchild (Eds.), Crowdsourcing Geographic Knowledge SE - 14 (pp. 245–263). CHAP, Springer Netherlands. doi:10.1007/978-94-007-4587-2_14

Coleman, D. J., Georgiadou, Y., & Labonte, J. (2009). Volunteered geographic information: The nature and motivation of produsers. *International Journal of Spatial Data Infrastructures Research, 4*(1), 332–358.

Corcoran, P., Mooney, P., & Bertolotto, M. (2013). Analysing the growth of OpenStreetMap networks. *Spatial Statistics, 3*, 21–32. doi:10.1016/j.spasta.2013.01.002

DMTI Spatial Inc. (2015). *CanMap® Content Suite Support Manual.* Markham.

Dobson, M. (2013). VGI as a Compilation Tool for Navigation Map Databases. In D. Sui, S. Elwood, & M. Goodchild (Eds.), Crowdsourcing Geographic Knowledge SE - 17 (pp. 307–327). CHAP, Springer Netherlands. doi:10.1007/978-94-007-4587-2_17

Dorn, H., Törnros, T., & Zipf, A. (2015). Quality evaluation of VGI using authoritative data-a comparison with land use data in southern Germany. *ISPRS International Journal of Geo-Information, 4*(3), 1657–167110.3390/ijgi4031657

Eckle, M., & De Albuquerque, J. P. (2015). Quality assessment of remote mapping in OpenStreetMap for disaster management purposes. In L. A. Palen, T. Comes, M. Buscher, A. L. Hughes, & L. A. Palen (Eds.), *12th International Conference on Information Systems for Crisis Response and Management, ISCRAM 2015.* Retrieved from https://www.scopus.com/inward/record.uri?eid=2-s2.0-84947764602&partnerID=40&md5=772a654ed328051eca2e2591b6430aa9

El-Ashmawy, K. L. A. (2016). Testing the positional accuracy of OpenStreetMap data for mapping applications. *Geodesy and Cartography, 42*(1), 25–30. 10.3846/20296991.2016.1160493

Elwood, S. (2006). Critical issues in participatory GIS: Deconstructions, reconstructions, and new research directions. *Transactions in GIS, 10*(5), 693–708. doi:10.1111/j.1467-9671.2006.01023.x

Elwood, S., Goodchild, M., & Sui, D. (2013). Prospects for VGI Research and the Emerging Fourth Paradigm. In D. Sui, S. Elwood, & M. Goodchild (Eds.), Crowdsourcing Geographic Knowledge SE - 20 (pp. 361–375). CHAP, Springer Netherlands. doi:10.1007/978-94-007-4587-2_20

Estima, J., & Painho, M. (2013). Exploratory analysis of OpenStreetMap for land use classification. *Geocrowd, 13*, 39–46. doi:10.1145/2534732.2534734

Estima, J., & Painho, M. (2015). Investigating the Potential of OpenStreetMap for Land Use/Land Cover Production: A Case Study for Continental Portugal. In J. Jokar Arsanjani, A. Zipf, P. Mooney, & M. Helbich (Eds.), OpenStreetMap in GIScience SE - 14 (pp. 273–293). CHAP, Springer International Publishing. doi:10.1007/978-3-319-14280-7_14

Fan, H., Yang, B., Zipf, A., & Rousell, A. (2015). A polygon-based approach for matching OpenStreetMap road networks with regional transit authority data. *International Journal of Geographical Information Science, 30*, 1–17. 10.1080/13658816.2015.1100732

Fan, H., Zipf, A., Fu, Q., & Neis, P. (2014). Quality assessment for building footprints data on OpenStreetMap. *International Journal of Geographical Information Science, 0*(0), 1–20. doi:10.1080/13658816.2013.867495

Feick, R., & Roche, S. (2013). Understanding the Value of VGI. In D. Sui, S. Elwood, & M. Goodchild (Eds.), Crowdsourcing Geographic Knowledge SE - 2 (pp. 15–29). CHAP, Springer Netherlands doi:10.1007/978-94-007-4587-2_2

Forghani, M., & Delavar, M. (2014). A Quality Study of the OpenStreetMap Dataset for Tehran. *ISPRS International Journal of Geo-Information, 3*(2), 750–763. doi:10.3390/ijgi3020750

Friedman, A., & Kohler, B. (2003). Bidimensional regression: Assessing the configural similarity and accuracy of cognitive maps and other two-dimensional data sets. *Psychological Methods, 8*(4), 468–491. doi:10.1037/1082-989X.8.4.468 PMID:14664683

Getis, A., & Ord, J. K. (1992). The analysis of spatial association by use of distance statistics. *Geographical Analysis, 24*(3), 189–206. doi:10.1111/j.1538-4632.1992.tb00261.x

Girres, J. F., & Touya, G. (2010). Quality Assessment of the French OpenStreetMap Dataset. *Transactions in GIS, 14*(4), 435–459. doi:10.1111/j.1467-9671.2010.01203.x

Goodchild, M. F. (2007). Citizens as sensors: The world of volunteered geography. *GeoJournal, 69*(4), 211–221. doi:10.1007/s10708-007-9111-y

Goodchild, M. F. (2009). What problem? Spatial autocorrelation and geographic information science. *Geographical Analysis, 41*(4), 411–417. doi:10.1111/j.1538-4632.2009.00769.x

Goodchild, M. F. (2013). The quality of big (geo)data. *Dialogues in Human Geography*, *3*(3), 280–284. doi:10.1177/2043820613513392

Goodchild, M. F., & Hunter, G. J. (1997). A simple positional accuracy measure for linear features. *International Journal of Geographical Information Science, 11*(3), 299–306. doi:10.1080/136588197242419

Goodchild, M. F., & Li, L. (2012). Assuring the quality of volunteered geographic information. *Spatial Statistics*, *1*, 110–120. doi:10.1016/j.spasta.2012.03.002

Graser, A., Straub, M., & Dragaschnig, M. (2014). Towards an open source analysis toolbox for street network comparison: Indicators, tools and results of a comparison of osm and the official Austrian reference graph. *Transactions in GIS*, *18*(4), 510–526. doi:10.1111/tgis.12061

Gröchenig, S., Brunauer, R., & Rehrl, K. (2014). Estimating completeness of VGI datasets by analyzing community activity over time periods. In C. Granell, S. Schade, & J. Huerta (Eds.), *17th AGILE Conference on Geographic Information Science, AGILE 2014* (pp. 3–18). Kluwer Academic Publishers. doi:10.1007/978-3-319-03611-3_1

Haklay, M. (2010). How good is volunteered geographical information? A comparative study of OpenStreetMap and ordnance survey datasets. *Environment and Planning. B, Planning & Design*, *37*(4), 682–703. doi:10.1068/b35097

Haklay, M. (2013). Citizen Science and Volunteered Geographic Information: Overview and Typology of Participation. In D. Sui, S. Elwood, & M. Goodchild (Eds.), Crowdsourcing Geographic Knowledge SE - 7 (pp. 105–122). CHAP, Springer Netherlands. doi:10.1007/978-94-007-4587-2_7

Haklay, M., Basiouka, S., Antoniou, V., & Ather, A. (2010). How Many Volunteers Does it Take to Map an Area Well? The Validity of Linus Law to Volunteered Geographic Information. *The Cartographic Journal*, *47*(4), 315–322. doi:10.1179/000870410X12911304958827

Haklay, M., & Weber, P. (2008). OpenStreet map: User-generated street maps. *IEEE Pervasive Computing / IEEE Computer Society [and] IEEE Communications Society*, *7*(4), 12–18. doi:10.1109/MPRV.2008.80

Hashemi, P., & Ali Abbaspour, R. (2015). Assessment of Logical Consistency in OpenStreetMap Based on the Spatial Similarity Concept. In J. Jokar Arsanjani, A. Zipf, P. Mooney, & M. Helbich (Eds.), OpenStreetMap in GIScience SE - 2 (pp. 19–36). CHAP, Springer International Publishing. doi:10.1007/978-3-319-14280-7_2

Hayakawa, T., Imi, Y., & Ito, T. (2012). Analysis of quality of data in OpenStreetMap. In *2012 IEEE 14th International Conference on Commerce and Enterprise Computing, CEC 2012* (pp. 131–134). doi:10.1109/CEC.2012.29

Hecht, R., Kunze, C., & Hahmann, S. (2013). Measuring Completeness of Building Footprints in OpenStreetMap over Space and Time. *International Journal of Geo-Information, 2*, 1066–1091. 10.3390/ijgi2041066

Helbich, M., Amelunxen, C., & Neis, P. (2012). Comparative Spatial Analysis of Positional Accuracy of OpenStreetMap and Proprietary Geodata.*Proceedings of GI_Forum 2012: Geovisualization, Society and Learning*, 24–33. doi:10.1080/13658816

Hochmair, H. H., & Zielstra, D. (2013). Development and Completeness of Points Of Interest in Free and Proprietary Data Sets: A Florida Case Study. *Gi_Forum 2013: Creating the Gisociety*, 39–48. 10.1553/giscience2013s39

Hochmair, H. H., Zielstra, D., & Neis, P. (2015). Assessing the Completeness of Bicycle Trail and Designated Lane Features in OpenStreetMap for the United States and Europe. *Transactions in GIS, 19*(1), 63–81. doi:10.1111/tgis.12081

Hunter, G. J. (1999). New tools for handling spatial data quality: Moving from academic concepts to practical reality. *URISA Journal, 11*(2), 25–34.

Index of /planet/full-history. (2016). Retrieved July 4, 2016, from http://planet.openstreetmap.org/planet/full-history/

IRC - OpenStreetMap Wiki. (2015). Retrieved July 4, 2016, from http://wiki.openstreetmap.org/wiki/IRC

ISO, T. C. 211. (2002). *ISO 19113, Geographic Information–Quality Principles.* The International Organization for Standardisation.

Jackson, S., Mullen, W., Agouris, P., Crooks, A., Croitoru, A., & Stefanidis, A. (2013). Assessing Completeness and Spatial Error of Features in Volunteered Geographic Information. *ISPRS International Journal of Geo-Information*, 2(2), 507–530. doi:10.3390/ijgi2020507

Jiang, B. (2013). Volunteered Geographic Information and Computational Geography: New Perspectives. In D. Sui, S. Elwood, & M. Goodchild (Eds.), Crowdsourcing Geographic Knowledge SE - 8 (pp. 125–138). CHAP, Springer Netherlands. doi:10.1007/978-94-007-4587-2_8

Jilani, M., Corcoran, P., & Bertolotto, M. (2014). Automated highway tag assessment of openstreetmap road networks. In M. Schneider, M. Gertz, Y. Huang, J. Sankaranarayanan, & J. Krumm (Eds.), *22nd ACM SIGSPATIAL International Conference on Advances in Geographic Information Systems, ACM SIGSPATIAL GIS 2014* (*Vol. 04–07*–Nove, pp. 449–452). Association for Computing Machinery. doi:10.1145/2666310.2666476

Jokar Arsanjani, J., & Bakillah, M. (2014). Understanding the potential relationship between the socio-economic variables and contributions to OpenStreetMap. *International Journal of Digital Earth, 8*(10), 1–16. doi:10.1080/17538947.2014.951081

Jokar Arsanjani, J., Barron, C., Bakillah, M., & Helbich, M. (2013). Assessing the Quality of OpenStreetMap Contributors together with their Contributions.*16th AGILE Conference.*

Jokar Arsanjani, J., Helbich, M., Bakillah, M., Hagenauer, J., & Zipf, A. (2013). Toward mapping land-use patterns from volunteered geographic information. *International Journal of Geographical Information Science, 27*(12), 2264–2278. doi:10.1080/13658816.2013.800871

Jokar Arsanjani, J., Helbich, M., Bakillah, M., & Loos, L. (2015). The emergence and evolution of OpenStreetMap: a cellular automata approach. *International Journal of Digita Earth, 8*(1), 74–88. 10.1080/17538947.2013.847125

Jokar Arsanjani, J., Mooney, P., Helbich, M., & Zipf, A. (2015). An Exploration of Future Patterns of the Contributions to OpenStreetMap and Development of a Contribution Index. *Transactions in GIS, 19*(6), 896–914.10.1111/tgis.12139

Jokar Arsanjani, J., Mooney, P., Zipf, A., & Schauss, A. (2015). Quality Assessment of the Contributed Land Use Information from OpenStreetMap Versus Authoritative Datasets. In J. Jokar Arsanjani, A. Zipf, P. Mooney, & M. Helbich (Eds.), OpenStreetMap in GIScience SE - 3 (pp. 37–58). CHAP, Springer International Publishing.doi:10.1007/978-3-319-14280-7_3

Jokar Arsanjani, J., & Vaz, E. (2015). An assessment of a collaborative mapping approach for exploring land use patterns for several European metropolises. *International Journal of Applied Earth Observation and Geoinformation, 35*, 329–337. doi:10.1016/j.jag.2014.09.009

Jokar Arsanjani, J., Zipf, A., Mooney, P., & Helbich, M. (2015). An Introduction to OpenStreetMap in Geographic Information Science: Experiences, Research, and Applications. In J. Jokar Arsanjani, A. Zipf, P. Mooney, & M. Helbich (Eds.), OpenStreetMap in GIScience SE - 1 (pp. 1–15). CHAP, Springer International Publishing.doi:10.1007/978-3-319-14280-7_1

Kalantari, M., & La, V. (2015). Assessing OpenStreetMap as an Open Property Map. In J. Jokar Arsanjani, A. Zipf, P. Mooney, & M. Helbich (Eds.), OpenStreetMap in GIScience SE - 13 (pp. 255–272). CHAP, Springer International Publishing.doi:10.1007/978-3-319-14280-7_13

Keßler, C., Theodore, R., & De Groot, A. (2013). Trust as a Proxy Measure for the Quality of Volunteered Geographic Information in the Case of OpenStreetMap. In *Geographic Information Science at the Heart of Europe* (pp. 21–37). Springer; doi:10.1007/978-3-319-00615-4_2

Koukoletsos, T., Haklay, M., & Ellul, C. (2012). Assessing Data Completeness of VGI through an Automated Matching Procedure for Linear Data. *Transactions in GIS, 16*(4), 477–498. doi:10.1111/j.1467-9671.2012.01304.x

Levenshtein, V. I. (1966). Binary codes capable of correcting deletions, insertions and reversals. *Soviet Physics, Doklady, 10*, 707.

Lin, W. (2013). When Web 2.0 Meets Public Participation GIS (PPGIS): VGI and Spaces of Participatory Mapping in China. In D. Sui, S. Elwood, & M. Goodchild (Eds.), Crowdsourcing Geographic Knowledge SE - 6 (pp. 83–103). CHAP, Springer Netherlands.doi:10.1007/978-94-007-4587-2_6

Ludwig, I., Voss, A., & Krause-traudes, M. (2011). A Comparison of the Street Networks of Navteq and OSM in Germany. In *Advancing Geoinformation Science for a Changing World* (Vol. 1, pp. 65–84). Berlin: Springer-Verlag; doi:10.1007/978-3-642-19789-5_4

Mailing lists - OpenStreetMap Wiki. (2016). Retrieved July 4, 2016, from http://wiki.openstreetmap.org/wiki/Mailing_lists

Mashhadi, A., Quattrone, G., & Capra, L. (2015). The Impact of Society on Volunteered Geographic Information: The Case of OpenStreetMap. In J. Jokar Arsanjani, A. Zipf, P. Mooney, & M. Helbich (Eds.), OpenStreetMap in GIScience SE - 7 (pp. 125–141). CHAP, Springer International Publishing. doi:10.1007/978-3-319-14280-7_7

McMaster, R. B. (1986). A statistical analysis of mathematical measures for linear simplification. *The American Cartographr, 13*(2), 103–116. doi:10.1559/152304086783900059

Meier, J. C. H. (2015). *An Analysis of Quality for Volunteered Geographic Information*. JOUR.

Miller, H. J., & Goodchild, M. F. (2015). Data-driven geography. *GeoJournal, 80*(4), 449–461.10.1007/s10708-014-9602-6

Mohammadi, N., & Malek, M. (2015). Artificial intelligence-based solution to estimate the spatial accuracy of volunteered geographic data. *Journal of Spatial Science, 60*(1), 119–135.10.1080/14498596.2014.927337

Mondzech, J., & Sester, M. (2011). Quality analysis of OpenStreetMap data based on application needs. *Cartographica: The International Journal for Geographic Information and Geovisualization, 46*(2), 115–125. doi:10.3138/carto.46.2.115

Mooney, P., & Corcoran, P. (2012a). Characteristics of Heavily Edited Objects in OpenStreetMap. Future Internet, 4(4), 285–305.doi:10.3390/fi4010285

Mooney, P., & Corcoran, P. (2012b). The Annotation Process in OpenStreetMap. *Transactions in GIS, 16*(4), 561–579. Journal Article.10.1111/j.1467-9671.2012.01306.x

Mooney, P., & Corcoran, P. (2014). Analysis of Interaction and Co-editing Patterns amongst OpenStreetMap Contributors. *Transactions in GIS, 18*(5), 633–659. doi:10.1111/tgis.12051

Mooney, P., Corcoran, P., & Winstanley, A. C. (2010). Towards quality metrics for OpenStreetMap. *Proceedings of the 18th SIGSPATIAL International Conference on Advances in Geographic Information Systems GIS 10*, 514–517.doi:10.1145/1869790.1869875

Moovit Company Overview. (2014). Retrieved July 3, 2016, from http://moovitapp.com/wp-content/uploads/2014/07/Moovit-At-A-Glance-USA-4-14.pdf

Mullen, W. F., Jackson, S. P., Croitoru, A., Crooks, A., Stefanidis, A., & Agouris, P. (2015). Assessing the impact of demographic characteristics on spatial error in volunteered geographic information features. *GeoJournal, 80*(4), 587–605.10.1007/s10708-014-9564-8

Natural Resources Canada. (2015). *National Road Network (NRN) - Full metadata*. Retrieved April 23, 2016, from http://geogratis.gc.ca/api/en/nrcan-rncan/ess-sst/c0d1f299-179c-47b2-bcd8-da1ba68a8032.html#full-screen-meta

Neis, P., & Zielstra, D. (2014). Recent developments and future trends in volunteered geographic information research: The case of OpenStreetMap. *Future Internet*, *6*(1), 76–106. doi:10.3390/fi6010076

Neis, P., Zielstra, D., & Zipf, A. (2011). The Street Network Evolution of Crowdsourced Maps: OpenStreetMap in Germany 2007–2011. *Future Internet*, *4*(4), 1–21. doi:10.3390/fi4010001

Neis, P., & Zipf, A. (2012). Analyzing the Contributor Activity of a Volunteered Geographic Information Project — The Case of OpenStreetMap. *ISPRS International Journal of Geo-Information, 1*(3), 146–165. doi:10.3390/ijgi1020146

OSMstats - Statistics of the free wiki world map. (2016). Retrieved July 3, 2016, from http://osmstats.neis-one.org/

Parr, D. A. (2015). *The production of volunteered geographic information: A study of OpenStreetMap in the United States*. Academic Press.

Pourabdollah, A., Morley, J., Feldman, S., & Jackson, M. (2013). Towards an Authoritative OpenStreetMap: Conflating OSM and OS OpenData National Maps Road Network. *ISPRS International Journal of Geo-Information*, *2*(3), 704–728. doi:10.3390/ijgi2030704

Quality assurance - OpenStreetMap Wiki. (2016). Retrieved July 3, 2016, from http://wiki.openstreetmap.org/wiki/Quality_assurance

Quattrone, G., Mashhadi, A., Quercia, D., Smith-Clarke, C., & Capra, L. (2014). Modelling growth of urban crowd-sourced information. *Proceedings of the 7th ACM International Conference on Web Search and Data Mining - WSDM '14*, 563–572.doi:10.1145/2556195.2556244

Raymond, E. S. (2001). *The Cathedral & the Bazaar: Musings on linux and open source by an accidental revolutionary*. O'Reilly Media, Inc.

Sehra, S. S., Singh, J., & Rai, H. S. (2015). Quality assessment of crowdsourced data against custom recorded map data. *Indian Journal of Science and Technology, 8*(33).10.17485/ijst/2015/v8i33/79884

Siebritz, L., Sithole, G., & Zlatanova, S. (2012). Assessment of the homogeneity of volunteered geographic information in South Africa. In M. Shortis & M. Madden (Eds.), *22nd Congress of the International Society for Photogrammetry and Remote Sensing, ISPRS 2012* (Vol. 39, pp. 553–558). International Society for Photogrammetry and Remote Sensing. Retrieved from https://www.scopus.com/inward/record.uri?eid=2-s2.0-84901710768&partnerID=40&md5=e65dd4e6feb4732bcd9d94f1c39a4e19

Statistics Canada. (2012). Focus on Geography Series, 2011 Census. Statistics Canada Catalogue no. 98-310-XWE2011004. Ottawa: Author.

Stein, K., Kremer, D., & Schlieder, C. (2015). Spatial Collaboration Networks of OpenStreetMap. In J. Jokar Arsanjani, A. Zipf, P. Mooney, & M. Helbich (Eds.), OpenStreetMap in GIScience SE - 9 (pp. 167–186). CHAP, Springer International Publishing.doi:10.1007/978-3-319-14280-7_9

Tenney, M. (2014). *Quality Evaluations on Canadian OpenStreetMap Data*. Retrieved from http://rose.geog.mcgill.ca/ski/system/files/fm/2014/tenney.pdf

Tobler, W. R. (1970). A computer movie simulating urban growth in the Detroit region. *Economic Geography*, *46*, 234–240. doi:10.2307/143141

Tobler, W. R. (1994). Bidimensional regression. *Geographical Analysis*, *26*(3), 187–212. doi:10.1111/j.1538-4632.1994.tb00320.x

Touya, G., & Brando-Escobar, C. (2013). Detecting Level-of-Detail Inconsistencies in Volunteered Geographic Information Data Sets. *Cartographica: The International Journal for Geographic Information and Geovisualization*, *48*(2), 134–143. doi:10.3138/carto.48.2.1836

Touya, G., & Reimer, A. (2015). Inferring the Scale of OpenStreetMap Features. In J. Jokar Arsanjani, A. Zipf, P. Mooney, & M. Helbich (Eds.), OpenStreetMap in GIScience SE - 5 (pp. 81–99). CHAP, Springer International Publishing.doi:10.1007/978-3-319-14280-7_5

Van Exel, M., Dias, E., & Fruijtier, S. (2010). The impact of crowdsourcing on spatial data quality indicators. *Proceedings of GiScience, 2011*, 1–4.

Van Oort, P. (2006). *Spatial data quality: from description to application. Publications on Geodesy 60*. DISS, WUR Wageningen UR. Retrieved from http://library.wur.nl/WebQuery/wdab/1788022

Vandecasteele, A., & Devillers, R. (2015). Improving Volunteered Geographic Information Quality Using a Tag Recommender System: The Case of OpenStreetMap. In J. Jokar Arsanjani, A. Zipf, P. Mooney, & M. Helbich (Eds.), OpenStreetMap in GIScience SE - 4 (pp. 59–80). CHAP, Springer International Publishing.doi:10.1007/978-3-319-14280-7_4

Vauglin, F. (1997). *Modèles statistiques des imprécisions géométriques des objets géographiques linéaires*. DISS.

Vaz, E., & Jokar Arsanjani, J. (2015). Crowdsourced mapping of land use in urban dense environments: An assessment of Toronto. *Canadian Geographer, 59*(2), 246–255.10.1111/cag.12170

Wang, M., Li, Q., Hu, Q., & Zhou, M. (2013). Quality analysis of open street map data. In E. Guilbert, B. Wu, & J. Shi (Eds.), *8th International Symposium on Spatial Data Quality* (2W1, Vol. 40, pp. 155–158). International Society for Photogrammetry and Remote Sensing. Retrieved from https://www.scopus.com/inward/record.uri?eid=2-s2.0-84929074534&partnerID=40&md5=cc6cb4934bc6f9cdeea8a05303363311

Yang, A., Fan, H., & Jing, N. (2016). *Amateur or professional: Assessing the expertise of major contributors in openstreetmap based on contributing behaviors. ISPRS International Journal of Geo-Information, 5(2)*. doi:10.3390/ijgi5020021

Zhang, X., & Ai, T. (2015). How to model roads in openstreetmap? A method for evaluating the fitness-for-use of the network for navigation. In Y. Leung & F. Harvey (Eds.), *16th International Symposium on Spatial Data Handling, SDH 2014* (*Vol. 19*, pp. 143–162). Springer Berlin.doi:10.1007/978-3-319-19950-4_9

Zhao, Y. J., Zhou, X. G., Li, G. Q., & Xing, H. F. (2016). *A Spatio-Temporal VGI Model Considering Trust-Related Information. ISPRS International Journal of Geo-Information, 5(2), 19*. doi:10.3390/ijgi5020010

Zhou, P., Huang, W., & Jang, J. (2014). Validation analysis of OpenStreetMap data in some areas of China. In J. Jiang & H. Zhang (Eds.), *ISPRS Technical Commission IV Symposium 2014* (4th ed.; *Vol. 40*, pp. 383–391). International Society for Photogrammetry and Remote Sensing.doi:10.5194/isprsarchives-XL-4-383-2014

Zielstra, D., Hochmair, H. H., & Neis, P. (2013). Assessing the effect of data imports on the completeness of openstreetmap - A United States case study. *Transactions in GIS, 17*(3), 315–334. doi:10.1111/tgis.12037

Zielstra, D., & Zipf, A. (2010). A comparative study of proprietary geodata and volunteered geographic information for Germany.*13th AGILE International Conference on Geographic Information Science 2010 Guimarães, Portugal, 1*, 1–15.

KEY TERMS AND DEFINITIONS

Attribute Accuracy: A measure of the correctness of attributes associated with geometrical shapes. It is also known as thematic accuracy.

Authoritative Dataset: A validated dataset published by governmental agencies or commercial organizations.

Buffer Analysis: A method of measuring positional accuracy. Buffers are created around objects from an authoritative dataset, and the percentages of objects from a test dataset that fall within the buffers are calculated.

Completeness: A measure of the comprehensiveness of a dataset. Data completeness not only reports how much data is missing, but also the amount of data that should be excluded.

Levenshtein Distance: A numerical value that reflects the differences of two strings, which is also known as edit distance.

Linus' Law: Errors can be easily identified with enough number of users in a software development environment.

OpenStreetMap: A crowd-sourced online platform that allows citizen scientists to edit maps of their neighborhood and the world using satellite images.

Positional Accuracy: A measure of the relative and absolute accuracy of coordinate values.

ENDNOTES

1 See https://mapzen.com/data/metro-extracts/

2 See https://mapzen.com/documentation/metro-extracts/overview/#choose-a-file-format

3 See http://geo2.scholarsportal.info/

APPENDIX

Table 10. Summary of recent literatures on quality analysis of OSM

Studies	Time	Study Areas	Reference Data Sources	Data Types			Quality Criteria or Methodology						
				POI	Line	Poly	C	PA	AA	LC	SA	TQ	L
Amelunxen (2010)	N/A	North Rhine-Westfalia, Germany	Geocoding service by Google	x				x					
Cipełuch, Jacob, Mooney, & Winstanley (2010)	2010	Ireland	Google Maps and Bing Maps		x		x		x	x			
Girres & Touya (2010)	2009	France	BD TOPO	x	x	x	x	x	x	x	x	x	x
Haklay (2010)	2007	England, UK	OS Meridian 2		x		x	x					
Haklay, Basiouka, Antoniou, & Ather (2010)	2007	London and England, UK	OS Meridian 2		x		Relationship between average positional error and number of contributors						
Mooney, Corcoran, & Winstanley (2010)	2010	European regions	N/A		x	x			x	x			
Zielstra & Zipf (2010)	2009	Germany^	Tele Atlas		x		x					x	
Antoniou (2011)	2009	England, UK	OS Meridian 2		x			x		x			
Ludwig, Voss, & Krause-traudes (2011)	2009	Germany^	Navteq		x		x	x	x				
Mondzech & Sester (2011)	N/A	Germany	ATKIS		x		Accessibility and length of simulated routes						
Neis, Zielstra, & Zipf (2011)	2007 to 2011	Germany^	TomTom		x		x			x		x	
Hayakawa, Imi, & Ito (2012)	2012	Japan and other regions	N/A	x	x	x	x						
Helbich, Amelunxen, & Neis (2012)	N/A	Germany	Tele Atlas		x			x					
Koukoletsos, Haklay, & Ellul (2012)	N/A	London and Newcastle, UK	OS ITN layer of MasterMap		x		x						
Mooney & Corcoran (2012a)	2011	UK and Ireland	N/A (User behavior)		x		Correlation between numbers of contributors and numbers of tags						
Mooney & Corcoran (2012b)	2011	UK, Ireland, Germany and Austria	N/A		x	x			x				
Siebritz et al. (2012)	2006 to 2011	South Africa	NMA	x	x							x	
Canavosio-Zuzelski, Agouris, & Doucette (2013)	2011	Purdue University, US	USGS National Map and TIGER/Line		x			x					
Corcoran, Mooney, & Bertolotto (2013)	2007 to 2011	Ireland	N/A		x							x	
Estima & Painho (2013, 2015)	2013	Portugal^	CLC	x					x				
Hecht, Kunze, & Hahmann (2013)	2011, 2012	Germany	Official building polygon dataset and ATKIS			x	x					x	
Hochmair & Zielstra (2013)	2012	Florida, US	TomTom, NAVTEQ, ESRI and TIGER/Line	x			x					x	
Jackson et al. (2013)	2011	Denver, US	ORNL	x			x	x					
Jokar Arsanjani, Barron, Bakillah, & Helbich (2013)	2012	Heidelberg, Germany	BKG		x		x	x			x		

continued on following page

Table 10. Continued

Studies	Time	Study Areas	Reference Data Sources	Data Types			Quality Criteria or Methodology						
				POI	Line	Poly	C	PA	AA	LC	SA	TQ	L
Jokar Arsanjani, Helbich, Bakillah, Hagenauer, & Zipf (2013)	2012	Vienna, Austria	GMESUA			x			x				
Keßler, Theodore, & Groot (2013)	2011	Münster, Germany	N/A (Data trust and vandalism)	x	x	x	Trustworthiness (e.g., versions, users, confirmations and tag corrections)						
Pourabdollah, Morley, Feldman, & Jackson (2013)	N/A	UK^	OS VMD		x				x				
Touya & Brando-Escobar (2013)	N/A	France	N/A	x	x	x	Level of Details						
Wang, Li, Hu, & Zhou (2013)	N/A	Wuhan, China	NavInfo		x		x	x	x				
Zielstra, Hochmair, & Neis (2013)	2006 to 2012	US^	TIGER/Line		x		x						
Barron, Neis, & Zipf (2014)	2007 to 2013	US, Spain, Cameroon	N/A (Framework)	x	x	x	x	x	x	x		x	
Fan, Zipf, Fu, & Neis (2014)	2013	Munich, Germany	ATKIS			x	x	x		x	x		
Forghani & Delavar (2014)	N/A	Tehran, Iran	Municipality of Tehran		x		x	x		x			
Jilani et al. (2014)	N/A	London and East Essex, UK	N/A		x				x		x		
Jokar Arsanjani & Bakillah (2014)	2013	Baden-Württemberg, Germany	N/A (User behavior)	x	x	x	Logistic regression relationship between highly contributed areas and socio-economic variables						
Quattrone, Mashhadi, Quercia, Smith-Clarke, & Capra (2014)	2007 to 2012	London, UK	N/A	x								x	
Tenney (2014)	N/A	Canada^	NRN (2011)		x		x	x	x				
Zhou, Huang, & Jang (2014)	N/A	China	National basic data		x		x	x	x	x			
Ballatore et al. (2015)	2015	Germany and UK	N/A (Framework)		x		Conceptual quality: accuracy, granularity, completeness, consistency, compliance and richness						
Camboim, Meza Bravo, & Sluter (2015)	2015	Brazil	IBGE		x	x	x		x			x	
Dorn, Törnros, & Zipf (2015)	2014	Rhine-Neckar, Germany	ATKIS			x	x		x				
Eckle & De Albuquerque (2015)	N/A	Germany	Map from expert mapper			x	x	x					
Hashemi & Ali Abbaspour (2015)	2014	Wörrstadt, Germany	N/A (Framework)	x	x	x				x			
Hochmair, Zielstra, & Neis (2015)	2013	Portland and Miami, US	Buehler & Pucher (2012)		x		x						
Jokar Arsanjani, Helbich, Bakillah, & Loos (2015)	2007 to 2012	Heidelberg, Germany	N/A	x	x	x						x	
Jokar Arsanjani, Mooney, Zipf, & Schauss (2015)	2013	Germany	GMESUA			x	x		x				
Jokar Arsanjani & Vaz (2015)	2013	European cities	GMESUA			x	x		x				
Kalantari & La (2015)	2013	Victoria, Australia	Victorian governmental data			x	x	x	x	x			

continued on following page

Table 10. Continued

Studies	Time	Study Areas	Reference Data Sources	Data Types			Quality Criteria or Methodology						
				POI	Line	Poly	C	PA	AA	LC	SA	TQ	L
Mashhadi, Quattrone, & Capra (2015)	2007 to 2012	London, UK	Navteq and Yelp	x			x					x	
Meier (2015)	N/A	Waterloo, Canada	NRN		x		x	x					
Mohammadi & Malek (2015)	2012	Tehran, Iran	N/A	x	x	x		x					
Mullen et al. (2015)	2011	Denver, US	ORNL	x			Non-spatial and spatial regression relationships between demographic characteristics and C and PA of OSM						
Parr (2015)	2006 to 2013	US^	US census and governmental data	x	x	x	The Activity-Context-Geography Model						
Sehra, Singh, & Rai (2015)	N/A	India	Ground data by smartphone		x		x	x	x				
Vaz & Jokar Arsanjani (2015)	2013	Toronto, Canada	DMTI Spatial Inc.			x			x				
El-Ashmawy (2016)	N/A	Saudi Arabia	Self-collected surveying data	x	x	x		x					
Yang, Fan, & Jing (2016)	2010 to 2014	Germany, France and UK	N/A (User behavior)	x	x	x	Use practice, skill and motivation as themes to identify the contributors' level of expertise						
Zhao, Zhou, Li, & Xing (2016)	2006 to 2014	Berlin, Germany and Pakistan	N/A (Data trust and vandalism)	x	x	x	Trustworthiness (e.g., contributor reputations)						

Note. ^: a national study; abbreviations:
AA Attribute Accuracy
ATKIS German Authority Topographic-Cartographic Information System
BD TOPO Topographic datasets from the French National Institute of Geography
BKG German Federal Agency for Cartography and Geodesy
C Completeness
CLC Corine Land Cover database
GMESUA Global Monitoring for Environment and Security Urban Atlas
IBGE Brazilian Institute of Geography and Statistics
ITN Integrated Transport Network
L Lineage
LC Logical Consistency
N/A Not Available
NMA National Mapping Agencies (South Africa)
NRN National Road Network (Canada)
ORNL Oak Ridge National Laboratory
OS Ordnance Survey, the national mapping agency for Great Britain
PA Positional Accuracy
SA Semantic Accuracy
TIGER/Line Topologically Integrated Geographic Encoding and Referencing (US)
TQ Temporal Quality
VMD Vector Map District

Chapter 3
Traditional vs. Machine-Learning Techniques for OSM Quality Assessment

Musfira Jilani
National University of Ireland, Ireland

Padraig Corcoran
Cardiff University, UK

Michela Bertolotto
University College Dublin, Ireland

Amerah Alghanim
University College Dublin, Ireland

ABSTRACT

Nowadays an ever-increasing number of applications require complete and up-to-date spatial data, in particular maps. However, mapping is an expensive process and the vastness and dynamics of our world usually render centralized and authoritative maps outdated and incomplete. In this context crowd-sourced maps have the potential to provide a complete, up-to-date, and free representation of our world. However, the proliferation of such maps largely remains limited due to concerns about their data quality. While most of the current data quality assessment mechanisms for such maps require referencing to authoritative maps, we argue that such referencing of a crowd-sourced spatial database is ineffective. Instead we focus on the use of machine learning techniques that we believe have the potential to not only allow the assessment but also to recommend the improvement of the quality of crowd-sourced maps without referencing to external databases. This chapter gives an overview of these approaches.

INTRODUCTION

While originally only *spatial professionals* (including cartographers, remote sensing scientists, photogrammetrists, etc.) were able to handle and exploit spatial data in standalone dedicated systems, nowadays a wide range of popular products and applications, used regularly by non-professionals, rely on such data (in particular maps). These include emergency response systems, location-based services, digital globes, etc. Many of these systems provide critical services and therefore require reliable and up-to-date map data. However, mapping is an expensive process which has, until recently, been carried out exclusively by official agencies and specialised companies. Because of the costs involved, authori-

DOI: 10.4018/978-1-5225-2446-5.ch003

tative maps are usually not entirely up-to-date as they are not very often edited to reflect the dynamic changes occurring, for example, in urban environments. Indeed, it has been argued that even in the most advanced countries such as the United States, the official maps produced by the US Census Bureau lack in detail and temporal accuracy primarily due to budget constraints and slow update cycles (Haklay, 2010). Furthermore, because these datasets are expensive to produce, they are often sold at high prices to users that want to incorporate them into their applications and products. This can be an impediment for many ordinary customers and companies.

It is precisely to overcome these problems that the OpenStreetMap (OSM) project started. Founded in July 2004 by Steve Coast with the aim of creating a freely accessible and editable map of the entire world (OSMwiki, 2015), OSM is considered as one of the most successful examples of Volunteered Geographic Information (VGI) (Goodchild, 2007). This new phenomenon, which refers to the crowd-sourcing of spatial (geographic) information, has been facilitated by the integration of cheap Global Positioning Systems (GPS) sensors in many popular mobile devices and the general adoption of advanced technologies such as the Web 2.0.

Similarly to other large crowd-sourced datasets collected and made available online to all, VGI repositories present both advantages and disadvantages. In particular, in addition to the lack of costs and ease of use of these datasets, benefits include the fact that the local knowledge of contributors allows them to add much finer detail (including vernacular names of places, specific features, etc.), which is typically not included by official agencies. Moreover, the global diffusion of crowd-sourcing technologies have the potential to create billions of volunteers (or human sensors) world-wide, thereby providing OSM (and other VGI projects such as wikimapia) with an unprecedented manpower that can update and refine the data in real time. Indeed, as remarked by many (Goodchild, 2007), these projects (OSM, wikimapia, etc.) are able to give rise to previously unknown applications.

However, the availability of these large amounts of data created by volunteers from all different types of backgrounds poses several different challenges. These range from the difficulty to efficiently store and manage this type of Big data, to the effective extraction of relevant information from it that matches the needs of users, to the reliability of this data for the application at hand. Indeed, while some parts of the OSM database have been mass imported from sources such as the TIGER database of the US Census Bureau (Zielstra et al., 2013), most of the map data is produced by local contributors which do not necessarily have any experience in mapping practices. As a consequence, many concerns regarding the quality and credibility of this type of geographic information arise.

Quality assessment is an important research issue because, in order to be able to rely on the use of a dataset for their applications, users want to know that it is of high quality. Even knowing that the data is not of very good quality is an important piece of information that provides an estimate of the confidence with which the data can be used. Therefore, many researchers have dedicated their efforts to tackling this issue. In particular, some authors (e.g., (Guptill and Morrison, 2013) and (Kresse & Fadaie, 2004)) have identified several dimensions of spatial data quality including geometric accuracy, attribute accuracy, temporal accuracy, lineage, logical consistency, and completeness. While the majority of studies have considered geometric accuracy in their approaches to quality assessment, other dimensions, including semantic accuracy did not receive the same level of attention.

The work presented in this chapter covers a brief overview of proposed approaches for analysing the quality of VGI, with particular emphasis on OSM which is the focus of the greatest majority of the literature. We do not claim to provide an extensive review of all the approaches, as this is covered in other chapters. Instead we focus more on approaches based on machine learning techniques, which have emerged recently as a viable alternative to more traditional techniques for data quality assessment. Indeed, it has been shown that in some cases, machine learning techniques can allow not only to automate the process of quality assessment but that they can also be used to improve the overall quality of crowd-sourced spatial data (Jilani et al., 2014; Jilani et al., 2016). This is an exciting research direction and one that seems very promising in providing viable solutions for several Big spatial data issues.

The remainder of the chapter is organised as follows. Section 2 briefly introduces the most popular issues and dimensions of spatial data quality. Section 3 discusses traditional approaches for the assessment of OSM quality, while the following two sections (Section 4 and Section 5) are dedicated to machine learning techniques and more recent methods for quality assessment that rely on the application of such techniques, respectively. Finally, in Section 6 we provide some concluding remarks.

SPATIAL DATA QUALITY

Quality assurance of a spatial dataset is necessary for assessing its suitability for certain applications. The quality of spatial data can be influenced by different factors. Given the spatial nature of the data, sometimes, even a small amount of noise can introduce unexplained phenomena within the dataset. Spatial datasets in general tend to suffer from the problem of preferential sampling (or mapping) which can cause the appearance of local phenomena in the data. Spatial data also suffers from the problem of variable information 'density'. Finally, spatial datasets are generated by onsite measurements which are associated with problems such as the accuracy and precision of the recording instrument, the knowledge of the person taking measurements, etc. The human factor becomes significantly important in the case of OSM data, as the knowledge, intention, and interest of the contributors can have a major impact on the overall quality and density of OSM.

Several dimensions of spatial data quality have been identified in the literature (see, for example, (Guptill & Morrison, 2013) and (Kresse & Fadaie, 2004)). The most commonly used ones assess the following aspects:

- Correctness of geometry of the data with respect to the ground truth (geometric accuracy);
- Quantitative and non-quantitative attributes (attribute accuracy);
- Omission of data and presence of excess (spurious) data (completeness);
- Internal consistency of the data with respect to geographic rules and constraints (logical consistency);
- Correspondence between the semantic information associated with the map objects and real world objects (semantic accuracy);
- Synchronicity of data with the changes in the actual world (temporal accuracy)
- Lineage of map data in terms of their capture and evolution (lineage)
- Fitness of use of the data for particular applications (usage)

While several efforts have focused on assessing the geometric accuracy and completeness of VGI in general and OSM data in particular, other aspects, such as the semantic accuracy, have been less studied. In the next section we discuss issues related to the quality of OSM data and some of the research efforts that have been carried out to tackle such issues.

A more general review of quality assessment methods, which covers not only OSM data but also other VGI repositories and subdivides the data as well as the corresponding assessment methods into types (map-based, text based and image-based), has recently been published by Hansi et al. (2016). In our discussion we only touch on some traditional approaches focussing on OSM data and distinguish the different methods based on the aspects of data quality they assess, to discuss some of the general conclusions they reached in relation to the quality of the OSM repository.

OSM QUALITY RESEARCH

Existing approaches for OSM data quality assessment have been classified as either extrinsic or intrinsic (Zhang and Ai, 2015), (Ali et al., 2015). Extrinsic approaches for OSM quality assessment involve comparisons of the OSM data with external datasets, while intrinsic approaches do not require the involvement of external sources of information, but rather an evaluation of the OSM data itself with respect to specific metrics.

Over the last few years, there has been a proliferation of attempts to analyse the quality of the OSM database, with particular emphasis on the geometry accuracy. However, the greatest majority of these approaches are extrinsic, i.e., they require referencing to accurate ground truth, which may be unavailable. And even if the ground truth is available, it can be argued that this process of comparing a crowd-sourced (heterogeneous) database with authoritative maps is ineffective. Indeed, an effective comparison is difficult because both datasets must first be registered, that is, correct correspondence between objects must first be determined before the objects in question can be compared (Koukoletsos et al., 2012).

In the following sections some of the existing approaches (both explicit and implicit) are reviewed with respect to the aspects of OSM data quality they consider.

Comparisons With External Datasets

Several studies have focused on the comparison of a subset of the OSM database with external datasets. Many of these studies use authoritative data (mostly produced by national mapping agencies) as the ground truth. For example, Haklay (Haklay, 2010) studied the geometric accuracy and completeness of OSM data for England by comparing it with the maps produced by the Ordnance Survey of Great Britain, while Koukoletsos et al. (2012) assessed the completeness and positional accuracy of OSM data in Liverpool using similar techniques.

Girres and Touya (2010) extended Haklay's study to the French OSM dataset and compared it with BD Topo and other official datasets. Their study involved several quality dimensions including geometric, attribute, semantic, temporal, logical, completeness, lineage, and usage. In general, these studies found that different aspects of accuracy and completeness increase proportionally with the number of contributors in the area and that there exists an urban rural divide. Likewise, in terms of temporal accuracy, they showed that the map is more up-to-date when the number of contributors increases. They also analysed logical consistency issues related from the lack of use of integrity constraints. With respect to usage, they found that the current usage was mainly limited to mapping and that owing to heterogeneity in accuracy, applications such as urban planning are not relying on OSM data. With respect to semantic accuracy, the need for a better standardization of the OSM tags was highlighted.

In the context of Iran, Forghani and Delavar (2014) compared the OSM of Tehran with official maps. They used fuzzy logic for evaluating the findings of their comparison. They concluded that while the quality of OSM data in the study area was fairly good, completeness had an uneven spatial distribution. Other comparisons of OSM data with official datasets reached conclusions similar to the ones discussed above. These include, for example, the positional and thematic accuracy and completeness of Greek OSM as compared to the official map service of Greece carried out by Kounadi (2009). Jackson et al. (2013) also used official datasets to assess the completeness and accuracy of some educational POIs (Points of Interest) in US OSM, while in (Siebritz, 2014), the accuracy of OSM data for Cape Town is compared with the official CD:NGI dataset.

In other studies, researchers used alternative data sources. For example, Zielstra and Zipf (2010) compared OSM data with Teleatlas data in Germany. Helbich et al. (2012) studied the positional accuracy for a German city by referencing to TomTom data (and other survey datasets). Ludwig et al. (2011) proposed an automated matching procedure between OSM and Navteq and compared the OSM data for Germany with Navteq. Ciepluch et al. (2010) compared the Irish OSM data with Google Maps and Bing Maps. Their comparison involved quality aspects such as coverage, currency, and positional accuracy.

While the studies discussed in this section (and indeed others not reported here) provide useful insights into some aspects of the quality of OSM data, the very process of assessing the quality of OSM data by comparing it with reference data presents from a number of drawbacks. As already pointed out, due to the high cost of the mapping process, official datasets are often outdated. Furthermore, one of the advantages of volunteered geographic information is that it includes detail that are not captured by official mapping agencies. This includes important semantic information which may be useful, for example, in map generalization, data integration, and data analysis. Finally, it has been pointed out that comparing an official (or other) dataset against data obtained from heterogeneous sources is complex and inefficient (Koukoletsos et al., 2012). In this context, alternative techniques for data quality assessment that do not require the availability of external datasets can be more useful. We discuss some of these approaches in the next section.

Alternative Approaches

Approaches to data quality assessment that do not involve the explicit comparison with external datasets are often referred to as implicit. Goodchild and Li (2012) identified three implicit mechanisms towards assuring the quality of volunteered geographic information and termed them as crowd-sourcing, social, and geographic approaches.

Similar to the studies discussed in the previous section (Haklay et al., 2010), some implicit assessment methods (Napolitano & Mooney, 2012; Neis & Zipf, 2012) have confirmed that there is a relationship between the quality of VGI data and the number of contributors. Other studies analysed the relationship between the number of edits and the quality of features in the OSM database (Scheider et al., 2011; Mooney & Corcoran, 2012a; Mashhadi et al., 2012). Mooney and Corcoran (2012b) and Barron et al. (2014) studied the OSM history data to verify if it can be linked to data quality. In this context, Mashhadi et al. (2013) found that the quality of OSM data does not depend on the number of times they have been edited or revised.

While the majority of work in this area has focussed on the geometric accuracy as the main dimension and indicator of VGI data quality, fewer studies have assessed the semantics of this data. However, data semantics is critical in many applications. On the other hand, semantic accuracy, or more precisely, semantic heterogeneity is one of the biggest challenges in crowd-sourcing projects. In particular, in OSM, semantic information is associated to data in the form of tags attached to geographic features. The association of tags to features is done in a quite unstructured way. In other words, while suggestions of possible tag values are provided to volunteers, they are free to ignore such suggestions and can create their own new tags. Furthermore, differently from many semantic web applications that might rely on deep hierarchical ontologies, the OSM semantics is very "shallow". For this reason, Ballatore and Bertolotto (2011) observed that OSM data is spatially rich but semantically poor. Their work developed an approach for extracting information implicit in the descriptions associated with data in the OSMwiki and on linking it to other online resources (Ballatore et al., 2013). The authors of (Scheider et al., 2011) point out that the current semantic taxonomy in OSM does not adequately support the identification of POIs. Baglatzi et al. (2012) proposed a method for finding structure with respect to POIs semantics by aligning the OSM tags with DOLCE Ultralite (Masolo et al., 2003).

Finally, several websites such as Wikipedia use trust hierarchies to assess the quality of their data (Adler et al., 2008), (Bishr and Kuhn, 2007), (Javanmardi et al., 2010). Following this idea, Van Exel et al. (2011) proposed the use of crowd quality while Kepler and Groot (2013) used the concept of trust for determining the quality of OSM data. Similarly, Neis et al. (2012) used a criterion based on 'user reputation' to develop a rule based system that identifies potential cases of vandalism in the OSM database. Other interesting intrinsic methods for OSM data quality assessment, to some extent similar to the rule based vandalism detection tool described above, rely on the application of machine learning which is the most popular and data-driven approach in the area of artificial intelligence. We discuss these methods in the next section, after introducing some of the most popular machine learning techniques.

MACHINE LEARNING: BACKGROUND

In this section we present some machine learning techniques used in the research contributions we discuss in later sections.

Machine learning has been popularly defined as a field of study that enables computers to learn with experience without being explicitly programmed (Mitchell, 1997). It incorporates approaches that can be broadly classified into three categories: supervised, unsupervised, and reinforcement learning. In supervised learning, a general hypothesis is produced by observing training instances with labels, which is then used to make predictions about future instances. Unsupervised learning aims at finding a structure or hidden regularities (e.g. clusters) or detecting anomalies in unlabelled data. Reinforcement learning tries to maximize cumulative reward: in other words, over time, the agents learn to perform the actions that maximize this reward. Unlike supervised learning, no training instances are presented to the agents.

In the next subsection we present the background material in supervised learning as this represents the most popular approach used and applied to the spatial domain. We focus on five popular learning techniques, namely, Naive Bayes, decision trees, support vector machines, neural networks, and random forests. The structured supervised learning is then introduced with details of the Conditional Random Fields (CRF), which forms the current state-of-the-art in structured learning.

Unstructured Supervised Learning

In the classical or unstructured supervised machine learning the goal is to develop a classifier that can learn from training instances to predict the class of new (or unseen) instances. This classifier is a function that maps an input **x** to an output *y*, where the inputs can be any kind of objects and the output is a real number (Daum'e III, 2006). The output represents a class in the case of classification and a value in the case of regression. The classifier function learns from training instances which are pairs (**x**,*y*), i.e., an input **x** and its associated known output *y*. The underlying assumption is that these training instances are drawn independently and identically (i.e., they come from an independent and identical distribution (*iid*)) from the joint distribution *P*(**x**,*y*): the inputs are not related to each other and there is no exploitation of the dependencies between related (neighbouring) inputs (Grimmett and Stirzaker, 2001). In the following, we introduce some of the most commonly used classical learning algorithms.

SVM is a large margin classifier (i.e., it tries to maximize the distance from the geometric margin or decision boundary of its classifications). It uses kernel functions to map the training instances into a feature space and separates the instances using a hyperplane that provides a large margin between the instances and the decision boundary (Cortes and Vapnik, 1995), (Platt et al., 1998). They have been especially effective with medical images (Torheim et al., 2014), text classification (Joachim, 1998), etc. While SVM can support the processing of massive datasets with millions of dimensions, their major drawback comes from the fact that they provide a kind of black-box model with the interpretation of the parameters of the model being difficult. Also, they do not directly support multi-class classification.

Decision Trees are an intuitive class of learning algorithms. The main idea is to build a tree whose nodes are as pure as possible (contain instances from a single class). The input space is repeatedly partitioned to reach this objective. Decision trees are not lazy learners because a tree is built before a target instance is encountered. Classification of target instance is performed by moving from top to bottom along the branches of the tree. Decision trees are found to be very effective especially with small datasets; large datasets result in complicated trees with huge memory requirements. The computational complexity also scales unfavourably with the number of dimensions. As these are greedy top-down learning techniques, decision trees often suffer from over-fitting especially if the tree is allowed to grow fully (Hastie et al., 2005), (Quinlan, 1996).

Random Forest is one of the most powerful learning techniques today. It is an ensemble approach to decision trees that helps overcome the overfitting problem. According to the ensemble learning technique (Dietterich, 2002), the overall classification accuracy increases by using a combination of multiple learning models. The bagging algorithm (Breiman, 1996) is also an ensemble approach to decision trees, where the output is produced by averaging the base learners (trees). Therefore, bagging is effective if the base learners are stronger (ie. they have high variance and little bias). In the random forest algorithm (Breiman, 2001), all the trees considered in the collection or ensemble have the same expectation values and are "de-correlated". Therefore, they are relatively more stable than bagging. Random forests can be used for both classification and regression.

Neural networks draw their inspiration from the connectivity of neurons in the biological nervous systems. They were first suggested by Turing (Turing, 1950) and are today one of the most commonly used approaches for classification. A neuron forms the basic computational unit of a neural network. It computes a weighted sum of its inputs and sometimes performs a non-linear function on this sum. Neural networks usually fail to give optimal results if the number of training instances is small. They can be used for both supervised and unsupervised learning.

Naïve Bayes is a simple probabilistic classifier that uses Bayes Theorem to calculate a probabilistic model over the training instances (John and Langley, 1995). It typically assumes that the features are independent from each other. Surprisingly, despite this assumption, which hardly correlates to the real world, Naïve Bayes classifiers are successfully used in several applications such as document classification and spam filtering (McCallum and Nigam, 1998). Often, the maximum likelihood estimate is used for parameter estimation.

Structured Learning

In a structured learning, the classifier is a function that maps inputs $\mathbf{x}$ to outputs y where, the inputs $\mathbf{x}$ can be any kind of objects, while the outputs y are *complex* or *structured* objects such as a joint labelling of all the streets in the given network, images, parse trees, etc. In our inherently connected and dependent world, several learning problems fall into the category of structured prediction. While the classical learning techniques can be used to solve them to a certain extent, they cannot and do not in general yield good performance because they fail to exploit the inherent dependencies within such data.

In the following we introduce probabilistic graphical models that allow us to encode the conditional dependencies between various variables in a given learning problem. In particular, we discuss the Conditional Random Field (CRF), by far the most extensively used structured learning algorithm in the machine learning practice. Later, we present an overview of the parameter learning and inference techniques in probabilistic graphical models.

Probabilistic Graphical Models

Murphy (2012) proposes that a probabilistic perspective is perhaps the best approach to solve modern machine learning problems. As the need to deal with *uncertainties* increases, the popularity of probabilistic models gains grounds. Graph theory provides a formal method to define complex probabilistic models (Tombone, 2015). A probabilistic graphical model is a graph-based structure consisting of nodes (or vertices) and edges (or links). The nodes in the model correspond to the random variables and the edges encode the (probabilistic) relationships between these variables (Koller and Friedman, 2009). The random variables (or the nodes) may correspond to discrete or continuous variables. However, in most problems the random variables correspond to discrete variables. There mainly exist two kind of graphical models: directed and undirected. In the *directed graphical models* or the *Bayesian Networks*, the edges of the graph are directed. Such models are specifically useful for modelling problems in which there exists a cause and effect relationship between the random variables (or nodes) of the model. The directed graphs used for modelling such relationships are specifically acyclic as no directed cycles are allowed. The *undirected graphical models*, also known as *Markov Random Fields (MRF)*, are graph structures in which there is no directionality attached to the edges of the graph. In essence, there is no specific causal relationship between the random variables and hence MRFs are more flexible and popular.

More details on probabilistic graphical models can be found in Bishop (2006), Koller and Friedman (2009), Nowozin and Lampert (2011). In particular, a particular probabilistic graphical model that is one of the most commonly used structured learning approaches is the Conditional Random Field (CRF) (Lafferty et al., 2001). It has found extensive applications in domains such as natural language processing (Mihalcea and Radev, 2011), computer vision (Blake et al., 2011), etc. CRFs are used to model the conditional distribution of an output y given an input x. Thus, CRF are discriminative models for which the input variables are not necessarily independent. More details can be found in Kschischang et al. (2001).

Learning and Inference

Graphical models provide a family of probabilistic distributions. Parameters in the model can help identify specific distributions of this family. Some of the parameter learning techniques include cross validation, Structured Support Vector Machines (Tsochantaridis et al., 2004), Maximum Likelihood Learning, Maximum Pseudolikelihood Learning, and Structured Perceptron.

Inference in a CRF model involves computing the posterior probability of a given output being generated by a given input. This problem may be intractable. However, several inference methods exist in the literature which can be used for obtaining an approximate solution. The interested reader is referred to (Koller and Friedman, 2009), (Bishop, 2006) for details of inference methods.

In the next section we present machine learning approaches applied to spatial data including OSM.

MACHINE LEARNING APPLIED TO SPATIAL DATA

Machine learning has been applied to the spatial domain particularly in the areas of aerial imaging and remote sensing. For example, Mantero et al. (2005) carried out partially supervised classification of remote sensing images by applying SVM-based methods for probability density estimation. Mnih (2013) used machine learning techniques for aerial image labelling, while Albert et al. (2014) applied multi-layer CRFs for simultaneously classifying land use and land cover information.

Some work has been carried out also in relation to vector map data. In particular, Lüscher et al. (2009) proposed a method to infer semantic type information corresponding to buildings. They used a Bayesian network (a directed graphical model) where the structure of the model is derived from a pre-defined ontology. They performed non-parametric learning using kernel density estimation and inference using belief propagation. The model uses geometrical and topological features defined by the ontology. Huang et al. (2013) proposed a markov random field model also for inferring the semantic type of buildings. Inference was performed using Gibbs sampling, while learning was not carried out as the model parameters were manually set. They used a set of features to represent the shape of building footprints.

Sester (2000) proposed a decision tree model for discriminating between houses, streets and land parcels. The model relied on geometrical and topological features. Keyes and Winstanley (2001), Keyes et al. (2003) performed supervised and unsupervised classification of buildings using the shape features of Fourier descriptors, moment invariants and scalar descriptors. Walter and Luo (2011) proposed a neural network model for discriminating between different types of geographical objects including streets and buildings. The model used geometrical features relating to the size and shape of objects. The results achieved were very promising but related to a very small dataset. Henn et al. (2012) used an SVM model to infer semantic type information corresponding to buildings in 3-dimensional city models. A large number of geometrical (e.g. area and width) and semantic (e.g. distance to nearest hospital) features were used.

In the following subsections we discuss unstructured learning applied to OSM data quality assessment as well as more recent methods that do not rely on the independent and identical distribution (iid) assumption and therefore take into account the inherent structure of spatial data, whereby entities are not independent but rather interrelated by means of connectivity relations.

Unstructured Learning Applied to OSM Data

In the context of OSM, Funke et al. (2015) developed a machine learning model for automatically completing missing street names in the OSM database. Giannopoulos et al. (2015) developed an SVM-based automated recommendation system for suggesting the categories of spatial entities. Hagenauer and Helbich (2012) proposed an artificial intelligence system for mining urban land use patterns from volunteered geographic information. Ali et al. (2014) developed a rule-based machine learning classifier to automatically distinguish between similar land use classes such as forest, meadow, and park. More recently they applied a similar approach to recommend the most appropriate class for OSM entities (Ali et al., 2016). They focus on overlapping grass-related classes (i.e., marsh vs. swamp, park vs. garden, etc.).

Using shape features similar to (Huang et al., 2013), Fan et al. (2014) proposed a rule-based approach to infer semantic type information corresponding to buildings. Werder et al. (2010) proposed a method for classifying buildings and city blocks in an unsupervised manner using a mixture of Gaussians model. The model used two sets of features which characterize individual buildings and city blocks.

A supervised classification method for automatically deriving road attributes by mining movement trajectories was developed by Van Wilden et al. (2016). They considered 8 different attributes (some of which are not really associated to OSM roads) and using a decision tree algorithm extracted these attributes for different roads from GPS data collected by users from Netherlands. Basiri et al. (2016) applied rule-based mining and clustering methods to GPS trajectories for quality assessment of OpenStreetMap data. The dataset used consisted of tracks collected by users from the towns of Hanover in Germany and Maynooth in Ireland.

Our own previous work in this area (Jilani et al. 2014) relates to the assessment and possible improvement of the semantic quality of street types in OSM. Indeed, streets are the most important feature in the OSM database and as such they have been extensively analysed. While the geometry of streets in OSM mainly comes from GPS trajectories, the semantics are entirely contributor dependent and hence even more prone to errors.

Toward addressing the problem of improving the semantic quality of street types in OSM, we proposed a machine learning based approach. The key idea is that if we can learn the semantic type of streets, we can predict them and make use of these predictions to assess the accuracy of existing semantics and also possibly fill the missing semantics. The semantic types of streets are recorded using the key 'highway' in the OSM database. Thus our focus is to learn the values associated with the key 'highway'. A street network consists of several street segments, where a street segment can be defined as the section of a street between two intersections, or between a dead-end and an intersection. A close observation of a street network suggests that the semantic type information of streets is implicitly contained within connected sets of street segments having uniform semantic type. For example, when comparing two connected sets of street segments belonging to two different street types in the city of London (UK), e.g. tertiary and residential, we observed that the connected set of segments corresponding to the residential type has more dead-ends as compared to the set corresponding to the tertiary type. Similarly, the overall linearity for the tertiary street is much higher than the residential street where linearity can be defined as the degree to which the shape of the street segments is similar to that of a straight line. Furthermore, a residential street is generally only connected to tertiary streets whereas tertiary streets are typically connected to both residential and secondary streets. This example suggests that the features of connected sets of segments of uniform type vary as a function of type.

Therefore, if we can measure the features enabling distinction between various semantic types of streets, we can subsequently use them for learning the semantic types of streets. However, these features are a function of a connected set of street segments where all elements of a particular set have the same semantic type. Such information relating to semantic type is implicitly (as opposed to explicitly) represented in the street network. Therefore, the measurement of such features requires two steps: firstly, determining the sets which are implicitly represented in the street network and secondly, measuring the features of these sets that may allow us to distinguish between different semantic types. In order to extract connected sets of streets of uniform type we constructed a multi-granular representation of the street network (Jilani et al. 2013).

Subsequently we identified features such as the geometrical and topological characteristics of connected sets of street segments (also referred to as streets in this chapter) that may allow us to discriminate between various semantic types of streets. Having measured such features, we applied several classical supervised machine learning classifiers including SVM, random forests, neural networks, and evaluated their performance in terms of classification accuracy, precision, recall, and f-scores. On the basis of this evaluation, a machine learning model to learn the features that characterise various semantic types of streets was developed. This model was used to assess and predict semantic types for streets in the OSM database. The approach produced interesting results and showed the potential of automatically learning the semantic types of streets given their geometrical and topological features.

In this section we discussed machine learning techniques applied to spatial data in general, and the OSM project in particular. Only a few works exist that use machine learning techniques for OSM data quality assessment. The majority of these make the assumption that map objects are independent and have an identical and independent distribution (iid), i.e., they are based on unstructured learning. However, any spatial data is inherently structured (i.e., consists of several spatial dependencies). Therefore, we argue that structured machine learning techniques are more suitable for assessing and improving the quality of spatial data. Towards this goal in the next section we present our work in this area.

Structured Learning for OSM Street Semantic Quality Assessment

In this section we discuss our recent work on the assessment of OSM data quality using structured learning techniques. Specifically, our focus is on the semantic type quality of streets in OSM.

The traditional machine learning framework described in the previous section considers each street as an identical and independent entity. However, street networks, being spatial in nature, consist of inherent spatial and topological dependencies. We, therefore, developed a structured prediction paradigm for inferring the semantic types of streets. A street network is a structured input as it consists of several streets, where not only the streets themselves contain information such as geometry, but also the way in which the streets are connected to each other is important. For such a structured input, we want to obtain a structured output of semantic types of streets covering all the streets in the network. We exploited the CRF framework for performing structured prediction. The CRF framework also allows us to incorporate prior knowledge available to us in the form of crowd-sourced semantics into the learning process. Figure 1 summarizes the overall approach described above using a flow diagram. The input boxes correspond to observations and the process boxes correspond to implementations. Details of this approach and the results obtained are reported in Jilani (2016) and Jilani et al. (2016).

Figure 1. Overall approach explained using a flow diagram where input boxes correspond to observations and the process boxes correspond to implementations

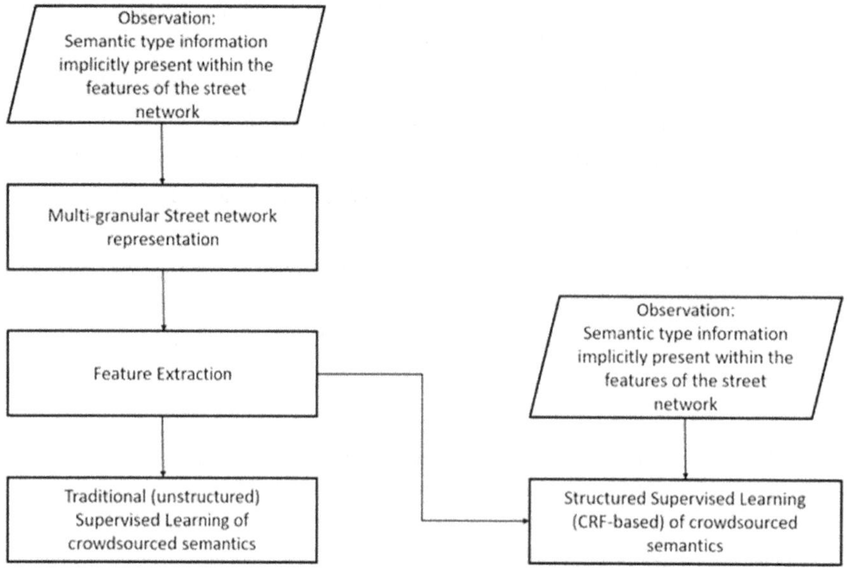

Our work produced very good results that showed the potential of this approach to not only assess the semantic quality of street networks but also to propose recommendations for improving it. In particular, the recommendations provided can be used for either changing the type of a street if deemed incorrect or for proposing a type when such information is missing (incomplete data).

CONCLUSION AND OUTLOOK

In this chapter we discussed issues related to the quality of OSM data. Several overview papers on the topic have been published in the literature (see for example, Senaratne et al, 2016). In particular, some researchers discussing the main directions for future research on OSM data in general and quality in particular pointed out recently that machine learning approaches are very promising (see for example, (Neis and Zielstra, 2014), (Jokar Arsanjani et al., 2015), (See et al., 2016)). Therefore, in this chapter we focussed particularly on efforts towards the applications of such approaches for OSM quality assessment.

Among others we presented our own work in the field. The primary goal of our work is to develop a learning framework that can be used to automatically predict the semantics of streets on a crowdsourced map. These automated predictions can then be used for assessing and improving the semantic quality of such maps. Supervised learning has been, by far the most widely used and successful learning strategy. Our approach also relies on the use of supervised learning techniques. In particular, we developed both unstructured (classical) and structured models with the latter performing significantly better. Although we obtained good results, there are still some open issues that require attention. In particular, our work focussed exclusively on streets while we completely disregard other types of features contained in OSM maps.

The potential of machine learning techniques for OSM quality assessment shown in this chapter has by no means been fully exploited. Even less has the potential of such techniques for automated quality improvement been exploited. This becomes more and more important as the data volumes continue increasing and the edits become more frequent. In the future, efforts should be devoted in continuing to develop structured learning techniques that exploit the inherent spatial relationships between geographic features.

REFERENCES

Adler, B. T., Chatterjee, K., De Alfaro, L., Faella, M., Pye, I., & Raman, V. (2008). Assigning trust to wikipedia content. In *Proceedings of the 4th International Symposium on Wikis*. ACM.

Albert, L., Rottensteiner, F., & Heipke, C. (2014). A two-layer conditional random field model for simultaneous classification of land cover and land use. *ISPRS International Archives of the Photogrammetry, Remote Sensing and Spatial Information Sciences, 1*, 17–24.

Ali, A. L., Schmid, F., Al-Salman, R., & Kauppinen, T. (2014). Ambiguity and plausibility: managing classification quality in volunteered geographic information. In *Proceedings of the 22nd ACM SIGSPATIAL International Conference on Advances in Geographic Information Systems*. ACM. doi:10.1145/2666310.2666392

Ali, A. L., Schmid, F., Falomir, Z., & Freksa, C. (2015). Towards rule-guided classification for volunteered geographic information. *ISPRS Annals of Photogrammetry, Remote Sensing and Spatial Information Sciences, 1*, 211–217.

Baglatzi, A., Kokla, M., & Kavouras, M. (2012). Semantifying OpenStreetMap.*Proceedings of the 5th International Terra Cognita Workshop*.

Ballatore, A., & Bertolotto, M. (2011). Semantically enriching VGI in support of implicit feedback analysis. In *Web and Wireless Geographical Information Systems* (pp. 78–93). Springer. doi:10.1007/978-3-642-19173-2_8

Ballatore, A., Bertolotto, M., & Wilson, D. C. (2013). Geographic knowledge extraction and semantic similarity in OpenStreetMap. *Knowledge and Information Systems, 37*(1), 61–81. doi:10.1007/s10115-012-0571-0

Barron, C., Neis, P., & Zipf, A. (2014). A comprehensive framework for intrinsic OpenStreetMap quality analysis. *Transactions in GIS, 18*(6), 877–895. doi:10.1111/tgis.12073

Basiri, A., Jackson, M., Amirian, P., Pourabdollah, A., Sester, M., Winstanley, A., & Zhang, L. et al. (2016). Quality assessment of OpenStreetMap data using trajectory mining. *Geo-spatial*[Taylor and Francis.]. *Information Science, 19*(1), 56–68.

Bishop, C. M. (2006). Pattern recognition. *Machine Learning*.

Bishr, M., & Kuhn, W. (2007). Geospatial information bottom-up: A matter of trust and semantics. In *The European information society* (pp. 365–387). Springer. doi:10.1007/978-3-540-72385-1_22

Blake, A., Kohli, P., & Rother, C. (2011). *Markov random fields for vision and image processing*. MIT Press.

Breiman, L. (1996). Bagging predictors. *Machine Learning*, *24*(2), 123–140. doi:10.1007/BF00058655

Breiman, L. (2001). Random forests. *Machine Learning*, *45*(1), 5–32. doi:10.1023/A:1010933404324

Ciepluch, B., Jacob, R., Mooney, P., & Winstanley, A. (2010). Comparison of the accuracy of OpenStreetMap for Ireland with Google Maps and Bing Maps.*Proceedings of the Ninth International Symposium on Spatial Accuracy Assessment in Natural Resources and Environmental Sciences*, 337.

Cortes, C., & Vapnik, V. (1995). Support vector machine. *Machine Learning*, *20*(3), 273–297. doi:10.1007/BF00994018

Daum'e, H., III. (2006). *Practical Structured Learning Techniques for Natural Language Processing* (PhD thesis). University of Southern California, Los Angeles, CA.

Dietterich, T. G. (2002). Ensemble learning. The handbook of brain theory and neural networks, 2, 110-125.

Fan, H., Zipf, A., & Fu, Q. (2014). Estimation of building types on openstreetmap based on urban morphology analysis. In *Connecting a Digital Europe Through Location and Place* (pp. 19–35). Springer. doi:10.1007/978-3-319-03611-3_2

Forghani, M., & Delavar, M. R. (2014). A quality study of the openstreetmap dataset for tehran. *ISPRS International Journal of Geo-Information*, *3*(2), 750–763. doi:10.3390/ijgi3020750

Funke, S., Schirrmeister, R., & Storandt, S. (2015). Automatic Extrapolation of Missing Road Network Data in OpenStreetMap.*2nd International Workshop on Mining Urban Data, ICML.*

Giannopoulos, G., Karagiannakis, N., Skoutas, D., & Athanasiou, S. (2015). Automatic recommendations of categories for geospatial entities. *LocalRec*, *15*, 1405.

Girres, J.-F., & Touya, G. (2010). Quality assessment of the French OpenStreetMap dataset. *Transactions in GIS*, *14*(4), 435–459. doi:10.1111/j.1467-9671.2010.01203.x

Goodchild, M. F. (2007). Citizens as sensors: The world of volunteered geography. *GeoJournal*, *69*(4), 211–221. doi:10.1007/s10708-007-9111-y

Goodchild, M. F., & Li, L. (2012). Assuring the quality of volunteered geographic information. *Spatial Statistics*, *1*, 110–120.

Google. (2015). Retrieved from http://www.google.com/mapmaker

Grimmett, G., & Stirzaker, D. (2001). *Probability and random processes*. Oxford University Press.

Guptill, S. C., & Morrison, J. L. (2013). *Elements of spatial data quality*. Elsevier.

Hagenauer, J., & Helbich, M. (2012). Mining urban land-use patterns from volunteered geographic information by means of genetic algorithms and artificial neural networks. *International Journal of Geographical Information Science*, *26*(6), 963–982. doi:10.1080/13658816.2011.619501

Haklay, M. (2010). How good is volunteered geographical information? a comparative study of openstreetmap and ordnance survey datasets. *Environment & Planning*, *37*(4), 682–703. doi:10.1068/b35097

Haklay, M., Basiouka, S., Antoniou, V., & Ather, A. (2010). How many volunteers does it take to map an area well? The validity of Linus law to volunteered geographic information. *The Cartographic Journal*, *47*(4), 315–322. doi:10.1179/000870410X12911304958827

Hastie, T., Tibshirani, R., Friedman, J., & Franklin, J. (2005). The elements of statistical learning: Data mining, inference and prediction. *The Mathematical Intelligencer*, *27*(2), 83–85. doi:10.1007/BF02985802

Helbich, M., Amelunxen, C., Neis, P., & Zipf, A. (2012). Comparative spatial analysis of positional accuracy of OpenStreetMap and proprietary geodata.*Proceedings of GI Forum*, 24–33.

Henn, A., Ro¨mer, C., Gro¨ger, G., & Plu¨mer, L. (2012). Automatic classification of building types in 3d city models. *GeoInformatica*, *16*(2), 281–306. doi:10.1007/s10707-011-0131-x

Huang, H., Kieler, B., & Sester, M. (2013). Urban building usage labeling by geometric and context analyses of the footprint data.*Proceeding of 26th international cartographic conference (ICC).*

Jackson, S. P., Mullen, W., Agouris, P., Crooks, A., Croitoru, A., & Stefanidis, A. (2013). Assessing completeness and spatial error of features in volunteered geographic information. *ISPRS International Journal of Geo-Information*, *2*(2), 507–530. doi:10.3390/ijgi2020507

Javanmardi, S., Lopes, C., & Baldi, P. (2010). Modeling user reputation in wikis. *SStatistical Analysis and Data Mining*, *3*(2), 126–139.

Jilani, M. (2016). *Structured and Unstructured Machine Learning for Crowdsourced Spatial Data* (PhD Thesis). University College Dublin, Ireland.

Jilani, M., Corcoran, P., & Bertolotto, M. (2013). Multi-granular street network representation towards quality assessment of OpenStreetMap data.*Proceedings of the Sixth ACM SIGSPATIAL International Workshop on Computational Transportation Science*, 19-24. doi:10.1145/2533828.2533833

Jilani, M., Corcoran, P., & Bertolotto, M. (2014). Automated highway tag assessment of OpenStreetMap road networks. Proceedings ACMSIGSPATIAL, 449-452. doi:10.1145/2666310.2666476

Jilani, M., Corcoran, P., & Bertolotto, M. (2016). Probabilistic graphical modelling for semantic labelling of crowdsourced map data. Intelligent Systems Technologies and Applications, 213-224. doi:10.1007/978-3-319-23258-4_19

Joachims, T. (1998). Text categorization with support vector machines: Learning with many relevant features. In *European conference on machine learning* (pp. 137-142). Springer Berlin Heidelberg. doi:10.1007/BFb0026683

John, G. H., & Langley, P. (1995). Estimating continuous distributions in bayesian classifiers. In *Proceedings of the Eleventh conference on Uncertainty in artificial intelligence*. Morgan Kaufmann Publishers Inc.

Jokar Arsanjani, J., Zipf, A., Mooney, P., & Helbich, M. (Eds.). (2015). *OpenStreetMap in GIScience: Experiences, Research, and Applications. Lecture Notes in Geoinformation and Cartography*. Springer. doi:10.1007/978-3-319-14280-7

Kepler, C., & Groot, R. T. A. d. (2013). Trust as a proxy measure for the quality of Volunteered Geographic Information in the case of OpenStreetMap. Geographic Information Science at the Heart of Europe, 21–37.

Keyes, L., & Winstanley, A. (2001). Using moment invariants for classifying shapes on large-scale maps. *Computers, Environment and Urban Systems*, *25*(1), 119–130. doi:10.1016/S0198-9715(00)00041-7

Keyes, L., Winstanley, A., & Hea, P. (2003). Comparing learning strategies for topographic object classification.*Geoscience and Remote Sensing Symposium, 2003. IGARSS'03. Proceedings. 2003 IEEE International*, 6, 3468– 3470. doi:10.1109/IGARSS.2003.1294824

Koller, D., & Friedman, N. (2009). *Probabilistic graphical models: principles and techniques*. MIT Press.

Koukoletsos, T., Haklay, M., & Ellul, C. (2012). Assessing data completeness of VGI through an automated matching procedure for linear data. *Transactions in GIS*, *16*(4), 477–498. doi:10.1111/j.1467-9671.2012.01304.x

Kounadi, O. (2009). *Assessing the quality of openstreetmap data. Msc geographical information science*. University College of London Department of Civil, Environmental And Geomatic Engineering.

Kresse, W., & Fadaie, K. (2004). *ISO standards for geographic information*. Springer Science & Business Media. doi:10.1007/978-3-662-08039-9

Kschischang, F. R., Frey, B. J., & Loeliger, H.-A. (2001). Factor graphs and the sum-product algorithm. *Information Theory. IEEE Transactions on*, *47*(2), 498–519.

Lafferty, J., McCallum, A., & Pereira, F. C. (2001). *Conditional random fields: Probabilistic models for segmenting and labeling sequence data*. Academic Press.

Ludwig, I., Voss, A., & Krause-Traudes, M. (2011). A Comparison of the Street Networks of Navteq and OSM in Germany. In *Advancing Geoinformation Science for a Changing World* (pp. 65–84). Springer. doi:10.1007/978-3-642-19789-5_4

Lu¨scher, P., Weibel, R., & Burghardt, D. (2009). Integrating ontological modelling and bayesian inference for pattern classification in topographic vector data. *Computers, Environment and Urban Systems*, *33*(5), 363–374. doi:10.1016/j.compenvurbsys.2009.07.005

Mantero, P., Moser, G., & Serpico, S. B. (2005). Partially supervised classification of remote sensing images through svm-based probability density estimation. *Geoscience and Remote Sensing. IEEE Transactions on*, *43*(3), 559–570.

Mashhadi, A., Quattrone, G., & Capra, L. (2013). Putting ubiquitous crowdsourcing into context. In *Proceedings of the 2013 conference on Computer supported cooperative work*. ACM.

Mashhadi, A., Quattrone, G., Capra, L., & Mooney, P. (2012). On the accuracy of urban crowd-sourcing for maintaining large-scale geospatial databases. In *Proceedings of the Eighth Annual International Symposium on Wikis and Open Collaboration*. ACM. doi:10.1145/2462932.2462952

Masolo, C., Borgo, S., Gangemi, A., Guarino, N., Oltramari, A., & Schneider, L. (2003). Dolce: a descriptive ontology for linguistic and cognitive engineering. *WonderWeb Project, Deliverable D17, 2,* 1.

McCallum, A., & Nigam, K. 1998, July. A comparison of event models for naive bayes text classification. In AAAI-98 workshop on learning for text categorization (Vol. 752, pp. 41-48).

Mihalcea, R., & Radev, D. (2011). *Graph-based natural language processing and information retrieval*. Cambridge University Press. doi:10.1017/CBO9780511976247

Mitchell, T. M. (1997). *Machine learning*. McGraw-Hill.

Mnih, V. (2013). *Machine learning for aerial image labeling* (PhD thesis). University of Toronto.

Mooney, P., & Corcoran, P. (2012a). The annotation process in OpenStreetMap. *Transactions in GIS*, *16*(4), 561–579. doi:10.1111/j.1467-9671.2012.01306.x

Mooney, P., & Corcoran, P. (2012b). Characteristics of heavily edited objects in OpenStreetMap. *Future Internet*, *4*(1), 285–305. doi:10.3390/fi4010285

Murphy, K. P. (2012). *Machine learning: a probabilistic perspective*. MIT press.

Napolitano, M., & Mooney, P. (2012). Mvp osm: A tool to identify areas of high quality contributor activity in openstreetmap. *The Bulletin of the Society of Cartographers*, *45*(1), 10–18.

Neis, P., Goetz, M., & Zipf, A. (2012). Towards automatic vandalism detection in OpenStreetMap. *ISPRS International Journal of Geo-Information*, *1*(3), 315–332. doi:10.3390/ijgi1030315

Neis, P., & Zielstra, D. (2014). Recent Developments and Future Trends in Volunteered Geographic Information Research: The Case of OpenStreetMap. *Future Internet*, *6*(1), 76–106. doi:10.3390/fi6010076

Neis, P., & Zipf, A. (2012). Analyzing the contributor activity of a volunteered geographic information project—The case of OpenStreetMap. *ISPRS International Journal of Geo-Information*, *1*(2), 146–165. doi:10.3390/ijgi1020146

Nowozin, S., & Lampert, C. H. (2011). Structured learning and prediction in computer vision. *Foundations and Trends* R *in Computer Graphics and Vision*, *6*(3–4), 185–365.

OSMwiki. (2015). *OpenStreetMap Wiki Community*. Retrieved from http://wiki.openstreetmap.org/wiki

Platt, J. (1998). *Sequential minimal optimization: A fast algorithm for training support vector machines*. Academic Press.

Quinlan, J. R. (1996). Bagging, boosting, and c4. 5. AAAI/IAAI, 1, 725–730.

Scheider, S., Keßler, C., Ortmann, J., Devaraju, A., Trame, J., Kauppinen, T., & Kuhn, W. (2011). Semantic referencing of geosensor data and volunteered geographic information. In *Geospatial semantics and the semantic web* (pp. 27–59). Springer. doi:10.1007/978-1-4419-9446-2_2

See, L., Mooney, P., Foody, G., Bastin, L., Comber, A., Estima, J., & Rutzinger, M. et al. (2016). Crowdsourcing, Citizen Science or Volunteered Geographic Information? The Current State of Crowdsourced Geographic Information, ISPRS Int. *J. Geo-Inf.*, *5*, 55.

Senaratne, H., Mobasheri, A., Loai Ali, A., Capineri, C., & Haklay, M. (2016). A review of volunteered geographic information quality assessment methods. *International Journal of Geographical Information Science*. DOI: 10.1080 13658816.2016.1189556

Sester, M. (2000). Knowledge acquisition for the automatic interpretation of spatial data. *International Journal of Geographical Information Science*, *14*(1), 1–24. doi:10.1080/136588100240930

Siebritz, L.-A. (2014). *Assessing the accuracy of openstreetmap data in south africa for the purpose of integrating it with authoritative data* (PhD thesis). University of Cape Town.

Tombone. (2015). Retrieved from http://www.computervisionblog.com/2015/04/deep-learning-vsprob-abilistic.html

Tsochantaridis, I., Hofmann, T., Joachims, T., & Altun, Y. (2004). Support vector machine learning for interdependent and structured output spaces. In *Proceedings of the twenty-first international conference on Machine learning*. ACM. doi:10.1145/1015330.1015341

Torheim, T., Malinen, E., Kvaal, K., Lyng, H., Indahl, U. G., Andersen, E. K., & Futsaether, C. M. (2014). Classification of dynamic contrast enhanced MR images of cervical cancers using texture analysis and support vector machines. *IEEE Transactions on Medical Imaging*, *33*(8), 1648–1656. doi:10.1109/TMI.2014.2321024 PMID:24802069

Turing, A. M. (1950). Computing machinery and intelligence. *Mind*, *59*(236), 433–460. doi:10.1093/mind/LIX.236.433

Van Exel, M., Dias, E., & Fruijtier, S. (2011). *The impact of crowdsourcing on spatial data quality indicators*. GiScience.

Walter, V., & Luo, F. (2011). Automatic interpretation of digital maps. *ISPRS Journal of Photogrammetry and Remote Sensing*, *66*(4), 519–528. doi:10.1016/j.isprsjprs.2011.02.010

Werder, S., Kieler, B., & Sester, M. (2010). Semi-automatic interpretation of buildings and settlement areas in user-generated spatial data. In *Proceedings of the 18th SIGSPATIAL international conference on advances in geographic information systems*, (pp. 330–339). ACM. doi:10.1145/1869790.1869836

van Winden, K., Biljecki, F., & van der Spek, S. (2016). Automatic Update of Road Attributes by Mining GPS Tracks. *Transactions in GIS*, *20*(5), 664–683. doi:10.1111/tgis.12186

Zhang, X., & Ai, T. (2015). How to model roads in openstreetmap? a method for evaluating the fitness-for-use of the network for navigation. In *Advances in Spatial Data Handling and Analysis* (pp. 143–162). Springer. doi:10.1007/978-3-319-19950-4_9

Zielstra, D., Hochmair, H. H., & Neis, P. (2013). Assessing the effect of data imports on the completeness of openstreetmap–a united states case study. *Transactions in GIS*, *17*(3), 315–334. doi:10.1111/tgis.12037

Zielstra, D., & Zipf, A. (2010). A comparative study of proprietary geodata and volunteered geographic information for germany. *13th AGILE international conference on geographic information science*.

Chapter 4
VGI in the Geoweb:
An Experiment to Test Data Reliability

Michael Buzzelli
The University of Western Ontario, Canada

Kenwoo Lee
The University of Western Ontario, Canada

David Brown
The University of Western Ontario, Canada

Justin Mullan
The University of Western Ontario, Canada

ABSTRACT

The advent of user-generated content, crowdsourcing and other forms of lay data generation have led to opposing arguments about the quality and reliability of data in the geoweb,. The main focus of this chapter is an 'experiment' to test the quality, validity and lay monitoring of volunteered geographic information (VGI) data. Given the growing importance of VGI, in particular its very different sources and potential uses, it is important that we also consider how this movement affects the ways in which we re-envision the pedagogy of geographic education. Accordingly, a sub-theme of this paper focuses on the manner in which the VGI experiment is undertaken: the experiment is run with students as a means of complementing their otherwise technical GIS training with primary research that exposes them to the wider social issues and debates relating to geographic data. We discuss the implications of this research project both for observers of the development of VGI and the pedagogy of GIS teaching and learning.

INTRODUCTION

This chapter is situated within a broad context of rapidly changing spatially referenced data infrastructures and technologies that substantially alter our conception of geographic information systems (GIS). A fundamental characteristic of these developments is that GIS have expanded beyond the expert domain and into everyday use. For example, 'locative media' forecast elections in near-real-time via web-based interfaces. Geographic location can now inform the search and purchase of products and services. And in the Web 2.0 world, georeferenced peer-to-peer content such as still images and video are continuously generated and exchanged. In general, geographic information is increasingly produced, accessed, disseminated and used via the Internet by lay citizens.

DOI: 10.4018/978-1-5225-2446-5.ch004

Of particular interest in this chapter is the rise of volunteered geographic information (VGI) in the geoweb because it fundamentally calls into question the 'authority' with which data is traditionally provided; by the state or other legitimate authority. The emergence of VGI - - information or data provided by ordinary citizens - - raises expert-or-lay tensions surrounding the origin, exchange, use and maintenance of geographic data. The main question of this chapter is: can we rely on the lay public to provide, maintain and monitor quality data, including geospatial data, in the Web 2.0 world? The main focus of this chapter is to address this question with an experiment of user-generated content in the Wikimapia web-mapping site. In answering this question, we recognise that the rise of VGI and related movements in the geoweb may be generating a 'neogeography' that necessitates a rethinking of GIS education. To address this issue, a secondary theme of this chapter is the pedagogical exercise used to undertake the main experiment. The pedagogical exercise is undertaken via a student-led project in which the pedagogy of GIS education is tooled to meet the new and qualitatively different nature of VGI. We conclude the chapter with a discussion of our experiment findings and what these signal for the ways in which we may reimagine GIS teaching and learning.

VGI IN THE GEOWEB: A DEBATE

VGI, a movement encompassing user generated content for the Internet, crowdsourcing and other forms of lay data and information generation, is both mirror and molder of wider social movements (Sui et al, 2012). We appear to be on the cusp of a fundamental shift in information collection and dissemination that increasingly incorporates volunteers or citizens as key actors (Elwood, 2011; Haklay, 2012). However these movements evolve, we argue that their importance will pivot around a key point: the extent to which volunteered data and information is 'reliable' in the broadest sense (Goodchild and Li, 2012). One side of this debate says that lay-volunteered data in the geoweb is championed by citizen sensors (Goodchild, 2007) who represent an opportunity for the geoweb to facilitate data provision, quality monitoring and citizen cartography. By contrast, others see the advent of user-generated data on the Internet as the rise of the 'the cult of the amateur' (Keen, 2007). This view refutes the 'wisdom of the crowd' and suggests that the lack of authority may usher in poor data quality and uncertainty around lay collection, dissemination and use of geodata.

We have begun to see examples of the kinds of answers that may arise in relation to our core question of VGI data quality and reliability. For instance, user-generated geoweb content may occasion mischief for some (Coleman et al., 2009). If true then there may be a need for some form of authority or official validation of volunteered data (Flanigan and Metzger, 2009). Yet as Cha et al. (2007) argued in relation to Youtube, it may be too soon to judge the reliability of user generated content given its novelty. Perhaps only the historian will be able to discern if and when volunteered information will have achieved widespread acceptance. We acknowledge that this might be true of Wikimapia (http://wikimapia.org/#lang=en&lat=43.650000&lon=-79.380000&z=11&m=b&search=toronto), the resource used here to test the sensor-versus-amateur debate, though we also argue as Elwood (2008) did several years ago that the advent of VGI bears ongoing research and interpretation.

Ever since Goodchild (2007) identified VGI, researchers have taken up the issue of data quality and reliability in various ways. Among the earliest examples is Haklay (2010) who undertook an analysis of data quality in the OpenStreetMap (OSM) system. By comparing OSM data with the Ordnance Survey for London and England as a whole, he found that OSM had produced 'fairly accurate' data. The popularity of OSM has generated many studies in this vein. For instance, Girres and Touya (2010) built upon and extended Haklay's work with a focus on France but, by contrast, found less certainty around site's data (focusing on a wider range of data, including object geometry, attributes, accuracy, logical consistency and usage among others). Similarly, Moony and Corcoran (2012b) found serious flaws in OSM data for Ireland, the UK, Germany and Austria based on what they termed 'highly edited' OSM objects (including tags and annotations edited at least 15 times over the object's history). They attribute this to the ways in which users interact with the OSM system but also to how user content is facilitated by the system itself. In a parallel study, Moony and Corcoran (2012a) showed that OSM's heavily edited objects exhibit constant change and suggest this is a data limitation for site users to be aware of. To address these kinds of issues, Grira et al. (2009) provided an early method in an attempt to impart confidence on the part of VGI data users by developing a classification scheme of data according to its intended use. More recently, others have adopted a different approach. Still using OSM, Koukoletsos et al. (2012) and Barron et al. (2014) aimed to develop methods to assess VGI data quality and reliability without the use of a reference data set, such as using OSM's own data history logs.

This string of studies helps us to identify two particular themes that are relevant in the present chapter. First, the focus on data quality and reliability suggests nuance within our broad theoretical crowd wisdom-versus-amateur tension. For instance, Haklay's (2008) early work points to the wisdom of the crowd yet Moony and Corcoran (2012b) echo Flanagan and Metzger (2009) in suggesting a need for moderated VGI systems. In fact a close read of these contributions points to nuance and overlap in the extent to which VGI is viewed as reliable and how it may be supported by various kinds of data mediators. From this follow a host of issues that will impact if, how and why VGI might come to be viewed as reliable and acceptable (e.g. Johnstone and Sieber, 2012; Leszczynski, 2012).

A second theme evident in these sorts of studies is the method by which VGI data may be assessed. For instance, Haklay (2010) found it necessary to compare OSM to the Ordnance Survey. And whereas Barron et al. (2014) pursued a strategy to bypass the need for comparison to a reference data set (as noted above), they nevertheless concluded (Barron et al., 2014, pp. 892): "absolute statements on data quality are only possible with a high quality reference dataset as a basis for comparison". Thus, a site's data history (if available) may only take us so far. Instead, a reference data set (along with moderation and perhaps also the aid of linus' law (see Haklay et al., 2010)) offer alternative approaches.

Undergirding these important issues of VGI data quality and reliability as well as the methods by which these are assessed is the data volunteer. What also becomes clear to the observer of these trends is that the user (who s/he is, what they provide and why they provide it) in user-generated-content is of vital importance (Goodchild 2009; Haklay et al., 2010). Amidst the theoretical debates and deluge of Internet geodata, we argue, is a growing need for observers of VGI to understand the origins of these data over and above their technical qualities and applications. Accordingly, our experiment of VGI data quality and reliability was run as a pedagogical exercise, namely an academic student-led research project in which the course instructor acted both as mentor and embedded group member. In place of - - or at least alongside - - the technical training traditionally core to the GIS field is the need for new approaches to teaching and learning necessitated by the advent of VGI. Thus we have parallel themes in this chapter: primary research undertaken via pedagogy aimed at coming to terms with societal issues and debates about lay-generated data in the geoweb.

METHODS AND DATA

We designed an experiment to answer our main research question: can we rely on the lay public to provide, maintain and monitor quality (geospatial) data in the Web 2.0 world? This exercise was undertaken in Spring 2010 using Wikimapia. This is a public site launched in 2006 and it combines Google Maps with the ability for users to 'contribute information by 'tagging' places anywhere on Earth. Users are able to create and edit point or polygon tags as well as add metadata in the form of plain text, web links and up to seven pictures per place.

At the time of this exercise Wikimapia had over 11 million unique places around the world and was growing at a rate of about 10,000 places per day. The site had a ranked user profile system totalling over 700,000 and was growing by over 1,000 users per day. According to the site (http://wikimapia.org/about/; accessed 15 September 2016):

Wikimapia data is wholly made by Internet volunteers, who contribute to the project on their free will. Marking places, adding descriptions provided with proof links, giving them appropriate categories and uploading photos to Wikimapia let people easily share information about the world with their friends or publicly. While contributing, user gains experience and gets access to more advanced tools (e.g., linear features for marking roads, rivers, railroads and ferry routes). A user can also share Wikimapia data to blogs, sites and social networks, as there is an option to put the map on your page.

The site's documentation page is linked on this same web page (i.e. http://wikimapia.org/docs/Main_Page) and provides a suite of information on the site itself, its policies, tools, the user interface, user profiles/accounts and the user community. At the time of this study, individuals could create a profile on Wikimapia through which data on the site could be added or manipulated. The user profiles were divided into three levels: UL0, UL1 and UL2. UL0 was reserved for new users, allowing them to add and edit places as well as use the site's personal messaging system. UL1 was assigned to most user profiles after a few days and, in addition, UL1 users were allowed to edit polygons (place outlines), delete places and contribute to the user forum. UL2 status was reserved only for a small number of users, and was granted solely at the discretion of the Wikimapia administrators. These users had the ability to delete large numbers of tags, add protections to certain tags and initiate the banning process of other users. Users also had the ability to vote for or against another user's contributions, thereby affecting that user's ranking on the website. At present, a full description of a longer list of possible user levels (nine numbered levels and four special categories) is found at this URL (http://wikimapia.org/docs/Community#User_levels) including how these levels are attained and potentially also lost in the course of users' activities (in the collection or loss of 'Experience points').

For this experiment, a small sample of user-generated content on the site was temporarily systematically modified and tracked to document and analyse the extent to which data were monitored and corrected by the site's volunteer users. Specifically, place names (such as a sports stadium or public institutions), geographies (neighbourhood boundaries or hiking trails) and metadata (descriptions and imagery of tagged places) were temporarily altered and then tracked in order to document the length of time needed for the user community to correct or further modify the data. This was done for a range of sites in several communities across Canada. It is critical to stress here that, as a condition for completion of this student-led course project, all data on Wikimapia were restored to their status prior to commencement of this experiment.

Relating to the latter point, the research methodology was undertaken primarily by students in Geography 4220, an undergraduate honours capstone course for students enrolled in the GIS programme at The University of Western Ontario (http://www.westerncalendar.uwo.ca/Archive/2011/pg915.html#39487). With the input and guidance of the course instructor, the students conceived, designed, and completed the project in partial fulfilment of the course requirements. This included an initial proposal and feedback stage, weekly meetings, a full-course presentation and final assessment. The project represented a significant share (65%) of students' overall assessment in the course. This course and assessment design was aimed at four broad objectives: to introduce students to original primary research; to do so by working collaboratively in an otherwise individualistic GIS curriculum; to do so by embedding the instructor within the group in a 'Socratic' teaching and learning approach to ensure research standards and feasibility (given the confined teaching term); in the case of this project topic, to expose students to the wider issues and debates of geodata beyond their otherwise technical GIS training.

RESULTS

The study commenced on 28 February 2010 with modification of place names, geographies and metadata across several Canadian communities. Multiple user accounts were created and used. The modifications totalled 135 sites and geographies, namely: 40 universities; 11 city halls; 42 landmarks and 42 sports venues. As shown in Table 1, 269 modifications were made across the above sites and geographies. Many of our data alterations were quickly corrected by users. In general, about half to two-thirds (in the case of meta-data) of our initial data changes were corrected by community users and nearly always back to their original form (one change saw a further error introduced by a community user). Thirteen community users were involved in the corrections to our site data modifications. On average, the community user corrections that were made were completed in four days. Figure 1 provides an example. Blue Mountain, a popular ski and recreation resort, was re-labelled by the project team as 'Gold Mountain' and occasioned the response shown from a community user. While providing an indication of the degree to which user-generated content is monitored and maintained by the community, what was most surprising is that this initial experiment phase came to an end when high-level Wikimapia users banned the research group's user accounts. A further strategy was necessary.

To address this issue and complete the exercise, further user accounts were added to Wikimapia and new and correct data of all types were added over a period of two weeks. The aim here was to build accounts that had a track record of acceptable data contributions. This might then permit further data alterations (including to the new data added by the group itself). In addition, selected neighbourhood names in Toronto and Vancouver were exchanged. Once again the incorrect modifications were introduced within a single day. In general, the same broad patterns were found: community users quickly

Table 1. Wikimapia site data alterations and corrections by community users, by data type, Spring 2010

	Geographic Polygon	*Metadata (description, etc.,.)*	*Place or feature name*	*TOTAL*
Change introduced by the group	52	93	124	269
Corrected by community user	21	60	65	146 (54%)

Figure 1. Example of a community user identifying incorrect data (place name "Gold Mountain" in place of the correct name Blue Mountain

corrected mistakes including those made to new correct data introduced by the team. For example, six of the eight neighbourhood name changes in both Toronto and Vancouver were corrected within one day. Figure 2 shows a community user's response when Rosedale, an affluent neighbourhood in Toronto, was re-labelled as Regent Park - - a nearby public housing neighbourhood. As with the example of Blue Mountain above and indeed many of the user corrections, reactions were swift and sometimes stern.

Figures 3 and 4 are further examples illustrating again the short community response time as well as the degree to which users feel the site data ought to be correct and protected. In the case of Winnipeg City Hall, the community user's incursion suggests that volunteers are not only monitoring labels and metadata but also the correct location of points of interest. In the case of Clarington municipality, a user appears to be monitoring data on a regular basis. S/he corrected their own spelling error after reintroduc-

Figure 2. A user comment on the labelling of Toronto's affluent Rosedale neighbourhood Rosedale (includes metadata describing the neighbourhood)

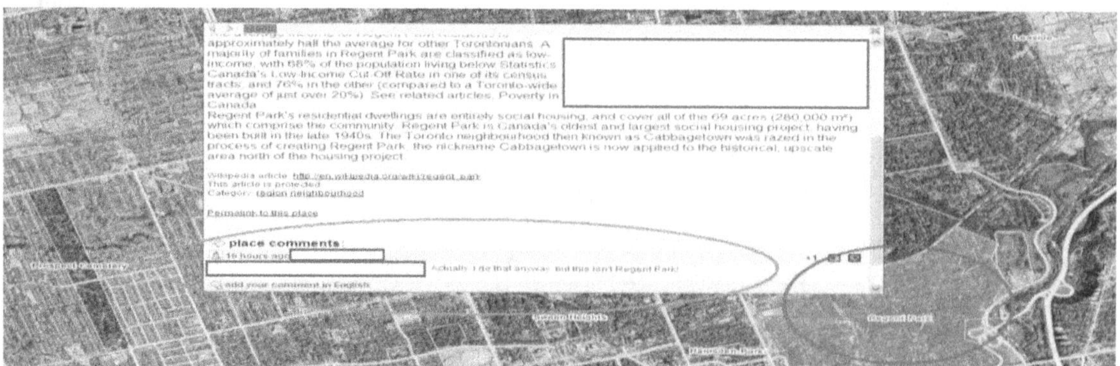

Figure 3. A highly ranked Wikimapia user questioning the placement of Winnipeg City Hall

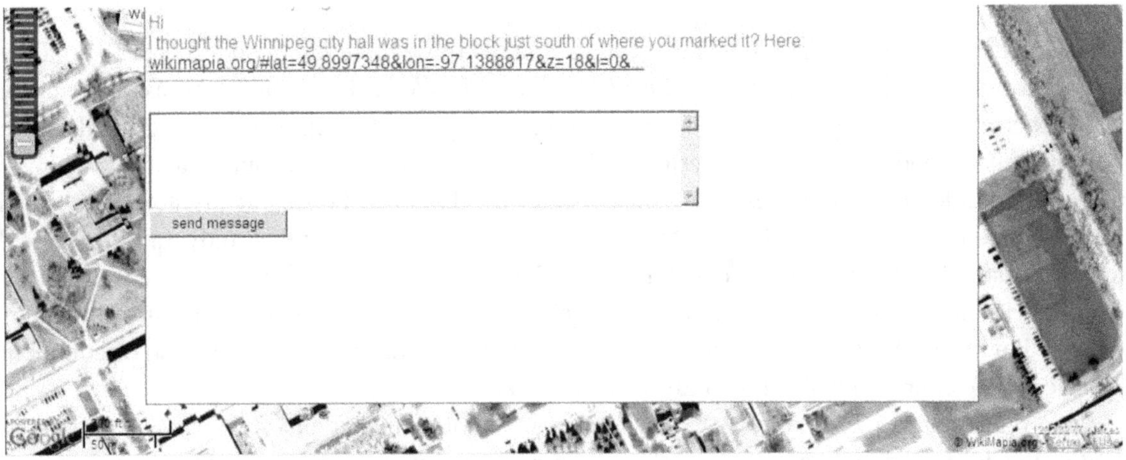

Figure 4. Community user feedback on an incorrectly labelled polygon

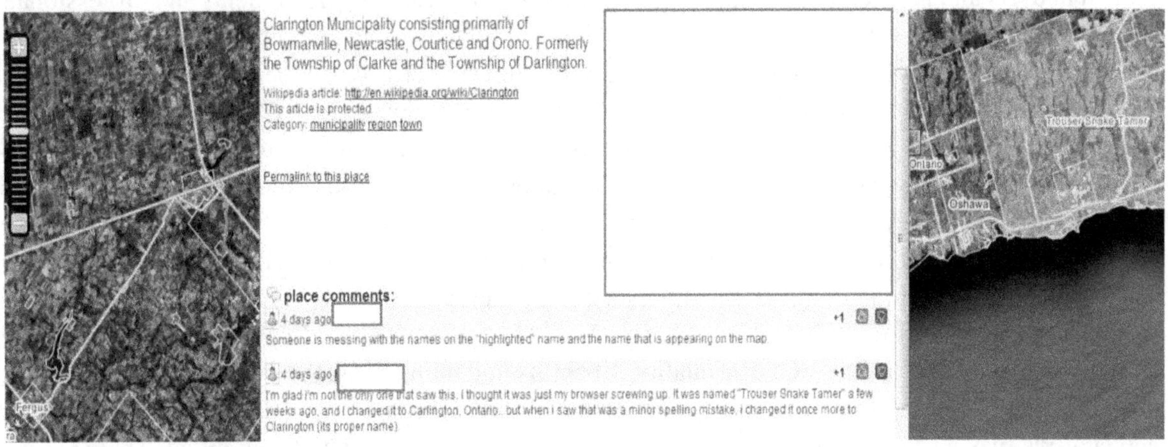

ing the correct city name. In this and other examples, we also see the community users communicating amongst themselves about the integrity of the various kinds of data on Wikimapia. These examples lead us to consider not only the potential technical merits of VGI and its ability to supplant some traditional 'authorised' sources of geodata, but also ponder the volunteer and her/his motivations. We return to this point in the discussion.

In summary, of the 269 temporary data modifications introduced overall (across the 135 sites), 146 or 54% were corrected by Wikimapia's community users by 3 April 2010. This ranged from 31% community user corrections made to university campus data to 89% for sporting venues. If corrections had not occurred by the 6-day mark, then the data usually remained intact as modified by the team until the end of the tracking period. Metadata was most likely to be corrected (65%) followed by place names (52%) and then polygons (40%). We re-state here that any remaining incorrect data introduced by the team at any time was returned to its original form as a condition of completion of the project for the course.

CONCLUSION AND DISCUSSION

This chapter has addressed the following question: can we rely on the lay public to provide, maintain and monitor quality (geospatial) data in the Web 2.0 world? We have sought to address this question via an experiment of user-generated content in the web-mapping site Wikimapia. It is critical to stress again the need to undertake future experiments like this with due care and caution and respecting the policies and guidelines of the sites used. This should include, for instance, our insistence that all site data be returned to their original form as they appeared at the beginning of this study. Indeed, while the group understood that it was introducing temporary data modifications of various kinds, we felt it was important even in a public domain data site that we abided by a pact to return the data to its original form by the end of our study (and, as mentioned earlier, ensured this was done as a condition of completion of the academic course in which this study was undertaken).

Based on this experiment, we may glean some insights into whether and in what ways we may rely on VGI. In addition, given the advent of VGI and in particular its very different nature (i.e. its sources and its data politics, as found in the present study and others), it is clear that we should also re-imagine what GIS education might have to adapt to in future. More specifically, if VGI continues to experience growth in users and in its ability to supply reliable data, we suggest that GIS students and professionals should be more aware: the genesis and nature of VGI data; the users and communities that are sprouting around this form of data provision; and perhaps also how the crowd may be engaged in discussions around data stewardship and use. We do not suggest here an easy conflation of VGI as GIS. Instead we want to emphasise that VGI is a new and qualitatively different movement within the broad ambit of GIS whose momentum must garner some attention as a matter of critical reflection for the GIS student and professional practitioner. Accordingly, we offer some comments on our second theme of GIS teaching and learning based on the pedagogical exercise that accompanied our main VGI experiment.

First, focusing on our main theme of VGI data quality and reliability, we can describe the provision, monitoring and protection of VGI on Wikimapia as swift and thorough but also incomplete. In the broad expert-versus-lay tension over VGI data quality, reliability and monitoring, this study suggests that some data may be reliable and that there exists great potential for VGI to become a dependable reservoir for geodata. We may also glean insights on the users themselves as engaged and committed. There is little in this experiment to support the view that VGI suffers 'amateur' treatment except perhaps with respect to the manner in which some community data volunteers publicly respond to data errors. If Wikimapia is any indication, it can be said that there is considerable potential for the growth of reliable VGI. Still, some issues remain that bear brief discussion here and further research.

If engaged volunteers can generate, disseminate and steward quality data, we should also consider the potential pitfalls that a growing reliance on VGI may mean for society as a whole. While the potential for VGI appears to be greatest in domains in which citizens invest energy and interest (e.g. environmental issues), the corollary suggests that the VGI ecosystem might unwittingly neglect themes and geographies of less concern today but of potential importance in future. Paradoxically, these kinds of data gaps would only grow and become more obvious with increasing reliance upon VGI.

A similar paradox may be identified with respect to data volunteers. The user community might not only be engaged data stewards: they may, at one extreme, be digitally marginalised or, at the other, unmediated gatekeepers. The digital divide might sort who may contribute and monitor data and therefore define and assign identity in the geoweb (Dunn, 2007; Curry, 1997). In a similar 'test' of VGI data reliability, Zielstra and Zipf (2010) found a digital divide between urban and rural data volunteers in terms of the quantity and quality of freely available data. At a different scale, consider what this might mean when a user generates a name, description or polygon relating to a place or territory of worship or religious significance. Or alternatively, how the growing cadre of so-called 'screeners' can influence what is tagged and said about particular places such as title-sponsored sports venues. One could imagine rather intense community forums and a range of ensuing issues. For instance, whereas a data authority (such as a municipal government) would normally follow due process in collecting and disseminating data, a data volunteer or community group may not. We would argue that these potential pitfalls are at least as significant as the potential for quality VGI data itself and it bears further research.

Future research can build upon and extend what is presented here in order to tackle conflicts that will inevitably arise in the VGI and GIS community. While guarding against gratuitous mischief, the answer to our core question necessitates research across diverse VGI sites, themes and data types. In addition, if we are to reasonably judge VGI as reliable, we need to know more about the data volunteers as well. More subtle still, research will be needed to document and understand how or even whether user and community disagreements can be resolved. VGI as a strategy for geodata development will depend upon whether citizen volunteers can self-organise to resolve data and information conflicts. If not, then even high quality reliable VGI data may come to a standstill when faced with intractable user and community conflicts.

This chapter has presented a small experiment that supports the view that VGI holds promise to grow and develop into a meaningful strategy for quality reliable data. Though the historian's gaze is likely needed, we might even say that we may be witnessing a momentous movement in the ways in which citizens engage one another. Some say this shifts the terrain of the GIS and geographical enterprise toward a 'neogeography' that is much less expert and more lay and citizen based (Haklay et al., 2008). To the extent that the technology is increasingly lay-based and less technocratic, the pedagogy of GIS curriculum must adapt. The fundamental pedagogical issue is that the data come with 'baggage' that the GIS teacher and learner have scarcely ever confronted. The pedagogical exercise used in this study was one attempt to expose and immerse the learner in the new, challenging and as yet unclear potential and pathways for VGI. If citizens volunteer their time and energy to provide a growing share of data on which we may rely in future, then we must at least consider alternative teaching and learning approaches to meaningfully acknowledge, understand and engage these changes.

REFERENCES

Barron, C., Neis, P., & Zipf, A. (2014). A comprehensive framework for intrinsic OpenStreetMap quality analysis. *Transactions in GIS*, *18*(6), 877–895. doi:10.1111/tgis.12073

Cha, M., Kwak, H., Rodriguez, P., Ahn, Y., & Moon, S. (2007). I tube, you tube, everybody tubes: analyzing the world's largest user generated content video system. *Proceedings of the 7th ACM SIGCOMM conference on Internet measurement*. Retrieved from http://portal.acm.org/citation.cfm?id=1298306.1298309

Coleman, D. J., Georgiadou, Y., & Labonte, J. (2009). Volunteered Geographic Information: The Nature and Motivation of Producer. *International Journal of Spatial Data Infrastructures Research*, *4*, 332–358.

Curry, M. (1997). The Digital Individual and the Private Realm. *Annals of the Association of American Geographers*, *87*(4), 681–699. doi:10.1111/1467-8306.00073

Dunn, C. (2007). Participatory GIS – a peoples GIS? *Progress in Human Geography*, *31*(5), 616–637. doi:10.1177/0309132507081493

Elwood, S. (2008). Volunteered geographic information: Future research directions motivated by critical, participatory, and feminist GIS. *GeoJournal*, *72*(3-4), 173–183. doi:10.1007/s10708-008-9186-0

Elwood, S. (2011). Geographic information science: Visualization, visual methods, and the geoweb. *Progress in Human Geography*, *35*(3), 401–408. doi:10.1177/0309132510374250

Flanagin, A. J., & Metzger, M. J. (2009). The credibility of volunteered geographic information. *GeoJournal*, *72*(3), 137–148.

Girres, J. F., & Touya, G. (2010). Quality Assessment of the French OpenStreetMap Dataset. *Transactions in GIS*, *14*(4), 435–459. doi:10.1111/j.1467-9671.2010.01203.x

Goodchild, M. (2007). Citizens as voluntary sensors: Spatial data infrastructure in the world of Web 2.0. *International Journal of Spatial Data Infrastructures Research*, *2*, 24–32.

Goodchild, M. (2009). NeoGeography and the nature of geographic expertise. *Journal of Location Based Services*, *3*(2), 82–96. doi:10.1080/17489720902950374

Goodchild, M., & Li, L. (2012). Assuring the quality of volunteered geographic information. *Spatial Statistics*, *1*, 110–120. doi:10.1016/j.spasta.2012.03.002

Grira, J., Bedard, Y., & Roch, S. (2009). Spatial data uncertainty in the VGI world. Going from consumer to producer. *Geomatica*, *46*(1), 61–71.

Haklay, M. (2010). How good is volunteered geographical information? A comparative study of OpenStreetMap and Ordnance Survey datasets. *Environment & Planning B*, *37*(4), 682–703. doi:10.1068/b35097

Haklay, M. (2012). Citizen scince and volunteered geographic information: Overview and typology of participation. In D. Sui, S. Elwood, & M. Goodchild (Eds.), *Crowdsourcing geographic knowledge: Volunteered geographic information and knowledge* (pp. 105–122). New York: Springer.

Haklay, M., Basiouka, S., Antoniou, V., & Ather, A. (2010). How Many Volunteers Does it Take to Map an Area Well? The Validity of Linus Law to Volunteered Geographic Information. *The Cartographic Journal*, *47*(4), 315–322. doi:10.1179/000870410X12911304958827

Haklay, M., Singleton, A., & Parker, C. (2008). Web mapping 2.0: The neogeography of the GeoWeb. *Geography Compass*, *2*(6), 2011–2039. doi:10.1111/j.1749-8198.2008.00167.x

Harvey, F., Kwan, M. P., & Pavlovskaya, M. (2004). Introduction: Critical GIS. *Cartographica*, *40*(4), 1–4. doi:10.3138/04L6-2314-6068-43V6

Johnson, P. A., & Sieber, R. E. (2012). Motivations driving government adoption of the Geoweb. *GeoJournal*, *77*(5), 667–680. doi:10.1007/s10708-011-9416-8

Kean, A. (2007). *The Cult of the Amateur: How Today's Internet is Killing Our Culture*. New York: Doubleday.

Koukoletsos, T., Haklay, M., & Ellul, C. (2012). Assessing Data Completeness of VGI through an Automated Matching Procedure for Linear Data. *Transactions in GIS*, *16*(4), 477–498. doi:10.1111/j.1467-9671.2012.01304.x

Leszczynski, A. (2012). Situating the geoweb in political economy. *Progress in Human Geography*, *36*(1), 72–89. doi:10.1177/0309132511411231

Moony, P., & Corcoran, P. (2012a). Characteristics of heavily edited objects in OpenStreetMap. *Future Internet*, *4*(4), 285–305. doi:10.3390/fi4010285

Moony, P., & Corcoran, P. (2012b). The annotation process in OpenStreetMap. *Transactions in GIS*, *1*(4), 561–579. doi:10.1111/j.1467-9671.2012.01306.x

Sui, D., Elwood, S., & Goodchild, M. (2012). *Crowdsourcing geographic knowledge: Volunteered geographic information and knowledge*. New York: Springer.

Zielstra, D., & Zipf, A. (2010). *A Comparative Study of Proprietary Geodata and Volunteered Geographic Information for Germany*. Paper presented at the 13th AGILE International Conference on Geographic Information Science, Guimarães, Portugal.

Chapter 5
Quality Assessment of Volunteered Geographic Information for Educational Planning

Hafiz Muhammad Muzaffar
National University of Sciences and Technology, Pakistan

Ali Tahir
National University of Sciences and Technology, Pakistan

Asmat Ali
Survey of Pakistan, Pakistan

Munir Ahmad
Survey of Pakistan, Pakistan

Gavin McArdle
University College Dublin, Ireland

ABSTRACT

Volunteered Geographic Information is the term used to describe the process of collecting spatial data using a network of volunteers. The approach collects spatial data to build maps which are often freely accessible. The maps and the underlying data can be used by the public, companies and government agencies for a variety of tasks such as route finding. Given that untrained volunteers may collect the spatial data, questions regarding the quality of VGI have been raised. Several studies have emerged to assess the quality (positional, semantic and thematic accuracy) of VGI by comparing the data to ground truth. This approach fails to capture the quality of VGI for domain specific tasks. In this chapter we examine the quality of VGI for an educational planning task in Islamabad, Pakistan, and show that while the data may be suitable for route finding tasks, they are insufficient for educational planning alone.

DOI: 10.4018/978-1-5225-2446-5.ch005

INTRODUCTION

Crowdsourcing of data involves volunteers collecting, processing and sharing data through specialized digital platforms or via ad-hoc methods. There are many examples of crowdsourcing projects. Wikipedia (encyclopedia), OpenSignal (cell tower information), Open Food Facts (food information) and RootMetrics (cell carrier quality metrics) are examples of crowdsourced repositories covering multiple domains. Goodchild (2007) introduced the term Volunteered Geographic Information (VGI) to describe the crowdsourcing of map and spatial data. OpenStreetMap (OSM) is perhaps the best known and most successful VGI project. The aim of the ongoing project is to map the world and to provide free access to the editable map and underlying data. The project builds on the local knowledge of volunteers to create detailed maps.

As crowdsourced data are often contributed by non-expert volunteers, questions regarding its quality have been raised. In the spatial domain many studies assess the quality of OSM data (Ciepłuch et al., 2011; Haklay 2010; Girres & Touya 2010; Ludwig et al., 2011; Ballatore & Zipf, 2015; See et al., 2016). The studies assess OSM data for geometric accuracy, overlap, attribute accuracy and completeness. The quality assessment generally involves comparing the volunteered spatial data to official data sources such as data collected by national mapping agencies or mapping companies. Such studies assume that the professionals who have trained in map generation and cartography have produced an accurate ground truth. Although, the studies indicate that the quality of OSM is acceptable with accurate positional precision for road networks, the results also indicate that there are spatial and regional differences in relation to the quality of OSM data.

With the exception of route planning, studies assessing the correctness of volunteered spatial data are typically domain independent and focus on assessing the overall quality of OSM without drawing on specific spatial analysis tasks. This can hide the true value of VGI data for certain domains. For example, questions about the usefulness and quality of VGI for urban planning, infrastructure planning and service delivery have not been widely assessed. While such analysis is a subset of the overall quality assessment measurements, there may be subsets of the data where quality is better or worse than that seen in the overall OSM dataset.

In this chapter we investigate this by specifically examining the quality of OSM data for tasks related to educational planning. We use the city of Islamabad in Pakistan as a case study. We take the classic approach of comparing the spatial, semantic and thematic data of educational facilities in Pakistan to data collected by the National mapping agency in Pakistan (Survey of Pakistan (SoP)). We augment this with other ground truth data which we collected as part of the study. We assess the volunteered geographic data for completeness, attribute accuracy and positional accuracy and report the results as they apply for educational planning in Pakistan.

BACKGROUND

This section considers existing research in the field of educational planning and the comparison of authoritative datasets with that of VGI. We establish the relationship between educational planning, GIS and the potential of VGI.

VGI and Educational Planning

Modern planning has multifaceted dimensions (Hall, 2002). Educational planning can be described as an exercise to establish an education system in order to accomplish the goals of policy makers. While educational planning is a broad subject, for the research presented in this chapter, educational planning concerns the setup of individual educational institutions in an area of interest by exploiting the power of spatial data. Public and private sectors need to establish educational institutions in areas where populations do not have easy access to the existing educational institutions. The main purpose behind establishing a new institution is to facilitate people and improve literacy rate. Okpala (2001) underpinned the importance of spatial information for practical and sustainable human settlement planning. Thus, planning to establish educational institution necessitates various spatial and non-spatial datasets. In Pakistan most of the spatial datasets held with public sector organizations are not shareable or accessible due to certain institutional and technical barriers like a lack of data policy, a lack of coordination, the high cost of data, the non-existence of data standards and infrastructure, non-availability of metadata as well as fear of exposure of data quality as identified by Ali & Ahmad, (2013). Moreover public sector data is not updated regularly due to updating cycle policies enacted in the country. This situation demands the use of free and accessible datasets like VGI in planning. There is immense potential of using VGI in planning (Elwood et al., 2012; Campagna et al., 2015) and its use in planning has been researched in several domain areas such as transportation, disaster management, environment, health, flood management and flood response.

In order to support transportation planners for better decision making, a novel community-driven wiki based route analysis tool has been introduced in (Masli et al., 2013). A large number of projects based on the potential use of VGI in disaster management have also been conducted. Poser & Dransch, (2010) elaborated on the opportunities and challenges associated with the use of VGI for disaster management especially in the context of response and recovery phases. The research was demonstrated with the help of a case study to rapidly estimate the flood damage using data collected from effected populations. Horita & Albuquerque, (2013) proposed a framework to integrate VGI and authoritative data for Spatial Decision Support System (SDSS) to support decision making in disaster management and to assist government organizations by providing mechanisms and information for better disaster management.

Similarly Hung et al., (2016) devised a method to estimate the credibility of VGI for disaster response. Datasets collected from two flood events in 2011 and 2013 from Brisbane, Australia and binary logistic regression were used to test the method. The result suggests the use of method for effective and rapid response as well as decision-making and coordination in emergency management whereas Degrossi et al., (2016) proposed a conceptual model that combines OSM, social media data and authoritative data in order to assess the quality of VGI. For the purpose of environmental monitoring Connors et al., (2012) conducted a case study in the Western United States to monitor a forest disease. The study makes use of a conceptual model to critically examine the OakMapper.org website that collects the spatial information pertaining to forest disease from volunteers. The authors argued that the proposed conceptual model can support environmental monitoring and help the public, researchers and policymakers collaborate to produce knowledge. Goranson et al., (2013) explore the value of VGI for public health research along with the potential challenges and ethical problems associated with the tools used for collecting VGI.

Bartoschek & Keßler, (2013) investigated whether students who engaged with VGI remains interested in the topic and remain involved in VGI communities after the conclusion of classes. The study makes use of a survey to collect information from students of past VGI classes and assemble a comprehensive overview of motivations and obstacles associated with different VGI platforms. The study recommends development of future curricula to include volunteered geographic information. It is evident from the above discussion that VGI has been used for planning and decision making in a variety of domains and applications, however VGI for educational planning is not evident from the literature. The same methods and models as endorsed in the above discussion can be utilized to benefit educational planning.

As we know, VGI is associated with certain accuracy and quality issues as reported by Elwood et al., (2012) and others. Thus to effectively use VGI for planning purposes, it is essential to assess its quality against certain aspects like completeness, logical consistency, positional accuracy, temporal accuracy, thematic accuracy, purpose, usage and lineage (Kresse & Fadaie, 2004); (Oort, 2006) before its incorporation in the planning process. The same is also required for educational planning in the context of city of Islamabad. Educational planning is selected as a domain area because in Pakistan the education sector is declining due to poor management and monitoring (Memon, 2007). Thus it is essential to use educational planning as a special case to boost the education sector and hence the economy of the country.

The main focus of the chapter is to assess the quality of VGI, in particular OSM, in the context of educational planning for the city of Islamabad, Pakistan. Data completeness, positional accuracy and attribute accuracy are the measure of quality selected for this case study. Data completeness is selected to provide detailed analysis about the coverage area of VGI in the city of Islamabad. Incomplete data does not reflect the true picture which is essential for educational planners to plan for the new educational facilities. Similarly, the attribute accuracy is important to check the consistency in the names and addresses of educational institutes. Finally, positional accuracy is selected in the context of educational planning to strategically plan the educational institutes in the city of Islamabad.

As for educational planning, complete picture of existing educational facilities is essential to plan for the new educational facilities, thus criterion for completeness will be that data from authoritative source and VGI source should match *100%* as far as the educational facilities within the dataset are concerned. For other features such as utilities network and services that are equally important to sustainably run as well as maintain the educational facility, data completeness may vary.

For the positional accuracy regarding educational facilities, < *50 meters* error in distance between the authoritative source and VGI source is acceptable, as < *50 meters* distance is sufficient to recognize the evidence of educational facility in the vicinity. The rule is equally applied to the point and polygon features.

For the attribute accuracy regarding educational facilities, > *70%* overlap in the attributes of educational facilities from authoritative source and VGI source is acceptable for the evidence of educational facility in the vicinity due to the fact that often volunteers collect attributes in short or abbreviated form.

Higher rate of data completeness, positional accuracy and attribute accuracy is desired for the features (such as utilities network and services) that are directly linked with the educational facilities as compared to the features that are distant from the educational facilities.

VGI and Authoritative Data

OSM is a much cited example of Volunteered Geographic Information (Mooney et al., 2010). OSM is a project which aims to create an editable and non-proprietary map of the world. It is achieving this through the collaboration of millions of volunteers who map their local area and share their results via an editable map along with the underlying spatial data. Using volunteers to produce spatial data about their locality can produce very fine grained data which builds on local knowledge. However, given that volunteers are typically untrained in cartographic principles and spatial data collection methods, questions regarding the quality of data output via OSM and VGI have been raised. In order to measure the quality of a VGI dataset the parameters against which to check the quality of the dataset need to be identified. A first perspective about the quality of VGI has been established by Haklay (2010) where the quality of OSM data is discussed. Haklay adopted a way to check the quality of OSM data by comparing it with authoritative data. Authoritative data is considered to be accurate and as per ground reality as it is gathered in a systematic manner by professionals and passed through a quality test in a standardized way. Haklay (2010) took the data from Ordnance Survey (OS), an authoritative mapping organization in UK, and compared it with freely available OSM dataset. He checked the positional accuracy of roads considering motorways (A and B category roads) first by using a buffer method. He created a buffer against OS roads dataset of a specific distance and checked the length of OSM roads within the buffer. He concluded that on average, OSM data is accurate to about *6 meters*. The study proposed to continue research on other aspects of VGI data quality.

Several other researchers tested the quality of OSM data against the authoritative data in other countries. For example Ludwig et al., (2011) compared the OSM dataset against Navteq data in Germany. The authors matched the OSM roads with Navteq roads dataset in an automated way. They further concluded that at the country level there exists high level of similarity in the datasets though at detailed level there exist a difference in the similarity of OSM and Navteq which further varies from town to town. Similarly, Ciepłuch et al., (2011) compared the quality of OSM data against authoritative data of Ordnance Survey Ireland (OSI). They calculated the length of all line features including roads, railways, rivers and trails etc. of OSM and the OSI dataset for Ireland. The same observations as Haklay (2010) were reported, De Leeuw et al., (2011) tried to investigate the role of local community in mapping the road network in Western Kenya. The authors used high resolution satellite imagery and got both the professionals and laymen to classify the roads with and without local territory knowledge. They observed that people with local area knowledge whether they were professional or not classified the roads with 92% accuracy. On the other hand, professionals and laymen who do not have local area knowledge classified roads data with 67.7% and 42.9% accuracy respectively. Hence, they concluded that data prepared by local community would be of higher quality other than that captured by general community.

Keßler et al., (2013) developed a concept to measure the quality of VGI data without authoritative data. They worked on the social part of VGI and assessed the number of people contributing in developing the VGI of a specific area and how often the data has been updated. They concluded that a high number of contributors and a high rate of altering the dataset meant the quality of data is good. However an increased number of corrections and revisions on a feature meant the quality of the data is lower. Similarly, D'Antonio et al., (2014) worked on the social aspect of data and tried to infer the quality of VGI data using VGI edit history. They presented a model to calculate the VGI user reputation and VGI feature trustworthiness as a standard to measure the VGI data quality. They concluded that data produced by a highly reputed user has a high quality.

Barron et al., (2014) presented a framework named iOSMAnalyzer which used all the history files of OSM data to evaluate the quality of OSM data in 25 different aspects. In order to evaluate the quality of OSM data based on the full history, iOSMAnalyzer divides its results in following use cases: General Information, Routing and Navigation, Address Search, POI Search, Map Application, User Information and Behavior.

Camboim et al., (2015) evaluated the quality of an OSM dataset in a heterogeneous area in Brazil. In the absence of a valid authoritative spatial dataset, the authors tried to compare the OSM data with a demographic dataset. They used the population count, average household income, Human Development Index (HDI) and other demographic datasets to evaluate the quality of OSM dataset in urban and rural areas. The results showed that the places where data is complete and is updated more frequently are usually the high density urban areas while the places where data do not exist are areas with a low population.

Methodology for Assessing VGI for Educational Planning

This section describes our methodology in order to deal with educational planning. The focus here is to make the optimal use of location which can help planners and decision makers. Using the geospatial technologies and VGI platforms, the decision making capability of stakeholders can be enhanced. For the proposed research, Islamabad, the federal capital of Pakistan has been selected as a study area. The design of Islamabad is considered as one of the best planned cities in Pakistan. It consists of five sectors starting from north to south in the order: E, F, G, H and I. Each sector is further divided into 16 subdivisions. The capital city is known as the hub of education in the country. The OSM map of Islamabad is shown in Figure 1 while the detailed spatial datasets are discussed in next section.

Figure 1. OSM study area- Islamabad

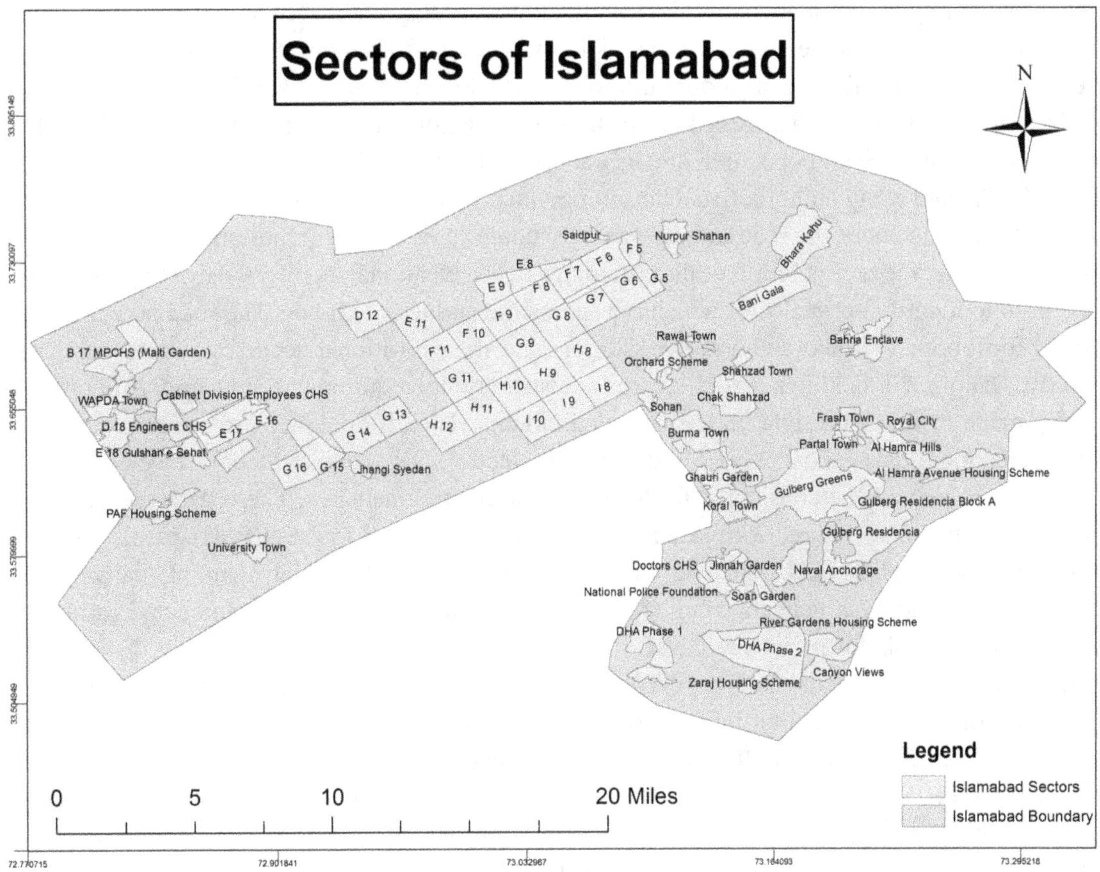

OSM is a free map of world contributed by volunteers under an open license while SoP is a national surveying and mapping organization. The main responsibility of SoP is to conduct topographical land surveys and produce map sheets on various map scales. Furthermore SoP maintains geodetic network for the entire country. These datasets are available to the public for both research and commercial purposes. Additionally, maps such as provincial, district, guide and street maps are available to use in national development projects.

Data for Educational Planning

An important dataset for educational planning is the location of existing educational institution. This is used to assess the relationship between space and educational requirements. In addition to location, the attribute information is also relevant. The number of schools, colleges, universities, type of facilities, accessibility, teachers' skills, population, land use, water availability, and administrative boundaries all play a role in educational planning. For example, administrative boundaries are a crucial spatial dataset for addressing educational challenges by showing the scope and jurisdiction of local government bodies responsible for educational services. This can assist in identifying the places where immediate steps are required to improve educational deficits. Mapping and spatial analysis related to education can be performed within the administrative boundary of a local council. We have found that accurate administrative boundaries can be obtained from both VGI and authoritative data sources.

Data concerning the population in a given geographical area is required in order to assess the number of schools with respect to the number of pupils in that area. Furthermore, the identification of type of settlement is also crucial for educational improvements. For instance, industrial areas do not require schools as opposed to residential areas. Furthermore, the categorization of low, medium and high density areas is also required. Such population data is not typically collected using crowdsourcing and is more reliable if collected using authoritative data sources such as a national census.

The information about the road network in a particular area is very important. The better the road network, the better the accessibility. While planning a new educational facility, the existing road network can be used to identify optimal sites which are accessible and near to roads. The road network data are obtained from both VGI and authoritative datasets and their positional accuracy is also compared in our study. The price of land in a particular area is also a crucial requirement. Educational decisions are typically made in line with tight budgetary requirements. The affordability further enriches the geospatial educational data repository. Utilities such as electric, gas, water, telecom, sewerage also form core datasets which can be considered for locating new educational sites. These datasets are normally authoritative and are not shared with general public. Other Points of Interest (PoIs) are also important. This data can contain hospitals, police stations, bus stops, mosques, hostels, restaurants, railway station, airport, etc. POI data is largely contributed by volunteers based on their experience and can be easily obtained from VGI platforms such as OSM.

Additionally, specific data relating to existing education facilities are required. This chapter compares OSM education facilities to authoritative sources. For this purpose, four datasets belonging to the capital city were obtained. Schools and colleges data were obtained from Federal Board of Intermediate Secondary Education (FBISE). This data contain the list of schools and colleges in the capital. The Universities list was obtained from the Higher Education Commission (HEC). Settlements and road network were acquired from SoP. All these datasets are used for spatial analysis and are compared to VGI data. FBISE has defined two categories for affiliated institutes. Either the institute is affiliated at Secondary School Certificate (SSC) level or at Higher Secondary School Certificate (HSSC) level.

The attribute information such as institution name, address, affiliation date, website etc. of affiliated schools and colleges are also present on FBISE website. FBISE has affiliated institutes in Pakistan and worldwide. A script was written which scrapes and parses data from the FBISE database. The flow of the script is shown in Figure 2.

After extracting the required data from the FBISE data source, the next step was to collect the similar data from OSM. There are a number of tools which facilitate OSM data download. These include *Geofabrik, Gisgraphy*, and *Thinkgeo*. We selected *Geofabrik* as this tool is more user-friendly. This is due to the fact that *Geofabrik* is updated on daily basis as opposed to *Gisgraphy* and *Thinkgeo*. Furthermore, Geofabrik provides data in multiple formats (pbf and shp) while *Gisgraphy* only provides pbf support. *Thinkgeo*, on the other hand is not completely free.

The downloaded data size for Pakistan is 28.1 MB as per (12-05-2016). Figure 3 outlines the steps of data extraction from *Geofabrik* into a local database for analysis.

Assessing VGI for Educational Planning

In this chapter educational data is assessed across three principle axis, namely: completeness, attribute accuracy and positional accuracy. Figure 4 illustrates the framework for quality assessment used in this study. Data completeness is an important element in the quality assessment of VGI for educational planning. VGI often do not contain complete coverage of existing educational infrastructure and therefore do not reflect the true picture for educational planners. For data completeness, the authoritative data of schools, colleges and universities are obtained from FBISE and HEC and the overlap between this ground truth and OSM is calculated. The concept is highlighted in Figure 5.

Figure 2. Scrape data

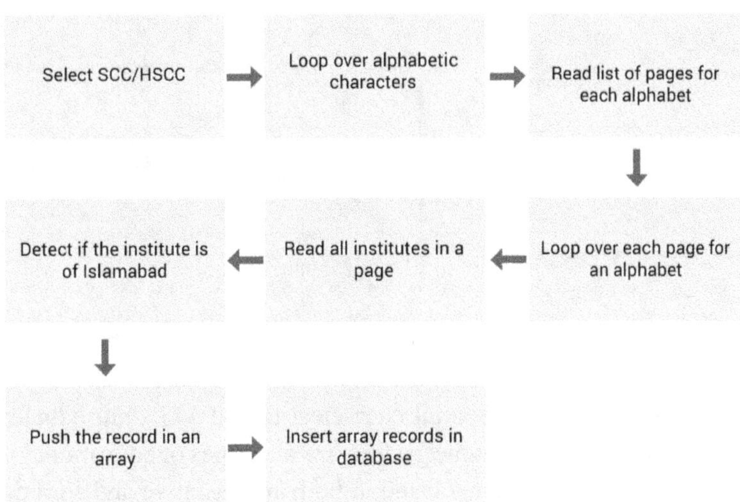

Figure 3. OSM data extraction

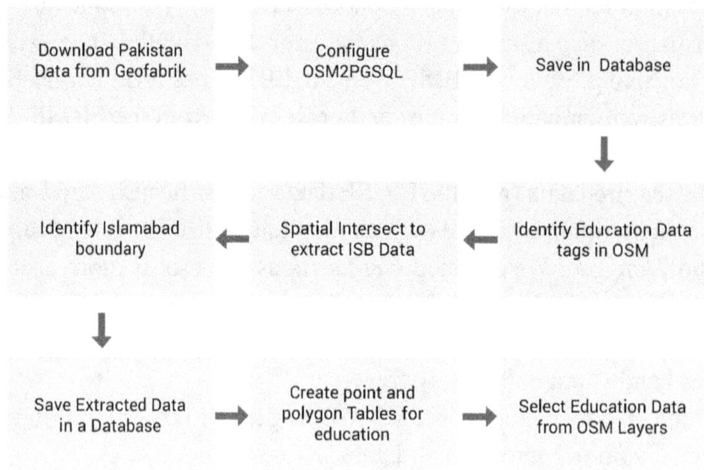

Figure 4. OSM quality framework w.r.t educational planning

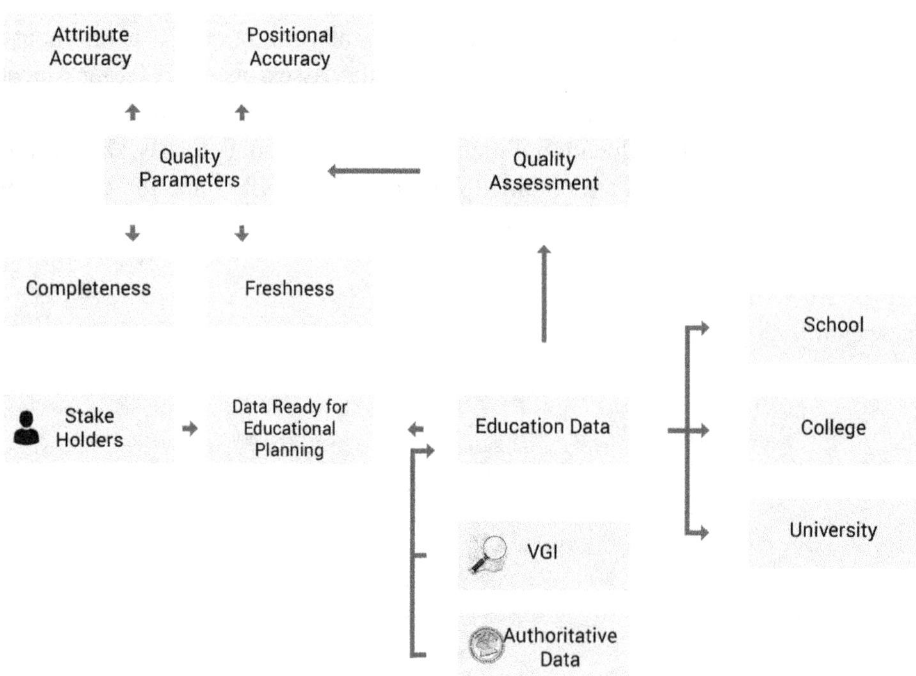

Attribute accuracy concerns the non-spatial characteristics of VGI. For educational datasets, it is important to check the consistency in the names, addresses and types of educational institutes. Therefore a comparison is made between the attributes listed in both authoritative and VGI data. The flow of attribute accuracy can be seen in Figure 6.

For educational planning, we assess positional accuracy in three dimensions as shown in Figure 7. Firstly, the positional accuracy of roads is compared against authoritative data. The roads are essential

Figure 5. The Concept of data completeness

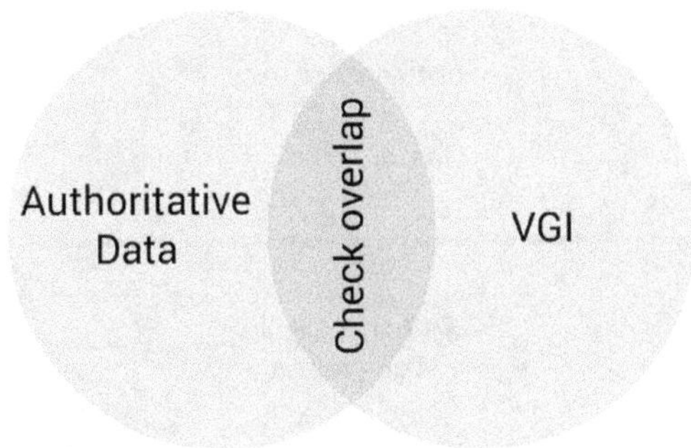

Figure 6. Attribute accuracy

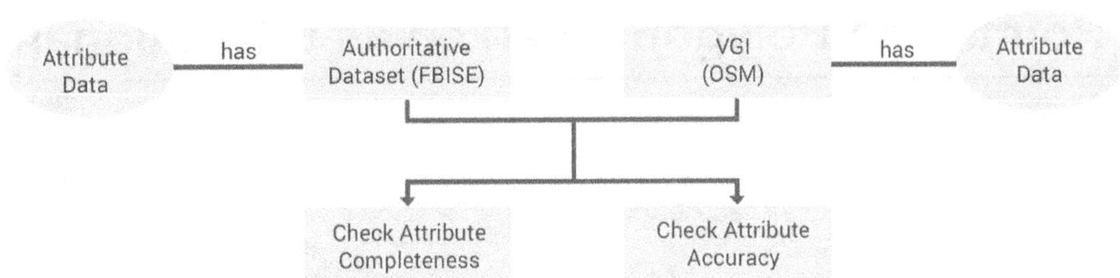

components in educational planning and facilitate accessibility and routing. Secondly, positional accuracy of schools and colleges is compared with OSM. Finally, the case of a single university is assessed as a measure of positional accuracy of individual building footprints within educational institutes. Ground truth for positional accuracy is also determined by a GPS survey to identify the boundaries of educational buildings.

ASSESSMENT AND RESULTS

Data Completeness

For this study, OSM has been used as an example of VGI dataset. In the OSM dataset for Islamabad, *181* features were tagged as educational. These include schools, colleges and universities. *64%* of these were polygons while the remaining *36%* were marked as points (see Figure 8). Interestingly, in some cases (approximately 20 times), the same educational institute is marked as point and polygon at the same time. This is an interesting research dimension to explore where an amenity is marked either as point, line or polygon. No educational points were marked as linear features in the given study area. Overall, in addition to data completeness, these factors become crucial when planning an educational move. For example, there are spatial analyses which are applicable on point features only. Similarly, the boundary or the area of each educational institute is measured using polygons.

Figure 7. Positional accuracy

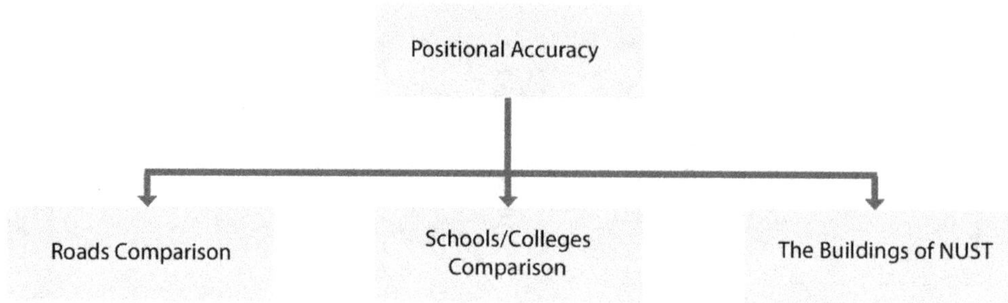

Figure 8. Point and polygons marked as educational institutes

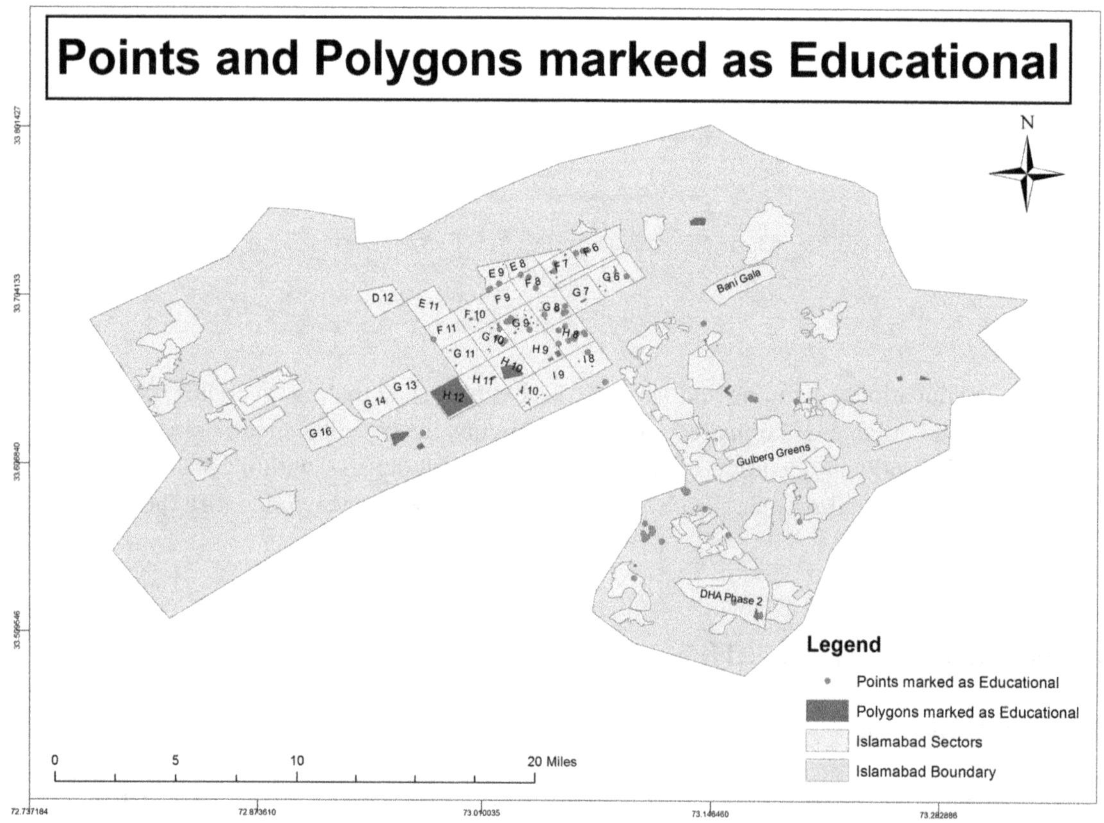

In the case of a specific University, data completeness is also discussed in this section. *19* polygon features of OSM were tagged as Universities while only *2* Universities were tagged as points. As per the authoritative data, *21* universities were affiliated with HEC. No occurrence was found where a University was marked as both point and polygon. There were other observations such as almost *53%* polygons were representing a true picture of Universities while the rest were the departments of respective Universities. Similarly, there was *50%* point coverage of Universities. The overall percentage of marked Universities on OSM is 52% as shown in Table 1.

Table 1. Data completeness of polygon and point data tagged for universities

	Number of Features in OSM	Number of Features in OSM matching Authoritative Data	Percentage Completeness
Polygons	19	10	53%
Points	2	1	50%
Total	21	11	52%

A case for data completeness of buildings within NUST, one of the largest Universities of the country is presented. In order to extract the buildings from OSM, a spatial query was created in *ArcGIS,* which performs the spatial intersection on all polygons within the NUST boundary. In the OSM schema, the polygons have an attribute called *building* which accepts "yes" or "no" as a Boolean value. Figure 9 shows the extracted buildings of NUST. As per the authoritative data, NUST contains a total of *40* buildings. In OSM, there was almost *90%* coverage which is very good. However these include buildings which are geographically present but their building attribute is marked as "no" or they were empty. This is due to a lack of awareness and knowledge amongst volunteers while marking these spatial features.

For the schools and colleges coverage, *97* polygons and *63* points in OSM were tagged as schools and colleges respectively. As per the authoritative data of FBISE, there are total of *424* educational institutes. On *20* occasions both points and polygons were representing the same school or college. Table 2 shows the percentages of data completeness. *38%* of polygons were matched in OSM as compared to FBISE while *27%* of points were matched. The overall accuracy percentage of data completeness was found to be *34%.* With this low percentage of completeness in OSM, the outcome can affect decision making. Out of a total *424* schools and colleges as per the FBISE list, only *160* were present in OSM. Furthermore, only *54* matched to FBISE. This is equal to *13%* only which shows a poor coverage of schools and colleges in OSM for the study region.

Attribute Accuracy

As described in the previous section, there were only *54* educational institutes present in OSM matching the authoritative data. To assess attribute accuracy, the total number, names and addresses of the records present were measured. Moreover, it has been checked if the correct attributes are reflected in OSM. When compared, it was found that 88.8% of the names matched while 37.0% of the addresses matched. The details are shown in Table 3. Achieving the attribute accuracy can be a challenging task. It depends on volunteers and what attributes they are specifying while creating a geographic feature. In the

Table 2. Data completeness of polygon and point data tagged for schools and colleges

	Number of Features in OSM	Number of Features in OSM matching Authoritative Data	Percentage Completeness
Polygons	97	37	38%
Points	63	17	27%
Total	160	54	34%

Figure 9. OSM polygons in NUST

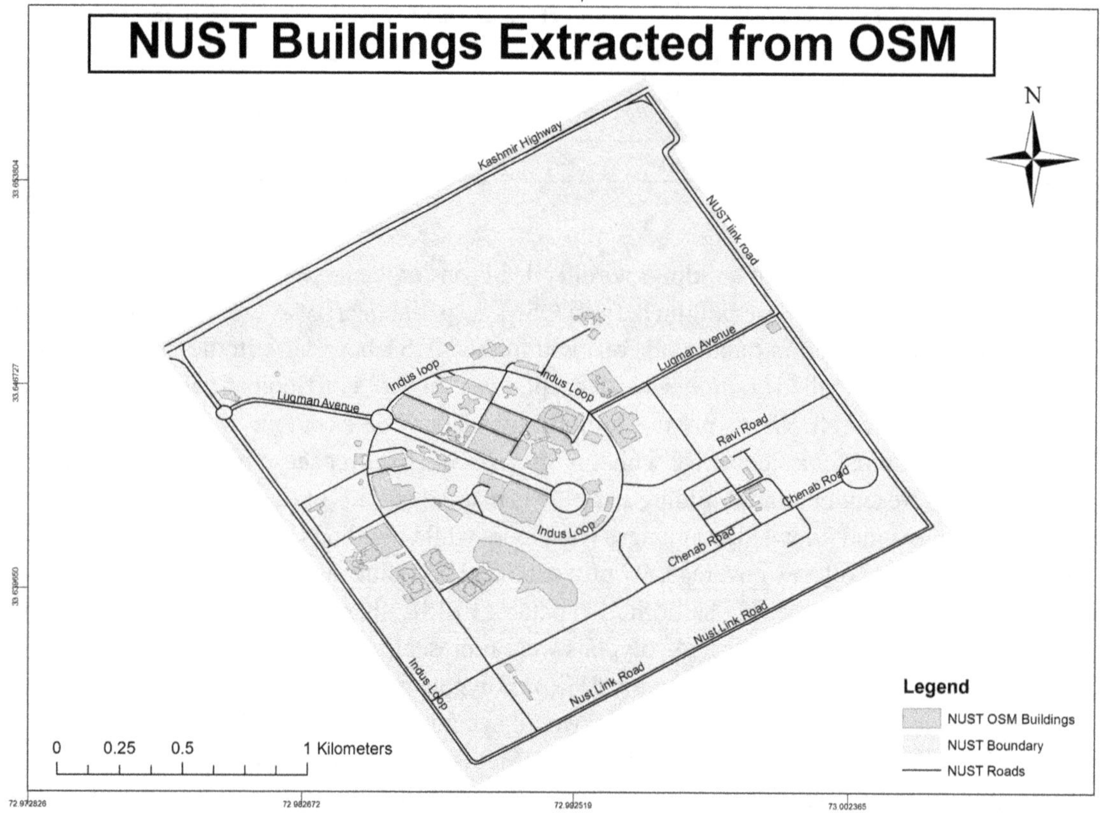

Table 3. Correct names and addresses percentage

Total	Name Present	Correct Name	Address Present	Correct Address
54	48	35	20	8
Percentage	88.80%	72.90%	37%	40%

case of attribute accuracy for educational institutes, some names were missing while for the addresses, street names were often omitted and so a complete match could not be obtained. Attribute accuracy is required for efficient educational planning. We consider attribute accuracy as inaccurate when at least a single word is missing.

Positional Accuracy

A comparison of OSM and SoP roads of *8* sectors of Islamabad was performed (see Figure 10). The authoritative data is considered as highly accurate prepared by professionals. These datasets normally go through rigorous quality measures and checks. It was observed that SoP considers the single line object for a two-way road while in OSM a two-way road is represented as two features. Goodchild & Hunter,

Figure 10. Flowchart of positional accuracy for roads

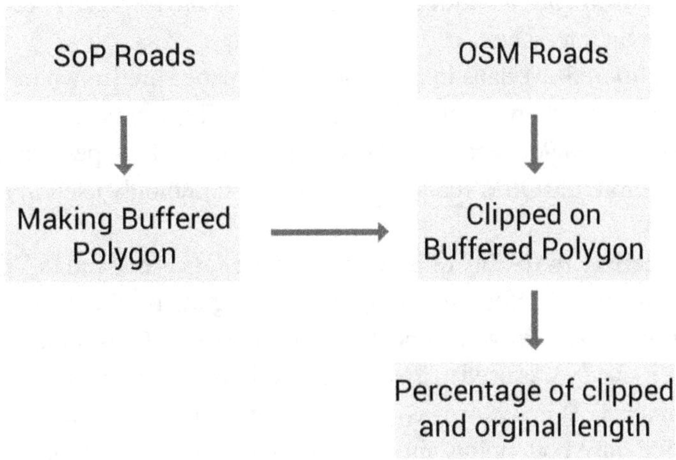

(1997) proposed to create a buffer in order to measure the positional accuracy of linear features in two datasets. Therefore, a buffer is created against OSM. This helps determine the positional accuracy of OSM. Figure 11 illustrates this concept where a buffer is drawn on OSM.

ArcMap was used to apply the buffer for positional accuracy of OSM and SoP roads. Firstly, the roads were separated for all *8* selected sectors of both OSM and SoP data. It was visually observed that both OSM and SoP roads for each sector represents almost the same area and length. Both the datasets were imported into ArcMap. A buffer of *15 meters* was taken against SoP dataset using the built-in buffer tool and the results were saved as a polygon shapefile. After buffering SoP dataset, the clip tool was used to select those parts of OSM roads which fall in the buffered polygon. The clipped roads of OSM were saved as a separate shapefile. The length of clipped roads were calculated and then divided by the length of original OSM roads. This provided the percentage of OSM roads which fall within the *15 me-*

Figure 11. Buffered Methodology to assess the positional accuracy of roads dataset (Haklay, 2010)

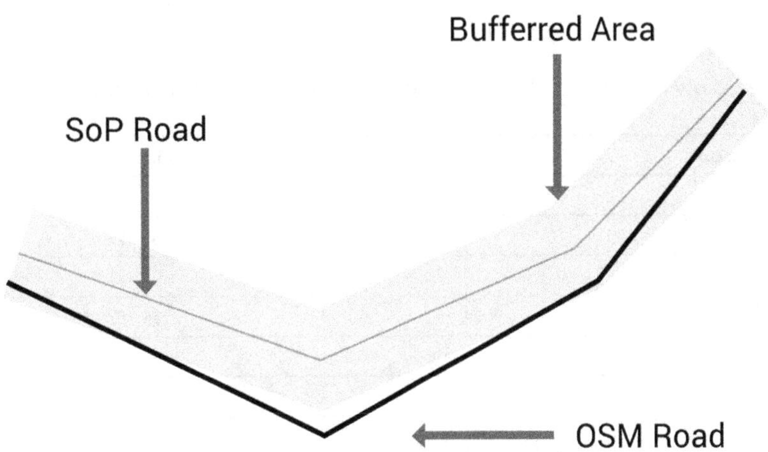

ters buffer of SoP roads. *15 meters* buffer was selected due to the fact that the widest road of Islamabad (called The Kashmir Highway) has a width of about *30 meters* on average. Therefore it was selected in order to cater the maximum tolerance.

The percentages of buffered-overlaps in *8* sectors of Islamabad are shown in Table 4. It can be seen that the lowest percentage is *84.2%* while the highest is *96.1%*.The average is *90.2%*. It can be concluded that OSM road network is up to *90%* accurate. This is reasonably a high percentage and corresponds to other jurisdictions. Therefore the OSM road network is very commonly used in routing studies even in developing countries.

Positional accuracy defines the reality of how accurate the data is marked in VGI with respect to their actual position in real world. Therefore, to strengthen this argument, the positional accuracy of OSM is measured by taking some samples while checking their positions (coordinates) using a GPS survey.

A Garmin based GPS survey was used to assess the positional accuracy of schools and colleges across the G-9 sector of Islamabad. This sector was selected as it contains the maximum number of schools and colleges. Since OSM have both points and polygons marked as schools and colleges, the positional accuracy of points is not considered. This is due to the fact that within the boundary of an educational institute, a point is marked anywhere. Therefore it is challenging to measure the positional accuracy of a point. Figure 12 shows the polygons which are selected for a survey. The GPS location was marked along the boundaries of these educational institutes near the main entrance gates. Figure 12 also illustrates the marked surveyed points along the institutions boundaries. The distance between the GPS points and the respective boundaries of the polygons are calculated using Haversine Formula (Sinnott, 1984).

As GPS points are marked near the main gates along the boundaries of institutions, a perpendicular point against the marked GPS point is selected from the OSM buildings. Considering the latitude and longitude as *osm_lat* and *osm_lng*, while GPS latitude and longitude as *gps_lat* and *gps_lng*. In order to calculate the distance, the following equations are used:

$$\alpha = osm\_lng - gps\_lng \text{ equation} \quad (1)$$

$$\beta = osm\_lat - gps\_lat \text{ equation} \quad (2)$$

$$a = sin^2(\beta/2) + cos(gps\_lng) * cos(osm\_lat) * sin^2(\alpha/2) \text{ equation} \quad (3)$$

Table 4. Percentage of overlap of OSM and SoP datasets for various sectors of Islamabad

Sector	Percentage of Overlap
G9	90.2%
F9	87.5%
H9	95.7%
G8	85.3%
F8	89.5%
H8	96.1%
G7	84.2%
F7	93.6%

$c = 2 * \tan^{-1}(\text{sqrt}(a), \text{sqrt}(1-a))$ equation (4)

$d = R * c$ (where R is the radius of the Earth) equation (5)

Hence d provides the distance between the GPS points and the points marked from OSM buildings. The result of survey's distance from the institution is shown in Table 5 along with the GPS accuracy. It can be seen that the minimum value for accuracy is *1.3 meters* while maximum value is *10.6 meters* with an average of *5.23 meters*. Keeping in mind the GPS average accuracy of *4.43 meters,* it can be said that the positional accuracy of schools and colleges lie between *0.8 meters* to *9.66 meters* which is acceptable.

The positional accuracy is further evaluated inside the NUST campus. OSM building polygons were extracted from downloaded data while the settlement data was obtained from SoP. The settlement data provides some information of NUST buildings. There were *13* NUST buildings which were part of SoP data and therefore all NUST buildings are not represented (see Table 6). A spatial query was performed in *PostGIS* to calculate the Hasudorff distance (Hausdorff, 1914). *PostGIS* provides a function called *ST_HausdorffDistance* to calculate the distance between two features. The function returns the Hausdorff distance between two geometries. The units returned are as per the spatial reference system of the geometries. The results of the query are shown in Table 6. It can be seen that OSM buildings are on average *4.2 meters* within SoP buildings while there is a variation from *3.1 meters* to *5.6 meters* overall.

Figure 12. The selected polygons and surveyed points for positional accuracy

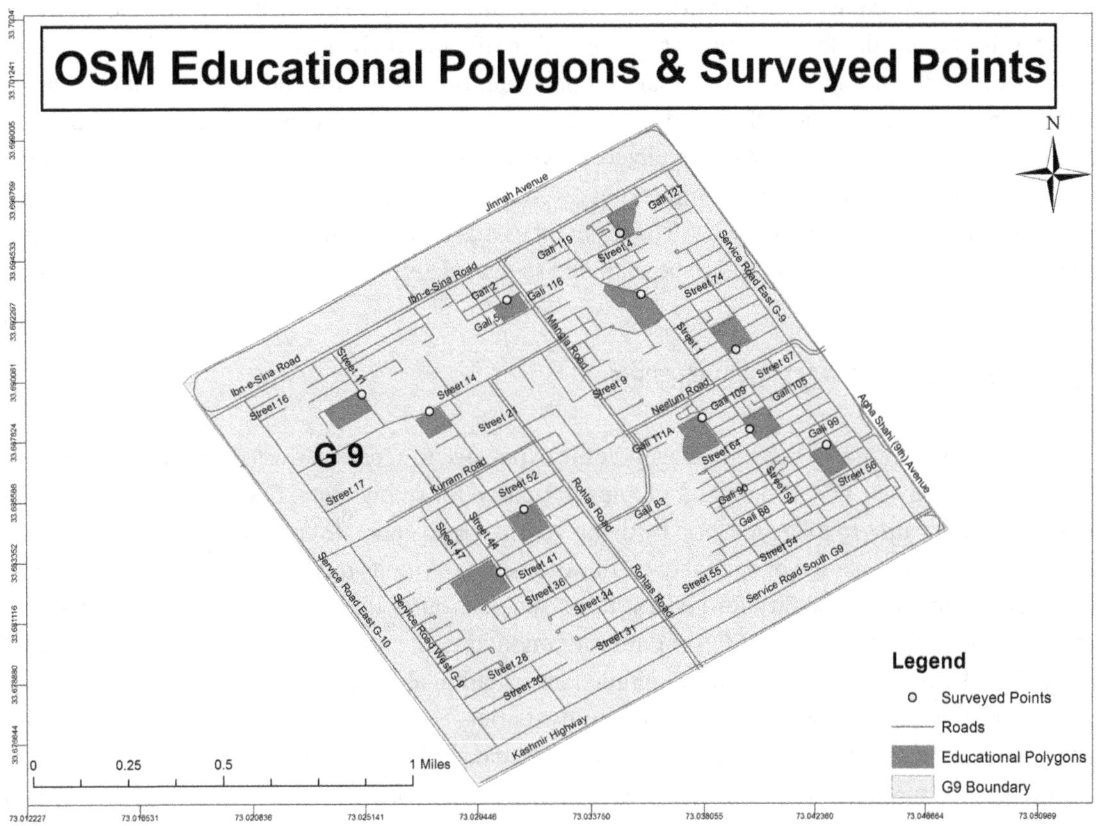

Table 5. Distance between GPS point and OSM boundary with GPS accuracy at each point

Sample #	Distance (meter)	GPS Accuracy (meter)
1	5.9	3.75
2	5.2	4.1
3	10.6	4
4	1.6	4.5
5	6.8	7
6	2.9	3.5
7	4.2	2.89
8	9.7	4.3
9	2.1	3.7
10	7.3	8.32
11	1.3	2.75

Table 6. Buildings name of NUST and the Hausdorff distance between OSM and SoP buildings

NUST Building	Hausdorff Distance (meters)
Admin Office	5.2
Habib Bank Limited	3.5
Rumi Hostels	4.9
Masjid-e-Rehmat	5.6
IGIS	3.9
Concordia 2(C2)	3.7
SCME	4.6
Concordia 1(C1)	3.5
NUST Villas	5.3
ISRA Apartments	3.3
IQRA Apartments	3.7
Masjid-e-Taqwa	3.1
RIMMS	4.8

The positional accuracy is valuable in educational planning. Questions such as where are new schools, colleges and buildings within a large University needed? What and where are the optimal routes for students to commute to educational institutes? If digitization is carried out with care and the high resolution imageries are available, an accurate representation of data can be achieved.

The results for positional accuracy described here reinforce previous studies which have focused on analyzing the accuracy of the OSM road network. A small offset between the location of educational institutes in OSM and ground truth was detected, however the anomaly would not affect educational planning. When attribute accuracy and completeness were assessed, the results indicate a severe mismatch between authoritative datasets and OSM. The differences make OSM insufficient for educational planning in Islamabad and other datasets need to be included to provide reliable results for educational planning.

CONCLUSION AND FUTURE WORK

When official spatial data from national agencies is expensive to access, slow to change and presented in difficult to use formats, VGI is seen as a viable alternative for spatial data. OSM is the most prominent VGI project which aims to map the world by using volunteers to generate spatial data. Given the increase in the use of OSM as a source of spatial data, questions regarding its quality have been raised. Previous studies have examined its quality for general tasks such as route finding and positional accuracy. This chapter has assessed the quality of OSM data for conducting the specific task of educational planning in Islamabad, Pakistan. The task requires detailed information about the current location, attributes and accessibility of existing schools, colleges and universities. Such information was extracted using the semantic descriptions found in OSM and compared to data collected from the websites of the authorities which regulate education institutes and SoP. Additionally, the positional accuracy of buildings was tested using GPS ground truth data which were collected for this study.

The quality of the OSM for a planning task in the study region was tested in several dimensions. Completeness, positional accuracy and attribute accuracy were chosen as key requirements for educational planning. For completeness, there are large gaps in the VGI for educational institutes. For example, only 35% of schools are present in the OSM data. Similarly, the names and addresses of educational institutions were not always correctly recorded in the OSM data and in some cases no attribute data was recorded for educational institutes. Finally, the positional accuracy of institutes and individual buildings within them was assessed. The offset was found to be insignificant and a high degree of overlap with official data was discovered.

While the positional accuracy of the institutes found in OSM is satisfactory, the measures of completeness and attribute data are not sufficient for an educational planning task. In order to determine the location for a new educational institute, complete knowledge of existing educational institutes is required along with reliable attribute information describing the name and type of each educational institute. The results support previous studies which highlight the positional and topographical accuracy of OSM is generally acceptable, however this chapter has highlighted the problems with using OSM for specific tasks, in this case educational planning.

The study presented here did not include an assessment of the temporal accuracy of the OSM data. This will be assessed in future studies. It is anticipated that crowdsourcing will be effective at capturing changes to local areas faster than official mapping sources which are not updated frequently and require rigorous testing before publishing. However, without completeness, this metric is irrelevant for education planning.

Finally, OSM was the only VGI source used to test the viability of VGI for educational planning while a single region of Pakistan was the only area analyzed. Further VGI projects, such as *Wikimapia* (http://wikimapia.org) will be explored to determine if additional information can be collected to support planning in regions where OSM data is missing or erroneous. The research will also be extended to other areas to compare the quality of VGI for educational planning in other cities. The knowledge created through this research will be used to develop an online platform for interactively planning educational institutions in Pakistan.

REFERENCES

Ali, A., & Ahmad, M. (2013). Geospatial data sharing in Pakistan: possibilities and problems? *Proceedings of Global Geospatial Conference 2013.*

Ballatore, A., & Zipf, A. (2015, October). A conceptual quality framework for volunteered geographic information. In *International Workshop on Spatial Information Theory* (pp. 89-107). Springer International Publishing. doi:10.1007/978-3-319-23374-1_5

Barron, C., Neis, P., & Zipf, A. (2014). A comprehensive framework for intrinsic OpenStreetMap quality analysis. *Transactions in GIS*, *18*(6), 877–895. doi:10.1111/tgis.12073

Bartoschek, T., & Keßler, C. (2013). VGI in education: From K-12 to graduate studies. In Crowdsourcing Geographic Knowledge (pp. 341-360). Springer Netherlands.

Camboim, S. P., Bravo, J. V. M., & Sluter, C. R. (2015). An Investigation into the Completeness of, and the Updates to, OpenStreetMap Data in a Heterogeneous Area in Brazil. *ISPRS International Journal of Geo-Information, 4*(3), 1366–1388. doi:10.3390/ijgi4031366

Campagna, M., Floris, R., Massa, P., Girsheva, A., & Ivanov, K. (2015). The role of social media geographic information (SMGI) in spatial planning. In Planning Support Systems and Smart Cities (pp. 41-60). Springer International Publishing.

Ciepłuch, B., Mooney, P., & Winstanley, A. C. (2011). *Building generic quality indicators for OpenStreetMap. In 19th annual GIS Research UK*. GISRUK.

Connors, J. P., Lei, S., & Kelly, M. (2012). Citizen science in the age of neogeography: Utilizing volunteered geographic information for environmental monitoring. *Annals of the Association of American Geographers, 102*(6), 1267–1289. doi:10.1080/00045608.2011.627058

D'Antonio, F., Fogliaroni, P., & Kauppinen, T. (2014). VGI edit history reveals data trustworthiness and user reputation.*17th AGILE Conference on Geographic Information Science.*

de Albuquerque, J. P., Fan, H., & Zipf, A. (2016). A conceptual model for quality assessment of VGI for the purpose of flood management.*19th AGILE Conference on Geographic Information Science.*

Elwood, S., Goodchild, M. F., & Sui, D. Z. (2012). Researching volunteered geographic information: Spatial data, geographic research, and new social practice. *Annals of the Association of American Geographers, 102*(3), 571–590. doi:10.1080/00045608.2011.595657

Girres, J. F., & Touya, G. (2010). Quality assessment of the French OpenStreetMap dataset. *Transactions in GIS, 14*(4), 435–459. doi:10.1111/j.1467-9671.2010.01203.x

Goodchild, M. F. (2007). Citizens as sensors: The world of volunteered geography. *GeoJournal, 69*(4), 211–221. doi:10.1007/s10708-007-9111-y

Goodchild, M. F., & Hunter, G. J. (1997). A simple positional accuracy measure for linear features. *International Journal of Geographical Information Science, 11*(3), 299–306. doi:10.1080/136588197242419

Goodchild, M. F., & Li, L. (2012). Assuring the quality of volunteered geographic information. *Spatial Statistics, 1*, 110–120. doi:10.1016/j.spasta.2012.03.002

Goranson, C., Thihalolipavan, S., & di Tada, N. (2013). VGI and public health: possibilities and pitfalls. In *Crowdsourcing Geographic Knowledge* (pp. 329–340). Springer Netherlands. doi:10.1007/978-94-007-4587-2_18

Haklay, M. (2010). How good is volunteered geographical information? A comparative study of OpenStreetMap and Ordnance Survey datasets. *Environment and Planning. B, Planning & Design, 37*(4), 682–703. doi:10.1068/b35097

Hall, P. (2002). *Cities of tomorrow: an intellectual history of urban planning and design in the twentieth century*. Malden: Blackwell.

Hausdorf, F., & Aumann, J. R. (1914). *Grundzüge der mengenlehre*. Veit.

Horita, F. E., & De Albuquerque, J. P. (2013, May). An approach to support decision-making in disaster management based on volunteer geographic information (VGI) and spatial decision support systems (SDSS).*Proceedings of the 10th International Conference on Information Systems for Crisis Response and Management*, 1215.

Hung, K. C., Kalantari, M., & Rajabifard, A. (2016). Methods for assessing the credibility of volunteered geographic information in flood response: A case study in Brisbane, Australia. *Applied Geography (Sevenoaks, England)*, *68*, 37–47. doi:10.1016/j.apgeog.2016.01.005

Keßler, C., & de Groot, R. T. A. (2013). Trust as a proxy measure for the quality of volunteered geographic information in the case of OpenStreetMap. In *Geographic Information Science at the Heart of Europe* (pp. 21–37). Springer International Publishing. doi:10.1007/978-3-319-00615-4_2

Kresse, W., & Fadaie, K. (2004). *ISO standards for geographic information*. Springer Science & Business Media. doi:10.1007/978-3-662-08039-9

Leeuw, J. D., Said, M., Ortegah, L., Nagda, S., Georgiadou, Y., & DeBlois, M. (2011). An assessment of the accuracy of volunteered road map production in Western Kenya. *Remote Sensing*, *3*(2), 247–256. doi:10.3390/rs3020247

Ludwig, I., Voss, A., & Krause-Traudes, M. (2011). A Comparison of the street networks of Navteq and OSM in Germany. In *Advancing Geoinformation Science for a Changing World* (pp. 65–84). Springer Berlin Heidelberg. doi:10.1007/978-3-642-19789-5_4

Masli, M., Bouma, L., Owen, A., & Terveen, L. (2013, February). Geowiki+ route analysis= improved transportation planning. In *Proceedings of the 2013 Conference on Computer Supported Cooperative Work Companion* (pp. 213-218). ACM.

Memon, G. R. (2007). Education in Pakistan: The key issues, problems and the new challenges. *Journal of Management and Social Sciences*, *3*(1), 47–55.

Mooney, P., Corcoran, P., & Winstanley, A. C. (2010). Towards quality metrics for OpenStreetMap. In *Proceedings of the 18th SIGSPATIAL International Conference on Advances in Geographic Information Systems* (pp. 514-517). ACM.

Okpala, D. C. I. (2001, October). Spatial Information: The basic tool for Sustainable human settlements development planning and management. In *International Conference on Spatial Information for Sustainable Development.*

Poser, K., & Dransch, D. (2010). Volunteered geographic information for disaster management with application to rapid flood damage estimation. *Geomatica*, *64*(1), 89–98.

See, L., Mooney, P., Foody, G., Bastin, L., Comber, A., Estima, J., & Liu, H. Y. et al. (2016). Crowdsourcing, citizen science or Volunteered Geographic Information? The current state of crowdsourced geographic information. *ISPRS International Journal of Geo-Information*, *5*(5), 55. doi:10.3390/ijgi5050055

Shumaker, B. P., & Sinnott, R. W. (1984). Astronomical computing: 1. Computing under the open sky. 2. Virtues of the haversine. *Sky and Telescope*, *68*, 158–159.

van Oort, P. A. (2006). *Spatial data quality: from description to application* (Doctoral dissertation). Publications on Geodesy 60, NCG.

KEY TERMS AND DEFINITIONS

ArcGIS: A well-known desktop commercial software developed by ESRI used primarily for all maps related tasks.

Attribute Accuracy: A parameter used in quality assessment to analyze how accurate the spatial information of features is as compared to the actual one.

Authoritative Data: Data produced by an authorized or well-known company which can be considered to be as per the ground reality.

Data Completeness: A parameter used in quality assessment to quantify how much ground reality has been mapped in user generated content.

Geofabrik: A website from where you can download update OSM data of any specific area of the world in various formats.

Hasudorff distance: A method used to calculate how far the two lines or polygon are from each other.

Haversine Formula: An equation which can be used to calculate the distance between two points having latitude and longitude information.

Higher Education Commission (HEC): An institution that controls all higher education related issues in Pakistan.

OSM2PGSQL: A command line tool which can be used to convert OSM dataset into a PostGIS enabled PostgreSQL database.

Positional Accuracy: A parameter used in quality assessment to analyze that how accurate the location information of a ground feature has been captured as compared to its actual location.

PostGIS: An extension of PostgreSQL database which enables storing, retrieving and numerous other functions of geometry type dataset.

Survey of Pakistan (SoP): A National Mapping Agency of Pakistan which is responsible for conducting topographic land surveys and maps production.

Web Scrapping: A technique used to extract data from a website in a systematic way.

Chapter 6
Emotional and Subjective Volunteered Geographical Information

Jiri Panek
Palacky University in Olomouc, Czech Republic

ABSTRACT

Crowdsroucing of emotional information can take many forms, from social networks data mining to large-scale surveys. The author presents the case-study of emotional mapping in Ostrava's district Ostrava-Poruba, Czech Republic. Together with the local administration, the author crowdsourced the emotional perceptions of the location from almost 400 citizens, who created 4,051 spatial features. Additional to the spatial data there were 1,244 comments and suggestions for improvements in the district. Furthermore, the author is looking for patterns and hot-spots within the city and if there are any relevant linkages between certain emotions and spatial locations within the city.

INTRODUCTION

The mapping of emotions and sentiments was once a neglected practice in cartography, nevertheless with the rise of volunteered geographical information (VGI), citizen science, neocartography and crowdsourcing it is growing in popularity among researchers, representatives of the local administration and community leaders. They all understand that places are strongly linked with emotions (Mody, Willis, & Kerstein, 2009) and can be considered safe or unsafe, pleasant or ugly, popular or boring, among other responses (Korpela, 2002). Subjective perceptions of the city can be a valuable source of information for city planners and local administration, and citizens can become engaged in community building processes. Emotional mapping can therefore be considered as an active form of public participation in local decision making processes.

Participatory mapping and qualitative GIS (sometimes also called GeoParticipation) allow city planners and decision makers to deploy new tools and methods that can collect qualitative and quantitative data about cities, their dynamics and the people who live there (Kloeckl, Senn, Di Lorenzo, & Ratti, 2011).

DOI: 10.4018/978-1-5225-2446-5.ch006

Most geospatial applications rely on objective data only, although there can be a discussion concerning the extent to which GIS data are objective, as there is always a level of generalisation, uncertainty and authorial bias (Pickles, 1995). The call for a more humanised and participatory approach to geospatial information and technologies has been heard since the publication of the *Ground Truth* more than twenty years ago (Pickles, 1995), and it currently echoes humanistic alternatives to GIS (Byrne & Pickard, 2016; Knowles, Westerveld, & Strom, 2015).

The author acted as an action researcher and, together with the local administration from the Ostrava-Poruba[1] neighbourhood, collected and analysed VGI subjective emotional data divided into seven categories. The data were collected as both paper and electronic emotional maps and the outcomes can be seen as a version of a Gould-style mental map (Gould, 1986; Gould & White, 1974).

The aim of this chapter is to present the methodology and deployment of the stand-alone web-based application for collecting VGI emotional data. The tool PocitoveMapy.cz is designed as a web-application based on Leaflet library. Similar to other web-based tools for crowdsourced mapping, it allows users to collect spatial data on a slippery map background. Unlike Ushahidi, Umap, ArcGIS Online and many others, PocitoveMapy.cz does not require the registration or installation of any specific software, plug-in or virtual server. The data collected through the crowdsourcing tool are saved in the GeoJSON format, which can be later exported to shapefile.

The tool used in the case-study allows for the duplication of the research methodology in other regions, as well as its implementation in various topics such as safety mapping (Pánek, Pászto, & Marek, 2017), the analyses of citizens' preferences for bicycle routes and neighbourhood revitalisation plans (Pánek & Pászto, 2017).

Emotions on Maps

In the past ten years several projects have dealt with georeferenced emotions and the methods used to gather emotional data can be divided into three technological groups: (1) biometric measurements (Bergner, Zeile, & Papastefanou, 2011; Nold, 2009), that use a Galvanic Skin Response (GSR) recorder, which is a simple indicator of emotional arousal, in conjunction with a user's geographical location. The outcome (map) visualises points of high and low emotional arousal, but gives no explanation as to what happened at each location. (2) extraction from user generated content, often called sentiment analysis, gathers data from services such as Twitter, Flickr, Facebook, etc. (Biever, 2010; Bollen, Mao, & Zeng, 2011; Mislove, Lehmann, Ahn, Onnela, & Rosenquist, 2010; Quercia, O'Hare, & Cramer, 2014; Quercia, Schifanella, & Aiello, 2014) and uses language processing and text analysis to identify and extract subjective/emotional information from source materials. This form of research often uses a lexicon-based approach (Taboada, Brooke, Tofiloski, Voll, & Stede, 2011) to assign values to words extracted from the source and to assess their emotional background. (3) surveys (Huang, Gartner, Klettner, & Schmidt, 2014; MacKerron & Mourato, 2010; Mody et al., 2009). The author's approach can be considered to be a survey.

Griffin and Mcquoid (2012) distinguish three categories when talking about maps and emotions. These categories are (1) maps of emotions, (2) using maps to collect emotional data, and (3) emotions while using maps. The case study described in this chapter is a combination of the first two categories. Maps were used to collect the emotional information, and also to visualise the emotional data. Emotions are one of the defining characteristics of every human being and yet their presence on maps and spatial data is uncommon (Griffin & Mcquoid, 2012). Several authors (Barrett, 1976; Reeve, 2014; Russell, 1980) have described emotions as a two dimensional structure with the axis being pleasant-unpleasant and high arousal-low arousal. This division does not fit the needs of local planning and public participation as citizens want to have a broader range of possible answers to describe their living environment. Geographers, as well as the author in this chapter, describe emotions as subjective, relational flows between places and people (Smith, Bondi, & Davidson, 2012).

While historically, cartography was mainly focused on treating that which is visible or can be mapped (including for example air temperature and wind speed) (Wilson, 2011), critical and feminist cartographers have always advocated mapping a space as people experience it, with emotions (Pearce, 2008). Hauthal & Burghardt (2016, p. 2) argued that *...mappers of georeferenced emotions are almost exclusively researchers...* using emotional maps in various fields such as tourism (Mody et al., 2009), navigation (Gartner, 2012; Huang et al., 2014) and city planning (Raslan, Al-hagla, & Bakr, 2014). This research is one example which shows that the argument of Hauthal and Burghardt may no longer be true. Applications such as Maptionnaire, SprayCan, and PocitoveMapy.cz open a wide range of opportunities to the general public in the area of GeoParticipation and in the use of maps as tools for negotiation and facilitation.

METHODOLOGY

As mentioned above, the author approached the research as an action researcher, with the aim of achieving practical benefits and generating practical information from the research (Brydon-Miller, Greenwood, & Maguire, 2003). The methodology involved sketchable maps with questions (Jankowski, Czepkiewicz, Młodkowski, & Zwoliński, 2015) in an analogue as well as a digital interface, in order to find the perceived centre of Ostrava-Poruba, and pins[2], analogue and digital, for collecting point-like locations. The decision to use points, even for areas that could theoretically be marked as polygons, was based on the research of Brown and Pullar (2012), who suggested using points as basic input data features for creating and analysing perception maps. The decision was also based on the author's previous experience with the creation of emotional maps. The only exception was the question regarding the perceived centre of Ostrava-Poruba. Based on a discussion with the city representatives, who were interested in a spatial definition of the "city centre", the author decided to use the polygon feature to mark the answers to this question. The combination of collection methods was selected because in ordinary participatory methods face-to-face meetings combined with time and space commitments narrow down the number of participants and lead to the exclusivity of a single opinion (Kahila-Tani, Broberg, Kyttä, & Tyger, 2015). The involvement of crowdsourcing and Computer-Assisted Web Interviews (CAWI) is in line with the concept of a participatory planning support system as defined by Kahila and Kyttä (2009).

During the meetings with citizens, two maps were available; (1) map of all Ostrava-Poruba with a scale of 1:10 000, printed on paper size A0 and (2) map of the centre of Ostrava-Poruba with a scale of 1:10 000, printed on paper size A2. Both maps were based on OpenStreetMap data and printed via the MapWebbing[3] service. When it was observed that the map of the centre was no longer readable due to the polygons people had drawn on the map, it was replaced[4] by a new map of the same size and scale. Respondents had an unlimited number of attempts to mark places and areas on the map and were able to skip questions if they wished.

The second option for collecting the data, as opposed to classical paper-based mapping, was the crowdsourcing online tool PocitoveMapy.cz. Responders had the opportunity to toggle between maps, as can be seen in figure 4. The default basemap for the application is OpenStreetMap, but users were free to choose between OpenStreetMap, OpenStreetMap toner, Esri topographic map, National aerial map and a National map of the Czech Republic (State Administration of Land Surveying and Cadastre, 2014). The skills required to use the tool effectively are similar to those of the average Internet user.

The crowdsourced data are saved in MySQL database, which is usually available on every hosting, and therefore there is no need to have specialised hosting or own server with geodatabase installed. Each entry in the database contains unique user ID (randomly generated), question indentificator, number of points/lines/polygons marked and geometry. These entries are later merged by GeoJSON PHP library script (Mikola, 2015), which allows GeoJSON to maintain data from multiple users. Furthermore, simple SQL queries can be used to filter data, based on user ID, type of question, etc.

Users of the online tool were asked the same seven questions as those using the paper-based map, with options to use points for six of the questions and polygon-freehand for their perception of the centre of Ostrava-Poruba. Once the feature was created, a pop-up window appeared which asked users to mark the level of their answers (as can be seen on Figure 4). The representatives of the local administration decided to ask citizens seven spatial questions:

- What do you consider to be the centre of Ostrava-Poruba?
- Where do you spend free time?
- Where do you have trouble getting to by public transport?
- Where do you feel unsafe?
- Where do you have problems with parking?
- Where is it ugly?
- Where do you like to shop?

CASE STUDY

Study Location

Poruba is one of 23 city districts of the city of Ostrava, the Czech Republic (Figure 1). Historically, it was a separate village and it became part of Ostrava in 1957. The heavy industrialization of Ostrava in the second half of the 19th century didn't have a major impact on Poruba and it retained its agricultural character. Several factories were built in Poruba at the beginning of the 20th century (Duží & Jakubínský, 2013). The major event in the history of Ostrava-Poruba was the decision of the communist authorities to build a large complex of apartment buildings intended to house the workers who came to Ostrava from various parts of the country in large numbers after World War II. Construction began in the early 1950s (Hosák & Šrámek, 1980) and at its peak almost 100,000 people lived in Ostrava-Poruba. Today the population is about 69,000 and the area of Ostrava-Poruba is 13.18 km^2 (Ostrava-Poruba, 2016).

Figure 1. Location of Ostrava-Poruba within the Czech Republic on Google Maps

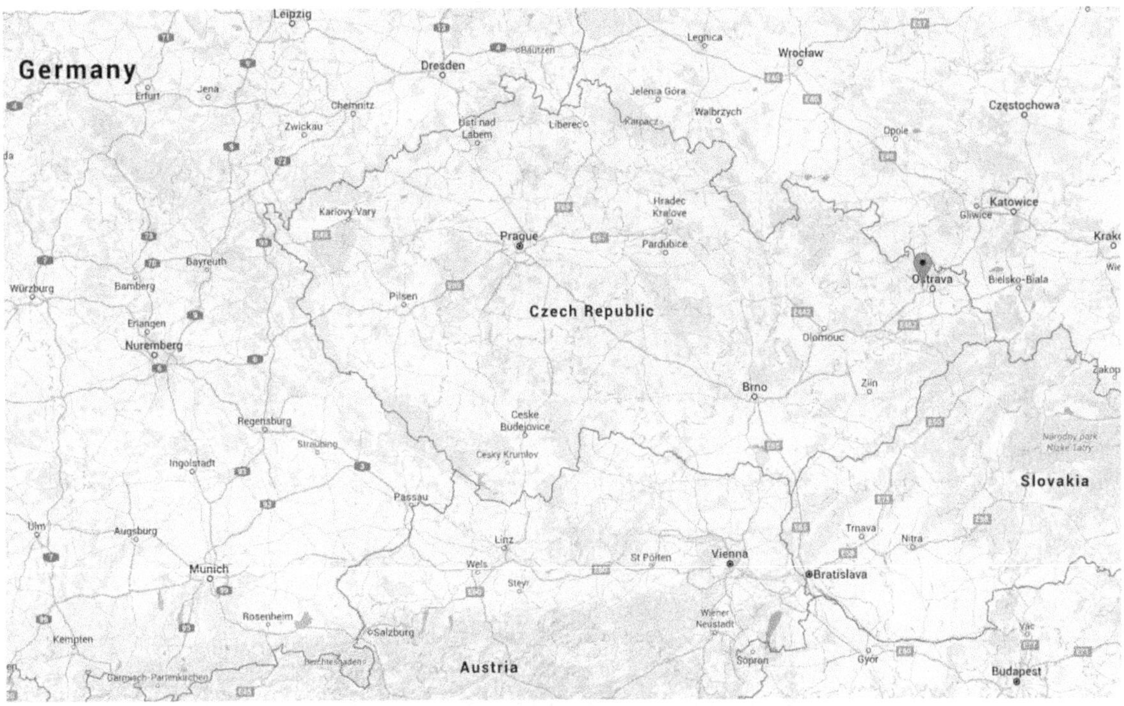

Collection of Data

UN Agenda 21 serves (IDRC, 1996), among other things, as a cornerstone for the global implementation of sustainable development. This 24 year-old document was adopted as a voluntary initiative and follows the main dimensions of sustainable development and its implementation on a global, national and local level. Its local and community strategy version (Local Agenda 21) emphasizes the central role of local authorities and calls upon them to develop local strategies for sustainable development. The main reason for this is the added value of the emphasis on communication, participation and cooperation among various entities at the local level. In the light of the ideas of sustainable development and Local Agenda 21, the administration of Ostrava-Poruba organised a meeting with its citizens, the purpose of which was to discuss issues relating to the district and to start creating an emotional map of the area. The meeting was held on 15th March 2016 and about 130 citizens participated in the discussion (Figure 2).

The collection of the data was in analogue form (paper map with coloured pins – see Figure 3) during the meeting and via the crowdsourcing online mapping tool PocitoveMapy.cz (Figure 4). The online mapping was available for the period of one month from 15th March to 13th April 2016. Overall, citizens recorded 4,051[5] points and polygons and made 1,244 comments regarding the quality of life in Ostrava-Poruba. The majority of responses were acquired via the online tool[6] as can be seen in Table 1. In total, 345 citizens answered via the online tool. The number of citizens who used the paper map during the meeting cannot be calculated, but it is expected to have been approximately 50 people.

Figure 2. Meeting with the citizens of Ostrava-Poruba

Figure 3. The analogue way of creating the emotional map of Ostrava-Poruba

Results of the Case Study

The main results of the case study are based on the dataset collected during the crowdsourcing exercise and the participatory mapping with citizens. In total, the dataset comprises 4,051 features. Additional to the spatial features are 1,244 comments and suggestions for improvements in the district of Ostrava-Poruba. The most answered question was "Where do you spend free time?", with 919 points crowdsourced; while the least answered was "Where do you have trouble getting to by public transport?", with 258 points. On average, people marked 1.5 places per question, which is not a high number, nevertheless it still gives a

Figure 4. The online tool for creating the emotional map of Ostrava-Poruba

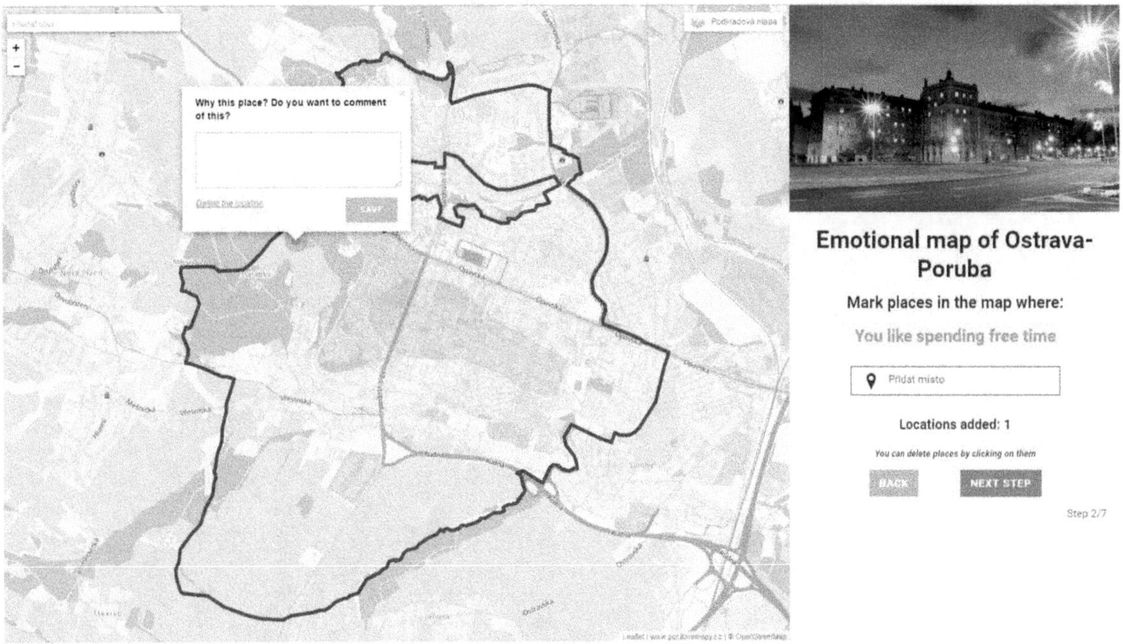

Table 1. Basic statistics for crowdsourced data

	What do you consider to be the centre of Ostrava-Poruba	This is where I spend free time	I have trouble getting here by public transport	I feel unsafe here	I have problems parking here	It is ugly here	I like shopping here
All answers	**400**	**919**	**258**	**619**	**584**	**689**	**582**
- From online tool	373	794	242	569	561	629	544
- From paper map	27	125	16	50	23	60	38
Average number of points per person (only for online tool)	1.1	2.3	0.7	1.6	1.6	1.8	1.6

good picture of the citizens' perceptions within the district of Ostrava-Poruba. The data were analysed in QGIS[7] as vector, but visualised as heatmap (Figure 5) with a radius of 7mm[8] and no transparency level.

Where Do You Like Spending Free Time?

The question about favourable locations to spend free time was the most answered and as can be seen on Figure 5 the answers provided several hot-spots. The main hot-spot on the western edge of the location is the forest, which locals very often use for walks, runs and other outdoor activities. Furthermore, the swimming pool (hot-spot on the east of the location) and the city park (hot-spot on the north) were among the popular answers. Most comments were related to walking areas, parks, forest and ponds.

Figure 5. Heat-map visualisation of the emotional mapping results for Ostrava-Poruba: Top-left: Where do you like spending free time; Centre-left: I have trouble getting here by public transport; Bottom-left: I like shopping here; Top-right: I have problems parking here, Centre-right: I feel unsafe here, Bottom-right: It is ugly here.

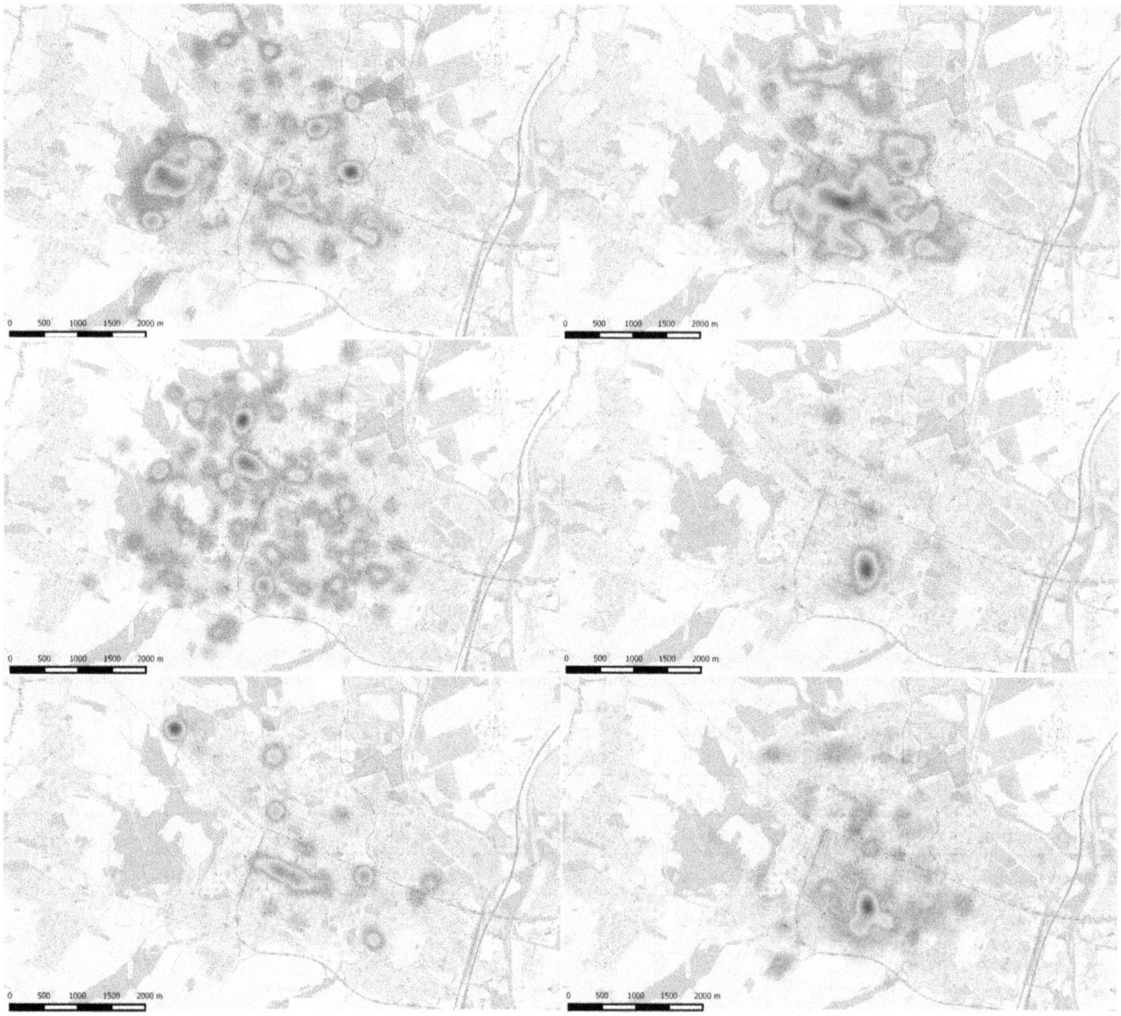

Where Do You Have Trouble Getting to by Public Transport?

On the other hand, the least answered question was about the inaccessibility of some areas to public transport. The map has several smaller hot-spots and one very strong hot-spot (but this one is outside the administration boundaries of the district). Comments on this question were mainly related to the distance to the closest tram/bus stops.

Where Do You Like to Shop?

A popular question with 582 answers and one very strong hot-spot – the shopping centre in the north-west corner of the district. The main street – centre of the map, which is also considered as a centre of the district (see Figure 6), also gained several answers. The majority of comments mentioned the names of shopping centres and the name of the main street.

Where Do You Have Problems With Parking?

Parking is a burning issue in many Czech cities and Ostrava is one of them. There is basically one large hot-spot which covers most of the densely populated centre of Ostrava-Poruba. Citizens did not mark many spatial comments for this question, but when they did, they often mentioned areas around the local hospital.

Where Do You Feel Unsafe?

Mapping of safety and fear of criminality has become a modern trend in various fields such as cartography, GIScience, environmental ecology, behavioural geography, urban planning and psychology. In the Czech/Slovak context only a few authors have dealt with the mapping of safety (Jíchová & Temelová, 2012; Sessar & Sirotek, 2001; Stasíková, 2011), but only (Pánek et al., 2017) by using emotional or mental maps to collect and visualise the data. Němečková (2014) used subjective data to analyse safety

Figure 6. Cumulative answer of 400 respondents on questions: What do you consider a centre of Ostrava-Poruba

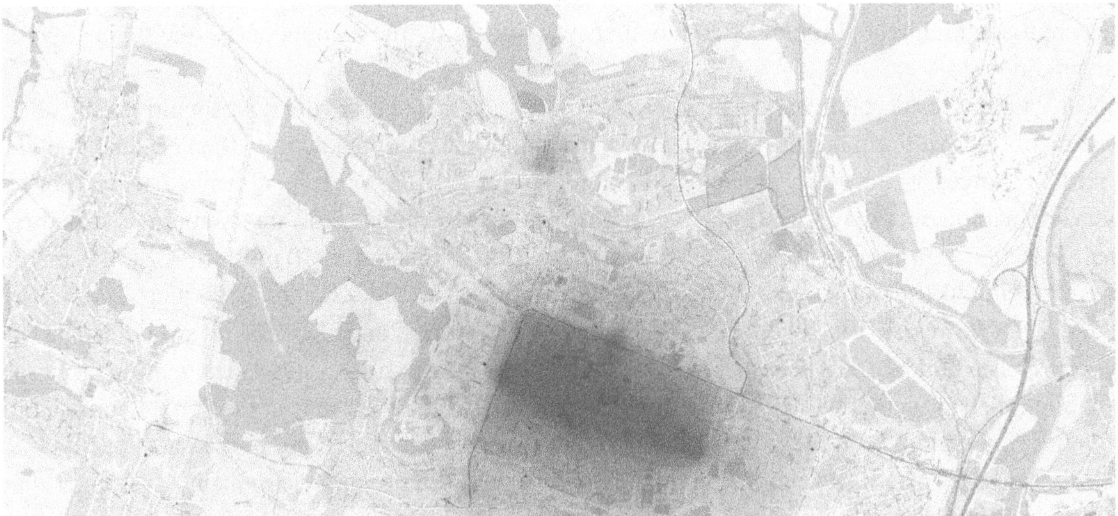

in the city, but the spatial element of the data was not even explored in her research. Some researchers (M. Brown & Polk, 1996; Clemente & Kleiman, 1977; Oc & Tiesdell, 1997) argued that fear of criminality could be as dangerous as criminality itself, therefore the application of GIScience became trendy in analysing spatial patterns of criminality and how it affects people's behaviour within an urban environment (Doran & Burgess, 2011; Chainey & Ratcliffe, 2013; Leitner, 2013; Santos, 2012). In the case of Ostrava-Poruba, there is only one hot-spot related to this question. The hot-spot is in the neighbourhood around "Dělnická" street and was marked by more than 54% of respondents. "Dělnická" is a socially excluded locality (Kvasnička, 2010) with a population of about 750 inhabitants, mainly of Roma[9] origin. Unlike other excluded localities in Ostrava, "Dělnická" has nice flats, is close to the centre of Ostrava-Poruba, has enough shops and schools, and people have access to public transport (Rusková, 2012), but it still has a bad reputation among the Ostrava-Poruba citizens. The other places have a very small density of points, hence they do not appear on the heat-map visualisation.

Where It is Ugly?

The question about the aesthetic quality of the place has one similar hot-spot to the previous question – the neighbourhood of "Dělnická" street, which is often called a *ghetto* by locals.

What Do You Consider a Centre of Ostrava-Poruba?

Because Ostrava-Poruba was an independent village that was incorporated into the city of Ostrava in 1957 and most of its inhabitants live in a large complex of apartment buildings built during the communist era, mainly in the 1950s, it has no historical centre. The representatives of the local administration therefore were interested in exploring citizens' answers about the centre of Ostrava-Poruba. During the research the author collected 400 polygons representing the perceived centre of the district. The visualisation (Figure 6) was created in QGIS using a single symbol method and setting the transparency level at 95%. This approach highlighted the most common answers given by the respondents – the main street and surrounding areas.

The data and results gathered during the case study were presented to the representatives and administration of Ostrava-Poruba. The outcomes from the emotional mapping activity then became one of the important sources for the newly created document – *Study of Public Space in Ostrava-Poruba* (personal interview with deputy-mayor of Ostrava-Poruba). Within the study[10], 252 interventions[11] scaled from small-size interventions (short term – up to two years, with a budget up to 10 million[12] CZK), to medium size interventions (up to five years with a budget up to 50 million[13] CZK) and large interventions (long term – over five years with a budget over 50 million CZK) were planned. Small interventions are often projects such as building new pedestrian crossings, re-cultivating public spaces and increasing the number of public toilets or public waste baskets. Most medium-sized interventions are new road constructions for better connections with the rest of the city or new car parks to increase the parking capacity within the area[14]. Large scale interventions are mainly linked with the urban development of the area – new commercial sites, or larger infrastructure constructions.

DISCUSSION

In total, about 400[15] respondents marked 4,051 points and polygons on seven topics. The author used the choice of sample size method (Agresti & Finlay, 2009; Raosoft, 2004) with a margin of error of 5% and a confidence level of 95% (z=1.96) to calculate the size of the population needed for the case study to be representative. The result of the calculation was that the representative sample, considering the settings described above, is 383 respondents. The author is aware of a few methodological gaps in the research that could have been addressed better, nonetheless due to certain restraints it was not possible, therefore only comments are provided below.

Although the size of the sample was calculated as being representative, the administration of the district did not want to collect any personal data, so it is not possible to (a) validate whether respondents really exist, (b) crosscheck the demographic structure of respondents with the demographic structure of the district, (c) draw any deeper geostatistical analysis, such as age distribution of answers, gender differences in answers, etc.

There is a discussion as to whether data gathered from paper-based maps can be combined with data gathered via online crowdsourcing. The author believes that accessing users[16] who would not fill in the online form is worth the risk of combining the data. When they were analysed further, the author realised that the data from the paper map are almost identical to the answers gathered via the crowdsourcing platform. The author is also aware of the disadvantages of paper-based map research as, for example, paper maps (a) cannot zoom and pan, (b) are not searchable, (c) and are done in a group, so people can be affected by the answers of other respondents.

CONCLUSION

Emotional mapping and participatory planning play an important role in the processes of community building and participatory development, as well as in the creation of knowledge about common spaces and the environment. Crowdsourcing of emotional perceptions allows users to get involved in the process of collecting information related to their emotional links with their environment. The idea is grounded in the GeoParticipationary approach – using spatial tools in order to involve citizens in community participation and decision-making processes related to public spaces (Pánek, Hrubeš, Kubásek, Valůch, & Zahumenská, 2014). The outcome of this case-study combines data from a traditional analogue paper-based map and from a participatory web-application PocitoveMapy.cz. The author sees the potential in using the PocitoveMapy.cz tool for crowdsourcing emotional data; to aid local administrations in making spatially-aware decisions. The tool has been developed and tested in various scenarios, both topic-wise as well as location-wise. The main obstacle for the further implementation of any participatory planning support system crowdsourcing platform is the readiness and preparedness of politicians. The future outlook for the research includes spatial analysis based on different age groups of respondents, and based on their relationship with the city. Citizens who live in the city may have different perceptions to citizens who only commute to the city for work or study purposes. Nevertheless, the first analysis proved that the inclusion of qualitative and subjective data can produce relevant spatial data, which can be used in local planning and community development. The combination of subjective layer with "official" data, for example from the police department, can enhance the information richness of the dataset and reveal new relationships among various datasets.

REFERENCES

Agresti, A., & Finlay, B. (2009). Choice of sample size. In Statistical Methods for the Social Sciences (p. 126). Academic Press.

Barrett, A. M. (1976). Some reflections on those who taught me a personal view. *The Journal of Pharmacy and Pharmacology*, *28*(4s), 338–341. doi:10.1111/j.2042-7158.1976.tb04176.x PMID:6736

Bergner, B., Zeile, P., & Papastefanou, G. (2011). Emotional Barrier-GIS – A new approach to integrate barrier-free planning in urban planning processes. In *Proceedings REAL CORP* (pp. 247–257). Retrieved from http://realcorp.at/archive/CORP2011_27.pdf

Biever, C. (2010). Twitter mood maps reveal emotional states of America. *New Scientist*, *207*(2771), 14. doi:10.1016/S0262-4079(10)61833-7

Bollen, J., Mao, H., & Zeng, X. (2011). Twitter mood predicts the stock market. *Journal of Computational Science*, *2*(1), 1–8. doi:10.1016/j.jocs.2010.12.007

Brown, G., & Pullar, D. V. (2012). An evaluation of the use of points versus polygons in public participation geographic information systems using quasi-experimental design and Monte Carlo simulation. *International Journal of Geographical Information Science*, *26*(2), 231–246. doi:10.1080/13658816.2011.585139

Brown, M., & Polk, K. (1996). Taking Fear of Crime Seriously: The Tasmanian Approach to Community Crime Prevention. *Crime and Delinquency*, *42*(3), 398–420. doi:10.1177/0011128796042003004

Brydon-Miller, M., Greenwood, D., & Maguire, P. (2003). Why Action Research? *Action Research*, *1*(1), 9–28. doi:10.1177/14767503030011002

Byrne, D., & Pickard, A. J. (2016). Neogeography and the democratization of GIS: A metasynthesis of qualitative research. *Information Communication and Society*, 1–18. doi:10.1080/1369118X.2015.1125936

Chainey, S., & Ratcliffe, J. (2013). *GIS and crime mapping*. John Wiley & Sons.

Clemente, F., & Kleiman, M. B. (1977). Fear of Crime in the United States: A Multivariate Analysis. *Social Forces*, *56*(2), 519–531. doi:10.2307/2577738

Doran, B. J., & Burgess, M. B. (2011). *Putting fear of crime on the map: investigating perceptions of crime using geographic information systems*. Springer Science & Business Media.

Duží, B., & Jakubínský, J. (2013). Brownfield dilemmas in the transformation of post-communist cities: A case study of Ostrava, Czech Republic. *Journal of Studies and Research in Human Geography*, *72*(2), 53–64. doi:10.5719/hgeo.2013.72.53

Gartner, G. (2012). *Putting Emotions in Maps–The Wayfinding Example*. Retrieved from http://www.mountaincartography.org/publications/papers/papers_taurewa_12/papers/mcw2012_sec3_ch08_p061-065_gartner.pdf

Gould, P. (1986). *Mental Maps*. Taylor & Francis.

Gould, P., & White, R. (1974). *Mental Maps*. Harmondsworth, UK: Pelican Books. doi:10.4324/9780203163757

Government of the Czech Republic. (2016). *Report about the Roma minority in the Czech Republic in 2015*. Retrieved from http://www.vlada.cz/cz/ppov/zalezitosti-romske-komunity/dokumenty/zprava-o-stavu-romske-mensiny-za-rok-2015-146576/

Griffin, A. L., & Mcquoid, J. (2012). At the Intersection of Maps and Emotion : The Challenge of Spatially Representing Experience. *Kartographisch Nachrichten*, (6), 291–299.

Hauthal, E., & Burghardt, D. (2016). Mapping Space-Related Emotions out of User-Generated Photo Metadata Considering Grammatical Issues. *The Cartographic Journal*, *53*(1), 78–90. doi:10.1179/1743277414Y.0000000094

Hosák, L., & Šrámek, R. (1980). *Místní jména na Moravě a ve Slezsku [Local names in Moravia and Silesia]* (Vol. 2). Brno: Academia. (In Czech)

Huang, H., Gartner, G., Klettner, S., & Schmidt, M. (2014). Considering Affective Responses towards Environments for Enhancing Location Based Services. ISPRS-International Archives of the Photogrammetry. *Remote Sensing and Spatial Information Sciences*, *1*, 93–96.

IDRC. (1996). *The Local Agenda 21 Planning Guide. The International Council for Local Environmental Initiatives (ICLEI), The International Development Research Centre (IDRC), The United Nations Environment Programme*. UNEP.

Jankowski, P., Czepkiewicz, M., Młodkowski, M., & Zwoliński, Z. (2015). Geo-questionnaire: A Method and Tool for Public Preference Elicitation in Land Use Planning. *Transactions in GIS*. 10.1111/tgis.12191

Jíchová, J., & Temelová, J. (2012). Kriminalita a její percepce ve vnitřním městě: případová studie pražského Žižkova a Jarova [Criminality and its perception in the inner city: case studies of Žižkov and Jarov in Prague]. *Geografie, 3*(117), 329–348. Retrieved from http://geography.cz/sbornik/wp-content/uploads/downloads/2012/11/g03_2012_5jichova_temelova.pdf

Kahila, M., & Kyttä, M. (2009). SoftGIS as a Bridge-Builder in Collaborative Urban Planning. In S. Geertman & J. Stillwell (Eds.), *Planning Support Systems Best Practice and New Methods* (Vol. 95, pp. 389–411). Dordrecht: Springer Netherlands; doi:10.1007/978-1-4020-8952-7_19

Kahila-Tani, M., Broberg, A., Kyttä, M., & Tyger, T. (2015). Let the Citizens Map—Public Participation GIS as a Planning Support System in the Helsinki Master Plan Process. *Planning Practice and Research*, 1–20. doi:10.1080/02697459.2015.1104203

Kloeckl, K., Senn, O., Di Lorenzo, G., & Ratti, C. (2011). Live singapore!-an urban platform for real-time data to program the city. In Computers in Urban Planning and Urban Management (Vol. 4, pp. 1–18). CUPUM. Retrieved from http://senseable.mit.edu/papers/pdf/20110708_Kloeckl_etal_LiveSingapore_ComputersUrban.pdf

Knowles, A. K., Westerveld, L., & Strom, L. (2015). Inductive Visualization: A Humanistic Alternative to GIS. *GeoHumanities*, 1–33. 10.1080/2373566X.2015.1108831

Korpela, K. (2002). Children's environment.Handbook of Environmental Psychology, 363–373.

Kvasnička, R. (2010). *Description of socially excluded locations in Ostrava region*. Ostrava.

Leitner, M. (2013). *Crime modeling and mapping using geospatial technologies* (Vol. 8). Springer Science & Business Media. doi:10.1007/978-94-007-4997-9

MacKerron, G., & Mourato, S. (2010). *Mappiness*. Retrieved September 12, 2015, from http://www.mappiness.org.uk/

Mikola, J. (2015). *GeoJSON implementation for PHP*. Retrieved from https://github.com/jmikola/geojson

Mislove, A., Lehmann, S., Ahn, Y.-Y., Onnela, J.-P., & Rosenquist, J. N. (2010). *Pulse of the Nation: U.S. Mood Throughout the Day inferred from Twitter*. Retrieved September 12, 2015, from http://www.ccs.neu.edu/home/amislove/twittermood/?utm_campaign=Facebook+Page&utm_content=Pulse+of+the+Nation+US+Mood+Throughout+the+Day+inferred+from+Twitter&utm_medium=postitat&utm_source=facebook

Mody, R. N., Willis, K. S., & Kerstein, R. (2009). WiMo: location-based emotion tagging. In *Proceedings of the 8th international Conference on Mobile and Ubiquitous Multimedia* (p. 14).

Němečková, M. (2014). *Komparace vnímání bezpečnosti ve městech Most a Litvínov* [Comparison of safety perception in cities Most and Litvínov]. Masarykova Univerzita. Retrieved from http://is.muni.cz/th/363798/fss_m/dp_nemeckova.pdf

Nold, C. (2009). *Emotional cartography: technologies of the self*. Retrieved from http://emotionalcartography.net/EmotionalCartography.pdf

Oc, T., & Tiesdell, S. (1997). Safer City Centres: Reviving the Public Realm. *Sage (Atlanta, Ga.)*.

Ostrava-Poruba. (2016). *Demography of Ostrava-Poruba* [in Czech]. Retrieved from https://poruba.ostrava.cz/cs/o-porube/zakladni-udaje/demograficke-udaje

Pánek, J., & Pászto, V. (2017). Emotional Mapping in Local Neighbourhood Planning: Case Study of Příbram, Czech Republic. *International Journal of E-Planning Research*, *6*(1), 1–22. doi:10.4018/IJEPR.2017010101

Pánek, J., Pászto, V., & Marek, L. (2017). Mapping emotions: spatial distribution of safety perception in the city of Olomouc. In I. Ivan, A. Singleton, J. Horák, & T. Inspektor (Eds.), *Lecture Notes in Geoinformation and Cartography: The Rise of Big Spatial Data* (pp. 211–224). Ostrava, Czech Republic: Springer International Publishing; doi:10.1007/978-3-319-45123-7_16

Pánek, J., Hrubeš, M., Kubásek, M., Valůch, J., & Zahumenská, V. (2014). GeoParticipace - jak používat prostorové nástroje v rozhodování o lokalitách ve kterých žijeme? [GeoParticipation - how to use spatial tool in decision making process about the locations we live in?]. Olomouc: Univerzita Palackého v Olomouci.

Pearce, M. W. (2008). Framing the days: Place and narrative in cartography. *Cartography and Geographic Information Science*, *35*(1), 17–32. doi:10.1559/152304008783475661

Pickles, J. (1995). *Ground Truth: The Social Implications of Geographic Information Systems* (1st ed.). The Guilford Press.

Quercia, D., O'Hare, N. K., & Cramer, H. (2014). Aesthetic capital. In *Proceedings of the 17th ACM conference on Computer supported cooperative work & social computing - CSCW '14* (pp. 945–955). New York: ACM Press. doi:10.1145/2531602.2531613

Quercia, D., Schifanella, R., & Aiello, L. M. (2014). The shortest path to happiness. In *Proceedings of the 25th ACM conference on Hypertext and social media - HT '14* (pp. 116–125). New York: ACM Press. doi:10.1145/2631775.2631799

Raosoft. (2004). *Sample Size Calculator by Raosoft*. Retrieved February 12, 2016, from http://www.raosoft.com/samplesize.html

Raslan, R., Al-hagla, K., & Bakr, A. (2014). Integration of Emotional Behavioural Layer "EmoBeL" in City Planning. In REAL CORP 2014 (Vol. 8, pp. 309–317).

Reeve, J. (2014). *Understanding motivation and emotion*. John Wiley & Sons.

Rusková, L. (2012). *Socially excluded localities in Ostrava*. Masaryk University Brno.

Russell, J. A. (1980). A circumplex model of affect. *Journal of Personality and Social Psychology*, *39*(6), 1161–1178. doi:10.1037/h0077714

Santos, R. B. (2012). Crime analysis with crime mapping. *Sage (Atlanta, Ga.)*.

Sessar, K., & Sirotek, J. (2001). Společnost v období transformace a strach z kriminality. *Sociologicky Casopis*, *37*(1), 7–22. Retrieved from http://www.jstor.org/stable/41131597?seq=1#page_scan_tab_contents

Smith, M., Bondi, L., & Davidson, J. (2012). *Emotional geographies*. Ashgate Publishing, Ltd.

Stasíková, L. (2011). Relevantnost výskumu strachu z kriminality v urbánnej geografii. *Geograficky Casopis*, *63*(4), 325–343.

State Administration of Land Surveying and Cadastre. (2014). *Fundamental Base of Geographic Data of the Czech Republic*. Retrieved September 21, 2015, from http://geoportal.cuzk.cz/(S(z2w1esrjen1dfactx0vyepfg))/Default.aspx?mode=TextMeta&text=dSady_zabaged&side=zabaged&menu=24

Taboada, M., Brooke, J., Tofiloski, M., Voll, K., & Stede, M. (2011). Lexicon-based methods for sentiment analysis. *Computational Linguistics*, *37*(2), 267–307. doi:10.1162/COLI_a_00049

Wilson, M. W. (2011). Training the eye: Formation of the geocoding subject. *Social & Cultural Geography*, *12*(4), 357–376. doi:10.1080/14649365.2010.521856

ENDNOTES

1. Poruba is an administrative district of the city of Ostrava, Moravian-Silesian region in the Czech Republic.
2. Six different coloured pins were used, each representing a different question.
3. http://print.mapwebbing.eu/_

4 In total 3 small maps of the city centre were used.

5 3,651 points and 400 polygons

6 The link for the survey was distributed by the administration representatives via local newspaper, Facebook profile, mailing lists, etc.

7 Version 2.14.0 with OpenLayers plugin.

8 Radius is the area around each point that will be used to calculate the `heat` a pixel receives. 7mm distance was set-up based on previous tests for the location of Ostrava-Poruba.

9 Roma people are one of the minorities in the Czech Republic. According to the last census from 2011, the Roma nationality was reported to be a total of 13,150 inhabitants in the Czech Republic. However, in the same census, 40, 370 inhabitants reported the Roma language as their language and, according to estimates by the Czech government, the size of the population is between 250,000 and 300,000. (Government of the Czech Republic, 2016)

10 Study of public space in Ostrava-Poruba.

11 155 small-scale interventions, 62 medium-scale interventions and 35 large-scale interventions.

12 Approximately 350k EUR.

13 Approximately 1.8 million EUR.

14 Direct response to the question „Where do you have problems with parking?“

15 The number is just an estimation as it is not possible to count how many people marked the community meeting map, nevertheless 373 users used the online tool.

16 Seniors or people without Internet connection.

Chapter 7
Using OpenStreetMap to Create Land Use and Land Cover Maps:
Development of an Application

Cidália Costa Fonte
University of Coimbra, Portugal & INESC Coimbra, Portugal

Joaquim António Patriarca
INESC Coimbra, Portugal

Marco Minghini
Politecnico di Milano, Italy

Vyron Antoniou
Hellenic Military Geographical Service, Greece

Linda See
International Institute for Applied Systems Analysis, Austria

Maria Antonia Brovelli
Politecnico di Milano, Italy

ABSTRACT

OpenStreetMap (OSM) is a bottom up community-driven initiative to create a global map of the world. Yet the application of OSM to land use and land cover (LULC) mapping is still largely unexploited due to problems with inconsistencies in the data and harmonization of LULC nomenclatures with OSM. This chapter outlines an automated methodology for creating LULC maps using the nomenclature of two European LULC products: the Urban Atlas (UA) and CORINE Land Cover (CLC). The method is applied to two regions in London and Paris. The results show that LULC maps with a level of detail similar to UA can be obtained for the urban regions, but that OSM has limitations for conversion into the more detailed non-urban classes of the CLC nomenclature. Future work will concentrate on developing additional rules to improve the accuracy of the transformation and building an online system for processing the data.

INTRODUCTION AND BACKGROUND

OpenStreetMap (OSM) is a well-known collaborative mapping project that involves volunteers from all over the world in the creation of a free, global geospatial database. With more than 2.8M registered contributors at the time of writing (July 2016) (http://wiki.openstreetmap.org/wiki/Stats), OSM is one of the most popular projects exemplifying the concept of Volunteered Geographic Information or VGI (See

DOI: 10.4018/978-1-5225-2446-5.ch007

et al., 2016). The availability of the OSM database under a fully open license, which allows anyone to use the data freely and produce derived products, has attracted the interest of a multitude of end users such as industry, professionals, governments and humanitarian organizations (Haklay, Antoniou, Basiouka, Soden, & Mooney, 2014; Soden & Palen, 2014; Olteanu-Raimond et al., 2015; Mooney & Minghini, in press). The success of the project has also attracted the attention of the academic community (Jokar Arsanjani, Zipf, Mooney, & Helbich, 2015), and OSM is now considered to be a research topic on its own.

There are a number of factors that can account for the popularity and massive exploitation of OSM data. The large number of contributors over time has ensured that OSM data have reached a high degree of quality. Many studies exist that have compared OSM data with authoritative datasets and showed that they are of a comparable quality, at least in urban areas where more contributors are active (see e.g. Girres & Touya, 2010; Haklay, 2010; Ciepłuch, Jacob, Mooney, & Winstanley, 2010; Ludwig, Voss, & Krause-Traudes, 2011; Fan, Zipf, Fu, & Neis, 2014; Zheng & Zheng, 2014; Brovelli, Minghini, Molinari, & Zamboni, 2016). The OSM database is also extremely rich, as it includes a variety of thematic layers (with attribute information) that are not traditionally available in other official or authoritative datasets. Lastly, the OSM database is constantly updated and enriched by contributors, and each new version is immediately available for use. In contrast, there are a number of problems related to OSM including an inconsistent spatial coverage (see e.g. Haklay, 2010; Zielstra & Zipf, 2010; Hecht, Kunze, & Hahmann, 2013; Fram, Chistopoulou, & Ellul, 2015; Ribeiro & Fonte, 2015; Brovelli, Minghini, & Molinari, 2016) and positional and thematic inconsistencies, where the latter is due to the relative freedom provided to the contributors in defining object attributes (Ballatore & Mooney, 2015).

Despite these known issues with OSM, the thematic richness of this dataset means that it has great potential for land use and land cover (LULC) mapping. LULC maps are fundamental inputs to many applications ranging from habitat monitoring to ecosystem accounting, among others. These maps are usually generated through the classification of satellite imagery. However, their creation is time consuming and therefore their release, even when at greatly detailed level, is only every few years. For example, CORINE Land Cover (CLC) has been produced for 2000, 2006 and 2012, where the 2012 product is still being validated in some countries. The Urban Atlas (UA) is another example of a detailed LULC map but is only available for cities in the European Union. There are also a number of global land cover products available, e.g. GLC-2000 (Mayaux et al., 2006), MODIS (Friedl et al., 2010) and the recently produced GlobeLand30 (Chen et al., 2015), where the latter product has been compared with other authoritative land cover products in Italy and Germany, and good overall agreement was found (Brovelli et al., 2015; Jokar Arsanjani et al., 2015). Other studies have shown that when these products are compared with one another, there are large spatial disagreements between them (Fritz et al., 2011). The Geo-Wiki crowdsourcing tool has been developed as one way of involving citizens in collecting data on land cover using Google Earth imagery to improve global land cover maps (See et al., 2015). OSM provides an alternative, relatively unexploited source of LULC information that could also be used to generate, verify and validate LULC maps.

Some initial investigations have already been undertaken to convert OSM into LULC maps (Jokar Arsanjani, Helbich, Bakillah, Hagenauer, & Zipf, 2013; Estima & Painho, 2015; Jokar Arsanjani, Mooney, Zipf, & Schauss, 2015; Jokar Arsanjani & Vaz, 2015; Martinho & Fonte, 2015). For example, OSM has been converted to the UA classes for several cities in Europe in order to undertake a comparison of the two products (Jokar Arsanjani et al., 2013; Jokar Arsanjani, Mooney, et al., 2015; Jokar Arsanjani & Vaz, 2015). The results of this comparison varied between 53.6% and 86.2% in terms of overall agreement depending upon the city considered. Similar positive results were obtained by Martinho & Fonte

(2015) in validating the UA with OSM data and by Estima & Painho (2015) when comparing OSM with CLC in Portugal.

Despite these encouraging outcomes, one of the first challenges when doing this comparison is the need to convert the OSM features into the corresponding nomenclature of the UA and CLC products. This harmonization process is not straightforward because of the way in which the data are collected in OSM. For example, there are frequently overlapping features, which means that these features should be assigned to different LULC classes at the same location. This can be a result of features that contain or cross other features, e.g. a bridge that crosses a water feature, or due to problems with positional accuracy, which results in partial overlapping of features that are inconsistent, e.g. a building footprint that overlaps with a road feature, or the assignment of different tags to the same region due to the subjectivity of the volunteers.

Thus, the creation of LULC maps using OSM data requires a solution for dealing with these types of inconsistencies as well as the harmonization of features found in OSM and the classes of LULC maps such as the UA and CLC. Although many of the aforementioned papers discuss these problems, none of them offer automated solutions for dealing with them. Thus, the overall aim of the chapter is to present an automated methodology for converting OSM features into LULC classes using a hierarchical approach, a set of decision rules and geoprocessing within a GIS environment. This procedure will automatically solve a number of inconsistencies found in OSM data when converting the features into LULC classes. The methodology is applied to parts of the cities of London, United Kingdom, and Paris, France, and the results are compared to available LULC maps with different levels of detail, i.e. the UA and CLC.

The remainder of this chapter is structured as follows. Section 2 provides a more technical description of the OSM database, the data model used and how the data are collected. It also offers an overview of the CLC and UA LULC nomenclatures. Section 3 explains the automated methodology developed for OSM conversion into a LULC map, while Section 4 describes the case study where the procedure was tested. The results are then provided in Section 5 and discussed more generally in Section 6, where the conclusions consider outstanding issues and recommendations for future work.

INPUT DATASETS

OSM Data

The OSM project was initiated in 2004 and has continued towards the goal of creating a free and openly-licensed map of the world (Jokar Arsanjani et al., 2015). OSM contains geospatial vector data, i.e. nodes (points), ways (polylines and polygons) and relations (logical collections of two or more nodes, ways or other relations), as well as the associated attributes or tags. Each tag consists of a key and value. Although the main focus of mapping by much of the OSM community has been on capturing road and building features and points of interest, which can be specified by different key value combinations, e.g. "building=residential", there are also natural and land use key value combinations that can be used to map other land cover and land use features. Thus it is possible to extract land cover and land use information from OSM.

Much of the research on OSM has been on assessing the quality of this data source in comparison to authoritative or reference data, including the study areas in this paper. For example, Haklay (2010) examined the positional accuracy of road features in London and found that many features were within

1-2 m of the authoritative data, with average accuracy for different districts in London varying between 3 and 8 m. In another study, Haklay et al. (2010) found that the positional accuracy of OSM in London was high, exceeding 80% overall and with one in four roads having a positional accuracy of between 95 and 100%. A similar exercise was undertaken by Girres and Touya (2010) to compare OSM to authoritative data for France. They found that positional accuracy varied between 5 and 15 m but peaked between 5 and 10 m, similar to that found in London. OSM data have also been compared to LULC maps in previous studies. For example, Jokar Arsanjani et al. (2015) compared OSM with GlobeLand30 and found an overall accuracy of 74%, with higher accuracies of 97% for agricultural areas in Germany. However, the authors also showed that OSM is more accurate than GlobeLand30 in areas where they disagreed using very high resolution satellite imagery. Thus, based on numerous accuracy assessments of OSM in the literature, the authors feel that OSM is accurate enough for LULC mapping.

The Urban Atlas (UA) and CORINE Land Cover (CLC)

The UA, which is officially known as the GMESUA (Global Monitoring for Environment and Security Urban Atlas), provides LULC data for urban areas in all EU member countries as part of the Copernicus Land Monitoring Service (EEA, 2011). The product was developed so that cities across the EU could be compared using a common LULC nomenclature. The UA is currently available for 305 cities for the reference year 2006, where the criterion for inclusion was a population greater than 100K. The minimum mapping unit (MMU) is 0.0025 km^2 in urban areas and 1 ha in rural areas, the minimum width of linear elements is 100m, and the positional accuracy is ±5m (EEA 2011). Satellite imagery with a resolution of 2.5m was used for classification, supported by numerous other datasets including topographic maps, navigation data from commercial providers and very high resolution imagery for verification, among others (EEA, 2011). The data are freely available for downloading from the European Environment Agency's (EEA) website (http://www.eea.europa.eu/data-and-maps/data/urban-atlas#tab-gis-data). In May 2016, the 2012 product was made available for downloading (http://land.copernicus.eu/local/urban-atlas/urban-atlas-2012), increasing the number of cities to 695 by including all urban areas with a population greater than 50K. Validation efforts for this product are still ongoing. Table 1 provides the nomenclature for the UA, up to level 3. Level 1 is the least detailed, e.g. all urban areas are contained in one class, while in level 3 the artificial surfaces are described using 12 classes. A fourth level is still available, but is not considered in this study.

The CLC product is also part of the Copernicus Land Monitoring Service and provides a consistent, comparable, pan-European land cover product (http://land.copernicus.eu/pan-european/corine-land-cover). The first CLC product was developed for the reference year 1990, with subsequent updates in 2000, 2006 and 2012. There are 44 land cover classes in the most detailed level (Table 1), where the nomenclatures of the UA and CLC are clearly compatible. The main difference is that more detailed urban classes can be found in the UA (even without considering the fourth level) and other land cover classes in CLC are more detailed than in the UA, which reflects differences in their overall purpose. The MMU for CLC is larger than the UA at 0.25 km^2 for areal objects but the minimum width of 100 m for linear objects is the same. The positional accuracy is 100m and the overall thematic accuracy is greater than 85%. Time series maps of land change are available at a finer MMU of 0.05 km^2. The products are generated in both vector and raster format at resolutions of 100m and 250m, respectively. CLC is freely available for downloading from the EEA website (http://land.copernicus.eu/pan-european/corine-land-cover/clc-2012).

Table 1. Nomenclatures of UA and CLC

Urban Altas nomenclature			CORINE Land Cover nomenclature		
Level 1	**Level 2**	**Level 3**	**Level 1**	**Level 2**	**Level 3**
1.Artificial Surfaces	1.1 Urban Fabric	1.1.1 Continuous urban fabric 1.1.2 Discontinuous urban fabric 1.1.3 Isolated Structures	1.Artificial Surfaces	1.1 Urban Fabric	1.1.1 Continuous urban fabric 1.1.2 Discontinuous urban fabric
	1.2 Industrial, commercial, public, military, private and transport units	1.2.1 Industrial, commercial, public, military and private units 1.2.2 Road and rail network and associated land 1.2.3 Port areas 1.2.4 Airports		1.2 Industrial, commercial, public, military, private and transport units	1.2.1 Industrial or commercial units 1.2.2 Road and rail network and associated land 1.2.3 Port areas 1.2.4 Airports
	1.3 Mine, dump and construction sites	1.3.1 Mineral extraction and dump sites 1.3.3 Construction sites 1.3.4 Land without current use		1.3 Mine, dump and construction sites	1.3.1 Mineral extraction 1.3.2 Dump sites 1.3.3 Construction sites
	1.4 Artificial non-agricultural vegetated areas	1.4.1 Green urban areas 1.4.2 Sports and leisure facilities		1.4 Artificial non-agricultural vegetated areas	1.4.1 Green urban areas 1.4.2 Sports and leisure facilities
2. Agricultural, semi-natural areas, wetlands			2.Agricultural areas	2.1 Arable land	2.1.1 Non-irrigated arable land 2.1.2 Permanently irrigated land 2.1.3 Rice fields
				2.2 Permanent crops	2.2.1 Vineyards 2.2.2 Fruit trees and berry plantations 2.2.3 Olive groves
				2.3 Pastures	2.3.1 Pastures
				2.4 Heterogeneous agricultural areas	2.4.1 Annual crops associated with permanent crops 2.4.2 Complex cultivation patterns 2.4.3 Land principally occupied by agriculture, with significant areas of natural vegetation 2.4.4 Agro-forestry areas
3. Forests			3. Forest and semi natural areas	3.1 Forests	3.1.1 Broad-leaved forest 3.1.2 Coniferous forest 3.1.3 Mixed forest
				3.2 Scrub and/ or herbaceous vegetation associations	3.2.1 Natural grasslands 3.2.2 Moors and heathland 3.2.3 Sclerophyllous vegetation 3.2.4 Transitional woodland-shrub
				3.3 Open spaces with little or no vegetation	3.3.1 Beaches, dunes, sands 3.3.2 Bare rocks 3.3.3 Sparsely vegetated areas 3.3.4 Burnt areas 3.3.5 Glaciers and perpetual snow
-------			4. Wetlands	4.1 Inland wetlands	4.1.1 Inland marshes 4.1.2 Peat bogs
				4.2 Maritime wetlands	4.2.1 Salt marshes 4.2.2 Salines 4.2.3 Intertidal flats
5. Water			5. Water	5.1 Inland waters	5.1.1 Water courses 5.1.2 Water bodies
				5.2 Marine waters	5.2.1 Coastal lagoons 5.2.2 Estuaries 5.2.3 Sea and ocean

PROCEDURES AND TOOLS

Methodology

OSM features are converted into a LULC map via a sequence of steps as follows:

Step 1: Associate the key=value combinations available in OSM with the LULC classes in the LULC product of interest, in this case the UA or CLC
Step 2: Choose any user defined values that are necessary for the processing
Step 3: Run the conversion process
Step 4: Eliminate inconsistencies such as overlapping regions assigned to different classes
Step 5: If a MMU is to be considered, then generalize the map so that all regions with smaller areas are merged with neighboring features

The first step listed above requires that all OSM features to be converted must be associated with corresponding LULC classes. This process can be done for the features that are listed in the OSM Map Features Wiki page (http://wiki.openstreetmap.org/wiki/Map_Features), which represents the official OSM mapping and tagging guide. However, since contributors are free to use keys and values other than the recommended ones, an analysis must always be done for each study area, as different key=value combinations may be available. Moreover, if LULC maps are to be created for the same region using OSM data obtained from different dates (since the data may be changing continuously), an analysis of the available data also needs to be undertaken, so that all of the available data and the changes over time are captured.

The second step refers to the choice of the parameters that may be necessary for the conversion process. These may be required for a diversity of processes. For example, the linear features that represent objects such as roads, railways or water regions need to be converted into areas, which is typically done by creating buffer zones. Three types of approach can be considered for this conversion:

1. Create a buffer with a predefined width, which is available in the application, for all the lines with the same key=value association. This value is used whenever no user specified values are provided. An example of this may be the use of the typical width of a railway line.
2. Create a buffer with a width chosen by the user for all the lines with the same key=value association. These parameters need to be provided by the user in step 2 above, prior to the processing. An example of this may be the use of typical widths for the types of roads in a particular region, such as primary or residential roads.
3. Create buffers with different widths for the segments that form a line, with values obtained using spatial analysis. This requires the separation of lines into segments and the identification of the buffer width for each segment, and is executed already as part of step 3 above. An example of this approach is the computation of the distance of each segment forming a road axis to the building polygons. These distances would be associated with each segment, and buffers would then be created for each segment if the distance to the building is smaller than a predefined or user defined value. The obtained buffers would then be merged.

Additional parameters may also be considered, such as the definition of a minimum segment length for specific key=value associations.

The processing of polygonal features may also require the use of parameters inserted by the user. Three types of approach can also be considered in this case, namely:

1. Simple selection of the polygons with the identified combination of key=value to insert into a LULC class.
2. Selection of the polygons with the identified combination of key=value that satisfies predefined or user defined restrictions (such as the polygon area) to insert into a LULC class. An example of this case is the insertion of regions with the key=value association natural/forest to the class Forest only if the area of the polygon is larger than a predefined or chosen value.
3. Selection of the polygons with an identified combination of key=value that satisfies a set of rules related to their neighborhood or topological relations, such as contain or is contained, to insert into a LULC class. Examples of this type of rule are, for example, polygons corresponding to green areas that are associated with recreation areas if they contain equipment such as sports fields (Loai Ali, Schmid, Falomir, & Freksa, 2015).

Step 3 refers to the conversion process that includes the creation of polygons corresponding to linear features, selection of all polygonal features for each LULC class and their conversion into a set of polygons for each class. Figure 1 illustrates the conversion process for class 1.2 (Industrial, commercial, public, military, private and transport units) for both the UA and CLC.

Figure 1. Typical workflow for converting OSM features into UA class 1.2

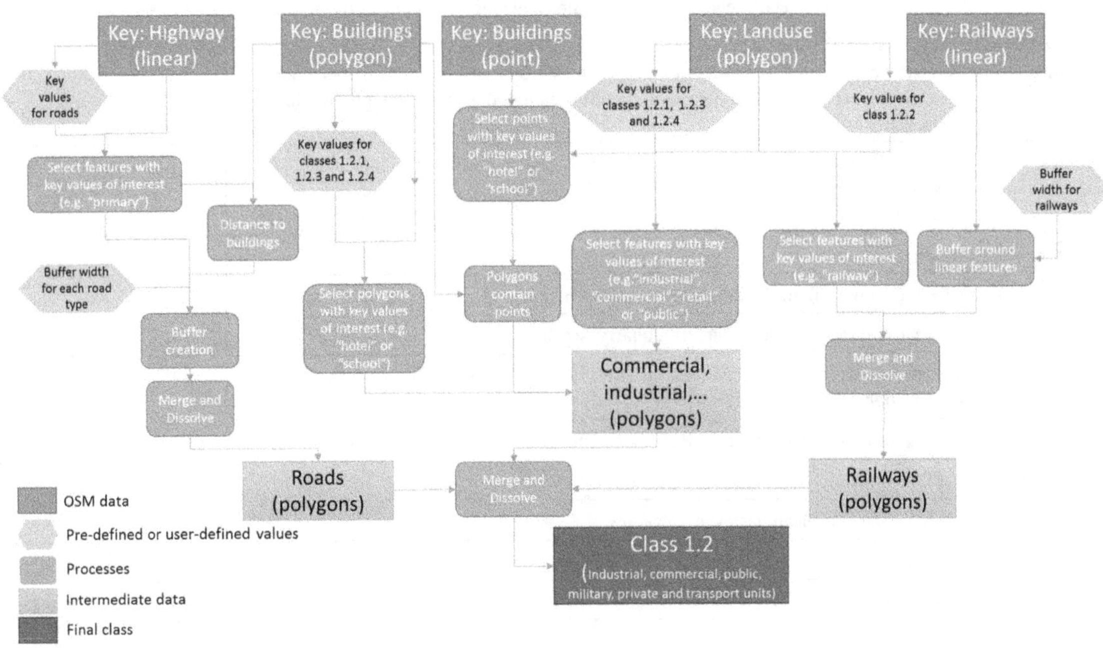

Similar workflows were created for the other classes. Table 2 shows which key=value combinations were considered for each class of both the UA and CLC nomenclatures for a subset of the map features listed in the OSM Map Features Wiki in the present version of this application. The processing approach uses Points of Interest (POI) associated with the keys "amenity" and "building". These were used to assign the buildings that contain them to the appropriate class, taking into consideration the type of use, such as a school, hospital or train station, whenever this information has not been associated with the building polygon (see Figure 1). In Table 2 the association with the classes of the UA is always done to level 2, even though the classes 2, 3 and 5 are not subdivided into additional classes in level 2 compared to level 1. For CLC classes, it can be seen that in some cases it is not possible to assign OSM tags to the more detailed classes in level 2, since it is not possible to determine if, for example, a polygon tagged as farmland corresponds to "Arable land" (class 2.1) or "Permanent crops" (class 2.2) or "Heterogeneous agricultural areas" (class 2.4) using only the information provided by the key and the value. For this reason, the output maps for level 2 with the CLC nomenclature are, at this stage of development, a

Table 2. OSM keys and values used to obtain UA and CLC classes

Key	Key values	UA class	CLC class
amenity (polygon)	animal_shelter, arts_centre, bank, bar, brothel, cafe, car_rental, car_wash, casino, cinema, clinic, college, community_centre, courthouse, crematorium, crypt, dentist, dive_centre, driving_school, embassy, fast_food, ferry_terminal, fire_station, food_court, fuel, grave_yard, gym, hospital, internet_cafe,	1.2	1.2
amenity (point)	bar, hospital, pub, restaurant, school, university	1.2	1.2
building (polygon)	apartments, house, garage, garages, residential, yes, "null"	1.1	1.1
	cathedral, chapel, church, civic, commercial, hangar, hospital, hotel, industrial, kiosk, mosque, office, public, retail, school, shrine, stadium, synagogue, temple, train_station, transportation, warehouse	1.2	1.2
building (point)	hotel, office, commercial,hospital, industrial, retail, warehouse, cathedral, chapel, church, mosque, temple, synagogue, school, stadium, train_station, transportation, public, shrine, civic, hangar, kiosk	1.2	1.2
highway (line)	living_street, motorway, motorway_link, pedestrian, primary, primary_link, residential, road, secondary, secondary_link, service, tertiary, tertiary_link, trunk, trunk_link, unclassified	1.2	1.2
landuse (polygon)	residential	1.1	1.1
	cemetery, commercial, industrial,military, railway, retail	1.2	1.2
	brownfield, construction, depot, landfill, quarry, scrub	1.3	1.3
	beer-garden, flowers, grass, greenfield, meadow, plants, recreation_ground, recreational_area, recreational, village_green	1.4	1.4
	allotments, farm, farmland, farmyard, grass, greenhouse-horticulture, meadow	2.0	2.0
	orchard, vineyard	2.0	2.2
	allotments, farm, farmland, farmyard, meadow	2.0	2.3
	grass, meadow	2.0	3.2
	forest (if area >10 000m2)	3.0	3.1
	reservoir, pond	5.0	5.1

continued on following page

Table 2. Continued

Key	Key values	UA class	CLC class
leisure (polygon)	adult_gaming_centre, amusement_arcade, dance, hackerspace, ice_rink, sports_centre, stadium, swimming_pool	1.2	1.2
	dog_park, garden, park, golf_course, miniature_golf, pitch, playground, summer_camp, track, track, water_park	1.4	1.4
	marina	1.4	1.2
	Nature_reserve	2.0	3.0
	beach_resort	2.0	3.3
	swimming_area	5.0	5.0
natural (polygon)	forest (if area <10,000 m^2), grassland, park	1.4	1.4
	grassland	2.0	2.1
	grassland	2.0	2.3
	grassland	2.0	2.4
	fell, grassland, heath, scrub	2.0	3.2
	beach, sand, shingle, bare_rock, scree, glacier	2.0	3.3
	mud, wetland	2.0	4.0
	forest (if area >10,000 m^2), wood	3.0	3.1
	bay, water	5.0	5.0
	riverbank	5.0	5.1
railway (line)	rail	1.2	1.2
waterway (line)	river, riverbank, stream	5.0	5.1
	dock	1.2	1.2

mixture of level 1 and level 2 classes, and whenever it is not possible to differentiate between the level 2 classes, a zero is indicated in the second digit of the used codes for the classes (Table 2). This type of difficulty was found for all classes of CLC except the Artificial Surfaces class (class 1), showing that, without additional spatial analysis and ancillary data, OSM appears to be more appropriate for creating detailed maps of urbanized regions.

Once the conversion of OSM features into each class is complete, overlapping of different classes may occur. This is due to the nature of OSM and may occur for several reasons, such as erroneous assignment of regions to the different keys or values by the volunteers, partial overlapping of features due to lack of accuracy in the definition of boundaries or inaccuracies in the conversion of the linear features to polygons. To eliminate this invalid information in the current version of this application, a hierarchical approach is considered in step 4 listed above, which assigns levels of priority to the classes based on their relative importance, size and most common spatial and topological relations. The priorities used here are listed in Tables 3 and 4 for levels 1 and 2 of the UA and CLC nomenclatures, respectively.

Although the use of such a list of priorities solves the problems of inconsistencies, it is not sufficient for solving all types of problems, as the choice of one class over another may change based on the context, spatial relations of features and their relative size. For example, it is appropriate to give priority

Table 3. Priority associated with level 1 classes when overlapping regions with different classes exist

Level of priority	UA Class	UA class name	CLC class	CLC class name
1	1	Artificial surfaces	1	Artificial surfaces
2	5	Water	5	Water
3	2	Agricultural, semi-natural areas, wetlands	4	Wetlands
4	3	Forests	2	Agricultural areas
5	---	---	3	Forests

Table 4. Priority associated with level 2 classes when overlapping regions with different classes exist

Level of priority	UA Class	UA class name	CLC class	CLC class name
1	1.2	Industrial, commercial, public, military, private and transport units	1.2	Industrial, commercial, public, military, private and transport units
2	5.0	Water	5.1	Inland water
3	1.4	Artificial non-agricultural vegetated areas	5.2	Marine water
4	1.3	Mine, dump and construction sites	4.1	Inland wetlands
5	1.1	Urban Fabric	4.2	Maritime wetlands
6	2.0	Agricultural, semi-natural areas, wetlands	1.4	Artificial non-agricultural vegetated areas
7	3.0	Forests	1.3	Mine, dump and construction sites
8	---	---	1.1	Urban Fabric
9	---	---	2.2	Permanent crops
10	---	---	2.1	Arable land
11	---	---	2.4	Heterogeneous agricultural areas
12	---	---	2.3	Pastures
13	---	---	3.2	Scrub and/or herbaceous vegetation associations
14	---	---	3.3	Open spaces with little or no vegetation
15	---	---	3.1	Forests

to water when it cuts a polygon that represents a park (i.e. a water line within the park), but if there is an island within a lake inside the park, then priority will be given to whatever exists on that island over the water feature.

Step 5 indicated above may need to be used when a LULC map extracted from OSM is to be created with a specific MMU. Although this process is necessary in order to make a direct comparison of the LULC maps extracted from OSM with other existing maps, this was not applied in the case studies presented in section 4. This step requires the creation of an additional workflow and more rules, so that the classes are appropriately merged and small regions with no data in OSM, which may be assigned to neighboring classes, are eliminated.

Application

The procedure for the creation of the LULC maps has been implemented in a fully free and open source environment. Several algorithms for converting OSM data into LULC classes were written in Python, where the flexibility of this programming language allowed different tools for manipulating vector data and the processing of OSM data to be integrated. These tools included GRASS GIS 7, the GDAL/OGR libraries (http://www.gdal.org/) and PostgreSQL 9.5 (https://www.postgresql.org/) with PostGIS 2.2 (http://postgis.net/). The result is a GRASS GIS (https://grass.osgeo.org/) module, available with a Graphical User Interface. In addition, other tools specifically designed to work with OSM data were used such as the osm2pgsql tool (http://wiki.openstreetmap.org/wiki/Osm2pgsql), incorporated in PostgreSQL/PostGIS, and the Osmosis tool (http://wiki.openstreetmap.org/wiki/Osmosis) for the manual preparation of OSM data.

Figure 2 illustrates the transformation process and where each tool was used to obtain the final output. First, Osmosis was used to extract the data for a specific bounding box (London and Paris study areas - see next Section) from a regional OSM file downloaded in July 2016 from Geofabrik (http://www.geofabrik.de/). The preparation of the input data is one of the few tasks that was done manually, but in the future, the user will be able to choose between an XML file (.osm or.pbf) with only the data to be processed or an XML file with data that goes beyond the boundaries of the study area. This is relevant for the next step in which the OSM data are added to PostgreSQL by the osm2pgsql tool. However, if the XML file with OSM data covers a large area, this tool is not capable of executing this in an effective way.

In addition to the file with the OSM data, the script reads a file with the nomenclature to use (UA or CLC) and an SQL file that contains the mapping between each OSM feature and the corresponding LULC classes.

Figure 2. Application workflow for converting OSM features into a LULC map

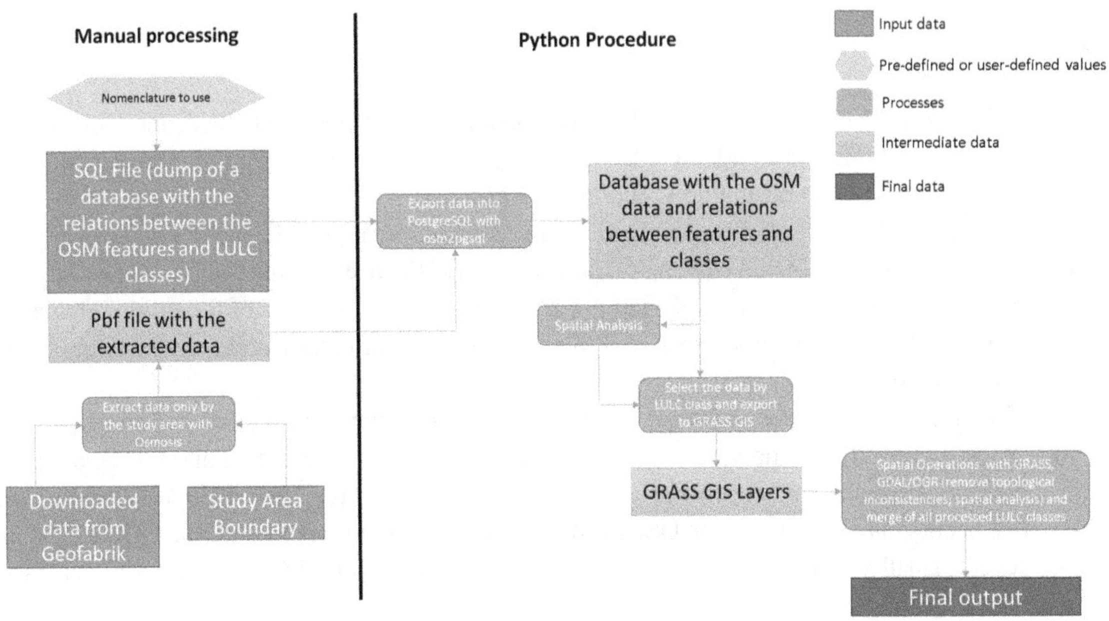

Once all of the inputs are ready, the procedure runs the osm2pgsql tool to add the OSM data into PostgreSQL. The script then reads the database uploaded from the SQL file to the PostgreSQL for each LULC class and obtains all related OSM features, exporting these geometries to GRASS GIS, where the geometric inconsistencies are solved and where some spatial analysis is made, also using tools from the GDAL/OGR Python Bindings. However, in some specific cases (e.g. the calculation of the distance between the roads and the buildings), the spatial operations are made with PostGIS instead of GRASS GIS or GDAL/OGR for reasons of efficiency. After processing all of the features related to each LULC class, these geometries are merged into a single layer and exported as an ESRI Shapefile.

CASE STUDY AREAS FOR DEMONSTRATION OF THE METHODOLOGY

London

Figure 3 shows the area within the city of London (UK), which was chosen as the first area to illustrate the methodology. The 10 km by 10 km bounding box shown in Figure 3 contains a highly urbanized section of London, but with some green space and water features. The spatial coverage of OSM features for the city of London is very good and a direct comparison with the UA will be most meaningful.

Paris

A second study area in Paris (France) was chosen, which is a section from the outskirts of the city; the 10 km by 10 km bounding box of the study area is shown in Figure 4. Like London, Paris also has very good spatial coverage of features in OSM, but in contrast, this portion of Paris has a more rural nature, with other land cover types present. This will allow us to compare the LULC map generated using the methodology outlined previously with both the UA and CLC products.

Results

Figures 5 and 6 show the results for the London study area that correspond to the UA and CLC nomenclatures, respectively, for levels 1 and 2. Figures 7 and 8 show the results obtained for the second study area, also for UA and CLC nomenclatures, and levels 1 and 2.

To assess the quality of the results obtained with the proposed approach, the obtained maps were overlapped with the original UA and CLC maps, and the areas of the overlapping regions were computed. The comparison of the level 2 classes associated with the overlapping regions of the original maps and the maps extracted from OSM can be seen in Tables 5 and 6 for both study areas, respectively for the UA and the CLC classes.

Tables 5 and 6 show that one of the best mapped classes is water, with match percentages higher than 90% in most cases. It can, however, be seen that for the CLC nomenclature and the London area several regions classified in OSM as water were assigned to other classes in CLC. As can be seen in Figure 6, this occurs mainly because the OSM extracted map was not converted to the CLC MMU and therefore several small water areas are present that are not mapped in CLC. The opposite occurs for

Figure 3. Location of the study area in the city of London, United Kingdom, in OSM
(data: © OpenStreetMap contributors)

the Paris region, where small land regions within the river are mapped in OSM but as they are smaller than the CLC MMU they are not in CLC, resulting in a match in relation to the CLC water area of 74%.

For the London area, which is a highly urbanized region, difficulties were found in separating agriculture from urban vegetation and dump sites without the application of additional rules. For the Paris region this was less evident, as there are large regions of agricultural areas.

The main findings from the results can be summarized as follows:

1. Overall there is good agreement between the features found in OSM and the UA, even though some classes are more difficult to differentiate, such as different types of vegetation.

Figure 4. Location of the study area in the city of Paris, France, in OSM
(data: © OpenStreetMap contributors)

2. There are some regions with missing data in OSM although these are only relatively small zones for both study areas. In some cases, data were associated with these regions but they did not provide any LULC information, e.g. administrative boundaries.
3. The region included in the black circle (see Figures 5 and 6) corresponds to a location where the Olympic Stadium was built for the 2012 Olympic Games that took place in London. The change in this region can be clearly identified in the data extracted from OSM compared to what was found in both the UA, which, according to the metadata associated with the London UA, was created using images from 2006 to 2009, and the 2006 CLC map.

Figure 5. London study area: On the left are the LULC maps obtained from OSM for the UA nomenclature, and on the right is the UA for the same region. Level 2 is represented above and level 1 below.

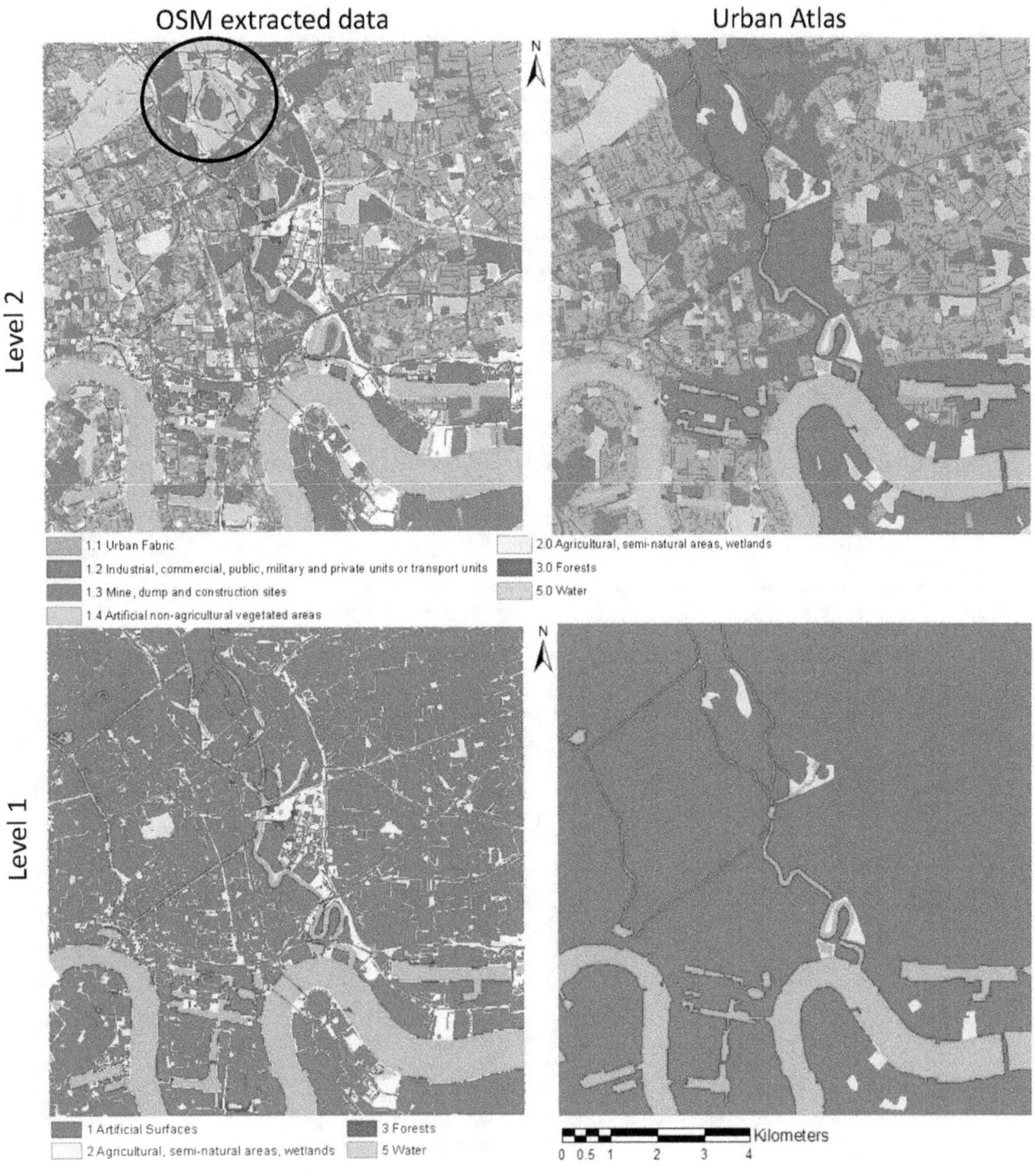

4. The level of detail for the urban regions is comparable to the UA. For a comparison with the CLC, the maps obtained from OSM need to be transformed to an appropriate MMU, mainly for the urban regions, while for rural areas this is less relevant.

Figure 6. London study area: London study area: On the left are the LULC maps obtained from OSM for the CLC nomenclature, and on the right is CLC for the same region. Level 2 is represented above and level 1 below.

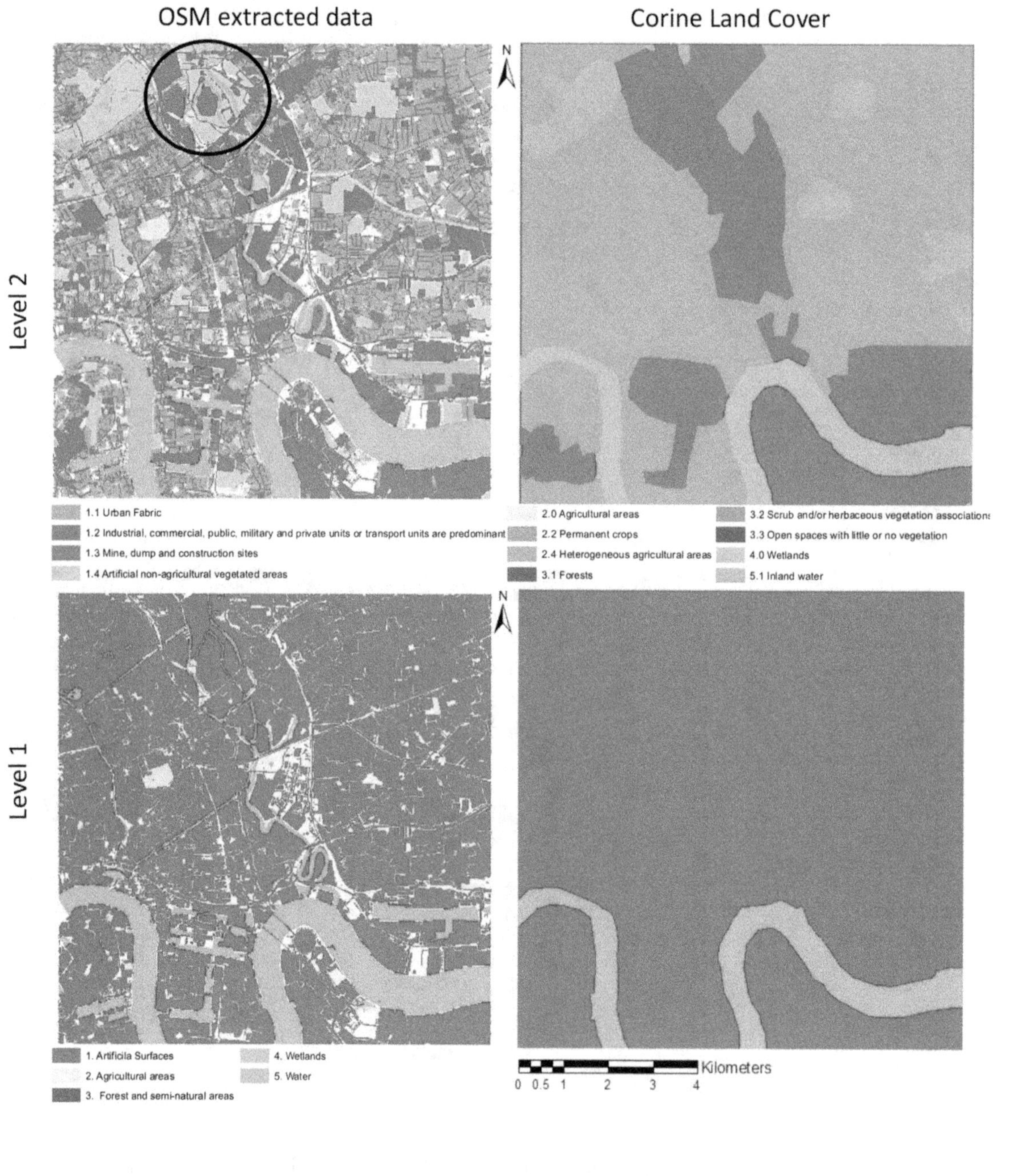

DISCUSSION, CONCLUSION, AND FUTURE DEVELOPMENTS

As already shown in previous research studies (Jokar Arsanjani et al., 2013; Estima & Painho, 2015; Jokar Arsanjani, Mooney, et al., 2015; Jokar Arsanjani & Vaz, 2015; Martinho & Fonte, 2015), OSM can be used to create LULC maps. However, several issues have been encountered during the process

Figure 7. Paris study area: On the left are the LULC maps obtained from OSM for the UA nomenclature, and on the right is UA for the same region. Level 2 is represented above and level 1 below.

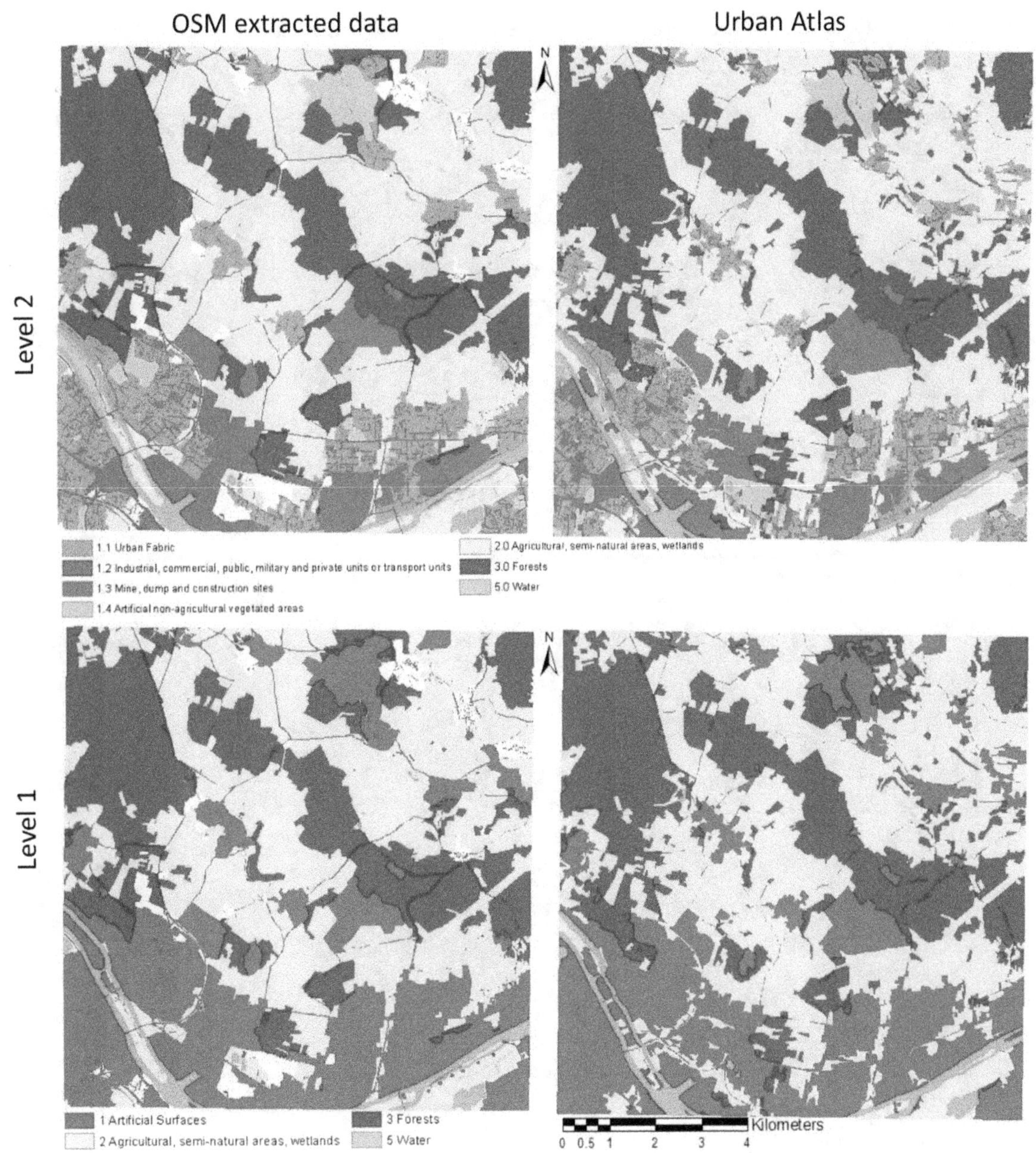

of conversion. These need to be documented and solved in order to develop an automated approach for converting OSM data into LULC maps. Here the authors have presented the first version of such an automated process, which uses a mix of predefined, feature-based and analysis-derived parameters.

The process was applied in two selected areas: a densely populated urban area (i.e. London) and a mixed urban/rural landscape (i.e. the outskirts of Paris). In both cases, our process delivered LULC maps that follow the nomenclature of two of the most prominent LULC products that are freely available for

Figure 8. Paris study area: On the left are the LULC maps obtained from OSM for the CLC nomenclature, and on the right is CLC for the same region. Level 2 is represented above and level 1 below.

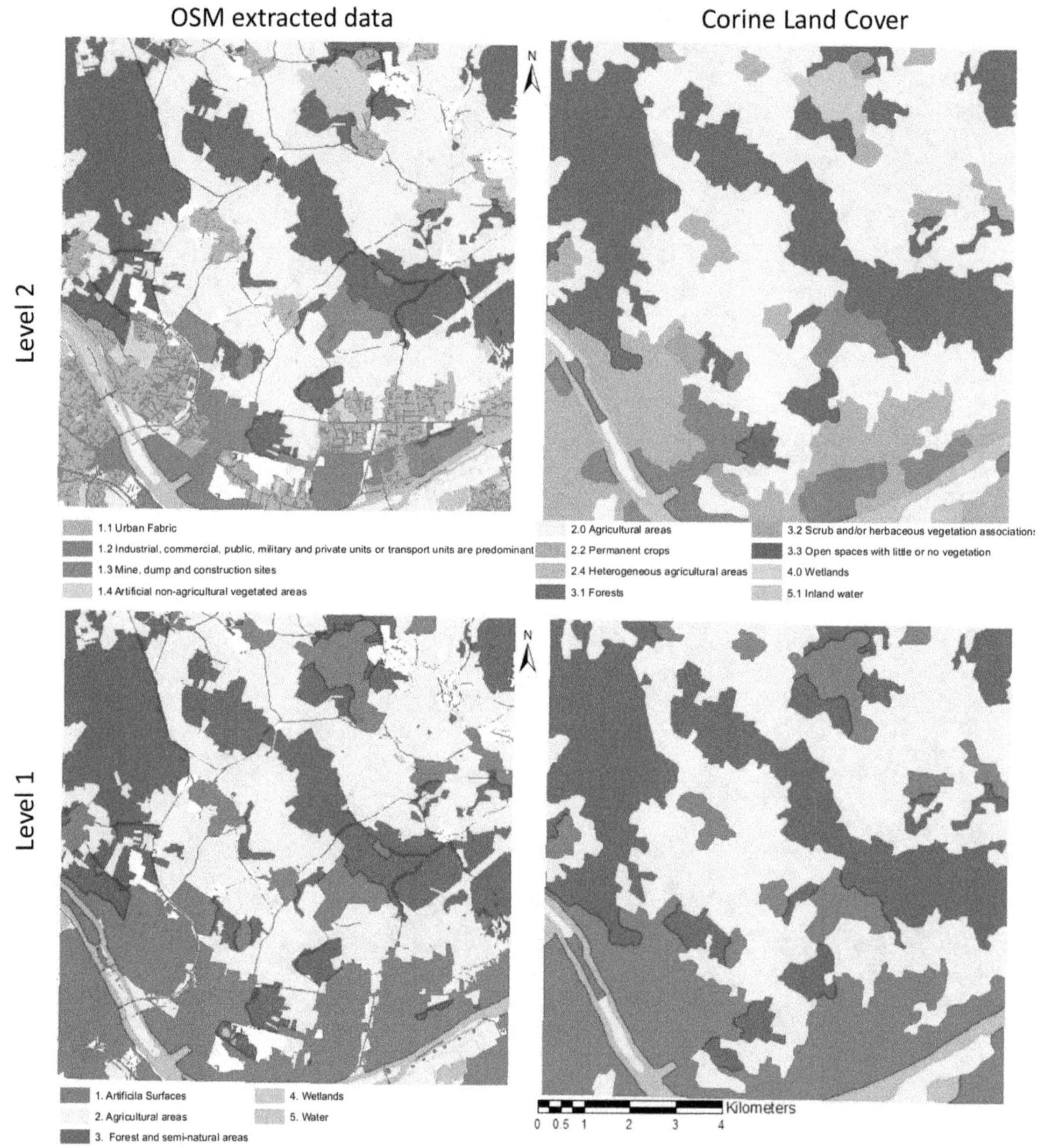

Europe: the UA and CLC. To achieve this, it was necessary to map the OSM key=values to the existing nomenclatures. Then the outcome was visually and quantitatively compared with the corresponding UA and CLC maps. The results show that, for both study areas, the maps obtained are very similar to the UA. The comparison with CLC shows that without a generalization to the MMU used in CLC, a direct comparison is not appropriate since OSM presents much more detail. From the perspective of the usefulness and usability of the LULC maps derived from OSM, this higher detail available in OSM represents

Table 5. Comparison of the area (ha) occupied by level 2 UA classes associated to the overlapping regions in the UA and the map extracted from OSM, for the London (a) and Paris (b) study areas

(a) London		Classes assigned to the overlapping regions in the OSM derived map							Area in UA (ha)	Match/Row Sum (%)	Match/Area in UA (%)
		11	12	13	14	20	30	50			
Classes assigned to the overlapping regions in UA	11	2346	796	16	86	8	2	21	3596	72	65
	12	525	2323	214	174	32	8	86	4023	69	58
	13	25	51	18	26	5	3	7	161	14	11
	14	19	111	5	644	17	5	18	891	79	72
	20	5	18	41	23	3	3	9	129	3	2
	30	0	0	0	0	0	0	0	0	---	---
	50	12	22	8	5	0	0	1107	1201	96	92
Match/Column Sum (%)		80	70	6	67	4	0	89			
(b) Paris		Classes assigned to the overlapping regions in the OSM derived map							Area in UA (ha)	Match/Row Sum (%)	Match/Area in UA (%)
		11	12	13	14	20	30	50			
Classes assigned to the overlapping regions in UA	11	967	106	1	11	50	24	1	1226	83	79
	12	186	640	37	20	50	13	3	972	67	66
	13	19	24	227	0	45	7	0	330	71	69
	14	56	26	0	161	57	6	5	341	52	47
	20	108	148	33	43	3545	124	10	4163	88	85
	30	21	28	11	44	138	2425	5	2724	91	89
	50	3	4	1	1	6	5	221	242	92	91
Match/Column Sum (%)		71	66	73	57	91	93	90			

a great advantage when LULC maps are to be derived in areas where no accurate LULC products are available, e.g. in underdeveloped or developing countries.

As indicated in the description of the application, the source of the OSM data used in the application was XML files. These files contain the original data created by the volunteers, and have been shown to contain more data than, for example, the shapefiles provided by Geofabrik (Fonte et al., 2016). Moreover, the organization of the data and the tags is different in the shapefiles compared with the.pbf files. For example, values such as "forest", which should have been associated with the "landuse" key according to the OSM Map Features Wiki, were found to be associated with the "natural" key in the shapefiles. The same thing occurred with "riverbank", which should have been associated with the key "waterway" but was also associated with the "natural" key in the Geofabrik shapefiles. Thus, to overcome these differences between the original OSM data and the data available in extracts, such as those from Geofabrik, the original XML files were used here instead of the already processed Geofabrik files.

In the case studies presented in this chapter, key=values associations that are not listed in the OSM Map Features Wiki were found, that had to be added to the list of tags, such as "landuse=flowers" or "amenity=conference_center". These types of problems are due to the fact that, despite guidelines existing

Table 6. Comparison of the area (ha) occupied by level 1 / 2 CLC classes associated to the overlapping regions in CLC and the map extracted from OSM, for the London (a) and Paris (b) study areas

(a) London		Classes assigned to the overlapping regions in the OSM derived map							Area in CLC (ha)	Match/Row Sum (%)	Match/Area in CLC (%)
		11	12	13	14	20	30	51			
Classes assigned to the overlapping regions in CLC	11	2685	1861	64	448	45	8	120	5880	51	46
	12	218	1258	228	171	17	13	374	2726	55	46
	13	12	77	11	29	2	2	2	156	8	7
	14	16	120	0	312	0	0	18	478	67	65
	20	0	0	0	0	0	0	0	0	0	0
	30	0	0	0	0	0	0	0	0	---	---
	51	2	6	0	1	0	0	740	760	99	97
Match/Column Sum (%)		92	38	4	32	0	0	59			

(b) Paris		Classes assigned to the overlapping regions in the OSM derived map								Area in CLC (ha)	Match/Row Sum (%)	Match/Area in CLC (%)
		11	12	13	14	20	31	32	51			
Classes assigned to the overlapping regions in CLC	11	1238	309	20	48	97	7	1	3	1762	72	70
	12	31	457	4	16	26	5	0	4	548	84	84
	13	0	16	167	0	31	9	0	0	224	75	75
	14	5	2	0	131	0	1	0	0	139	94	94
	20	52	132	9	37	3480	128	1	9	4109	90	85
	31	19	43	310	282	1	0	0	0	2763	0	0
	32	5	5	95	0	13	2	23	4	150	16	15
	51	9	11	1	7	45	3	0	223	303	75	74
Match/Column Sum (%)		91	47	54	46	90	0	88	91	91		

on how to tag in the OSM project, a complete freedom in the choice of tags is left to contributors. This presents considerable challenges to the creation of an application that can be widely used by non-experts. Each time the process is run, an analysis of available tags must be undertaken, and the ones that are not assigned to any class need to be identified and assigned to an appropriate class.

The application developed here is already able to produce results, which may be particularly useful for regions where no detailed LULC maps are available. A larger set of OSM tags are exploited than the ones mentioned in Fonte et al. (2016), and some POI have also been added to the procedure. However, additional developments are under preparation, which include:

1. The creation of spatial analysis procedures to enable the choice of the most appropriate class for some key=value associations. Examples of this are the association of a tag such as water to a lake, pond, marina or even the sea, or the association of grass to urban areas, agricultural or natural areas. With context information and information on the size and topological relations with other features, it will be possible to eliminate at least some of the possible choices.

2. Development of procedures using spatial analysis to enable the solution of inconsistencies. The hierarchy used to eliminate the existence of more than one class at the same location takes each particular situation and its surroundings into account. A set of rules can be developed to ensure that additional topological relations of features, such as inclusion and contain, context and size of features are also included.
3. Insertion of rules for generalization when a MMU is to be considered. This process will also enable the elimination of some small regions where no data are available in OSM.

The application under development will be made freely available under an open source license, enabling anyone to create a LULC map from OSM data for any part of the world and for different timestamps for the same regions, allowing for analysis of change. Future work will also concentrate on generalizing this approach to any LULC product and building an online processing system, e.g. a Web Processing Service (WPS), to accommodate user requests directly through the Web.

ACKNOWLEDGMENT

The authors would like to acknowledge the support and contribution of COST Actions TD1202 'Mapping and Citizen Sensor' http://www.citizen-sensor-cost.eu, IC1203 'ENERGIC' http://vgibox.eu/, the ERC CrowdLand project (No. 617754) and FCT project grant UID/MULTI/00308/2013.

REFERENCES

Ballatore, A., & Mooney, P. (2015). Conceptualising the geographic world: The dimensions of negotiation in crowdsourced cartography. *International Journal of Geographical Information Science*, *29*(12), 2310–2327. doi:10.1080/13658816.2015.1076825

Brovelli, M. A., Minghini, M., & Molinari, M. (2016). An automated GRASS-based procedure to assess the accuracy of the OpenStreetMap Paris road network. In The International Archives of the Photogrammetry, Remote Sensing and Spatial Information Sciences (Vol. XLI-B7, pp. 919–925). ISPRS Archives. doi:10.5194/isprsarchives-XLI-B7-919-2016

Brovelli, M. A., Minghini, M., Molinari, M., & Zamboni, G. (2016). Positional accuracy assessment of the OpenStreetMap buildings layer through automatic homologous pairs detection: the method and a case study. In The International Archives of the Photogrammetry, Remote Sensing and Spatial Information Sciences (Vol. XLI-B2, pp. 615–620). ISPRS Archives. doi:10.5194/isprsarchives-XLI-B2-615-2016

Brovelli, M. A., Molinari, M. E., Hussein, E., Chen, J., & Li, R. (2015). The First Comprehensive Accuracy Assessment of GlobeLand30 at a National Level: Methodology and Results. *Remote Sensing*, *7*(4), 4191-4212. http://doi:10.3390/rs70404191

Chen, J., Chen, J., Liao, A., Cao, X., Chen, L., Chen, X., & Mills, J. et al. (2015). Global land cover mapping at 30 m resolution: A POK-based operational approach. *ISPRS Journal of Photogrammetry and Remote Sensing*, *103*, 7–27. doi:10.1016/j.isprsjprs.2014.09.002

Ciepłuch, B., Jacob, R., Mooney, P., & Winstanley, A. (2010). Comparison of the accuracy of OpenStreetMap for Ireland with Google Maps and Bing Maps. *Proceedings of the Ninth International Symposium on Spatial Accuracy Assessment in Natural Resources and Environmental Sciences*, 337. Retrieved from http://eprints.nuim.ie/2476/

EEA. (2011). *Mapping guide for a European Urban Atlas.* Available from: http://www.eea.europa.eu/data-and-maps/data/urban-atlas

Estima, J., & Painho, A. (2015). Investigating the potential of OpenStreetMap for land use/land cover production: A case study for continental Portugal. In J. Jokar Arsanjani, A. Zipf, P. Mooney, & M. Helbich (Eds.), *OpenStreetMap in GIScience* (pp. 273–293). Cham: Springer International Publishing. doi:10.1007/978-3-319-14280-7_14

Fan, H., Zipf, A., Fu, Q., & Neis, P. (2014). Quality assessment for building footprints data on OpenStreetMap. *International Journal of Geographical Information Science*, *28*(4), 700–719. doi:10.1080/13658816.2013.867495

Fonte, C., Minghini, M., Antoniou, V., See, L., Brovelli, M. A., Patriarca, J., & Milčinski, G. (2016). An automated methodology for converting OSM data into a land use/cover map.*Proceedings of the 6th International Conference on Cartography & GIS.*

Fram, C., Chistopoulou, K., & Ellul, C. (2015). Assessing the quality of OpenStreetMap building data and searching for a proxy variable to estimate OSM building data completeness.*Proceedings of the 23rd GIS Research UK (GISRUK) Conference.*

Friedl, M. A., Sulla-Menashe, D., Tan, B., Schneider, A., Ramankutty, N., Sibley, A., & Huang, X. (2010). MODIS Collection 5 global land cover: Algorithm refinements and characterization of new datasets. *Remote Sensing of Environment, 114*(1), 168–182. doi:10.1016/j.rse.2009.08.016

Fritz, S., See, L., McCallum, I., Schill, C., Obersteiner, M., van der Velde, M., & Achard, F. et al. (2011). Highlighting continued uncertainty in global land cover maps for the user community. *Environmental Research Letters*, *6*(4), 44005. doi:10.1088/1748-9326/6/4/044005

Girres, J.-F., & Touya, G. (2010). Quality assessment of the French OpenStreetMap dataset. *Transactions in GIS*, *14*(4), 435–459. doi:10.1111/j.1467-9671.2010.01203.x

Haklay, M. (2010). How good is volunteered geographical information? A comparative study of OpenStreetMap and Ordnance Survey datasets. *Environment and Planning. B, Planning & Design*, *37*(4), 682–703. doi:10.1068/b35097

Haklay, M., Antoniou, V., Basiouka, S., Soden, R., & Mooney, P. (2014). Crowdsourced Geographic Information Use in Government. London: UCL-Geomatics.

Hecht, R., Kunze, C., & Hahmann, S. (2013). Measuring completeness of building footprints in OpenStreetMap over space and time. *ISPRS International Journal of Geo-Information*, *2*(4), 1066–1091. doi:10.3390/ijgi2041066

Jokar Arsanjani, J., Helbich, M., Bakillah, M., Hagenauer, J., & Zipf, A. (2013). Toward mapping land-use patterns from volunteered geographic information. *International Journal of Geographical Information Science, 27*(12), 2264–2278. doi:10.1080/13658816.2013.800871

Jokar Arsanjani, J., Mooney, P., Zipf, A., & Schauss, A. (2015). Quality assessment of the contributed land use information from OpenStreetMap versus authoritative datasets. In J. Jokar Arsanjani, A. Zipf, P. Mooney, & M. Helbich (Eds.), *OpenStreetMap in GIScience* (pp. 37–58). Cham: Springer International Publishing. doi:10.1007/978-3-319-14280-7_3

Jokar Arsanjani, J., & Vaz, E. (2015). An assessment of a collaborative mapping approach for exploring land use patterns for several European metropolises. *International Journal of Applied Earth Observation and Geoinformation, 35*(B), 329–337. 10.1016/j.jag.2014.09.009

Jokar Arsanjani, J., Zipf, A., Mooney, P., & Helbich, M. (2015). An introduction to OpenStreetMap in Geographic Information Science: Experiences, research, and applications. In J. Jokar Arsanjani, A. Zipf, P. Mooney, & M. Helbich (Eds.), *OpenStreetMap in GIScience* (pp. 1–18). Cham: Springer International Publishing. doi:10.1007/978-3-319-14280-7_1

Loai Ali, A., Schmid, F., Falomir, Z., & Freksa, C. (2015). Towards rule-guided classification for Volunteered Geographic Information. In ISPRS Annals of the Photogrammetry, Remote Sensing and Spatial Information Sciences (Vol. II-3/W5). La Grande Motte, France: ISPRS Annals. doi:10.5194/isprsannals-II-3-W5-211-2015

Ludwig, I., Voss, A., & Krause-Traudes, M. (2011). A comparison of the street networks of Navteq and OSM in Germany. In S. Geertman, W. Reinhardt, & F. Toppen (Eds.), *Advancing Geoinformation Science for a Changing World* (Vol. 1, pp. 65–84). Berlin: Springer Berlin Heidelberg. doi:10.1007/978-3-642-19789-5_4

Martinho, N., & Fonte, C. C. (2015). *Assessing the applicability of OpenStreetMap data to validate Land Use/Land Cover Maps* (Working Paper No. 5). Coimbra, Portugal: Instituto de Engenharia de Sistemas e Computadores de Coimbra.

Mayaux, P., Eva, H., Gallego, J., Strahler, A. H., Herold, M., & Agrawal, S. ... Roy, P. S. (2006). Validation of the global land cover 2000 map. *IEEE Transactions on Geoscience and Remote Sensing, 44*, 1728–1737. Retrieved from http://www.scopus.com/scopus/inward/record.url?eid=2-s2.0-33746330734&partnerID=40

Mooney, P., & Minghini, M. (in press). A review of OpenStreetMap data. In G. Foody, L. See, V. Antoniou, C. C. Fonte, S. Fritz, P. Mooney, & A.-M. Olteanu (Eds.), *Mapping and the Citizen Sensor*. London, UK: Ubiquity Press.

Olteanu-Raimond, A.-M., Hart, G., Foody, G. M., Touya, G., Kellenberger, T., & Demetriou, D. (2015). The scale of VGI in map production: A perspective of European National Mapping Agencies. *Transactions in GIS*. doi:10.1111/tgis.12189

Ribeiro, A., & Fonte, C. C. (2015). Methodology for assessing OpenStreetMap degree of coverage for purposes of land cover mapping. In *ISPRS Annals of the Photogrammetry, Remote Sensing and Spatial Information Sciences* (Vol. WG II/4). La Grande Motte, France: ISPRS Annals. Retrieved from http://www.isprs-ann-photogramm-remote-sens-spatial-inf-sci.net/II-3-W5/297/2015/isprsannals-II-3-W5-297-2015.pdf

See, L., Fritz, S., Perger, C., Schill, C., McCallum, I., Schepaschenko, D., & Obersteiner, M. et al. (2015). Harnessing the power of volunteers, the internet and Google Earth to collect and validate global spatial information using Geo-Wiki. *Technological Forecasting and Social Change*, *98*, 324–335. doi:10.1016/j.techfore.2015.03.002

See, L., Mooney, P., Foody, G., Bastin, L., Comber, A., Estima, J., & Rutzinger, M. et al. (2016). Crowdsourcing, citizen science or Volunteered Geographic Information? The current state of crowdsourced geographic information. *ISPRS International Journal of Geo-Information*, *5*(5), 55. doi:10.3390/ijgi5050055

Soden, R., & Palen, L. (2014). From crowdsourced mapping to community mapping: The post-earthquake work of OpenStreetMap Haiti. In C. Rossitto, L. Ciolfi, D. Martin, & B. Conein (Eds.), *COOP 2014 - Proceedings of the 11th International Conference on the Design of Cooperative Systems, 27-30 May 2014, Nice (France)* (pp. 311–326). Springer International Publishing. Retrieved from http://link.springer.com/chapter/10.1007/978-3-319-06498-7_19

Zheng, S., & Zheng, J. (2014). Assessing the completeness and positional accuracy of OpenStreetMap in China. In T. Bandrova, M. Konecny, & S. Zlatanova (Eds.), *Thematic Cartography for the Society* (pp. 171–189). Springer International Publishing. doi:10.1007/978-3-319-08180-9_14

Zielstra, D., & Zipf, A. (2010). *A comparative study of proprietary geodata and volunteered geographic information for Germany*. Presented at the 13th AGILE International Conference on Geographic Information Science 2010, Guimarães, Portugal. Retrieved from http://www.agile-online.org/Conference_Paper/CDs/agile_2010/ShortPapers_PDF/142_DOC.pdf

KEY TERMS AND DEFINITIONS

CORINE Land Cover: LULC pan-European land cover product. The first CLC product was developed for the reference year 1990, with subsequent updates in 2000, 2006 and 2012.

Feature Inconsistencies: When features in OSM are converted to LULC maps, there are inconsistencies in the data due to overlapping of features, freeform tagging by volunteers that may result in different areas with the same features being tagged in different ways, and due to positional accuracies. These feature inconsistencies need to be handled in order to convert OSM data to a LULC map.

GRASS: A free and open source GIS package.

Harmonization of Nomenclatures: The mapping of features in one space (e.g. OSM) to features in another space (e.g. classes in the Urban Atlas and CORINE Land Cover) in order to allow for conversion between nomenclatures.

Land Use/Land Cover Map (LULC map): Land use and land cover (LULC) maps provide important information about urban and rural landscapes and are used in numerous applications from habitat monitoring to ecosystem accounting. Authoritative LULC maps are produced in a semi-automated, top down manner, using satellite imagery, classification algorithms and rigorous validation.

OpenStreetMap (OSM): OSM is a crowdsourced global map of the world created by a community of volunteers. OSM was started by Steve Coast in 2004.

PostGIS: Spatial database extender for PostgreSQL object-relational database.

PostgreSQL: Open-source object-relational Database Management Systems.

Python: A high level computer programming language.

Urban Atlas: LULC data for urban areas in all EU member countries developed for 305 cities for the reference year 2006, where the criterion for inclusion was a population greater than 100K.

Volunteered Geographic Information (VGI): Coined by Goodchild in 2007, VGI is georeferenced data that are contributed by volunteers, e.g. mapping of features in OSM.

Chapter 8
Mapping Regional Landscape by Using OpenstreetMap (OSM): A Case Study to Understand Forest Patterns in Maya Zone, Mexico

Di Yang
University of Florida, USA

ABSTRACT

A forest patterns map over a large extent at high spatial resolution is a heavily computation task but is critical to most regions. There are two major difficulties in generating the classification maps at regional scale: large training points sets and expensive computation cost in classifier modelling. As one of the most well-known Volunteered Geographic Information (VGI) initiatives, OpenstreetMap contributes not only on road network distributions, but the potential of justify land cover and land use. Google Earth Engine is a platform designed for cloud-based mapping with a strong computing power. In this study, we proposed a new approach to generating forest cover map and quantifying road-caused forest fragmentations by using OpenstreetMap in conjunction with remote sensing dataset stored in Google Earth Engine. Additionally, the landscape metrics produced after incorporating OpenStreetMap (OSM) with the forest spatial pattern layers from our output indicated significant levels of forest fragmentation in Yucatan peninsula.

INTRODUCTION

Incorporating Road Networks into Forest Landscape Mapping

Forest ecosystems are one of the most important ecosystems. Forests cover one-third of the surface land and contain nearly 80% of terrestrial biodiversity. Forests also provide a broad range of ecosystem goods and services for human well-being (Aerts & Honnay 2011; Costanza et al., 2006). Forests contribute significantly to the terrestrial carbon cycle and sequester massive amounts of anthropogenic CO_2 emissions. In landscape ecology perspective, forest landscape is defined as a spatial mosaic of arbitrary

DOI: 10.4018/978-1-5225-2446-5.ch008

boundaries containing forest patches that are able to interact functionally (Turner 1989). Also, forests are amongst the most biologically rich terrestrial systems and provide broad ranges of goods and services for human well-being (Costanza et al., 2006). However, human activities are negatively transforming and altering the global forest patterns and structures. Human activities, such as road-building and resource extraction (e.g., timber harvesting and timber production), alter forest patterns and landscapes, thereby fragmenting continuous forests into smaller and more isolated patches (Jaeger et al., 2007). Understanding forest landscape patterns is critical in ecological research and forest land management. Remote sensing and GIS science are powerful tools that make monitoring forest dynamics over decades at regional scale possible. However, interpreting remote-sensing-derived images is not a new technology since the first aerial photographs (Carls, 1947) and satellite images (Estes et al., 1980 and Landsat 1). The recent increase of satellite sensing prompts the rapid development of open geoscience and helps us see the world in a more complete way.

Moreover, the updates for most land cover map projects are far from frequent, particularly at regional scale due to the uneven distribution of data resources. Most developed countries have a well-established infrastructure to provide high quality, large scale and comprehensive dataset on land cover/land-use for a variety of purposes. North America and Europe have well-done QA/QC assessments, such as COoRdination of Information on the Environment (CORINE), National Land Cover Database (NLCD) and Coast Change Analysis Program (C-CAP). Although land cover mapping techniques are well studied and have been applied to map most of the developed areas; however, there are still difficulties mapping land cover and land changes at a large scale with high in spatial and temporal resolution, especially for areas with high heterogeneity. In the developing world, such high-quality datasets are limited, and most developing countries lack the resources to collect extensive ground data.

Mapping forest fragmentation by incorporating road network is a fundamental basis of forest landscape ecological analysis (Haddad et al., 2015, Abdullah et al., 2007). A variety of forest fragmentation results have been characterized from the regional to global scale (Espírito et al., 2014, Haddad et al., 2015). McGarigal et al., investigated the landscape structure changes of the San Juan Mountains from 1950-1993 and found that roads had a more significant impact on landscape structure than logging (McGarigal et al. 2001). Wickham examined forest changes in cumulative distribution based on patch sizes for the Chesapeake Bay, Virginia (Wickham et al. 2007). However, there is persistent difficulty in delineating ecologically relevant patches based only on remote sensing images. In landscape ecology perspective, the significant impacts of road networks are continuously disregarded, not to mention the ignorance of small roads. In other cases, the systematic mapping of roads excludes small roads (e.g. logging roads and trails). Moreover, the effects of small secondary roads remain largely unknown. We therefore urgently need a more comprehensive road network database to evaluate the effects of road-caused fragmentation. As human activities make forestland more accessible, road building itself becomes a domino effect increasing secondary roads.

Roads inherently cut the existing forest patches into pieces, which is one of the most obvious reasons for forest fragmentations. Road networks are also treated as one of the most important human factors that cause forest fragmentation (add more citations here). Roads and the adjacent area contribute upwards of 20% of American total geographical land, as Forman et al., analyzed in 2000 (Forman et al. 2000). In some heavily forested areas, road and water features are hiding under canopy, so we have difficulty discovering and mapping them using regular classification techniques, especially when we want to explore the forest patterns. Also, there is persistent difficulty in delineating ecologically relevant patches based only on remote sensing images. Besides, road networks have significant secondary impacts on ecosystems and the distributions of species. The assessment of road databases is, therefore, imperative/crucial, and roads must be mapped as accurately as possible. Otherwise, the impacts end ecological effects of roads on the landscape might be underestimated. OpenStreetMap created the best map when considering temporal change trajectories by providing "up-to-date" data.

Volunteered Geographic Information (VGI) and OpenstreetMap

Volunteered Geographic Information (VGI) opens a new era in mapping and visualizing our world (Elwood et al., 2008, Haklay et al., 2008). OpenStreetMap (OSM) is one of the well-supported VGIs, and it has been studied and applied in multiple disciplines, but there are still large amounts of information needing to be investigated. OSM has its unique advantages, which might fit a wider range of applications over the official geographic databases. From the geographic perspective, scaling has always been a prominent and prevailing issue. Because of the growing coverage of spatial big data and cloud computing, OSM attracted our attention to its suitability in regional and continental scale mapping (Zook et al., 2010). Recent studies focused on land cover and land-use change accuracies by incorporating OSM (Estima et al, 2013 and 2015; Johnson 2016, Jokar et al., 2013, See et al., 2013 and 2015). In this study, we aim to incorporate OSM into regional land cover databases to investigate and quantify landscape fragmentation. Spatializing the distribution of the forest patches is a concrete way to measure the landscape structure and understand the forest patterns. The accuracy of VGI and OSM data are studied (Hakley et al, 2010). Moreover, the combining of remote-sensing data and OSM constitutes a powerful tool to monitor, characterize and quantify land cover and forest-patch distributions. The product will be a major source of information for researchers and policy-makers for investigating land parcels.

For the objective of optimizing our framework, we chose the Maya Zone of the Yucatan Peninsula, Mexico as a case study. The Maya Zone (Zona Maya) is a geographic area extending from north Quintana Roo to the borders of Belize, and it is known for its large forest coverage area and wide diversity of civilizations (Scarborough 2010). The Maya Zone of the Yucatan Peninsula is home to the most densely forested region in Mexico, which has a massive amount of forest cover and is also well known as native Mayan land. Current road networks have strongly influenced the distribution of forest patches in the Yucatan Peninsula. In the Maya Zone, local communities manage a large amount of forest. Fieldwork included in this study explores nine different sites covering a large part of the Quintana Roo state and includes at least three different types of land tenures (*Ejido*, Private land and conservation land). Spatializing the distribution of the forest patches is a concrete way to measure the landscape structure and understand the forest patterns.

In this study, we proposed a framework to improve the potential of land cover mapping based on selected features in OpenstreetMap in Google Earth Engine. Also, we were able to build a database demonstrating the near real-time road effects on forest-pattern spatial distributions. In addition, we validated and assessed the potentials of mapping land cover and land change at a regional scale with high heterogeneity.

STUDY AREA AND SOCIAL BACKGROUNDS

The Maya Zone (Zona Maya) in our study is a geographic area extending from North Yucatán state to the borders of Belize locating on the Yucatan Peninsula, and it is known for its wide diversity of civilizations (Bray et al., 2004). Yucatan Peninsula is a region dominated by semi-deciduous, tropical dry forest with distinct dry seasons (October - May) and wet seasons (June – September). Average forest height of the dry forest is 10-20m at maturity status and a lot of waterbodies are hiding under canopy because of its distinctive karstic topography (Flores et al., 1994, White and Hood. 2004). Our study area contains two states: Quintana Roo and Yucatan (Figure 1). These two adjacent states of interest were selected for comparison since their social backgrounds and different forest management practices. Also, the locations of these two regions were chosen because of the dominant land covers. In Yucatán state, the land cover is dominated by cropland and smallholder agriculture and Quintana Roo state is dominated by forest and secondary forest.

Yucatan Peninsula (the Maya Zone) has an organized combination of pre-colonial indigenous Mayan people and post-colonial development (Turner et al., 2001). In forest management and land-use practice perspective, the structure of land tenure and resource use rights essentially obligates a community-based approach in Mexico. At least three types of land tenure exist in the study area: *ejido* land (largely usufruct), private land, and conservation land. An *ejido* is a uniquely Mexican form of communal land tenure that covers large portions of the rural landscape in many states. The areas that covered the land tenure types including *ejido* land, private land and conservation land.

This uniquely Mexican form of communal land tenure covers large portions of the rural landscape in many states, and such land property is called an "*ejido*". *Edijos* and the local communities have maintained their traditional methods of managing the forest sustainably for thousands of years.

For cropland cover, local Mayan people keep the tradition of shifting cultivation known as milpa managed by *ejidos* (Klepeis and Vance 2003). Milpa croplands include the crop types of maize, squash, yoka and beans. In 1992, residents were given rights for the land based on revised Article 27 of the Constitution (Geoghegan et al., 2001). So we hypothesized that the privatization of the *ejidos* might change the land-use and forest management practice strategy that cause the land cover change which can be observed by remotely sensed satellite images. This study seeks to determine a way in which the forest cover patterns in the Quintana Roo state can be understood and the how the characteristics of this heavily forested land can be evaluated at a regional scale. This type of broader view enables the identification of the interactions between local scale (communities) and regional scale (forest management).

METHODS

In this section, we introduce the framework of producing a land cover dataset by embedding the features from OSM using Random Forest (RF) machine learning techniques. Also, a landscape ecology analysis will be performed by incorporating road networks into regional land cover maps to investigate the landscape pattern matrix. In this case study, we divided our methods into two parts: 1) Mapping land cover by exploring the feature points from OSM by using machine learning classification techniques. 2) Incorporate OSM road networks into the classification map, which produced from part 1, to quantify road-caused forest fragmentations.

Conceptual Model

To examine how road networks affect the forest pattern distributions in the Maya Zone, we proposed a measurement to refine the forest map by OpenStreetMap. Road-caused forest fragmentation was measured

Figure 1. Study region of the Maya Zone in Yucatan Peninsula, Mexico

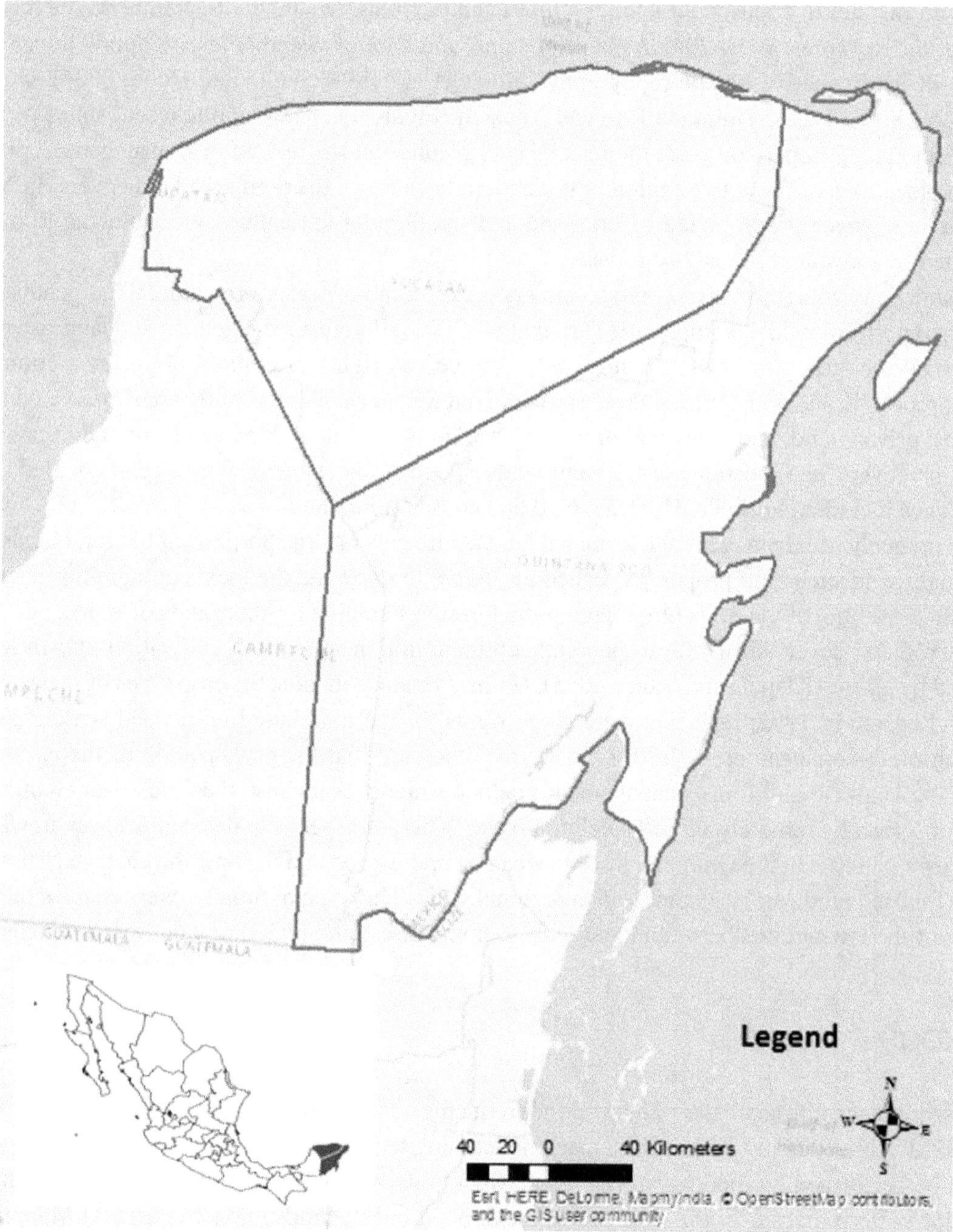

on the land cover maps after superimposing full national road and street maps from open source OSM roads database. We incorporated updated roads data into our forest land cover map so that the maps include additional classes of roads such as private access roads and driveways in rural areas, small service roads or alleys in urban areas, and forest access roads. When superimposing road maps, we converted all the forest pixels that contained at least one road segment to non-forest pixels (30-m spatial resolution based). No distinctions were drawn between the types of road, traffic volume, or other factors. By superimposing our resulting land cover maps with road networks data, the refined forest patches map will be compared and the spatial distribution will be analyzed. Set our classification strategy with appropriate numbers of training points and their weights.

We chose random forest (RF) as our classier because it is now a widely used algorithm for remote sensing image classification. It has an excellent ability to handle high dimensional raster database, which perfectly fit our research framework. Random Forest is a classifier that produces many classification and regression-like trees, where each tree is grown according to the split training data. RF also provides an accuracy assessment called "out-of-bag" (OOB) error to avoid bias during the validations.

As our computing platform, Google Earth Engine (GEE), anewly developed data analysis environment, enables large-scale spatial analysis by its special infrastructure and automatically parallelized computation techniques (Padarian et al., 2015, Patel et al., 2015). By storing and bringing together large amounts of earth observations data on Earth Engine's server, GEE is able to analysis big data "on-the-fly" (Yu et al., 2012). We can also upload our private and personal database into the server as assets to achieve time series generating, zonal statistics, spectral analysis and many machine learning classification techniques. All the data we used in this study as inputs for the classifier are completely embedded in GEE and we concluded java program codes for any repeatable analysis of the other regions. With the embedding of GEE, the regional land cover mapping can be done in a much simpler way, as many spatial and temporal covariates are already available in GEE server and the calculation itself is parallelized automatically to optimize the computing time cost. By uploading our own training sample from OSM, we need to generate a fusion table representation the geometry including class types of our training sample extracted from OSM.

The four-step procedure for mapping forest patterns and qualifying road-caused fragmentation is summarized below, and the flow chart of the process is shown in Figure. 2:

1. Extract the datasets of "places", "land-use", and "nature" from OSM, set the land cover classification strategy based on research objective, reclassify and generate a training dataset. (In this study, we reclassify the points into urban, waterbody, forest, cropland and cropland and bare land five classes.)
2. Validate with Google Earth both on time and space for the training dataset.
3. Collaborate the spatial and temporal covariates as inputs for Random Forest classifier, and run Random Forest classifier by using Google Earth Engine API, and analyzed the classification accuracy for each classification type.
4. Extract the forest land from the resulting map at step 3, superimpose rasterized OSM road networks and run cluster analysis to generate forest pattern map.1

Mapping Land Cover Using GEE and OSM

OpenStreetMap (OSM) is one of the most popular and well-supported Volunteered Geographic Information (VGI) (Mooney et al., 2010). All OSM data was downloaded from the Geofabrik website http://download.geofabrik.de. The main reason why we used OSM was to create the fully free and openly accessible map of road networks. Community volunteers collect geographic information and submit it to the global OSM database (Ciepluch et al., 2009). The impact of the OSM has been assessed and is significant. Girres and Touya (2010) performed a spatial analysis of the quality of OSM street network representations in the UK and France respectively through a comparison to ground-truth data obtained

Figure 2. Flowchart illustrating the procedure of mapping forest land and quantify forest patterns

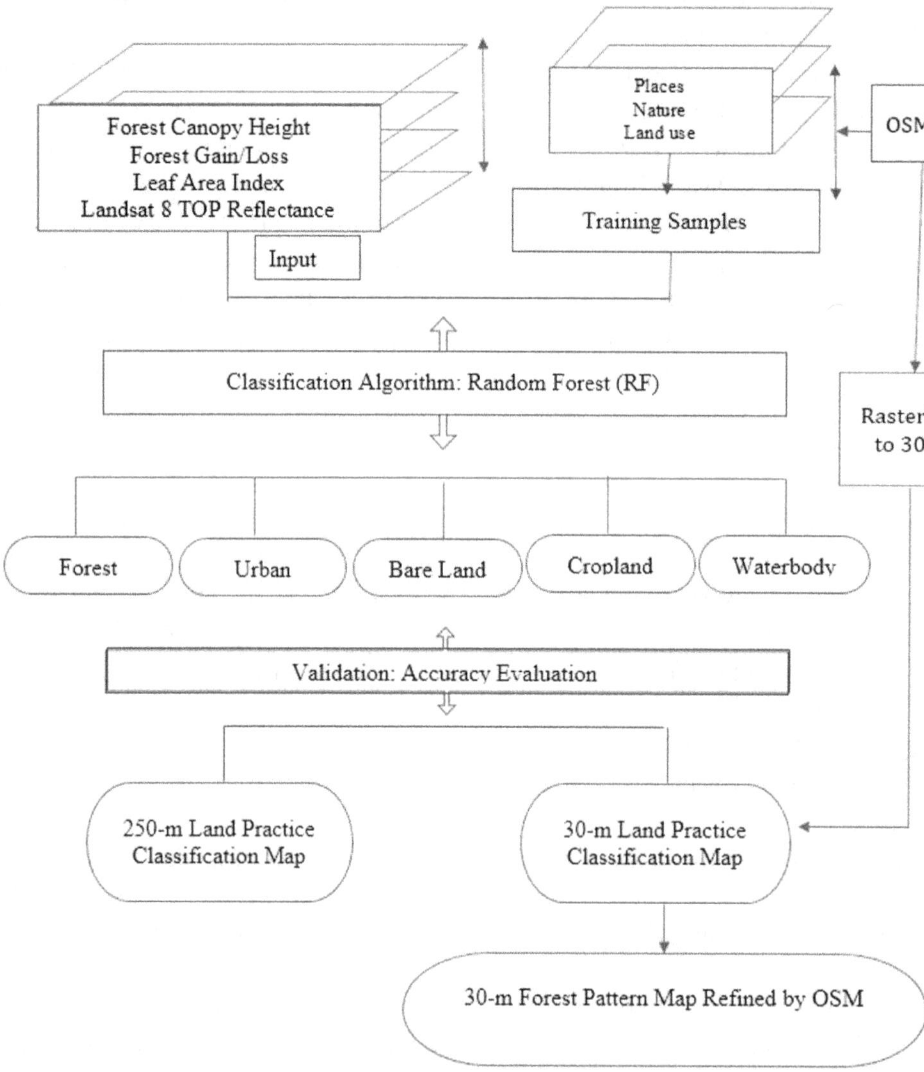

from the corresponding national mapping agency. Both studies found that on average the quality of the data was reasonably good but exhibited significant spatial heterogeneity. Neis et al 2012 analyzed how the quality of the OSM street network in different regions of Germany changed between the years 2007–2011.

Our study area of the Maya Zone is composed of cropland and forestland covering around 95% of the whole Yucatan Peninsula. The OSM database has a full coverage of our study area with the following sub-datasets: places, points, railways, roads, waterways, buildings, land-use and nature areas. It's hard to obtain accurate ground information that covers a broad study area from the field. We have to rely on the visual interpretation of high spatial resolution remotely sensed data for ground information; however, it is time consuming and labor intensive. To facilitate the virtual interpretation, we selected two publicly available datasets, OpenStreetMap features and Google Earth virtual global, and integrated them as our training and validation data resources. The OpenStreetMap feature classes we applied for the thematic map including "land-use", "natural", "places", and "waterway". The "land-use" is a polygon class identified forests, residential areas, industrial areas, etc. In the class "natural", all the water body, parks, and recreation areas are represented by polygons. The point class "places" includes urban and suburb landmarks and attractions. The "waterway" is different with water body, it is a polyline feature indicated rivers, canals, streams, ditches, etc. All of these classes were converted to point features and we used a stratified randomly distributed sample design to extract training and validation points. Then we integrated the selected points with Google Earth virtual globe for visual interpretation. Google Earth virtual globe offers substantially high-resolution imagery and enabled us to assess the five land cover classes in all scenes. The reason we used Google Earth is its higher spatial resolution with a time span of more than 20 years. So we will get more spatial information, what types of land practice occur and the land cover type during this spatiotemporal validation. More details about the training sample setting strategy will be represented as a flow chart. We converted the randomly selected points into a Keyhole Markup Language (KML) file. Each selected point centered at a 30×30-meter polygon, one-pixel image size. The selected points and associated polygons overlaid in Google Earth virtual globe for visualization. We identified and assigned the classes of selected points based on the OSM feature types and their surrounding areas. The selected points were split into training and validation data sets. One-third of points were randomly selected as validation data. With a stratified random sampling design, the points were selected from five strata. The main focus of the thematic map was the forest and forest is the dominant domain.

The training samples we extracted from OSM were re-classified into five classes: Urban and built-up, waterbody, forest, cropland and bare land. Figure 3 showed the spatial distribution of our training samples of each class. Feature points density is virtualized in Figure.3. It represented the spatial distribution of the training points geographically and also illustrated the degree of urbanization.

The spatial and temporal covariates used in the land cover classification include NDVI-based reflectance, forest canopy height, forest gain and loss, and leaf area index (LAI). Those spatial and temporal covariates are significant indicators of forest ecosystems, and those remotely sensed inputs are consistently and accurately derived. All the data we used for our land cover map were up-to-date and subset to the boundaries of our study area of Yucatan peninsula and resampled to 30-m spatial resolution. In Table 1, we list our existing data for mapping task. All the data we used as spatial and temporal covariates cover our study regions and can be extended to a global scale. Therefore, it is repeatable for the other study regions if there are similar research objectives.

1. Landsat 8 Top-of-atmosphere (TOA) (Chander et al., 2009) of 2016: This composite image reflects the greenest pixel during the whole time period based on NDVI. (Google earth engine ID: LANDSAT/LC8_L1T_ANNUAL_GREENEST_TOA)
2. Global forest canopy height version of 2005 (Simard et al): This dataset represents the canopy height on the global level by incorporating Geoscience Laser Altimeter System (GLAS) and ancillary data. (Google earth engine ID: NASA/JPL/global_forest_canopy_height_2005)
3. Hansen et al., global forest change dataset V1.2 (Version of 2015): Hansen et al., global forest change database is the first database describing and monitoring the forest changes at a global-scale. We extracted 3 layers of "treecover2000" "forestloss" and "forestgain". Here the layer of "treecover2000" is defined as canopy closure for all the vegetation taller than 5m and covers a percentage

Figure 3. Distribution of the OSM-derived training data for creation of the forest pattern map in Yucatan Peninsula, Mexico

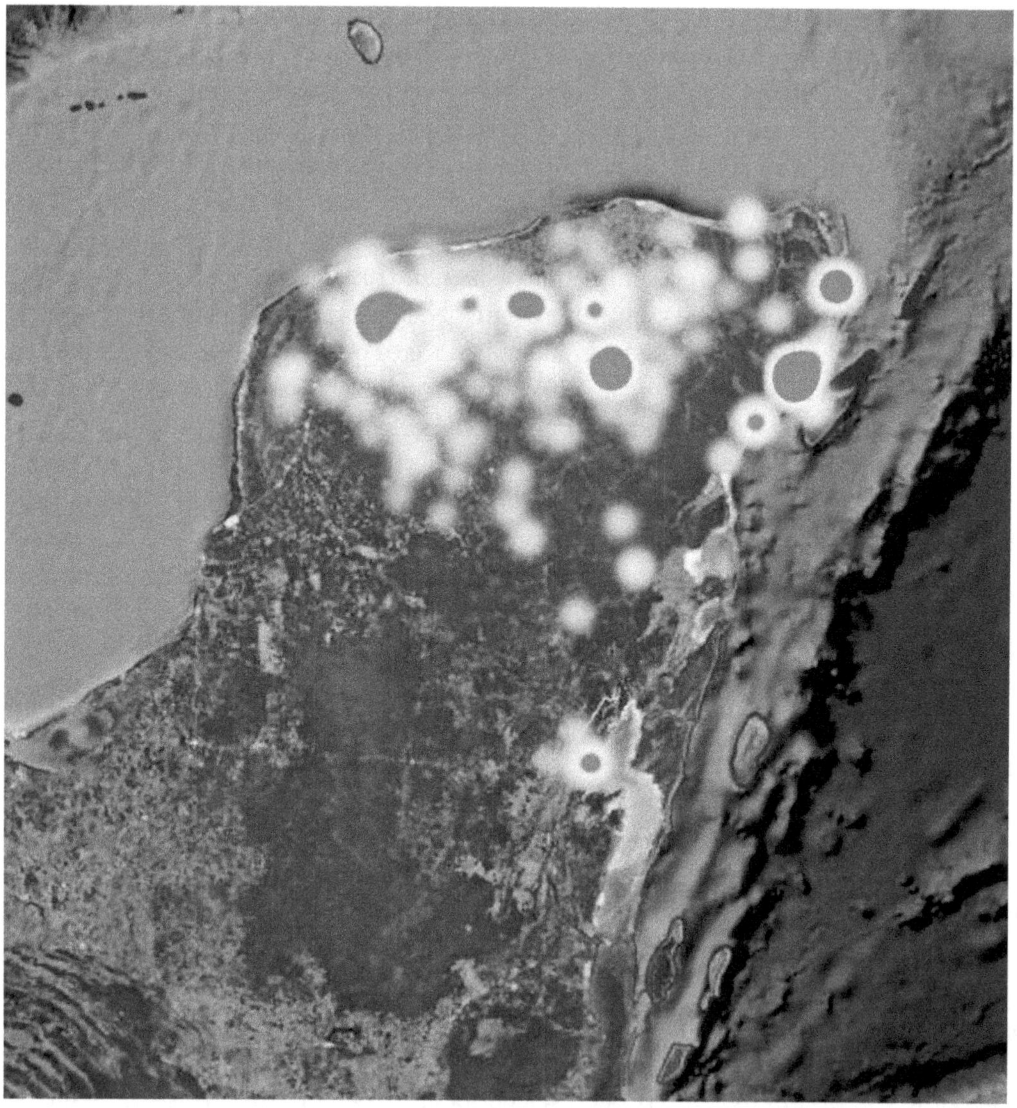

Table 1. Spatial and temporal covariates for random forest classifier

Database Name	*Objective*	*Google Earth Engine ID*	*Spatial resolution*	*Temporal resolution*
Landsat 8 TOP Reflectance	Reflectance	LANDSAT/LC8_L1T_ANNUAL_GREENEST_TOA	30m	16-day
Forest Canopy Height	Canopy Height	NASA/JPL/global_forest_canopy_height_2005	1km	Composite
Hansen et al., global forest change dataset V1.2	Forestland changes	UMD/hansen/global_forest_change_2015	30m	Composite
MODIS 15 Leaf Area Index	LAI	MODIS/006/MCD15A3H	500m	4-day

Source: Google Earth Engine

in the analysis unit. "Forestloss" during the period of 2000-2014 is defined as a stand-replacement disturbance, or a change from a forest to non-forest, represented as either 1 (loss) or 0 (no loss). "Forestgain" during the period 2000-2014 is defined as the inverse of loss, or a non-forest to forest change (Hansen et al., 2013). (Google earth engine ID: UMD/hansen/global_forest_change_2015)

4. MODIS 15 Leaf Area Index global 500m: The MCD15A3H version 6 MODIS Level 4, Combined Fraction of Photosynthetically Active Radiation (FPAR), and Leaf Area Index (LAI) product is a 4-day composite database with 500-meter pixel size. (Google Earth Engine ID: MODIS/006/MCD15A3H)

Quantifying Road-Caused Forest Fragmentation

Uncertainty assessment is a critical component of any earth observation derived classification-mapping techniques. Usually, the road feature is hiding due to the dense forest canopy, whereas we were able to incorporate the road layer into the land cover map as a way of achieving our forest pattern map's assessments. The combining of remote sensing and Geographic Information Systems (GIS) is a powerful tool to monitor, characterize and quantify forest fragmentation in both time and space (Ritters et al., 2002, Heilman et al., 2002). The modeled land cover map cover those two states was re-classified into forest/nonforest dummy variable map to quantify road-caused fragmentation effects on forest patches. The non-forest class includes cropland, bare land, residential and water body. All data sets were projected into the coordinate system Albers Conical Equal Area to follow the North America mapping criteria. To characterize forest fragmentation, the OSM database was superimposed onto a 30-m spatial resolution raster vision. Comparing with our analytical extent, we assumed the interaction between urban forest and rural forest would be minimal. Additionaly, we also generated near-real time "Openroad density" map from OSM, which included all types of roads.

Choosing the unit and scale of analysis is important in quantifying forest fragmentation, as smaller landscapes are more sensitive to finer-scale patterns (Riitters et al. 2002). The effects of both spatial and temporal scale should be considered in forest fragmentation, because forestland is heterogeneous, and the structures, functions and ecological processes are themselves scale-dependent. We calculated the patch size-distributions at both the local and regional scales. The spatial distribution represented the diversity and heterogeneity of road-caused forest fragmentation; and the patch-size class map elucidated the landscape composition and configurations. Therefore, in this study, we produced the dataset at local and regional scale at a spatial resolution of 30m (Landsat spatial resolution) and 250m (MODIS spatial resolution), respectively.

RESULTS

In this section, we present the results obtained by applying the conceptual model and the methodology described in Methodology section.

Analysis of Refined Land Cover Map

The map of the dominant land cover types in Yucatan Peninsula is shown in Figure 4, and the error matrix is Table 2. The landscape of Yucatan Peninsula, which consisted 2,222km^2, is comprised of 45%

forestland, 45% cropland, 12% urban areas. In the random forest classifier, we used to estimate the classification error without bias. In our study, the OOB error is 12% yielding a total classification accuracy of 91% of our random forest classifier building

In Table 2, we made the comparison between the forest cover and cropland cover between those two states. Quintana Roo is about 31.4% higher Yucatan at forest cover and 31% lower at cropland cover. Yucatán state has a higher average road density of 4.53km/km^2 than Quintana Roo state 2.55 km/km^2 derived from OSM road feature. This huge difference is not only just because the degree of urbanized, but also there is a big difference between the land tenure, tourism policies, and timber rights in those two states.

We constructed an error matrix from the results of our land practice map, and calculated the commission, omission and the overall errors (Table.3). The error matrix is shown in Table. 2. Rows represent the reference data and columns represent the classification results. Producer's accuracy and user's accuracy are listed in the last row and column, respectively. The overall accuracy of the thematic map was 92%, and the kappa coefficient was 77%. Overall 96% of the forest pixels were correctly identified as forest, and 95% of the forest area are successfully classified as forest. Bare land shows the lowest user's accuracy of 72%, which is because of the light weight in the training samples, and VGI contributors define bare land differently. Moreover, the training samples cannot fully represent the surrounding land cover types, for example, a small conservation park in urban regions that was marked "forest".

Forest Patterns Map

By proposing the framework by incorporating OSM road networks, we have the capability to spatially access the potential of forest fragmentation according to forest patch sizes and its surrounding land covers and road densities. Figure.5 shows the refined spatial distribution of forest patterns based on patch size. We can see clearly that there is the big difference between Yucatan and Quintana Roo state. In Yucatan forest pattern map, the visualization result tells us a large proportion of forest patches ranges from 6-100 ha which is relatively small and has a high risk for further deforestation and fragmentation. Moreover, in Quintana Roo state, the forest is less fragmented than Yucatán state, but we can still easily see the forest pattern was separated and divided by roads. In general, as the road networks increase, the forest fragmentation will increase by a wide margin. After incorporating the road network, there are 603 forest patches in the state of Quintana Roo. A total number of 477 forest patches are under the area of 500 Ha.

To account for the complexity spatial patterns at regional scale, frequency distributions are used heavily in quantifying the total number of patches and their patch sizes. Combining with the forest pattern map from Figure 6, a size-frequency distribution has the capability to represent locations where has a high risk of deforestation and forest fragmentation. As shown in Figure. 6, we calculated the size-frequency distribution of forest patches over the whole Yucatan and Quintana Roo state. Those total 15,000 forest

Table 2. Comparison of Forest and Cropland Cover between Quintana Roo and Yucatán states

	Forest Cover	Cropland Cover	Average Road Density (km/km^2)
Quintana Roo	86.20%	11.10%	2.55
Yucatan	44.80%	53.00%	4.53

Figure 4. Spatial distribution of dominant land cover types of Yucatan Peninsula

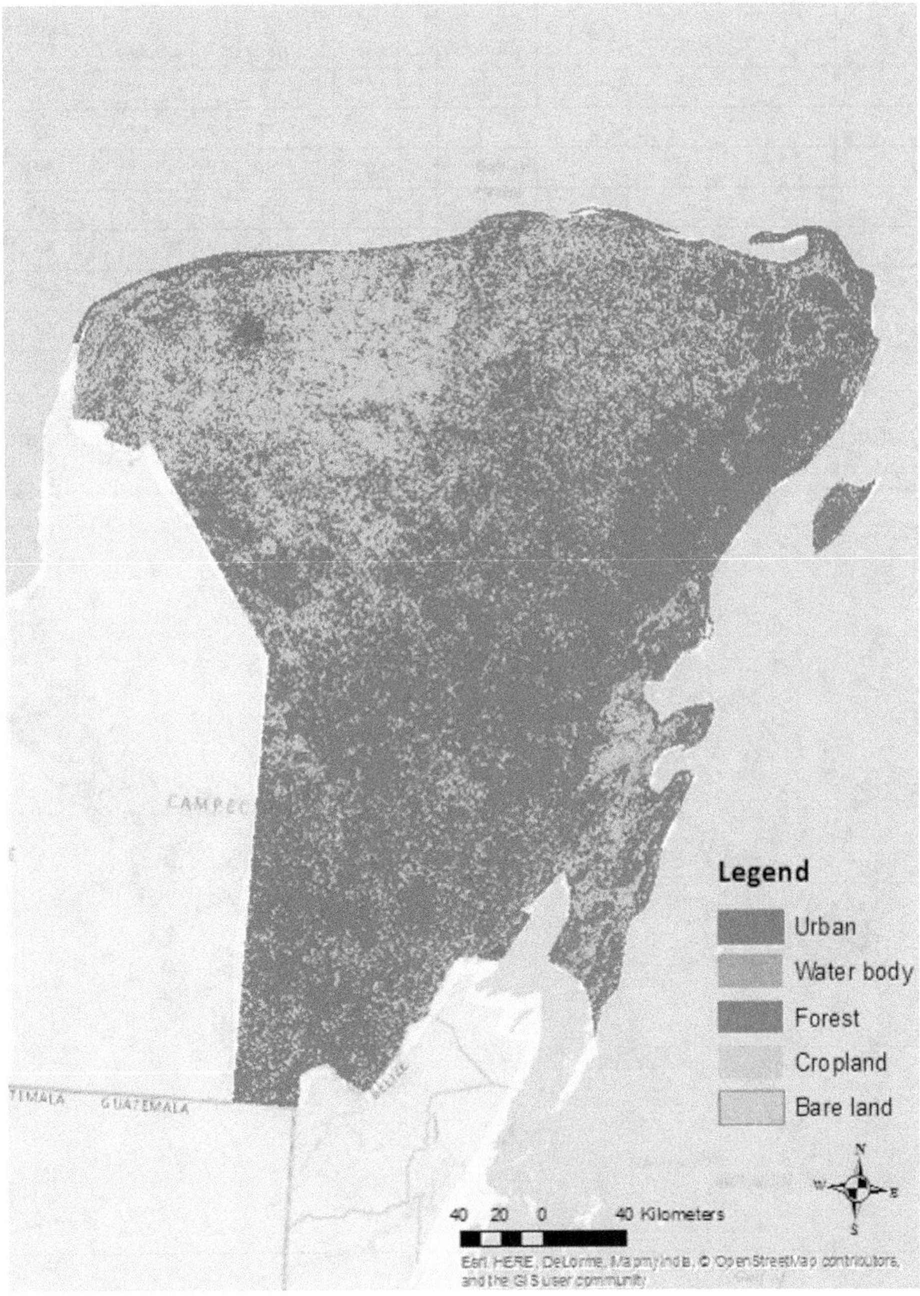

patches in Yucatán state just cover 3% of the forest cover, which can be understood with a high potential to get further fragmented. And the spatial patterns map tells us where those patch located and we can provide some advises to the land managers or decision makers. In Quintana Roo state, no more major road building will be recommended within the large continuous patches with size more than 10,000 ha.

Table 3. Classification accuracy from random forest classifier OOB of each land cover type

	Urban	Water	Forest	Cropland	Bare land	Points	User's acc. (%)
Urban	36	0	7	2	0	45	80
Water	0	56	10	7	0	73	77
Forest	4	15	1280	40	12	1351	95
Cropland	0	4	28	147	1	180	82
Bare land	0	0	4	10	36	50	72
Total	40	75	1329	206	49	1699	
Producer's acc. (%)	90	75	96	71	73		

Figure 5. Spatial patterns classified by patches area of Yucatán state and Quintana Roo state

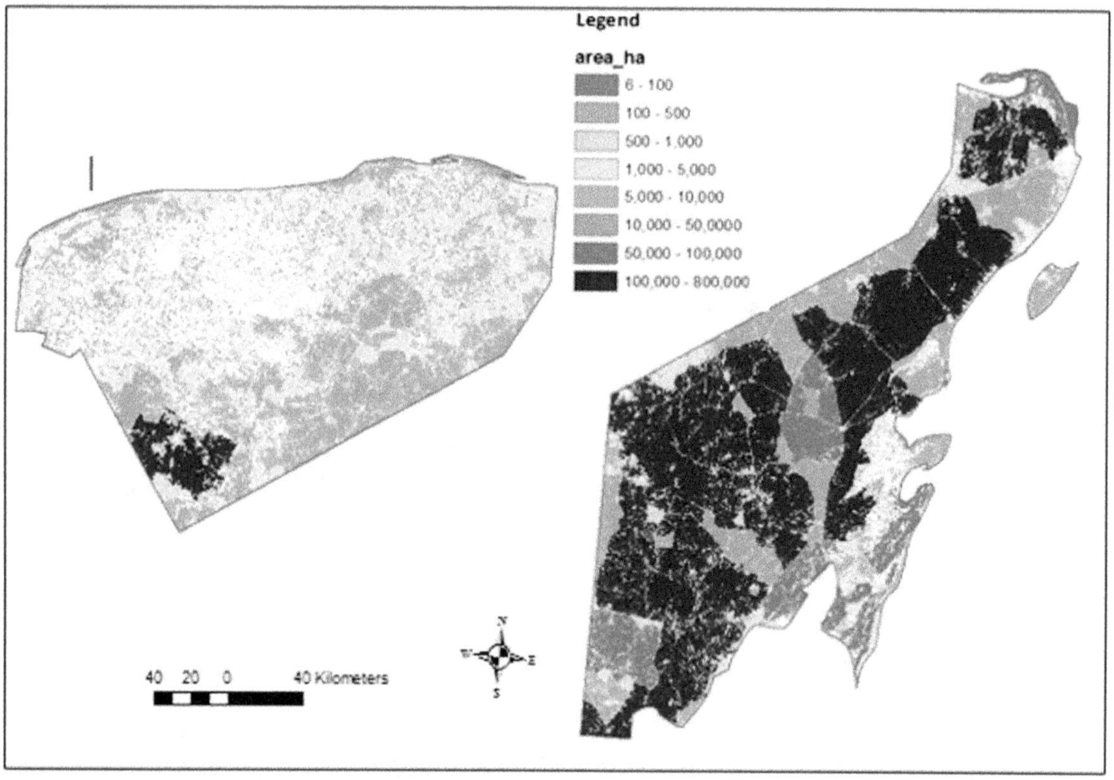

DISCUSSIONS

This study described the potential that OSM's ability to mapping forest patterns, and its critical rules in setting training samples. We incorporated the OSM into remote sensing satellite layers to classify land and forest practices. We then compared the dominant land cover types between Quintana Roo and Yucatán states.

Figure 6. Forest Patches Size-frequency Distribution of Yucatan and Quintana Roo state

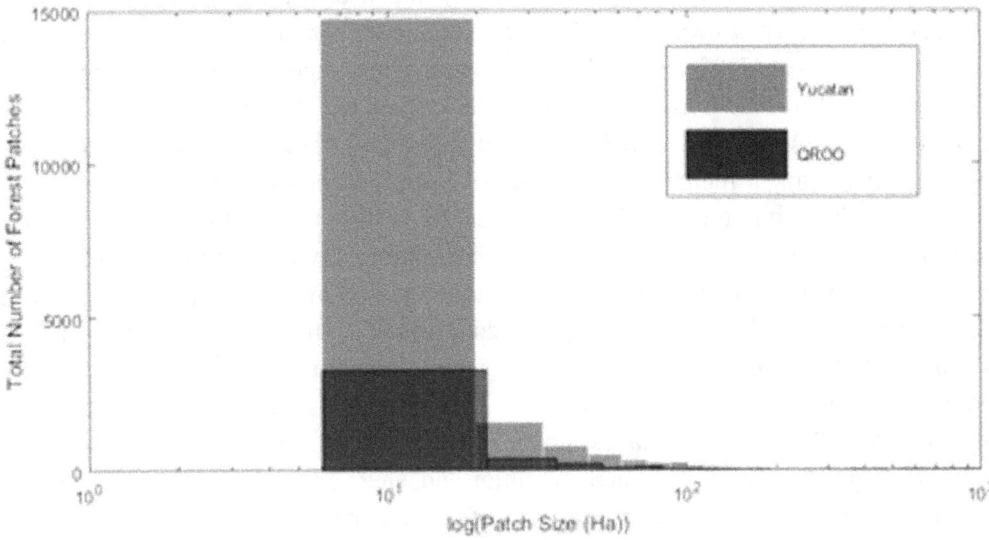

OpenstreetMap Road Density

At the regional scale, road density is an overall term used to describe the road networks. It has been proposed as a broad index of roads' ecological effects in a landscape. We calculated OpenStreetMap derived road density of our study area based on our land cover map including the land transition matrix.

Ownership Affecting Forest Patterns Distributions

To support our map, to prove the by combining the satellite images and VGI data have the potential to tell us what is going on over the ground and the landscape. We also need people in situ testimonies to monitor and explain the ground truth because the land and forest management practices on Maya Zone is unique and cannot be repeatable by the others according to the historical and climate environments (Corbera et al., 2011). We took nine places in Quintana Roo state and the state border between Yucatan and Quintana Roo state as the sub-case study. First of all, we concluded those places and classified them based on the management ownership types. In this research, land ownership in the state of Quintana Roo was separated into three types: *Ejido* and local communities land; private land; and conservation areas.

- ***Ejido* Land**: Chan Santa Cruz, Petcacab, Betania, Nuevo Durango
- **Private Land**: Don's land, and agriculture land in Jose Maria Morelos
- **Conservation Land**: Siijil Nol Ha, Punta Laguna, and Coba

Forest ownership is a vital link between human and environmental factors, particularly at the re. At regional scale, forest ownership patterns explain different types of land management practices and trajectories of land cover change (Turner 1996). Turner et al., also similarly showed a correlation between land ownership categories and land cover change patterns in North Carolina and Washington (Turner et al., 1996). Albers and Ando found that federal ownership of conservation lands created crowding-in and crowding-out effects of private ownership (depending on the states), thereby also indirectly influencing fragmentation (Albers et al, 2008). But in Mexico, not many researches have been done.

Much of Quintana Roo land is assigned to the ejidal system, which is managed by community-based planting and logging. In many *ejidos*, mahogany-a lucrative forest product, is used as the guide species to design forest management regimes. In the *ejido* of Petcacab, the total forested area is divided into five-year blocks with these in turn divided into annual harvesting blocks. In a given year, members of the Petcacab *ejido* are authorized to harvest a specified volume of wood. Within each harvesting block, only trees with a diameter at breast height (DBH) larger than 55cms are supposed to be harvested. In contrast, few people are involved in forest management practice per management unit in privately owned land. Within private land ownership systems, there is more freedom in forest management practices but less freedom regarding planting and harvesting. Conservation areas, on the other hand, are a form of passive management. These forested lands are "area where natural processes and natural disturbance regimes can develop without management intervention and where ecological and societal goals are given primacy" (Duncker et al., 2012). The characteristics of forest management practices in the Quintana Roo state vary by land ownership type, which in turn may have implications for the level of forest fragmentation associated with each of these land tenure systems. To examine the different implications of the land ownership types on forest fragmentation and hence biodiversity conservation, zonal statistics were calculated for each of the land tenures types and their forest management systems.

CONCLUSION

By using the OSM trained forest pattern map as the forest reference data, we produced the OSM refined forest parcels and analyzed the spatial size distribution of forest patterns. We successfully demonstrated the application of OSM as training sample and assessments layer embedding in land cover maps at a regional scale. The framework got a satisfactory performance of generating the maps with a total accuracy of 91% and Kappa coefficient of 77% and also did an excellent job incorporating OSM road features in characterizing forest patterns. In addition, by incorporating the OSM, the secondary and logging roads are shown to play an important role in causing fragmentation of the remnant forestlands. Landscape metrics for the Quintana Roo state were produced. Spatial frequency distributions of forest patches were calculated and analyzed. The results of the analysis showed significant levels of forest fragmentation throughout the both Quintana Roo and Yucatán state.

All platforms in this study are in active developments, GEE is building more machine learning classification techniques into the program, and increasing the computational limit for the goal of large-scale mapping. Moreover, more researchers and land managers are digitalizing their rural area maps into OpenstreetMap, which fills the gaps of VGI in rural areas.

OSM is an exciting new platform, which could be implemented in land cover mapping and assessing, because of its geo-located features and marks. The combination of Google Earth Engine with OpenstreetMap features will be a tool for large scale mapping and will present significant opportunities in the creation ideas because of its unique workflow to fulfill specific mapping needs.

FUTURE RESEARCH DIRECTIONS

Conserving forest biodiversity is a prerequisite for the long-term flow of forest ecosystem services. Forest management has evolved considerably in the last two decades, demonstrating significant positive impacts for conserving biodiversity and delivering economic benefits to local communities.

According to the temporal availability of our covariates, these up-to-date forest patterns maps can be extended to a multi-temporal changes database. It also has the capability to make predictions based on the spatiotemporal forest pattern changes and patch size distributions. There is still a strong need for a continental coverage land cover map with a higher temporal resolution, which will reflect the land cover transition processes at shorter intervals. This issue is urgent; in order to abate the rapidly increasing construction of road networks and the associated ecological impacts, we need better road network designs that combine both top-down and bottom-up processes, which would retain/preserve large forest patches and decelerate the parcelization rate. The use of open data source will be open our mind of exploring the area that doesn't have an official classification database. More information on the effects of land ownership and forest management, combined with the detailed road network and a continental coverage land cover maps can aid in thwarting further forest fragmentation by promoting more thoughtful road planning by land planners and decision makers.

REFERENCES

Abdullah, S. A., & Nakagoshi, N. (2007). Forest fragmentation and its correlation to human land-use change in the state of Selangor, peninsular Malaysia. *Forest Ecology and Management*, *241*(1), 39–48. doi:10.1016/j.foreco.2006.12.016

Aerts, R., & Honnay, O. (2011). Forest restoration, biodiversity and ecosystem functioning. *BMC Ecology*, *11*(1), 29. doi:10.1186/1472-6785-11-29 PMID:22115365

Albers, H. J., Ando, A. W., & Batz, M. (2008). Patterns of multi-agent land conservation: Crowding in/out, agglomeration, and policy. *Resource and Energy Economics*, *30*(4), 492–508. doi:10.1016/j.reseneeco.2008.04.001

Arsanjani, J. J., Mooney, P., & Zipf, A. (2015). Quality assessment of the contributed land-use information from OpenStreetMap versus authoritative datasets. OpenStreetMap in GIScience, 37-58.

Baskent, E. Z., & Jordan, G. A. (1995). Characterizing spatial structure of forest landscape. *Canadian Journal of Forest Research*, *25*(11), 1830–1849. doi:10.1139/x95-198

Bray, D. B., Ellis, E. A., Armijo-Canto, N., & Beck, C. T. (2004). The institutional drivers of sustainable landscapes: A case study of the 'Mayan Zonein Quintana Roo, Mexico. *Land Use Policy*, *21*(4), 333–346. doi:10.1016/j.landusepol.2003.11.001

Carls, N. (1947). *How to read aerial photographs for census work.* US Government Printing Office.

Chander, G., Markham, B. L., & Helder, D. L. (2009). Summary of current radiometric calibration coefficients for Landsat MSS, TM, ETM+, and EO-1 ALI sensors. *Remote Sensing of Environment*, *113*(5), 893–903. doi:10.1016/j.rse.2009.01.007

Ciepluch, B., Mooney, P., Jacob, R., & Winstanley, A. C. (2009). Using openstreetmap to deliver location-based environmental information in Ireland. *SIGSPATIAL Special*, *1*(3), 17–22. doi:10.1145/1645424.1645428

Cihlar, J. (2000). Land cover mapping of large areas from satellites: Status and research priorities. *International Journal of Remote Sensing*, *21*(6-7), 1093–1114. doi:10.1080/014311600210092

Corbera, E., Estrada, M., May, P., Navarro, G., & Pacheco, P. (2011). Rights to land, forests and carbon in REDD+: Insights from Mexico, Brazil and Costa Rica. *Forests*, *2*(1), 301–342. doi:10.3390/f2010301

Corcoran, P., Mooney, P., & Bertolotto, M. (2013). Analysing the growth of OpenStreetMap networks. *Spatial. Statistics*, *3*, 21–32.

Costanza, R., Fisher, B., Ali, S., Beer, C., Bond, L., Boumans, R., & Snapp, R. et al. (2007). Quality of life: An approach integrating opportunities, human needs, and subjective well-being. *Ecological Economics*, *61*(2), 267–276. doi:10.1016/j.ecolecon.2006.02.023

Duncker, P. S., Raulund-Rasmussen, K., Gundersen, P., Katzensteiner, K., De Jong, J., Ravn, H. P., & Spiecker, H. (2012). How forest management affects ecosystem services, including timber production and economic return: Synergies and trade-offs. *Ecology and Society*, *17*(4), 50. doi:10.5751/ES-05066-170450

EEA Corine Land Cover. (n.d.). Available online: http://www.eea.europa.eu/publications/COR0-landcover

Ellis, E., & Pontius, R. (2007). Land-use and land cover change. Encyclopedia of Earth.

Elwood, S. (2008). Volunteered geographic information: Key questions, concepts and methods to guide emerging research and practice. *GeoJournal*, *72*(3), 133–135. doi:10.1007/s10708-008-9187-z

Espírito-Santo, F. D., Gloor, M., Keller, M., Malhi, Y., Saatchi, S., Nelson, B., & Palace, M. et al. (2014). Size and frequency of natural forest disturbances and the Amazon forest carbon balance. *Nature Communications*, 5. PMID:24643258

Estima, J., & Painho, M. (2013). Exploratory analysis of OpenStreetMap for land-use classification. *Proceedings of the second ACM SIGSPATIAL international workshop on crowdsourced and volunteered geographic information*, 39-46. doi:10.1145/2534732.2534734

Estima, J., & Painho, M. (2015). Investigating the potential of OpenStreetMap for land-use/Land cover production: a case study for Continental Portugal. In *OpenStreetMap in GIScience* (pp. 273–293). Springer International Publishing. doi:10.1007/978-3-319-14280-7_14

Flores Guido, J. S., & Espejel Carvajal, I. (1994). *Tipos de vegetación de la Península de Yucatán* (No. Y/581.61 E8/3). Academic Press.

Foody, G. M., & Boyd, D. S. (2013). Using volunteered data in land cover map validation: Mapping West African forests. *IEEE Journal of Selected Topics in Applied Earth Observations and Remote Sensing*, *6*(3), 1305–1312. doi:10.1109/JSTARS.2013.2250257

Forman, R. T. (2000). Estimate of the area affected ecologically by the road system in the United States. *Conservation Biology*, *14*(1), 31–35. doi:10.1046/j.1523-1739.2000.99299.x

Fritz, S., McCallum, I., Schill, C., Perger, C., Grillmayer, R., Achard, F., & Obersteiner, M. et al. (2009). Geo-Wiki. Org: The use of crowdsourcing to improve global land cover. *Remote Sensing*, *1*(3), 345–354. doi:10.3390/rs1030345

Geoghegan, J., Villar, S. C., Klepeis, P., Mendoza, P. M., Ogneva-Himmelberger, Y., Chowdhury, R. R., & Vance, C. et al. (2001). Modeling tropical deforestation in the southern Yucatan peninsula region: Comparing survey and satellite data. *Agriculture, Ecosystems & Environment*, *85*(1), 25–46. doi:10.1016/S0167-8809(01)00201-8

Girres, J. F., & Touya, G. (2010). Quality assessment of the French OpenStreetMap dataset. *Transactions in GIS*, *14*(4), 435–459. doi:10.1111/j.1467-9671.2010.01203.x

Haddad, N. M., Brudvig, L. A., Clobert, J., Davies, K. F., Gonzalez, A., Holt, R. D., & Cook, W. M. et al. (2015). Habitat fragmentation and its lasting impact on Earths ecosystems. *Science Advances*, *1*(2), e1500052. doi:10.1126/sciadv.1500052 PMID:26601154

Haklay, M. (2010). How good is volunteered geographical information? A comparative study of OpenStreetMap and Ordnance Survey datasets. *Environment and Planning. B, Planning & Design*, *37*(4), 682–703. doi:10.1068/b35097

Haklay, M. (2010). How good is volunteered geographical information? A comparative study of OpenStreetMap and Ordnance Survey datasets. *Environment and Planning. B, Planning & Design*, *37*(4), 682–703. doi:10.1068/b35097

Haklay, M., Singleton, A., & Parker, C. (2008). Web mapping 2.0: The neogeography of the GeoWeb. *Geography Compass*, *2*(6), 2011–2039. doi:10.1111/j.1749-8198.2008.00167.x

Heilman, G. E., Strittholt, J. R., Slosser, N. C., & Dellasala, D. A. (2002). Forest fragmentation of the conterminous United States: Assessing forest intactness through road density and spatial characteristics. *Bioscience*, *5241*, 1–22.

Homer, C. G., Dewitz, J. A., Yang, L., Jin, S., Danielson, P., Xian, G., & Megown, K. (2015). Completion of the 2011 National Land Cover Database for the conterminous United states-Representing a decade of land cover change information. *Photogrammetric Engineering and Remote Sensing*, *81*(5), 345–354.

Jaeger, J. A., Schwarz-von Raumer, H. G., Esswein, H., Müller, M., & Schmidt-Lüttmann, M. (2007). Time series of landscape fragmentation caused by transportation infrastructure and urban development: A case study from Baden-Württemberg, Germany. *Ecology and Society*, *12*(1), 22. doi:10.5751/ES-01983-120122

Johnson, B. A., & Iizuka, K. (2016). Integrating OpenStreetMap crowdsourced data and Landsat time-series imagery for rapid /land cover (LULC) mapping: Case study of the Laguna de Bay area of the Philippines. *Applied Geography (Sevenoaks, England)*, *67*, 140–149. doi:10.1016/j.apgeog.2015.12.006

Jokar Arsanjani, J., Helbich, M., & Bakillah, M. (2013). Exploiting volunteered geographic information to ease mapping of an urban landscape. *ISPRS-International Archives of the Photogrammetry. Remote Sensing and Spatial Information Sciences*, *1*(1), 51–55.

Jokar Arsanjani, J., Helbich, M., Bakillah, M., Hagenauer, J., & Zipf, A. (2013). Toward mapping land-use patterns from volunteered geographic information. *International Journal of Geographical Information Science*, *27*(12), 2264–2278. doi:10.1080/13658816.2013.800871

Klepeis, P., & Vance, C. (2003). Neoliberal policy and deforestation in southeastern Mexico: An assessment of the PROCAMPO program. *Economic Geography*, *79*(3), 221–240. doi:10.1111/j.1944-8287.2003.tb00210.x

Lambin, E. F., Turner, B. L., Geist, H. J., Agbola, S. B., Angelsen, A., Bruce, J. W., & George, P. et al. (2001). The causes of land-use and change: Moving beyond the myths. *Global Environmental Change*, *11*(4), 261–269. doi:10.1016/S0959-3780(01)00007-3

McGarigal, K., Romme, W. H., Crist, M., & Roworth, E. (2001). Cumulative effects of roads and logging on landscape structure in the San Juan Mountains, Colorado (USA). *Landscape Ecology*, *16*(4), 327–349. doi:10.1023/A:1011185409347

Mooney, P., Corcoran, P., & Winstanley, A. C. (2010). A study of data representation of natural features in openstreetmap. Proceedings of GIScience, 150, 150-156.

Neis, P., Zielstra, D., & Zipf, A. (2012). The street network evolution of crowdsourced maps: OpenStreetMap in Germany 2007–2011. *Future Internet*, *4*(4), 1–21. doi:10.3390/fi4010001

OpenStreetMap in GIScience. (2015). *Experiences, Research, and Applications*. Springer.

OpenStreetMap Map Features. (2016). Retrieved from http://wiki.openstreetmap.org/wiki/Map_Features

Padarian, J., Minasny, B., & McBratney, A. B. (2015). Using Googles cloud-based platform for digital soil mapping. *Computers & Geosciences*, *83*, 80–88. doi:10.1016/j.cageo.2015.06.023

Patel, N. N., Angiuli, E., Gamba, P., Gaughan, A., Lisini, G., Stevens, F. R., & Trianni, G. et al. (2015). Multitemporal settlement and population mapping from Landsat using Google Earth Engine. *International Journal of Applied Earth Observation and Geoinformation*, *35*, 199–208. doi:10.1016/j.jag.2014.09.005

Scarborough, V. L., & Burnside, W. R. (2010). Complexity and sustainability: Perspectives from the ancient Maya and the modern Balinese. *American Antiquity*, *75*(2), 327–363. doi:10.7183/0002-7316.75.2.327

See, L., Comber, A., Salk, C., Fritz, S., van der Velde, M., Perger, C., & Obersteiner, M. et al. (2013). Comparing the quality of crowdsourced data contributed by expert and non-experts. *PLoS ONE*, *8*(7), e69958. doi:10.1371/journal.pone.0069958 PMID:23936126

See, L., Schepaschenko, D., Lesiv, M., McCallum, I., Fritz, S., Comber, A., & Siraj, M. A. et al. (2015). Building a hybrid land cover map with crowdsourcing and geographically weighted regression. *ISPRS Journal of Photogrammetry and Remote Sensing*, *103*, 48–56. doi:10.1016/j.isprsjprs.2014.06.016

Simard, M., Pinto, N., Fisher, J. B., & Baccini, A. (2011). Mapping forest canopy height globally with spaceborne lidar. *Journal of Geophysical Research: Biogeosciences*, *116*(G4), G04021. doi:10.1029/2011JG001708

Turner, B. L., Villar, S. C., Foster, D., Geoghegan, J., Keys, E., Klepeis, P., & Plotkin, A. B. et al. (2001). Deforestation in the southern Yucatán peninsular region: An integrative approach. *Forest Ecology and Management*, *154*(3), 353–370. doi:10.1016/S0378-1127(01)00508-4

Turner, M. G. (1989). Landscape ecology: The effect of pattern on process. *Annual Review of Ecology and Systematics*, *20*(1), 171–197. doi:10.1146/annurev.es.20.110189.001131

White, D. A., & Hood, C. S. (2004). Vegetation patterns and environmental gradients in tropical dry forests of the northern Yucatan Peninsula. *Journal of Vegetation Science*, *15*(2), 151–160. doi:10.1111/j.1654-1103.2004.tb02250.x

Wickham, J. D., Riitters, K. H., Wade, T. G., & Vogt, P. (2010). A national assessment of green infrastructure and change for the conterminous United States using morphological image processing. *Landscape and Urban Planning*, *94*(3), 186–195. doi:10.1016/j.landurbplan.2009.10.003

Yu, L., & Gong, P. (2012). Google Earth as a virtual globe tool for Earth science applications at the global scale: Progress and perspectives. *International Journal of Remote Sensing*, *33*(12), 3966–3986. doi:10.1080/01431161.2011.636081

Zook, M., Graham, M., Shelton, T., & Gorman, S. (2010). Volunteered geographic information and crowdsourcing disaster relief: A case study of the Haitian earthquake. *World Medical & Health Policy*, *2*(2), 7–33. doi:10.2202/1948-4682.1069

Chapter 9
Leveraging Volunteered Geographic Information to Improve Disaster Resilience:
Lessons Learned From AGORA and Future Research Directions

João Porto de Albuquerque
University of Warwick, UK

Flávio Eduardo Aoki Horita
University of São Paulo, Brazil

Livia Castro Degrossi
University of São Paulo, Brazil

Roberto dos Santos Rocha
University of São Paulo, Brazil

Sidgley Camargo de Andrade
Federal University of Technology – Paraná, Brazil

Camilo Restrepo-Estrada
University of São Paulo, Brazil

Werner Leyh
University of São Paulo, Brazil

ABSTRACT

Volunteered Geographic Information (VGI) has emerged as an important additional source of information for improving the resilience of cities and communities in the face of natural hazards and extreme weather events. This chapter summarizes the existing research in this area and offers an interdisciplinary perspective of the challenges to be overcome, by presenting AGORA: A Geospatial Open collaboRative Architecture for building resilience against disasters and extreme events. AGORA structures the challenges of using VGI for disaster management into three layers: acquisition, integration and application. The chapter describes the research challenges involved in each of these layers, as well as reporting on the results achieved so far and the lessons learned in the context of flood risk management in Brazil. Furthermore, the chapter concludes by setting out an interdisciplinary research agenda for leveraging VGI to improve disaster resilience.

DOI: 10.4018/978-1-5225-2446-5.ch009

INTRODUCTION

Disasters caused by natural hazards such as floods and landslides have caused significant material damage and human losses worldwide, and particularly affected countries in the Global South. Future extreme environmental events are likely to become more frequent and their patterns less predictable due to climate change. These events have immediate and longer-term socio-economic impacts on both an international and national scale (Jha, Barenstein, Phelps, Pittet and Sena, 2010). In 2013, disasters incurred financial losses of approximately US$135 billion worldwide and caused 20,500 fatalities. Moreover, 44% of these reported disasters, 49% of the fatalities, and 37% of the financial losses were linked to extreme hydrological events (Munich RE, 2013).

These facts show the need to carry out activities with a view to building resilience against disasters and extreme events, i.e. to enable communities to resist, change or adapt in face of a disaster (Norris, Stevens, Pfefferbaum, Wyche, & Pfefferbaum, 2008). Disaster management is an important adaptation strategy to achieve this goal, since it consists of a cycle of tasks that involve the mitigation of risks, preparatory measures, effective response and post-event recovery. Accurate and up-to-date information is essential for supporting the tasks of disaster management throughout this cycle. For instance, the continuous monitoring of a flood risk and the real-time mapping of flood hazards require local information such as rainfall data and water level observations to evaluate the risk of potential flooding. Moreover, the monitoring often requires an extensive spatial coverage to be efficient, which can be a considerable challenge for most cities in emerging and developing countries, like Brazil, where updated local data about the current state of the rivers is rarely available.

Volunteered Geographic Information (VGI) – i.e., user-generated information that is crowdsourced and relies on collaborative web platforms, specific web applications and/or mobile phone apps (Goodchild, 2007) – has emerged in the past few years as an important additional source of information for disaster management (Goodchild & Glennon, 2010; Horita, Degrossi, Assis, Zipf, & Albuquerque, 2013). The combination of VGI with traditional sources of information (e.g., environmental sensors) to explore complementarities, offers a great potential for building resilience against flooding, since they can be used for improving the availability, coverage and accuracy of the information needed for disaster management (Horita, Albuquerque, Degrossi, Mendiondo, & Ueyama, 2015).

However, there are still a number of serious challenges to be addressed to make VGI an effective means of supporting disaster management. With the aim of making sense of these challenges and investigating approaches to tackle them, this chapter presents AGORA - **A** **G**eospatial **O**pen collabo**R**ative **A**rchitecture for building resilience against disasters and extreme events (AGORA, 2016). This is a conceptual architecture that is being designed by a group of interdisciplinary projects hosted at the University of São Paulo at São Carlos, Brazil, and involving international partners at Heidelberg University (Germany), the University of Münster (Germany) and at the University of Warwick (UK). AGORA is aimed at improving flood risk management in Brazil and works in close collaboration with the National Center for Monitoring and Early Warning of Natural Disasters (CEMADEN). The conceptual architecture of AGORA structures the task of using VGI for disaster management into three layers: acquisition, integration, and application. In this chapter, we will explore the challenges and current state-of-the-art about each of these layers, as well as the lessons learned from the corresponding AGORA projects.

The remainder of this chapter is structured as follows: Section 2 outlines the theoretical background of this work, as well as related works and identifies gaps in research. The components of the AGORA framework are introduced in Section 3, while Section 4 describes the case studies carried out within the AGORA projects and the lessons learned. Finally, Section 5 concludes the chapter by recommending future directions for leveraging VGI as a means of improving disaster resilience.

BACKGROUND: CHALLENGES IN USING VGI FOR DISASTER PREPAREDNESS AND EMERGENCY RESPONSE

The term VGI has been used in the context of disaster management to refer to citizen-generated data from various information sources (Albuquerque, Eckle, Herfort, & Zipf, 2016): (i) social media – i.e., information sharing through platforms like Twitter, Facebook, Flickr etc.; (ii) crowd sensing – i.e., citizens on the Web or equipped with smartphones using dedicated applications to register and share observations (e.g., citizen observatories); (iii) collaborative mapping – i.e., the editing of geographic features to create internet-based interactive maps (e.g., OpenStreetMap). However, some researchers have criticized the use of the term "volunteered" to refer to information that is collected without the conscious knowledge and clear choice of the contributor, as is the case with georeferenced social media messages from Twitter (Harvey, 2013). Recently, some researchers have started using the term "Crowdsourced Geographic Information" to encompass both the more "active/conscious" and "passive/unconscious" information that is produced (Albuquerque et al., 2016; See et al., 2016). In this chapter, however, we employ a wider definition of VGI, as an equivalent of CGI, to encompass all forms of citizen-generated geographic information, while taking into account that they include very different ways of involving citizens.

Several works can be found in the literature on the use of VGI to support vulnerable communities and assist emergency agencies in preparing for and responding to disasters. The next sections attempt to point out the research gaps that can be found in the following areas: data integration, quality and credibility, visualization and decision-making support.

Data Integration

The first challenge when using VGI, concerns the way it can be integrated and shared with other available data, since VGI are found in different formats from different locations and time zones (e.g., countries and summer time) (Tu, Li, & Liu, 2009). Hence, the combination of VGI with an authoritative source raises some key issues.

Within this context, Schnebele, Oxendine, Cervone, Ferreira and Waters (2015) recommended an application to fill the gaps in remote sensing data during disasters by making use of data that was either collected from non-authoritative sources like VGI, or collected for purposes other than disaster assessment, such as that obtained from traffic cameras. Other research studies have also explored the spatio-temporal relationship between VGI provided by social media and disaster events (Sakaki, Okazaki, & Matsuo, 2010; Spinsanti & Ostermann, 2013). Furthermore, other works focus on the interoperability of environmental sensors that combine onboard storage with Web service capabilities (Kotsev, Pantisano, Schade, & Jirka, 2015). However, although these works incorporate sensor data and some forms of data produced by citizens, their approach fails to take account of the need to integrate the data provided by a citizen in different VGI platforms (e.g., OpenStreetMap, citizen observatories, and social media) with real-time environmental sensors, as a means of effectively supporting disaster management tasks.

Quality and Credibility

Despite the potential uses of VGI for disaster management, it is often accompanied by concerns about data quality (Flanagin & Metzger, 2008). These concerns arise from several factors: (i) false, inaccurate or incomplete information provided by volunteers; (ii) a lack of quality standards and systematic documentation; (iii) the limitations of a volunteer's spatial knowledge and expertise. In view of this, it is essential to assess the quality of VGI.

A quality assessment of VGI is an important stage in understanding whether an item of information is fit-for- purpose for a specific application domain (e.g., a flood management domain) (Ballatore & Zipf, 2015). Fitness-for-purpose means that an item of information should be suited to the purpose for which it will be used.

One way of evaluating VGI quality is based on manual assessment. However, manual manipulation for data validation and interpretation is time-consuming, labor-intensive and error-prone. Hence, automated methods are more suitable for handling both the amount of information and the speed with which it is provided when an extreme event occurs (e.g., a flood). In the literature, there are several methods for evaluating VGI quality (e.g., Ali & Schmid, 2014; Arsanjani, Mooney, Helbich, & Zipf, 2015; Barron, Neis, & Zipf, 2014). Among these methods, a common way of assessing the quality is to compare it by means of authoritative data. However, there are often constraints with regard to the availability or accuracy of reference data (Hung, Kalantari, & Rajabifard, 2016), as well as costs and licensing restrictions (Mooney, Corcoran, & Winstanley, 2010). An alternative solution is to use intrinsic data, as Barron et al. (2014) pointed out, or other volunteered information. With regard to the disaster management domain, Hung et al. (2016) proposed an automatic method "to assess the credibility of VGI for time-critical conditions, such as disaster response". This method combines authoritative data – including Digital Elevation Model (DEM), water resource areas, and flood risk zones – with other VGI reports provided in the same platform.

In our view, the next stage in the assessment of VGI quality is to consider other sources of volunteered information (e.g., Twitter, Flickr, Instagram, etc). This cross analysis could strengthen the credibility of the information. Thus, it is necessary to investigate how a cross-platform VGI could be used for quality assessment. It is worth noting that this is in itself a challenge since this kind of information is often regarded as insufficiently structured, documented and validated (Flanagin & Metzger, 2008). Moreover, further investigations are required on how real-time sensor data can be used to evaluate the quality of VGI. The relationship between sensor data and VGI can help to remove the uncertainty regarding the accuracy of the information provided by volunteers (Fava, 2015).

Visualization and Decision-Making Support

As well as the challenges mentioned above, most of the current visualization techniques employed for displaying VGI do not match the decision-makers' requirements (Dolif et al., 2013; Imran, Castillo, Lucas, Meier, & Rogstadius, 2014; Vieweg, Castillo, & Imran, 2014). This is largely because they are often concerned with providing a huge amount of data rather than focusing on the kind of data that really gives information about the overall situation in the areas at risk. Decision-making is also affected since there is not a clear link between the required information and the data provided by volunteers (Horita, Link, Albuquerque, & Hellingrath, 2014). This support could then provide a better understanding of how VGI can be employed in each disaster management task.

One line of inquiry in this area concerns the use of VGI in the decision-making required for disaster preparedness and emergency response by providing valuable and updated information for supporting critical tasks and activities, e.g., evacuation plans (Chae et al., 2014), heightening situational awareness (MacEachren et al., 2011; Schnebele, Cervone, & Waters, 2014), resource allocation (Kongthon, Haruechaiyasak, Pailai, & Kongyoung, 2012; Kumar & Havey, 2013), or localized event forecasting (Hashimoto, Kuboyama, Chakraborty, & Shirota, 2012; Pohl, Bouchachia, & Hellwagner, 2012). Furthermore, VGI can be employed by digital analysts for conducting an analysis of the high volume of data from the affected areas (Imran et al., 2014). However, although the existing literature has made some advances in the use of VGI for decision-making in disaster management, it still lacks research studies that investigate the relationship between VGI and the decision-makers' tasks in a more systematic way, e.g., a humanitarian logistician that wants to decide the optimal route could use data provided by Twitter (condition of roads) or OpenStreetMap (road networks). Furthermore, the current works also fail to provide decision support systems that can properly process VGI to meet the decision-makers' requirements. This is because most of them are only concerned with creating algorithms for data processing and analysis that are not necessarily aligned to the decision-makers' requirements.

In addition, it is still a challenge to decide how to provide information about data quality to different types of users (e.g., citizens, decision-makers, hydrologists, etc). The description and communication of data quality are far from having achieved satisfactory outcomes (Hunter, Bregt, Heuvelink, Bruin, & Virrantaus, 2009). While decision-makers and hydrologists may understand the results of a spatio-temporal analysis, it may be difficult for ordinary citizens since most of them have not had a formal grounding in geographic information science. Furthermore, effective communication also requires visualization tools to support the data users. However, data quality information is not normally coupled with the data itself, and this subsequently prevents visualization (Hung et al., 2016).

Another line of inquiry seeks to analyze how VGI can be either incorporated into the hydrologic modeling systems or employed to obtain quantitative indicators that are useful to support decision-making in disaster management. For example, Smith, Liang, James, and Lin, (2015) modeled floods in Newcastle, UK, by using information from Twitter which included indicators such as the rainfall data and flood depth for the city in 2012. On the other hand, Albuquerque et al. (2015) employed a methodology to identify significant messages in flood management from social networks. The study shows a relationship between these messages and relevant data for the river basin of the Elbe (Germany) for a flood that occurred in 2013. They conclude that it is possible to obtain quantitative indicators extracted from social networks, which can serve for the management of flood crisis events. However, there are no reported studies that employ these data for the operation of hydrological models in real time, which would allow an effective means of forecasting floods. VGI data can emerge from a large number of sources and players, such as experts in acquiring and processing suitable information, or simply volunteers who are keen to help. Thus once acquired and validated, they can be used for calibration or the employment of hydrological models (Furquim et al., 2014; Hirata, Giannotti, Larocca, & Quintanilha, 2015).

AGORA: A GEOSPATIAL OPEN COLLABORATIVE APPROACH FOR BUILDING RESILIENCE AGAINST DISASTERS AND EXTREME EVENTS

In this chapter, a conceptual architecture called AGORA is introduced to tackle the problems raised in the previous section, in a structured way. The acronym AGORA derives from the Greek word agora, which referred to the spot in ancient Greek cities that was the center of its athletic, artistic, spiritual, and political life (Agora, 2016). The agora of Athens is thus regarded as the birthplace of democracy. Gibbons et al. (1994) used the term agora to refer to a new mode of making science, which involved experts from different areas and also stakeholder groups from the 'public'. In this sense, the agora is a principal domain for transdisciplinary knowledge, i.e., it enables research to be carried out that cuts across disciplinary boundaries and is strongly oriented towards and driven by problem-solving – and this is particularly suited to the characteristics of applied research in Information Systems (Albuquerque, Simon, Wahoff, & Rolf, 2008). Thus, the agora metaphor can be viewed here as an image for an envisaged transdisciplinary approach that can be adopted to gather heterogeneous sources of information from organizations and individuals, which can thus be dynamically analyzed and interpreted by different experts/researchers and other stakeholders, as illustrated in Figure 1.

As can be seen in Figure 1, AGORA is composed of three layers: (i) Acquisition, (ii) Integration, and (iii) Application. The acquisition is undertaken by components responsible for collecting data provided by static and mobile sensors, mobile phones, social media and collaborative mapping. The integration allows activities to be carried out that combine, manage and qualify the information shared by the acquisition. Finally, the application aims to support the work of the decision makers of government agencies

Figure 1. Main components of AGORA

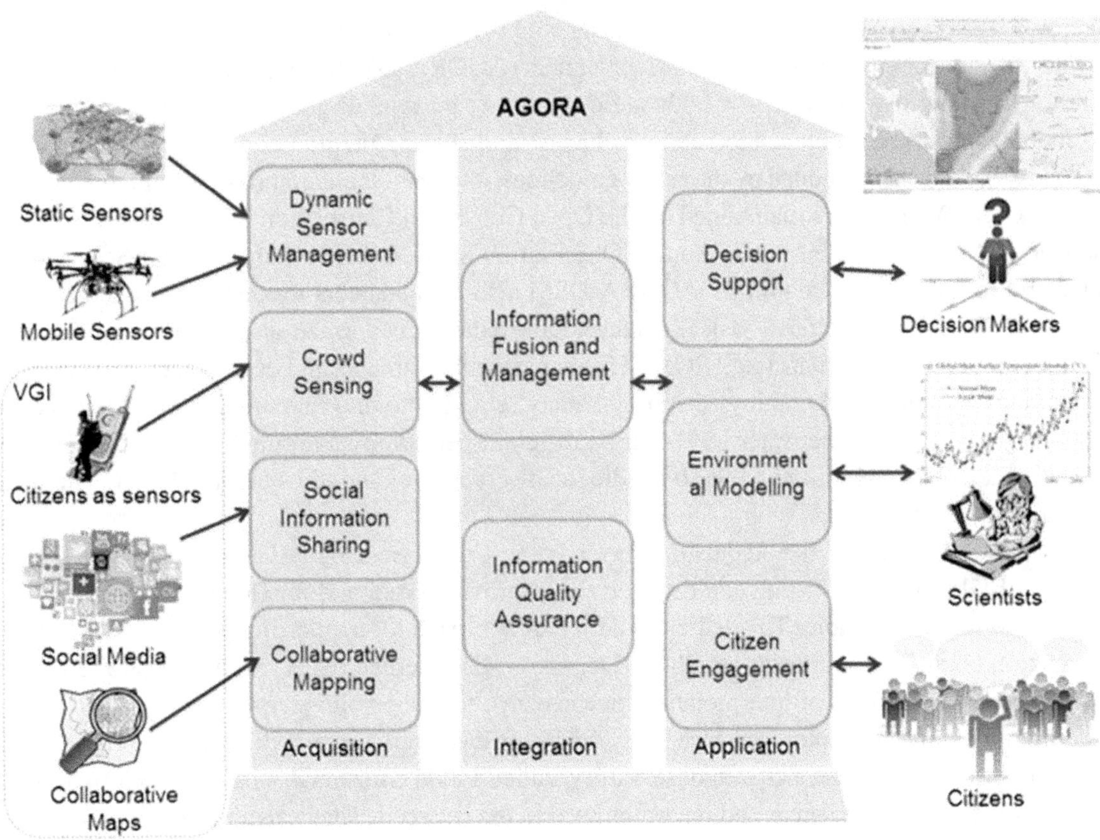

(e.g., civil defense and emergency services) and assist citizen engagement by creating risk maps, early warning systems, and social media for disaster response.

It is expected that, through this set of layers and components (which will be examined in detail in the next section), AGORA will be a knowledge management platform where key issues and data are provided by the relevant stakeholders. It will thus bring together their multiple perspectives from science, government (including local decision-makers, disaster management and emergency agencies), and the general population, to define the problems that must be solved. Specialists /researchers can also provide their expertise and interpretation of the gathered data in AGORA, on the basis of which decisions can be made. In our case, the decisions are related to building resilience against flooding, particularly with regard to funding preparatory schemes for risk mitigation, such as real-time flood monitoring, local early warning systems, and risk analysis supported by hazard, risk and situation maps.

Moreover, AGORA not only provides decision-making support interfaces for the relevant decision makers but also makes the value-added data openly available in a standardized fashion. In this manner, AGORA is able to provide a flexible ecosystem that can cope with the ever-changing needs of participatory environmental monitoring and resilience building against disasters and extreme events. The next sections will give more details about each of the layers, as well as their individual components.

Acquisition

The overall aim of the Acquisition layer is to gather data from several different data sources, and make these data available to other systems. It consists of four components: (1) dynamic sensor management, (2) crowd sensing, (3) social information sharing, and (4) collaborative mapping.

The objective of the Dynamic Sensor Management (AGORA-DSM) component is to create a middleware for handling data provided by both static sensors (e.g., rainfall gauges or hydrological stations) and mobile sensors (e.g., unmanned aerial vehicles) and sharing them in an interoperable manner. In achieving this, it is underpinned by the rules, guidelines, and open standard protocols of the Sensor Web Enablement (SWE) that are maintained by the Open Geospatial Consortium (OGC) (Bröring, Jirka, & Foerster, 2010), such as SensorML, Sensor Observation Service (SOS), Sensor Event Service (SES), and Web Processing Service (WPS). AGORA-DSM also uses adapters for encapsulating the required features of SWE. These adapters basically interpret the data source formats and convert them to uniform and interoperable data formats (e.g., SOS). This is particularly important because each data source has its own data structure and employs distinct formats (e.g., photo and numerical values) for the same observed property (e.g., the water level), and a different periodicity and timeliness of data provision. The use of adapters makes it possible to handle the discrepancies and idiosyncratic features in the data (Horita et al., 2015).

Another component of the Acquisition layer is the Crowd Sensing (AGORA-CS), which aims at gathering VGI to fill the existing gap between the information required for flood management and the information that is available. This led us to adopt an approach for supporting volunteers in supplying information about environmental variables (Degrossi, Albuquerque, Fava, & Mendiondo, 2014). The approach consists of a set of interpretative mechanisms, to assist volunteers in having a better understanding of the environmental variables (e.g., the water level in the riverbed), and a web-based citizen observatory. Volunteers use a citizen observatory, called Flood Citizen Observatory (FCO), to provide information about flooded areas and the water level in the riverbed. They carry this out by means of the interpretative mechanisms that are represented as categories in the citizen observatory. The information collected in the citizen observatory has a good chance of being an additional source of information source for AGORA-DSM, especially in areas where there is a lack of sensors (also known as ungauged areas) (Horita et al., 2015).

Social media platforms like Facebook, Twitter, Flickr, and Instagram, can also be used as a valuable source of information in areas with a lack of sensors. The Acquisition layer handles all these data by defining the Social Information Sharing (AGORA-SS) component. The purpose of this is to collect data from the social media platforms via their streaming APIs; for example, the Twitter Streaming API. These APIs include a set of functions that can be used by external systems to access the individual data of a system. Furthermore, the AGORA-SS employs different approaches for filtering the data collected from available social media platforms. This is particularly important because the huge volume of collected data makes it difficult to manually identify the relevant data to support decision-making (Vieweg et al., 2014). Finally, the filtered data is made available to other systems and components by means of a RESTful interface.

Lastly, the Collaborative Mapping (AGORA-CM) component seeks to gather information about the geographic features of disaster-affected or disaster-prone areas, using the OpenStreetMap (OSM) platform. A feature of this component is the collaborative collection of a very specific type of data, i.e., a georeferenced data about features like roads, buildings, paved areas, and points of interest (e.g., hospital, schools, warehouses etc.).

Integration

Since the objective of the Integration layer is to merge all the data provided by the Acquisition layer, it defines two essential components: (i) information fusion and management, and (ii) information quality assurance.

The first component, Information Fusion and Management (AGORA-IFM), seeks to propose ways to support the fusion of data obtained from the acquisition layer by using methods, techniques, and geospatial approaches to achieve a close relationship between two or more datasets related to the occurrence of the phenomenon. The integrated data are then shared either with another component of the integrating layer or with other systems. Moreover, AGORA-IFM also assists the fusion process by automating the activities, and making the process flexible and adaptable, so that it can be modified through service chaining. There is a wide range of possible activities in the AGORA-IFM component that include, exploiting and determining the types of relation between the data, reducing inequalities between inputs and ensuring a common structure for processing, synchronization and conversion of data and geographic coordinates, among other factors.

The Information Quality Assurance (AGORA-IQA) is the second component that composes the Integration layer. This aims at assessing the quality of VGI provided by AGORA-CS. Since the information quality depends, to a great extent, on the application domain (Bordogna, Carrara, Criscuolo, Pepe, & Rampini, 2016), i.e., on the purpose for which the information will be used, here, the VGI quality is assessed for a specific purpose, e.g., for flood management. To achieve this goal, we adopt a novel approach that combines a cross-platform VGI with authoritative data (e.g., sensor data) as reference sources. The assessment of VGI quality is an important stage since it influences the reliability of the spatial analysis outputs and decision-making process (Bordogna et al., 2016). It is important to evaluate the quality of VGI from AGORA-CS since its data will be used for other external systems or components of AGORA.

Application

The objective of the Application layer is to provide visualization and interactive features to assist the decision-makers in carrying out their tasks. The components of this layer are threefold: (i) decision support, (ii) environmental modelling, and (iii) citizen engagement.

First of all, the Decision Support (AGORA-DS) component seeks to create a bridge between the decision-makers' requirements and the data provided by the integration layer (Figure 1). This includes designing and adopting approaches that employ innovative information visualization techniques, new models for describing real-world elements and their relationships, or modern architectures with decision support systems. To achieve these objectives, we employ modelling languages and formal standards such as Business Process Model and Notation (BPMN) (OMG, 2013) and Decision Model and Notation (DMN) (OMG, 2014) that are maintained by the Object Management Group (OMG), as well as the Observation & Measurements (O&M) from the Open Geospatial Consortium (OGC) (OGC, 2011). These languages and standards are particularly important for describing the decision-makers' requirements.

The visualization of the information is carried out to create digital (geo-) dashboards as a SDSS (Horita et al., 2015; Horita, Fava, Mendiondo, & Albuquerque, 2014). These dashboards are a type of information system, which is available for use on a single screen. They provide the most useful management information that is needed to support strategic activities, such as goal setting, planning and forecasting, and tracking performance in strict accordance with organizational objectives. In addition, they can be understood as comprising a type of information system that is usually used to support decisions at the strategic level of organizations. There are several essential features that these digital dashboards must include to achieve their objectives. First, they must be customizable and adaptive so that they can carry out the specific objectives of their users, e.g., the managers. Furthermore, they are characterized by their use of drill-down capabilities that provide quick access and allow them to identify the root cause of a problem (Vasiliu, 2006). Finally, they must provide an overview of everything that is currently going on in the business by means of simple charts, indicators, and forecasting capabilities (Vasiliu, 2006). The effectiveness of these approaches is related to the integrated data that are shared by the integration layer. This means that the adoption of approaches for integrating data is beyond the scope of this component, which is an essential feature of AGORA-IFM.

Secondly, the aim of the Environmental Modelling component (AGORA-EM) is to develop an approach for incorporating VGI in flood forecasting and early warning systems. This approach relies on traditional hydrological models for flood forecasting and uses VGI as input data. It is concerned with how VGI can be used to reduce the predictive uncertainty of hydrological models (Fava, Abe, Restrepo-Estrada, Mendiondo, & Souza, 2015). Hence, VGI is retrieved from two information sources, i.e. Flood Citizen Observatory and social media, which increases the amount of available information and allows decisions to be made quickly in highly frequent flood events, which affect the people living near to watersheds.

Finally, the Information Citizen Engagement (AGORA-CE) component aims to overcome some of these challenges by adopting a model-based approach to structure the collaborative mapping activities and to improve the situational awareness of the people in a crisis management domain. The focus of the AGORA-CE will be on OpenStreetMap (OSM), the most significant project in the field of collaborative mapping. OSM has emerged, particularly in developing countries, as a way of overcoming the problem of a lack of base maps, since georeferenced data – streets, roads, buildings, etc – are provided in the form of a map by the digital volunteers. According to Zook, Graham, Shelton and Gorman (2010), the greatest benefit of collaborative mapping is that a large number of maps can be produced in a short period of time. However, disaster managers are still often uncertain about what they can expect from this collaborative work, since it is not clear how volunteers can be organized and how they can become more involved.. In addition, the collaboration of volunteers – single individuals and volunteer technical communities (VTCs) – with traditional response agencies is of crucial importance if it is to have a real impact on disaster management. Thus, the AGORA-CE is the main feature of AGORA that can make it possible to understand how formal humanitarian organizations and different volunteer communities can collaborate with each other to support decision-making in disaster management.

CASE STUDIES AND LESSONS LEARNED

This section describes the case studies, which were carried out within the context of AGORA. It first examines the lessons that can be learned from the use of VGI and Wireless Sensor Networks (WSN) for flood risk monitoring in the city of São Carlos, Brazil (Case Study 1). Following this, it gives details about the use of social media messages for supporting rainfall detection in the city of São Paulo, Brazil (Case Study 2).

Case Study 1: The Use of VGI and WSN for Flood Risk Monitoring in Brazil

Study Area

For Case Study 1, we selected the city of São Carlos/SP as the study area. São Carlos is located 230 km away from São Paulo in Brazil and has been built around the Monjolinho catchment area. Figure 3 shows the creek of Monjolinho, Santa Maria Madalena, Mineirinho, and Tijuco Preto, which are all located in the town center of São Carlos. Two critical flood zones (the blue shaded areas in Figure 2) were highlighted at the intersection of the Santa Maria Madalena and Monjolinho creeks, and at the intersection of Mineirinho and Monjolinho creeks. These regions have a large number of inhabitants living in residential buildings and houses, and are often affected by floods (Mendes & Mendiondo, 2007).

Owing to the problem of flood hazards, the areas are often monitored by a WSN installed along the riverbed (Hughes et al., 2011). The network has four sensor nodes that continuously provide measurements of the water level (i.e., the numeric value) from their location every five minutes. In a single day, this represents a volume of 288 measurements per sensor node and 1,152 measurements for the whole WSN. Furthermore, the citizens provided reports related to the water level from different parts of the study area by adopting a crowdsourcing-based approach (Figure 3) (Degrossi et al., 2014). This approach consists of a web-based citizen observatory – called "Flood Citizen Observatory" – for data collection and visualization, as well as interpretation mechanisms installed in the riverbed to enable the volunteers to have a better understanding of the environmental variables (Figure 4). The aim of the interpretative mechanisms is to enable volunteers to interpret an environmental variable, i.e. water level in the riverbed, since most volunteers are not specialists. They were designed by an interdisciplinary team comprising hydrologists and computer scientists. Three different scenarios were examined for their design: (i) a controlled point, (ii) a semi-controlled point and (iii) an uncontrolled point. The scenarios vary according to the interpretative mechanism – i.e., a water level ruler, a puppet similar to the human figure and multi-colored bands – that are available in the locality. The mechanisms were incorporated

Figure 2. The study area in São Carlos/SP, Brazil
(Adapted from Horita et al., 2015)

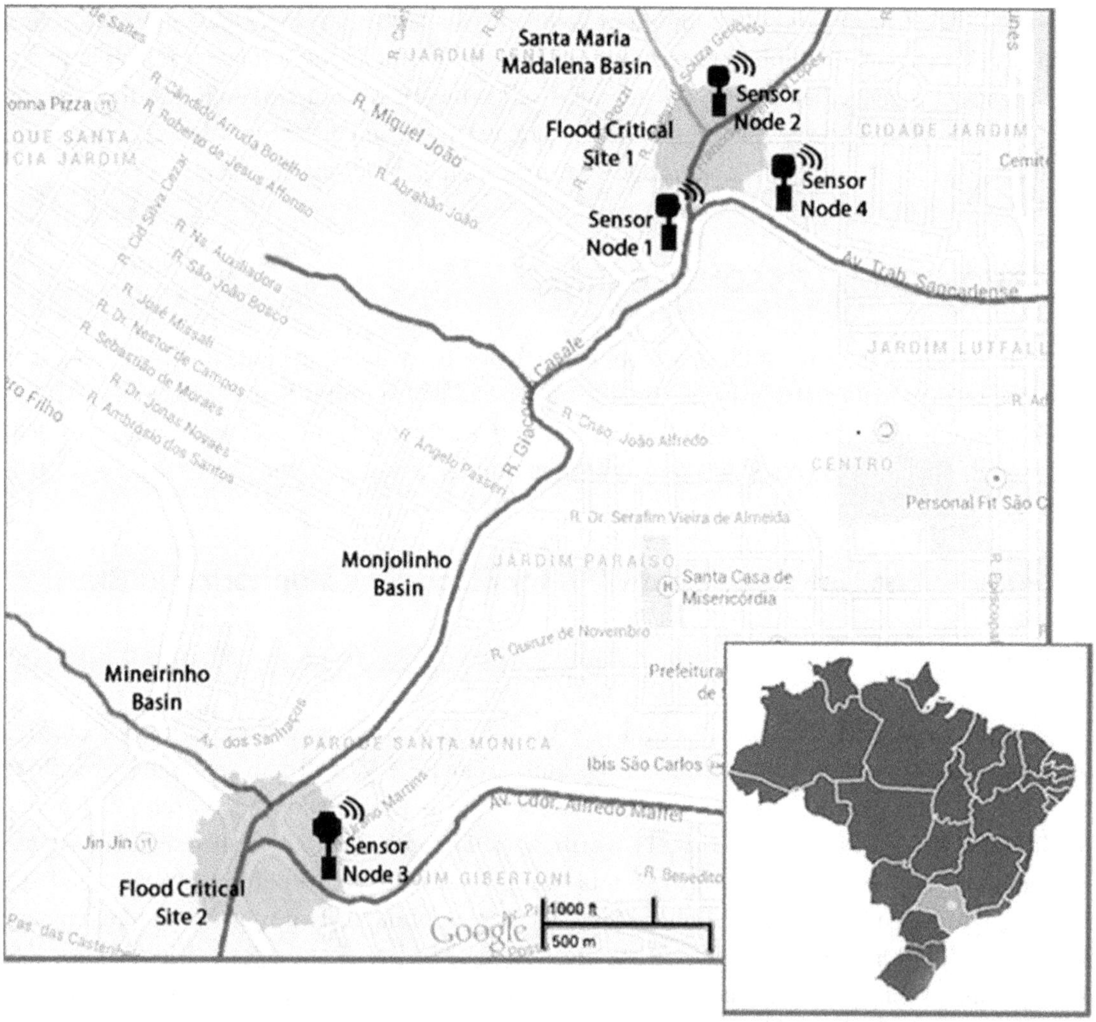

in the crowdsourcing platform, called a Flood Citizen Observatory (Figure 3), to facilitate the provision of information about flood risks.

On the basis of the visual analysis of the interpretative mechanisms, a citizen can share the water level at a location by sending a structured measurement (e.g., 60 cm), or semi-structured measurements (e.g., the knee of the puppet). Owing to doubts regarding VGI and to improve its credibility and quality, all the information is manually checked by the administrator of the Flood Citizen Observatory before it is shown on the Web or made accessible through the Ushahidi REST API (Ushahidi, 2016). The purpose of this task is to reduce the amount of false or inaccurate information disseminated to the public and emergency services.

Finally, a web-based decision support tool, called "GeoDashboard" (Figure 5), displays the integrated information provided by the volunteered information and the in-situ sensors on the map. The purpose of

Figure 3. Flood Citizen Observatory

this is to support flood risk management. Specialist models like "Hazard Index" (Rotava, Mendiondo, & Souza, 2013), are also used on the GeoDashboard with the aim of providing more valuable information that can assist environmental monitoring and thus enhance the decision-making.

Lessons Learned

In the AGORA-DSM, we developed an approach for near real-time integration of data provided by stationary and mobile sensors. This approach defines a dynamic sensor management component that aims at handling simple message protocols sent by the stationary or mobile sensors and delivering them to the integration layer via the sensor bus (Assis et al., 2016). By means of this component, the approach is able to provide a standardized form of interoperability, as well as a low latency for publishing data on the Web, which is crucial for applications requiring near real-time data publication such as disaster management. The evaluation that was carried out through realistic scenarios, provided evidence that the approach is efficient in terms of interoperability and performance. Furthermore, the AGORA-DSM was employed for integrating messages collected from social media platforms and sensor data streams

Figure 4. Interpretative mechanisms for assisting volunteers

Figure 5. GeoDashboard
(Adapted from Horita et al., 2015)

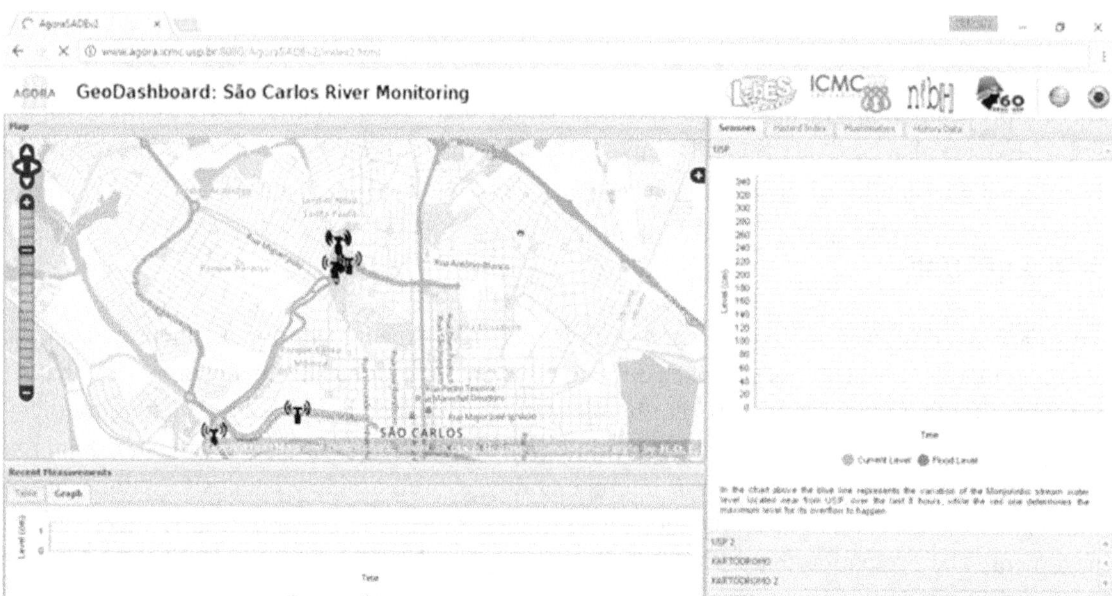

in near real-time scenarios, such as flood risk management (Assis, Herfort, Steiger, Horita, & Albuquerque, 2015). This integration was essential for prioritizing the collected messages and thus finding out those that were flood-related messages, which is particularly important when dealing with the huge volume of data produced by social media platforms. The findings of a case study in Brazil highlighted that AGORA-DSM is also useful for managing data provided by social media platforms (esp. Twitter) and sensor data streams (esp. rainfall gauges from Cemaden) in near real-time.

With regard to the results obtained from AGORA-CS, we developed a crowdsourcing-based approach for gathering useful volunteered geographic information in the flood management domain (Degrossi et al., 2014). An item of information is considered useful if hydrologists and decision-makers can use it in hydrological models and for decision-making, respectively. After several controlled experiments conducted, the results revealed that VGI is, on average, comparable to sensor data (Degrossi et al., 2014; Moreira, Degrossi, & Albuquerque, 2015). It can thus be concluded that VGI is a potential source of information for flood management. However, the use of manual manipulation for data validation and interpretation is time-consuming, labor-intensive and error-prone, especially during flood events, when a large amount of information is provided. In view of this, an automatic method is being developed for quality assessment.

We have also designed a service-oriented architecture (SOA), called AGORA-IFM (Castanhari, Rocha, Andrade, & Albuquerque, 2016), to support the integration of sensor data and VGI. This architecture relies on the AGORA-DSM to obtain sensor data, and the Flood Citizen Observatory to obtain VGI. Finally, these data are encoded in Observations and Measurements (O&M), a standard that specifies an XML implementation for the OGC conceptual model, and includes a schema for Sampling Features (OGC, 2011). The composition of the architectural components allows sensor data to be integrated with VGI by using several algorithms in a flexible and automated manner. The architecture was implemented by means of a prototype as a proof-of-concept and the results were used to create thematic maps. These maps can heighten flood risk awareness and support decision-making in the flood risk management domain.

The results of the case study also revealed the usefulness of integrating heterogeneous data sources (esp. WSN and VGI) to support decision-making in flood risk management. Indeed, VGI can provide information from areas of the river basin, which lack data since there is no appropriate station in the area. The evidence also showed that VGI could supplement authoritative data by expanding the monitoring of the vulnerable areas of the river basin. It is also worth mentioning that AGORA-DS is able to combine information provided by WSN and VGI, and provide useful information for supporting flood risk management.

Furthermore, there was evidence that showed that the connection of decision-making to VGI is an important means of understanding how this volunteered information can supply decision-makers with more appropriate information, and thus enhance decision-making in disaster management. For this reason, we developed a model-based framework that employs BPMN, DMN, and O&M as a basis for establishing a connection between the decision-makers' tasks to data sources. This model was called "observation-aware Decision Model and Notation" (oDMN) (Horita, Link, Albuquerque, & Hellingrath, 2016). By means of oDMN, decision-makers are able (i) to determine the impact of incoming observations on relevant tasks and decisions, (ii) to understand how emerging data sources are useful for their business processes, iii) to widen their understanding of existing activities, and (iv) to assist in targeted data collection. An initial evaluation case study focused on humanitarian logistical operations during the 2015 Nepal earthquake and the emergency response showed that oDMN is indeed able to connect tasks, decisions, information and data sources, and thus support better decision-making (Horita, Link, Albuquerque, & Hellingrath, 2016).

On the basis of the outcomes of the initial evaluation case, we are extending oDMN and applied it to the context of the monitoring and early-warning systems of Cemaden (Brazil) using data collected in a qualitative analysis (Horita, Albuquerque, Marchezini, & Mendiondo, 2016). Apart from being able to connect decision-making and heterogeneous data sources, the findings of this application case showed that oDMN enables decision-makers to identify critical decisions, as well as to understand how required information can be combined when seeking to identify a more suitable alternative. In other words, decision-makers can understand the data sources that are useful for their decisions. Furthermore, they are also able to find out if their decisions are based on a lack of appropriate data sources. Similarly, when a new data source emerges, they can also discover if it is important for a particular decision, and then communicate this to the people responsible for taking action.

Case Study 2: The Use of Social Media Messaging Patterns for Rainfall Detection in São Paulo

Study Area and Description of the Datasets

The case study was carried out in São Paulo, Brazil, where heavy rainfall often affects the city's infrastructure and citizens. Figure 6 shows the scenario with the spatial distribution of active rainfall gauges and rainfall-related Twitter messages in January 2016. The rainfall-related tweets are closer to the regions with rainfall gauge data than the regions without rainfall gauge data. In the last few years, there has been strong evidence of the spatial relationship between social media messages and the proximity of targeted events (Steiger, Albuquerque, & Zipf, 2015). However, we seek to address the problem of the temporal validity of social media data - that is, to establish the correlation between social media activity and the different phases of targeted events in real-time, within a short time-scale (e.g., 10 minutes). Thus, we investigated if there is a temporal relationship between the rainfall gauge time-series and the rainfall-related social media messages.

Our social media dataset contains 243,333 georeferenced tweets retrieved by Twitter Streaming API from January 1st to January 30th 2016. All the tweets were geotagged within the administrative borders of the city of São Paulo. As well as this, the geotagged tweets were filtered with the aid of keywords and synonyms (including stem words and wildcards). The keywords were in Brazilian-Portuguese and chosen from a list of keywords used in one of our previous studies (Assis et al., 2015). After this, we carried out a manual search for the subset of rainfall-related tweets to remove the false-positives. In this context, the false-positives are tweets that contain one or more keywords, but do not fall within the context. Furthermore, we removed all the retweet messages. In our case study, only 0.25% (620 tweets) were designated as rainfall-related tweets.

The rainfall data were obtained from Cemaden through a REST API. There were 81 active rainfall gauges in São Paulo during the data collection period. The rainfall data were measured in millimeters and with two sizes of temporal window: 10 minutes when it was raining and 60 minutes otherwise. We only used reliable rainfall gauges and removed those that supplied no data throughout the month and those, which registered negative values. In total, we removed 21 rainfall gauges.

We generated two time-series with a time-scale of 10 minutes, i.e., one for rainfall gauge data and another for rainfall-related tweets. The tweets from the rainfall-related time-series corresponded to the

Figure 6. Map of the distribution of active rainfall gauges and rainfall-related tweets in January 2016, in the city of São Paulo, Brazil

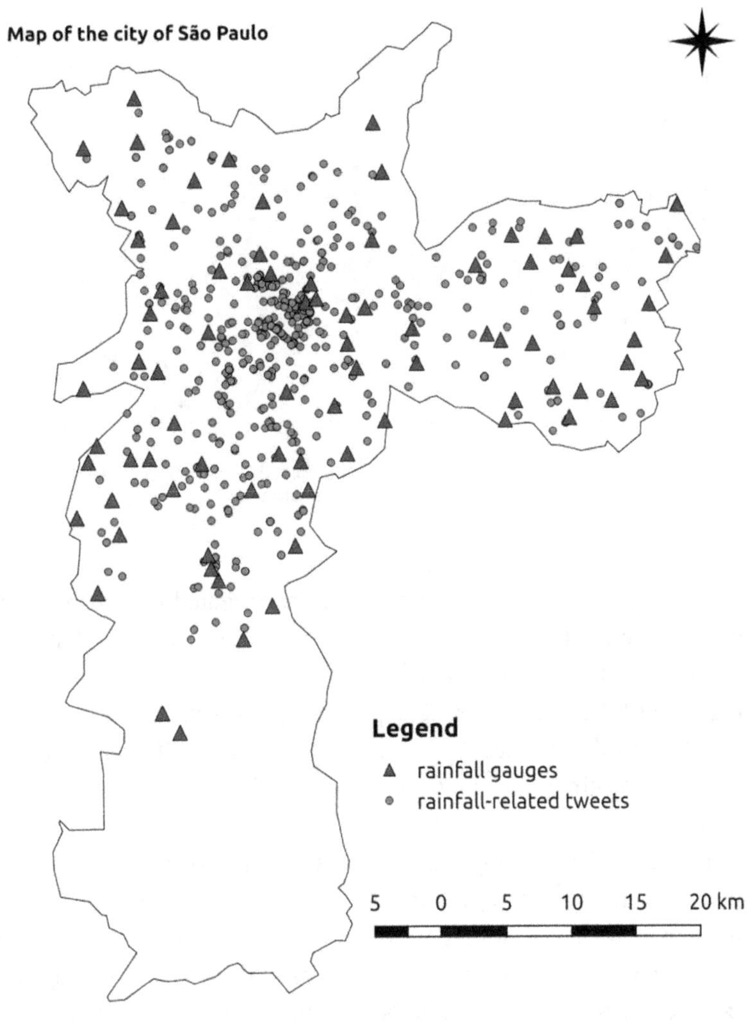

absolute frequency of rainfall-related tweets during the period of the case study, whereas the rainfall time-series were rainfall gauge data interpolated at the centroid of the sensors by means of the Inverse Distance Weighting method.

Lessons Learned

Social network users often report messages related to rainfall episodes. That is, they post texts, photos and videos, such as gray clouds, drizzle, rainfall and floods, and on rare occasions, give a forecast of rain. These posts appear at different times and thus their temporal scale and frequency are not known. Indeed, they may occur at the beginning, middle or end of the rainfall. On the other hand, there is a well-known temporal scale for rainfall data. Figure 7 depicts this temporal feature.

Figure 7. The pattern of increasing rainfall and frequency of rainfall-related tweets in January 2016

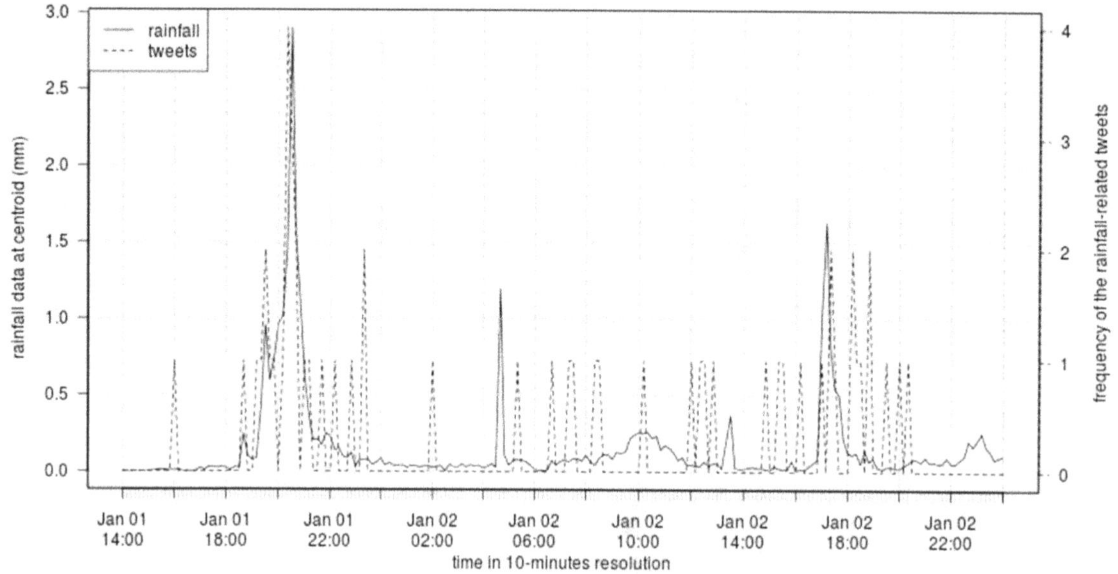

In the AGORA-IFM, spatio-temporal patterns were investigated on the basis of the authoritative data and social media messages. Thus, our assumption in the case study is that the temporal relationship between authoritative data and event-related social media messages can be explained by duration, frequency and the lag-time of the time-series. The preliminary results provide evidence that the rainfall gauge time-series and the rainfall-related tweets are occurring in different lag-times (Figure 8), but they are closely linked to a lag-time that ranges from -1 to +1 lag-times. In this case study, 1 lag-time represents 10 minutes.

In view of this, our results can be applied to detect patterns of rainfall events in real-time using authoritative data and social media messages, since the individual data sources can be represented as a single time-series. Moreover, the lag-times are very useful for monitoring activities where the time-scale and frequency of the data can change over time, such as rainfall events. For example, when the analyst chooses a short time-scale, the peaks of rainfall data and rainfall-related social media activity can be very close. Thus, it is not clear to a decision-maker whether or not the social network activity is useful for decision-making. However, our results show that the rainfall time-series patterns can be approximated by social network messages when full account is taken of the duration, frequency, and lag-time. In addition, the findings suggest that the lag-time can be useful to fit or approximate the best temporal scale between two time-series in any setting (e.g., real-time, near real-time and a post-hoc analysis).

FUTURE DIRECTIONS: AN INTERDISCIPLINARY RESEARCH AGENDA

With the aid of the conceptual architecture of AGORA and the case studies that were carried out, this chapter has been able to show the potential value of VGI for those who are coping with the effects of disaster events. Moreover, it also provided evidence that VGI can enable preventive measures to be taken that can make urban areas more resilient when faced with natural hazards and extreme events.

Figure 8. Cross-correlation between rainfall data time-series and rainfall-related tweets with a 10-minute time-scale in January 2016

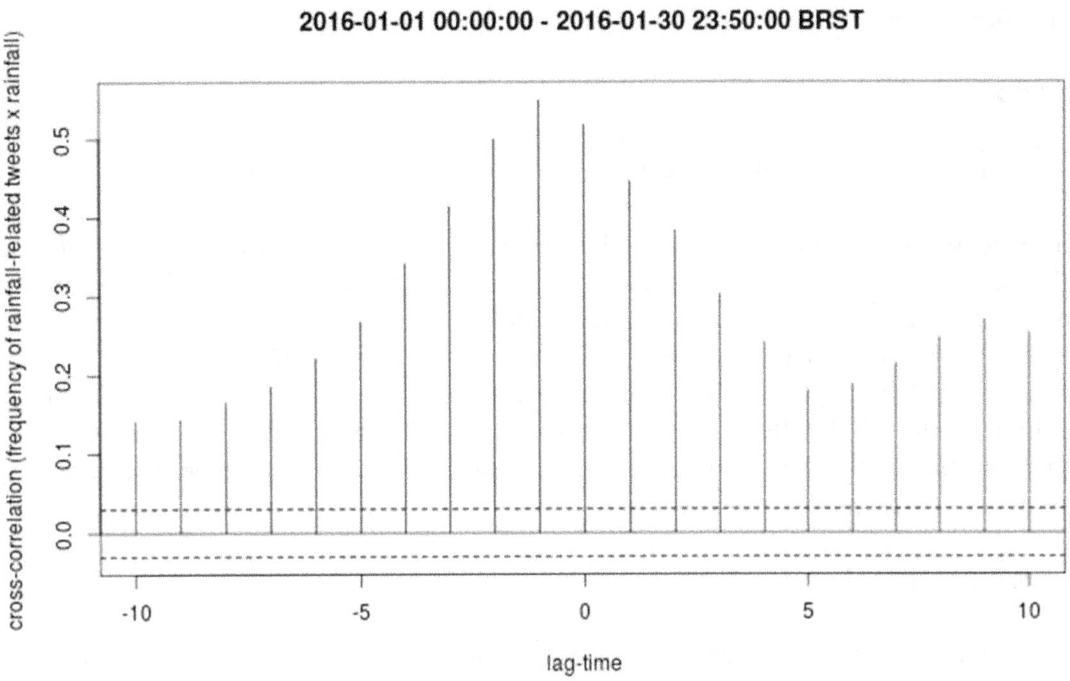

However, there are still important lines of inquiry to be pursued in the future to allow VGI to achieve its full potential in the context of disaster management. In the light of this chapter concludes by setting out a research agenda below that is structured into the three layers mentioned above: acquisition, integration and application.

Acquisition

Despite the advances made in this field, further investigations are still required on how to effectively obtain VGI (e.g., in citizen observatories or social media) to assist those involved in disaster management tasks. This is particularly challenging because of the sheer amount of data available and the lack of automated methods to extract relevant pieces of information. Thus, methods are still needed to automatically retrieve and filter VGI streams on the basis of the identified information required for disaster management tasks.

Another challenges facing the application of VGI in disaster management is how to encourage citizens to provide valuable information (Albuquerque et al., 2016). This will require approaches that go beyond considering citizens as "sensors", in the sense of being mere "data providers", but that fully take into account of their concerns by adopting genuinely participatory approaches.

Moreover, it is still necessary to extend the direction initiated by the AGORA-CM component so that data can be collected from OpenStreetMap. Currently, there are approaches that rely on different ways of extracting this data without any systematic method. The designing of an architecture that allows interoperability with other types of data and complies with open geospatial standards, as defined by the Open Geospatial Consortium (OGC, 2016), will be if great value, since it allows interoperability between different applications.

Integration

Information Fusion and Management

Further studies on understanding how VGI can be used individually or combined with authoritative data are still required. For this reason, there is a real need for, research into the relationships between these data sources in the light of future challenges, such as how to make the integration process more robust and flexible. Future works should thus go beyond simply conducting an exploratory analysis (e.g., like the one carried out in Section 4.2) and systematically investigate other social networking platforms and regions, to enable the AGORA-IFM approach to be generalized. Thus, a key future direction for AGORA-IFM is to adopt an approach that can integrate VGI with social media and citizen observatories so that it can be used a "virtual rainfall gauge".

Semantic Enrichment and Fusion

Since disaster-related scientific data are multidisciplinary by nature, and comprise entities from different data sources, there is a growing trend to adopt Linked Open Data (LOD) that can be used to semantically describe the meaning of the available data. This open up further opportunities for leveraging VGI in disaster management, particularly because this information is often shared and lacks any features that can describe its meaning (e.g., meta-data). An SDI-Node architecture composed of a GeoNode (GeoNode, 2016) and GeoNetwork (OSGeo, 2016) could be used for addressing this issue. GeoNode acts as a main data management gate and a metadata search tool. Furthermore, GeoNetwork is a standard decentralized spatial information management system, designed to enhance spatial information exchange and allow information to be shared between organizations and their audience.

Information Quality Assurance

Heterogeneous sources of volunteered geographic information have been used in disaster management to supplement or replace authoritative sources. However, owing to its nature, VGI raises concerns about its quality. Hence, different strategies are required to assess its quality because each source provides a different type of information (Craglia, Ostermann, & Spinsanti, 2012). We are developing an approach to assess the quality of VGI from citizen observatories for the purpose of flood management. This approach is based on the assessment of VGI quality based on information about the geographic context, e.g., distance to water resource areas. The baseline information comes from other VGI and authoritative sources (e.g., sensors). However, further research is still necessary to investigate how the integration of different types of volunteered information and authoritative data affects the quality of the resulting information.

Application

Decision Support

Further research is still necessary to find a way to improve the use of the volunteered data. First of all, data processing and visualization features of decision support systems should match the decision-makers' information requirements, i.e., the design of these systems must incorporate these requirements. Another line of inquiry should explore the effects of different visualization approaches to VGI in terms of "decision time" and "decision accuracy". This is because VGI tends to be shared without following any standard and this poses a number of challenges in data processing. At the same time, decision-makers often do not have a clear understanding of what information they might need to carry out a task until they face it. Finally, a set of design principles might provide a solid foundation for employing VGI more effectively as a useful data source for spatial decision support systems.

Environmental Modeling

In future studies, the use of VGI for supporting environmental modeling should seek to design a forecasting model that is able to handle the uncertainties arising from its variables and parameters (e.g., water flow, lag time, or watershed slope) on the basis of the information provided by volunteers. Depending on the environmental variables, VGI can provide more accurate and updated information than traditional sensors. For example, by visually observing the conditions of a river, citizens can share data about its water level. This in turn could be linked to a variable from a hydrological model (e.g., Probability Distributed Model (PDM)) and supplement the traditional data employed for forecasting a flood. The decision-makers would thus be supported with better forecasting information. Another line of inquiry might be the use of VGI for refining the estimated values of the hydrological model when addressing the question of the water level.

Citizen Engagement

Further research is still necessary about the best way to involve volunteers in supplying high-quality information for disaster management. For instance, how can we ensure that local communities are, to a meaningful extent, involved in different phases of a disaster. This can be achieved, for instance, by forming a community group or by providing training for volunteers on how to supply high-quality VGI. Moreover, different VGI categories – social media, crowd sensing, and collaborative mapping activities – require different strategies for encouraging citizen engagement. This study rests on the assumption that knowledge of the VGI categories is important for disaster managers as it can enable them to recruit users and encourage them to make use of VGI-systems. Thus, one important direction for future research endevors is to establish a holistic framework for fostering the engagement of digital volunteers in the context of disaster management, with a particular focus on floods.

ACKNOWLEDGMENT

The authors would like to express thanks for the financial support provided by CAPES (Grant No. 88887.091744/2014-01) and the FAPESP-IVA project "Assessment of Impacts and Vulnerability to Climate Change in Brazil and Strategies for Adaptation Options" (Grant no. 2008/58161-1). João Porto de Albuquerque acknowledges financial support from FAPESP (Grant no. 2011/23274-3), CAPES (Grant no. 12065-13-7), and Heidelberg University (Excellence Initiative II/Action 7). Flávio Eduardo Aoki Horita is grateful for the financial support from CNPq (Grant no. 202453/2014-6). Lívia Castro Degrossi is grateful for the financial support from CNPq (Grant no. 201626/2015-2). Roberto dos Santos Rocha is grateful for the financial support from European Community's Seventh Framework Programme: Marie Curie Actions/Initial Training Networks (Grant no. 317382). Sidgley Camargo de Andrade would like to thank the Araucária Foundation of Supports Scientific and Technological Development in the State of Paraná (FAPPR) and State Secretariat of Science, Technology and Higher Education of Paraná (SETI) for their financial support. Camilo Restrepo-Estrada is grateful for the financial support from CAPES. The authors are also grateful to the editors, Cláudio Campello, Michela Bertolotto, and Padraig Corcoran, as well as to the anonymous reviewers for their helpful suggestions.

REFERENCES

AGORA. (2016). *AGORA - A Geospatial Open collaboRative Architecture for Building Resilience against Disasters and Extreme Events*. Retrieved from http://www.agora.icmc.usp.br/site/agora/

Agora. (2016). *Wikipedia, The Free Encyclopedia*. Retrieved from https://en.wikipedia.org/w/index.php?title=Agora&oldid=746994959

Albuquerque, J. P., Eckle, M., Herfort, B., & Zipf, A. (2016). Crowdsourcing geographic information for disaster management and improving urban resilience: an overview of recent developments and lessons learned. In C. Capineri, M. Haklay, H. Huang, V. Antoniou, J. Kettunen, F. Ostermann, & R. Purves (Eds.), *European Handbook of Crowdsourced Geographic Information* (pp. 309–321). Ubiquity Press. doi:10.5334/bax.w

Albuquerque, J. P., Herfort, B., Brenning, A., & Zipf, A. (2015). A geographic approach for combining social media and authoritative data towards identifying useful information for disaster management. *International Journal of Geographical Information Science*, *29*(4), 667–689. doi:10.1080/13658816.2014.996567

Albuquerque, J. P., Simon, E. J., Wahoff, J.-H., & Rolf, A. (2008). The Challenge of Transdisciplinarity in Information Systems Research. In *Information Systems Research Methods, Epistemology, and Applications* (Vol. 28, pp. 88–102). IGI Global. https://doi.org/10.4018/978-1-60566-040-0.ch005

Ali, A. L., & Schmid, F. (2014). Data Quality Assurance for Volunteered Geographic Information. In *Geographic. Information Science*, *8728*, 126–141.

Arsanjani, J. J., Mooney, P., Helbich, M., & Zipf, A. (2015). An Exploration of Future Patterns of the Contributions to OpenStreetMap and Development of a Contribution Index. *Transactions in GIS*, *19*(6), 896–914. doi:10.1111/tgis.12139

Assis, L. F. F. G., Herfort, B., Steiger, E., Horita, F. E. A., & Albuquerque, J. P. (2015). Geographical prioritization of social network messages in near real-time using sensor data streams: an application to floods. In *Proc. of the XVI Brazilian Symposium on Geoinformatics (GEOINFO 2015)* (pp. 26–37). Campos do Jordão: MCTI/INPE.

Ballatore, A., & Zipf, A. (2015). A Conceptual Quality Framework for Volunteered Geographic Information. In *Proceedings of the XII Conference on Spatial Information Theory*. Santa Fe, NM: COSIT. doi:10.1007/978-3-319-23374-1_5

Barron, C., Neis, P., & Zipf, A. (2014). A Comprehensive Framework for Intrinsic OpenStreetMap Quality Analysis. *Transactions in GIS*, *18*(6), 877–895. doi:10.1111/tgis.12073

Bordogna, G., Carrara, P., Criscuolo, L., Pepe, M., & Rampini, A. (2016). On predicting and improving the quality of Volunteer Geographic Information projects. *International Journal of Digital Earth*, *9*(2), 134–155. doi:10.1080/17538947.2014.976774

Bröring, A., Jirka, S., & Foerster, T. (2010). The Sensor Bus – Integrating Geosensors and the Sensor Web. In *Proceedings of the 2nd Open Source GIS UK Conference (OSGIS)* (pp. 2–5). Nottingham, UK: OGC.

Castanhari, R. E. S., Rocha, R. dos S., Andrade, S. C., & Albuquerque, J. P. (2016). A Software Architecture to Integrate Sensor Data and Volunteered Geographic Information for Flood Risk Management. In *Proceedings of the 13th International Conference on Information Systems for Crisis Response and Management*. Rio de Janeiro, Brazil: ISCRAM.

Chae, J., Thom, D., Jang, Y., Kim, S., Ertl, T., & Ebert, D. S. (2014). Public behavior response analysis in disaster events utilizing visual analytics of microblog data. *Computers & Graphics*, *38*(1), 51–60. doi:10.1016/j.cag.2013.10.008

Craglia, M., Ostermann, F., & Spinsanti, L. (2012). Digital Earth from vision to practice: Making sense of citizen-generated content. *International Journal of Digital Earth*, *5*(5), 398–416. doi:10.1080/17538947.2012.712273

Assis, L. F. F. G., Behnck, L. P., Doering, D., de Freitas, E. P., Pereira, C. E., & Horita, F. E. A. … Albuquerque, J. P. de. (2016). Dynamic Sensor Management: Extending Sensor Web for Near Real-Time Mobile Sensor Integration in Dynamic Scenarios. In *Proceedings of the 30th International Conference on Advanced Information Networking and Applications (AINA 2016)* (pp. 303–310). IEEE. https://doi.org/10.1109/AINA.2016.100

Degrossi, L. C., Albuquerque, J. P., Fava, M. C., & Mendiondo, E. M. (2014). Flood Citizen Observatory: a crowdsourcing-based approach for flood risk management in Brazil. In *Proc. of the 26th Int. Conf. on Software Engineering and Knowledge Engineering (SEKE 2014)* (pp. 1–6). Vancouver, Canada: Knowledge Systems Institute Graduate School.

Dolif, G., Engelbrecht, A., Jatobá, A., da Silva, A. J. D., Gomes, J. O., & Borges, M. R. S. … de Carvalho, P. V. R. (2013). Resilience and brittleness in the ALERTA RIO system: A field study about the decision-making of forecasters. *Natural Hazards*, *65*(3), 1831–1847. https://doi.org/10.1007/s11069-012-0448-2

Fava, M. C. (2015). *Modelo de Alerta Hidrológico com Base Participativa usando Sistema de Informações Voluntárias para Previsão de Enchentes* (Masters Thesis). São Carlos School of Engineering.

Fava, M. C., Abe, N., Restrepo-Estrada, C., Mendiondo, E. M., & Souza, V. C. B. (2015). Redução de incertezas na previsão de enchentes com uso de dados colaborativos. In *Proceedings of the XXI Simpósio Brasileiro de Recursos Hídricos* (pp. 1–8). Brasilia, DF: ABRH.

Flanagin, A. J., & Metzger, M. J. (2008). The credibility of volunteered geographic information. *GeoJournal, 72*(3-4), 137–148. doi:10.1007/s10708-008-9188-y

Furquim, G., Neto, F., Pessin, G., Ueyama, J., Albuquerque, J. P., & Clara, M., … Braun, T. (2014). Combining wireless sensor networks and machine learning for flash flood nowcasting. In *Proceedings of the IEEE 28th International Conference on Advanced Information Networking and Applications Workshops, IEEE WAINA 2014* (pp. 67–72). Victoria, Canada: IEEE. doi:10.1109/WAINA.2014.21

GeoNode. (2016). *GeoNode*. Retrieved from http://geonode.org/

Gibbon, D. (1994). *The new production of knowledge: the dynamics of science and research in contemporary societies*. London, UK: SAGE Publications, Inc.

Goodchild, M. F. (2007). Citizens as sensors: The world of volunteered geography. *GeoJournal, 69*(4), 211–221. doi:10.1007/s10708-007-9111-y

Goodchild, M. F., & Glennon, J. A. (2010). Crowdsourcing geographic information for disaster response: A research frontier. *International Journal of Digital Earth, 3*(3), 231–241. doi:10.1080/17538941003759255

Harvey, F. (2013). To volunteer or to contribute locational information? Towards truth in labeling for crowdsourced geographic information. In *Crowdsourcing Geographic Knowledge: Volunteered Geographic Information (VGI) in Theory and Practice* (Vol. 9789400745, pp. 31–42). Dordrecht: Springer Netherlands. doi:10.1007/978-94-007-4587-2_3

Hashimoto, T., Kuboyama, T., Chakraborty, B., & Shirota, Y. (2012). Discovering topic transition about the east japan great earthquake in dynamic social media. In *Proceedings of the IEEE Global Humanitarian Technology Conference (GHTC 2012)* (pp. 259–264). IEEE. https://doi.org/10.1109/GHTC.2012.42

Hirata, E., Giannotti, M. A., Larocca, A. P. C., & Quintanilha, J. A. (2015). Flooding and inundation collaborative mapping - use of the Crowdmap/Ushahidi platform in the city of Sao Paulo, Brazil. *Journal of Flood Risk Management*. https://doi.org/10.1111/jfr3.12181

Horita, F. E. A., Albuquerque, J. P., Degrossi, L. C., Mendiondo, E. M., & Ueyama, J. (2015). Development of a spatial decision support system for flood risk management in Brazil that combines volunteered geographic information with wireless sensor networks. *Computers & Geosciences, 80*, 84–94. doi:10.1016/j.cageo.2015.04.001

Horita, F. E. A., Albuquerque, J. P., Marchezini, V., & Mendiondo, E. M. (2016). A qualitative analysis of the early warning process in disaster management. In *Proceedings of the 13th International Conference on Information Systems for Crisis Response and Management (ISCRAM)* (pp. 1–9).

Horita, F. E. A., Degrossi, L. C., Assis, L. F. F. G., Zipf, A., & Albuquerque, J. P. (2013). The use of Volunteered Geographic Information and Crowdsourcing in Disaster Management: a Systematic Literature Review. In *Proceedings of the Nineteenth Americas Conference on Information Systems* (pp. 1–10). Chicago, IL: AIS.

Horita, F. E. A., Fava, M. C., Mendiondo, E. M., Rotava, J., Souz, V. C., Ueyama, J., & Albuquerque, J. P. (2014). AGORA-GeoDash: A Geosensor Dashboard for Real-time Flood Risk Monitoring. In *Proceeding of the 11th International Conference on Information Systems for Crisis Response and Management*. ISCRAM.

Horita, F. E. A., Link, D., Albuquerque, J. P., & Hellingrath, B. (2014). A Framework for the Integration of Volunteered Geographic Information into Humanitarian Logistics. In *Proceedings of the the 20th Americas Conference on Information Systems (AMCIS 2014)*. Savannah, GA: AIS.

Horita, F. E. A., Link, D., Albuquerque, J. P., & Hellingrath, B. (2016). oDMN: An Integrated Model to Connect Decision-Making Needs to Emerging Data Sources in Disaster Management. In *Proceedings of the 49th Hawaii International Conference on System Sciences (HICSS 2016)* (pp. 2882–2891). IEEE. https://doi.org/10.1109/HICSS.2016.361

Hughes, D., Ueyama, J., Mendiondo, E., Matthys, N., Horré, W., & Michiels, S. ... Guan, S. U. (2011). A middleware platform to support river monitoring using wireless sensor networks. *Journal of the Brazilian Computer Society, 17*(2), 85–102. https://doi.org/10.1007/s13173-011-0029-3

Hung, K.-C., Kalantari, M., & Rajabifard, A. (2016). Methods for assessing the credibility of volunteered geographic information in flood response: A case study in Brisbane, Australia. *Applied Geography Journal*, *68*, 37–47. doi:10.1016/j.apgeog.2016.01.005

Hunter, G. J., Bregt, A. K., Heuvelink, G. B. M., De Bruin, S., & Virrantaus, K. (2009). Spatial Data Quality: Problems and Prospects. In G. Navratil (Ed.), *Research Trends in Geographic Information Science* (pp. 101–121). Berlin: Springer Berlin Heidelberg. doi:10.1007/978-3-540-88244-2_8

Imran, M., Castillo, C., Lucas, J., Meier, P., & Rogstadius, J. (2014). Coordinating human and machine intelligence to classify microblog communications in crises. In *Proceedings of the 11th International Conference on Information Systems for Crisis Response and Management (ISCRAM 2014)* (pp. 712–721). ISCRAM.

Jha, A. K., Barenstein, J. D., Phelps, P. M., Pittet, D., & Sena, S. (2010). *Safer Homes, Stronger Communities*. The World Bank. https://doi.org/10.1596/978-0-8213-8045-1

Kongthon, A., Haruechaiyasak, C., Pailai, J., & Kongyoung, S. (2012). The role of Twitter during a natural disaster: Case study of 2011 Thai Flood. In *2012 Proceedings of PICMET'12: Technology Management for Emerging Technologies* (pp. 2227–2232). Vancouver, Canada: PICMET.

Kotsev, A., Pantisano, F., Schade, S., & Jirka, S. (2015). Architecture of a Service-Enabled Sensing Platform for the Environment. *Sensors (Basel, Switzerland)*, *15*(2), 4470–4495. doi:10.3390/s150204470 PMID:25688593

Kumar, S., & Havey, T. (2013). Before and after disaster strikes: A relief supply chain decision support framework. *International Journal of Production Economics*, *145*(2), 613–629. doi:10.1016/j.ijpe.2013.05.016

MacEachren, A. M., Jaiswal, A., Robinson, A. C., Pezanowski, S., Savelyev, A., & Mitra, P., ... Blanford, J. (2011). SensePlace2: GeoTwitter Analytics Support for Situational Awareness. In *Proceedings of the IEEE Conference on Visual Analytics Science and Technology* (pp. 181–190). Providence, RI: IEEE. doi:10.1109/VAST.2011.6102456

Mendes, H. C., & Mendiondo, E. M. (2007). Histórico da Expansão Urbana e Incidência de Inundações : O Caso da Bacia do Gregório, São Carlos - SP. *Revista Brasileira de Recursos Hídricos, 12*(1), 17–27. https://doi.org/10.21168/rbrh.v12n1.p17-27

Mooney, P., Corcoran, P., & Winstanley, A. C. (2010). Towards Quality Metrics for OpenStreetMap. In *Proceedings of the 18th SIGSPATIAL International Conference on Advances in Geographic Information Systems* (pp. 514–517). San Jose, CA: ACM.

Moreira, R. de B., Degrossi, L. C., & Albuquerque, J. P. (2015). An experimental evaluation of a crowdsourcing-based approach for flood risk management. In *12th Workshop on Experimental Software Engineering* (pp. 393–403). Lima, Peru: SPC.

Munich RE. (2013). *Münchener Rückversicherungs-Gesellschaft*. Geo Risiko Forschung, NatCatSERVICE.

Norris, F. H., Stevens, S. P., Pfefferbaum, B., Wyche, K. F., & Pfefferbaum, R. L. (2008). Community Resilience as a Metaphor, Theory, Set of Capacities, and Strategy for Disaster Readiness. *American Journal of Community Psychology, 41*(1–2), 127–150. doi:10.1007/s10464-007-9156-6 PMID:18157631

OGC. (2011). *Observations and Measurements*. Retrieved from http://www.opengeospatial.org/standards/om

OGC. (2016). *Open Geospatial Consortium*. Retrieved from http://www.opengeospatial.org/

OMG. (2013). *Business Process Model and Notation (BPMN), Version 2.0 (formal/2013-12-09)*. Retrieved July 8, 2016, from http://www.omg.org/spec/BPMN/2.0/PDF/

OMG. (2014). *Decision Model and Notation (DMN) (dtc/2014-02-01)*. Retrieved from http://www.omg.org/spec/DMN/1.0/

OSGeo. (2016). *GeoNetwork opensource*. Retrieved from http://geonetwork-opensource.org/

Pohl, D., Bouchachia, A., & Hellwagner, H. (2012). Supporting crisis management via sub-event detection in social networks. In *Proceedings of the Workshop on Enabling Technologies: Infrastructure for Collaborative Enterprises, WETICE* (pp. 373–378). Toulouse, France: IEEE. doi:10.1109/WETICE.2012.58

Rotava, J., Mendiondo, E. M., & Souza, V. C. B. (2013). Simulação de instabilidade humana em inundações: primeiras considerações. In *Proceedings of the 20th Simpósio Brasileiro de Recursos Hídricos* (pp. 1–8).

Sakaki, T., Okazaki, M., & Matsuo, Y. (2010). Earthquake Shakes Twitter Users: Real-time Event Detection by Social Sensors. In *Proceedings of the 19th International Conference on World Wide Web* (pp. 851–860). New York, NY: ACM. doi:10.1145/1772690.1772777

Schnebele, E., Cervone, G., & Waters, N. (2014). Road assessment after flood events using non-authoritative data. *Natural Hazards and Earth System Sciences, 14*(4), 1007–1015. doi:10.5194/nhess-14-1007-2014

Schnebele, E., Oxendine, C., Cervone, G., Ferreira, C. M., & Waters, N. (2015). Using Non-authoritative Sources During Emergencies in Urban Areas. In M. Helbich, J. Jokar Arsanjani, & M. Leitner (Eds.), *Computational Approaches for Urban Environments* (pp. 337–361). Cham, Switzerland: Springer International Publishing. doi:10.1007/978-3-319-11469-9_14

See, L., Mooney, P., Foody, G., Bastin, L., Comber, A., & Estima, J. … Rutzinger, M. (2016). Crowdsourcing, Citizen Science or Volunteered Geographic Information? The Current State of Crowdsourced Geographic Information. *ISPRS International Journal of Geo-Information, 5*(5), 55. https://doi.org/10.3390/ijgi5050055

Smith, L., Liang, Q., James, P., & Lin, W. (2015). Assessing the utility of social media as a data source for flood risk management using a real-time modelling framework. *Journal of Flood Risk Management.* https://doi.org/10.1111/jfr3.12154

Spinsanti, L., & Ostermann, F. (2013). Automated geographic context analysis for volunteered information. *Applied Geography (Sevenoaks, England), 43*, 36–44. doi:10.1016/j.apgeog.2013.05.005

Steiger, E., Albuquerque, J. P., & Zipf, A. (2015). An Advanced Systematic Literature Review on Spatiotemporal Analyses of Twitter Data. *Transactions in GIS, 19*(6), 809–834. doi:10.1111/tgis.12132

Tu, Y., Li, Q., & Liu, R. (2009). A geospatial information portal for emergency management of natural disasters. In *Proceedings of the IEEE International Geoscience and Remote Sensing Symposium (IGARSS 2009)* (Vol. 2, pp. II-404–II-407). Cape Town, South Africa: IEEE. doi:10.1109/IGARSS.2009.5418100

Ushahidi. (2016). *Ushahidi.* Retrieved from https://www.ushahidi.com/

Vasiliu, A. (2006). *Dashboards and scorecards: Linking management reporting to execution.* Retrieved from http://dssresources.com/papers/features/vasiliu/vasiliu04302006.html

Vieweg, S., Castillo, C., & Imran, M. (2014). Integrating Social Media Communications into the Rapid Assessment of Sudden Onset Disasters. In *Proceedings of the 6th International Conference on Social Informatics* (pp. 444–461). Barcelona, Spain: Springer International Publishing. doi:10.1007/978-3-319-13734-6_32

Zook, M., Graham, M., Shelton, T., & Gorman, S. (2010). Volunteered Geographic Information and Crowdsourcing Disaster Relief: A Case Study of the Haitian Earthquake. *World Medical & Health Policy, 2*(2), 6–32. https://doi.org/10.2202/1948-4682.1069

KEY TERMS AND DEFINITIONS

Collaborative Mapping: Information about the geographic features of disaster-affected or disaster-prone areas, which is produced by volunteers using mapping platforms.

Crowd Sensing: Information collected from dedicated mobile applications and web platforms that are specifically aimed at producing information for disaster risk management.

Disaster Management: Creation of plans through which communities reduce vulnerability to hazards and cope with disasters.

Preparedness: Refers to a very concrete research based set of actions that are taken as precautionary measures in the face of potential disasters.

Response: Actions carried out immediately before, during, and after a hazard impact.

Social Media: Information produced by people about the event in popular social media platforms (e g. Twitter, Flickr, Instagram, Facebook).

VGI: User-generated geospatial content being created in these and many other sites to satisfy a variety of needs within industry, government, and social networking communities.

VTCs: Communities that apply and leverage their technical skills in collecting, processing and managing data in support of response efforts for disasters.

Chapter 10
Quality of Urban Life Index From Location-Based Social Networks Data:
A Case Study in Belo Horizonte, Brazil

Rodrigo Smarzaro
Federal University of Viçosa, Brazil

Tiago França Melo de Lima
Federal University of Ouro Preto, Brazil

Clodoveu Augusto Davis Jr.
Federal University of Minas Gerais, Brazil

ABSTRACT

Several indicators are developed to support the decision-making processes in public policy for urban planning. Some of them seek to measure the quality of urban life. For example, the city of Belo Horizonte developed and uses an index called Quality of Urban Life Index, which identifies inequalities within the city, and therefore, those areas that need more investment. This index is calculated by measuring the availability of various kinds of services (e.g. education, infrastructure) and their accessibility (based on travel time and mobility data). For that, data from several government sources must be collected and used, which can delay updates of index values. In this chapter, the authors describe how data from Location-Based Social Networks (LBSN) can be used to calculate urban indicators, and hence, how they could be used as an alternative data source for estimating quality of urban life with faster results to support urban planning policies.

DOI: 10.4018/978-1-5225-2446-5.ch010

INTRODUCTION

Since the last century, the world has experienced high rates of population migration from rural to urban areas. At the beginning of the 20th century, about 10% of the world population lived in urban areas. In 2007, urban population surpassed the number of people living in rural areas. In 2014, 53% of the world population were living in cities, and the predictions for 2050 indicate that urban population will reach 64% in developing countries and 86% in developed ones (The World Bank, 2015). In absolute numbers, urban population increases by 6 million people per month. It is as if a big city such as Washington (DC), Madrid or Belo Horizonte appeared every month in the world.

The fast urbanization process generates impacts on economy, health and environment, among others, and many challenges arise from them. For local governments, it is important to have indicators that summarize the evolution of citizens' quality of life, allowing the identification of regions that need further attention, assisting urban planning and supporting the allocation of resources to deal with current and future problems. Many metrics have been proposed over the years and there is no common agreement about how to define and measure quality of urban life. In 1990, the Human Development Index (HDI) was created by the United Nations Development Programme to show welfare status, aiming to be a more human-centered measure, instead of focusing only in economic issues. It is based on three dimensions: life expectancy, education and per capita income. HDI is used globally as an indicator for ranking countries, states and cities, but it fails to "reflect inequalities, poverty, human security, empowerment" (UNDP, 2015). The HDI uses the city, or municipality, as its smallest territorial unit.

Researchers and governments were stimulated by the creation and establishment of the HDI. Some of them started to propose and use their own indicators in order to measure urban quality of life in spatial units of different sizes, varying from country levels to metropolitan regions and municipalities. For instance, the Social Development Index (*Indice de Desenvolvimento Social* - IDS) was created in 1991 based on the HDI, but replacing per capita income with comfort and sanitation indicators (Rodrigues, 1991). The IDS is calculated for state-size geographic areas. The Living Conditions Index (*Indice de Condições de Vida* - ICV) is a city level index that was created to measure poverty and childhood conditions (Nahas, 2002). Several other efforts have been made to measure urban quality of life, with varying approaches and focusing on different aspects, including the evaluation of indexes (Badland et al., 2014; Blomquist, Berger, & Hoehn, 1988; Gavrilidis, Ciocănea, Niţă, Onose, & Năstase, 2016; Hagerty et al., 2001; Helliwell, 2003; Marans, 2015; Pissourios, 2013; Rezvani & Mansourian, 2013; Roback, 1982; Rogerson, 1999; Węziak-Białowolska, 2016).

One problem with these indicators is that in general they attempt to summarize the complexity of cities into a single number. The local government uses these indexes to monitor city performance as a whole, but cannot use them to identify problems in specific regions of the city. To address this situation, Nahas, (2002) describes the idea of a quality of urban life index (*Índice de Qualidade de Vida Urbana* - IQVU), which is composed by ten dimensions: food supply, culture, education, sports, housing, urban infrastructure, environment, health, urban services and urban security. It is based on quantifying the spatial inequality of services that are available and accessible to population, and can be used to show areas that need increased public investment. The index was created and is used in Belo Horizonte (BH), a 2.5-million-people city that is the capital of the Brazilian state of Minas Gerais. IQVU is calculated for a geographic subdivision of the city, called Unit of Planning (UP), and is used to support urban planning. The first component in the calculation of the IQVU is a Local Availability Index (*Índice de Oferta Local* - IOL) that measures the availability of services inside an UP. The IOL is then adjusted by an accessibility factor to deal with situations in which services from one UP are accessed by people from other UPs. The IOL, adjusted by accessibility, comprises the IQVU. Since its first published results in 1996, the IQVU has been used by the municipal government as a factor to balance public investment among the UPs. IQVU relies on official data sources provided by about 16 different governmental organizations.

Although the IQVU has been conceived to be easily and periodically recalculated, this is not what happened since its creation (PBH, 2014). We observe that the variety of temporal granularity among the several data sources may be a major cause for irregular updating of the indicator. Other factors may include difficulties in obtaining some indicator components, due to methodological changes implemented throughout the years by the organizations that provide information. Thereby, this chapter shows the results of an investigation into the use of data from social media as an alternative approach to measure the quality of urban life. To this end, data from two open Location-Based Social Networks (LBSN), Foursquare and Yelp, were collected and used. For each LBSN, its Application Programming Interface (API) was used for searching and gathering data of places (e.g. business) from Belo Horizonte city. Then, this data was used to estimate the IOL component of IQVU using its calculation methods as close as possible. Results show that the use of LBSN data to infer the quality of urban life is promising, with possible impacts in the formulation of similar indicators in the future.

The rest of this chapter is organized as follows. The Background section presents a brief summary of related work. The Methods section describes the procedures for gathering and processing data from LBSNs and how they were used to calculate the IOL. The results of using LBSN to support the measuring quality of urban life are presented and discussed in the Results and Discussion section. Finally, the last section presents final remarks and discusses future research opportunities.

BACKGROUND

This section presents some basic concepts related to accessibility, diversity and quality of urban life. Several indicators have been created aiming to measure quality of life and support urban planning. Some of them focus mainly on social and economic aspects, while others also include dimensions like health and environmental aspects. It also presents concepts about Location Based Social Networks, with special attention to Foursquare and Yelp.

Accessibility, Diversity, and Quality of Urban Life

The Death and Life of Great American Cities (Jacobs, 1961) is one of the most influential books in city planning. Jacobs states that death in urban areas is caused by the elimination of pedestrian activity from streets, while life means pedestrian activity at all times of the day (which leads to greater interaction with the neighborhood environment). Jacobs was a firm believer in that streets should experience a mix of activities of different purposes, which attracts pedestrian activity at all hours of the day. To achieve this, Jacobs presented four essential diversity criteria. The first one states that a district should have two or more primary uses. However, this mix of land uses should be maintained across an entire city in the street level. The second criteria requires city blocks to be shorter in order to create more alternate travel routes. Since larger blocks increase the walking distance, the presence of pedestrians in these blocks is decreased and the usage of vehicles such as cars is encouraged. Besides, small blocks produce more intersections, and thus slow down vehicles and shorten pedestrian walking distances. The third criteria mentions that buildings in a district need to be mixed with regard to age and shape in order to ensure diverse economic activity. The diversity of building types allows a diverse range of jobs. Finally, the last criteria states that the built environment should contain a sufficient concentration of buildings to attract people in both residential and nonresidential urban areas.

Sung et al. (2015) present the implementation of Jacobs's four conditions in the urban life of Seoul, South Korea. Instead of districts, authors analyzed data on *dongs*, the smallest administrative unit in Seoul. With this, the study found empirical evidence that Jacobs' specifications for sustaining urban diversity, such as mixed use and small block sizes, can play a significant role in urban life. However, the biggest problem of the methodology employed in that work lies on data collection. It took the authors 10 years to complete their work, due to the number of mechanical and repetitive processes for obtaining data.

Venerandi et al. (2015) present a methodology to measure urban deprivation from user-generated content. They quantitatively describe neighborhoods using a metric called Offering Advantage, which was then used to infer urban deprivation as well. Nevertheless, they only used data from Foursquare and OpenStreetMap (OSM) datasets, which have geographic and social biases. Also, they used the UK Index of Multiple Deprivation (IMD), a composite score that is calculated as the weighted means of seven distinct domains: income deprivation, employment deprivation, health deprivation, education deprivation, barriers to housing and services, crime, and living environment deprivation. As this index is not used by Brazilian governments, it cannot be used in this work since the necessary data sets are not readily available for analysis and comparison.

Yuan et al. (2012) propose a framework capable of identifying regions that performs different functions in urban areas. Discovering those functions for each region is important for urban planning and for predicting the development of a city. The solution presented in that work first creates partitions of the selected urban area based on its major roads (called regions). After this step, it uses human mobility data from GPS-equipped taxicabs. Such mobility data are used to infer each region's functionality (e.g., business, commercial, entertainment). One of the problems, however, comes from the evaluation of urban mobility, which only takes data from Beijing's GPS-tracked taxi trajectories, creating an important bias, as it is a too simplistic approach to represent overall mobility. Another issue is the difficulty in obtaining representative data on mobility.

Quercia and Saez (2014) focused on determining if data from social media offers an alternative data source for studying the relationship between resources and neighborhood deprivation. Authors gathered data from Foursquare users in the city of London and used classification algorithms to infer land-use information and to study the relationship between the presence of specific venues and socio-economic deprivation. The limitations of this work comprise its geographical and demographic biases, and the potential problems of replicating the methodology in other urban areas.

Shelton et al. (2015) present an extensive analysis of geotagged tweets from Louisville, Kentucky, a city which is becoming a reference when it comes to data-driven urban governance. The objective was to provide useful insight for urban planning and geographical research. The approach combines relational socio-spatial theory and GIScience to show issues like segregation between neighborhoods, mobility and inequality within city boundaries. This work, once again, has a considerable bias, since it just uses data from Twitter (limited to messages that included positional information) and its methodology was applied in a city with very specific features, having no guarantees if it would work for urban areas with different settings.

The IQVU-BH Index

The Quality of Urban Life Index of Belo Horizonte (IQVU) was developed by a multidisciplinary group from government and academia with the purpose to be a tool for improvement of distribution of public resources (Nahas, 2002). It is a multidimensional indicator that aims at quantifying the spatial inequality of services accessible to population, and consequently can be used to show areas that should be prioritized for public investment. The IQVU was created using variables that should be periodically updated (maximum of two years). However, IQVU calculation instances did not follow the same rule and were published without the expected regularity (1994, 2000, 2006, 2010, 2012).

The index is calculated on spatial regions called Units of Planning (UPs). It is based on the availability and accessibility of facilities and services, classified in ten dimensions (called variables): *food supply, culture, education, sports, habitation, urban infrastructure, environment, health, urban services and public security*. Each variable is calculated from the value of *components*, which in turn are obtained from *indicators*. For instance, *health centers* (the number of public health facilities in a UP per thousand inhabitants) is an indicator of the *health care* component of the *health* variable. The full list of variables, components and indicators used to calculate IQVU is presented in Table 2. Georeferenced data are used to calculate the indicators, which are then aggregated into the components and variables. The steps to calculate the IQVU are shown in Figure 1.

Figure 1. Steps to calculate the IQVU index
(adapted and translated from (PBH, 2014))

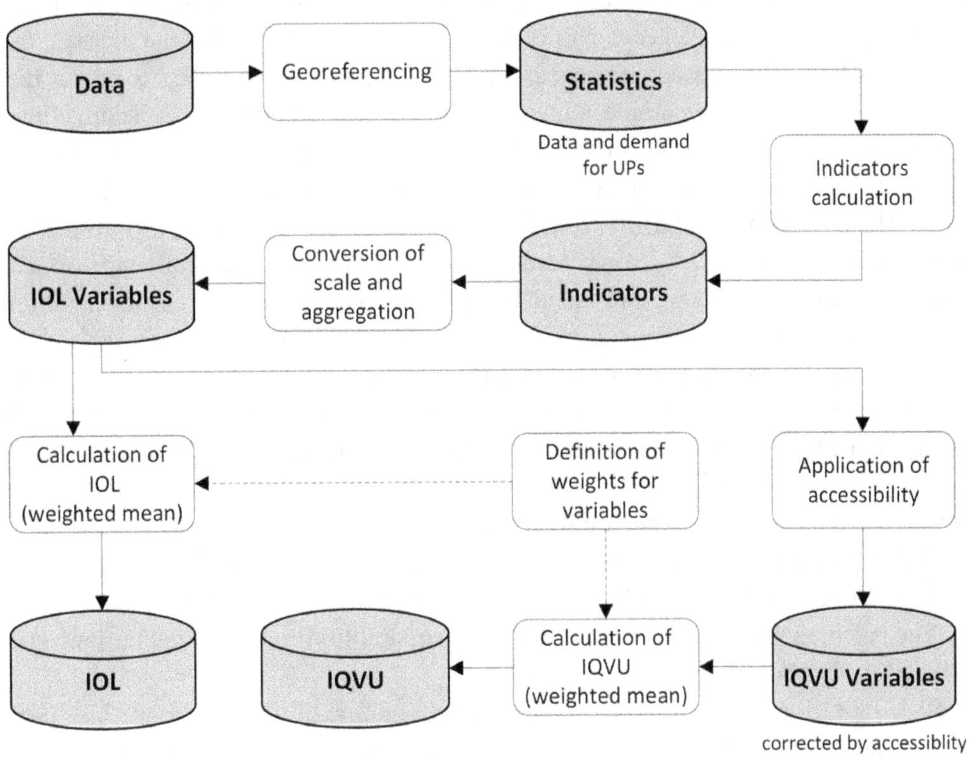

In simplified form, after data gathering and calculation of indicators, three main results are obtained:

- The Local Availability Index (IOL) is calculated for each dimension on each UP. First, each indicator's availability is calculated with a specific measure, which is, generally, a count per thousand inhabitants (e.g. number of hospitals / population x 1000). Then, the indicator values are normalized between 0 and 1 using equation:

$$I_c = 1 - e^{-(f \times v)}$$

where I_c is the normalized indicator value, v is the original indicator value, and f is given by:

$$f = \frac{-\ln(0.05)}{L_{ref}}$$

where L_{ref} is the reference value for the indicator, which is assumed to be the 95% percentile. This adjustment compensates for higher values, which cause less impact on the index (Lemos, Esteves, & Simões, 1995). The IOL of a dimension is given by the simple mean of the normalized value of its indicators, and the IOL for the UP region is given by the weighted average of all dimensions.

- The IQVU Variables (sectorial indices) are calculated by applying an accessibility index on the IOL. The accessibility index is based on an estimate of the displacement time of a citizen, using public transportation, between each pair of UPs. The accessibility index can increase or decrease the variable's IOL value, based on how easy it is for the population of a UP to access the services available in other UPs. For example, someone from UP 1 decides to go to a health clinic in UP 2, as her original UP does not offer this kind of service. If UP 2 is easily reached from UP 1, the low IOL value in UP 1 is partially compensated by the existence of services in UP 2, in an accessible location. The accessibility adjustment will then increase the local availability for UP 1, but will decrease the value for UP 2 by some amount. When we calculate the influence of accessibility over all variables for all UPs we can calculate the IQVU. The accessibility works in different ways for each variable, considering four accessibility modes: *immediate*, *near*, *average* and *remote*. An *immediate* accessibility measure means that the services corresponding to the variable should only be accessed inside the UP. A *near* accessibility measure needs the displacement time to be relatively short, for services that should be available near home. For instance, a person can go to a supermarket outside her UP, but it is less likely that she will visit a distant one. The values of the *average* and *remote* accessibility dimensions decay more slowly with distance than in the near accessibility measure. These are the cases of more complex services, for which traveling a relatively longer time makes little difference, such as a consultation with a medical specialist across town. The calculation procedure to incorporate accessibility measures is rather long and will not be detailed further here (see Lemos et al.(1995)).
- The IQVU index for each region is given by the aggregation of the IQVU variables (sectorial indexes) using a weighted average. The weight of each dimension was established by the specialists group responsible for the creation of the IQVU.

Once the IQVU results are available, the local government can use it to decide which regions will need more public investment. UPs with a large population and low IQVU are more likely to receive more investment than others with a smaller population and a higher IQVU. The IQVU has already been used to support decision processes for public investment in Belo Horizonte (PBH, 2014).

Location Based Social Networks (LBSN)

Online Social Networks (OSN), like Facebook and Twitter, enable users to create a network of friends (or people to follow or that will follow you, in case of Twitter) and share any content to all or a group of its contacts. Smartphones and other portable devices equipped with Global Positioning System (GPS) and connected to internet motivated the creation of Location-Based Social Networks (LBSN) (Pontes, Vasconcelos, Almeida, Kumaraguru, & Almeida, 2012). In LBSNs, geographical location can be shared along with the regular content of interest.

Zheng (2011) uses the role location plays on LBSNs to classify them into three groups. Geo-tagged-media-based, point-location-driven and trajectory-centric. *Geo-tagged-media-based* add the location to the media being shared. It is important to notice that the location can be added to the media after it is posted. Geotagged photos, videos, tweets are examples of georeferenced media. Flickr, Panoramio, Youtube and Twitter can all be considered LBSNs on this category. *Point-location-driven* services encourage users to share their current location in real time. For example, a user can arrive at a shopping mall and share his location. His contacts are informed and maybe they can meet at the place. The user also can share his opinion about the place. Foursquare and Yelp are examples of point-location-driven LBSN. *Trajectory-centric* services enable users to share a route, which is a sequential connection of point locations. Information like distance, speed, duration, altimetry and others the user can provide (such as tags, photos, opinions) are also shared. Sports logging services like Garmin Connect, Nike+, Strava are representative of this category. Our interest in this work is on point-location-driven LBSNs, more specifically Foursquare and Yelp.

Foursquare was launched in 2009. First, it was available in 100 cities worldwide, but in January 2010 it became global. In 2011, Foursquare reached about 7 million users. There are currently more than 50-million users (Foursquare, 2016). Initially, the user could only share her location with friends by making a "check in" at a venue. Later, more features were added, such as a user "to do list", which allows saving locations to be notified when nearby. Foursquare issues rewards based on the user's check-in history. For example, users who make 50 check-ins earn a "superstar" badge in their profiles. Another reward are venue mayorships. The user who makes more check-ins on a venue is considered its "mayor". Many venue owners offer advantages to Foursquare mayors, like discounts or free parking. Foursquare lets users post opinions about venues. People searching for a restaurant nearby can look at how Foursquare users reviewed that place. Good or bad reviews can have a great impact on business performance. In 2014, Foursquare was divided in two apps: Foursquare and Swarm. Social aspects like sharing locations with friends and finding out the location of friends migrated to Swarm, while Foursquare concentrates on venue recommendation.

Yelp was launched in 2004 with the mission to help people find good services and places. Like Foursquare, Yelp users can check-in at venues and share their location with friends. However, Yelp focuses on user reviews of business venues. Users can rate other user reviews so that Yelp can make better recommendations of places. In 2016, Yelp has over 23 million unique users on its website and 69 million using the service through a mobile app (Yelp, 2016).

METHODS

In this section, we describe the data sources and steps used for data gathering and analysis. Then, we present the steps used to calculate the IOL using data from LBSNs. The data were obtained from local government sources and from the social networks Foursquare and Yelp, as described on this section.

Official Data Sources

In this work, we used data from two official data sources: (i) the most recent IQVU-BH dataset, and (ii) Brazilian Census data.

The IQVU-BH dataset was obtained from Belo Horizonte's Web portal (PBH, 2015). It is available as a spreadsheet that contains calculated values of indicators, components and variables of IQVU for each UP in 1994, 2000, 2006, 2010 and 2012. The raw data and calculation formulas were not available on the spreadsheet, making it difficult to perform further analysis and comparison of results.

Besides the IQVU dataset, demographic information is also needed, since the calculation of almost all indicators use the population of each UP. Official data from the Brazilian census was used to reduce methodological issues by employing the same data used by previous IQVU calculations. The Brazilian Institute of Geography and Statistics (IBGE) (IBGE, 2015) performs the census of Brazilian population on a regular interval of ten years. The most recent was realized in 2010. A census sector is a minimum territorial unit, which has a continuous area located completely inside an urban or rural area and has an appropriate size for a census surveyor to manage on the field. Belo Horizonte has 3,936 census sectors, for which the full range of census data are available.

IBGE data provides the population count for each census sector, but the IQVU uses Units of Planning (UP). The UPs were created in 1993 by Belo Horizonte's administration as the spatial reference unit for public policies and urban planning. UPs have been defined as spatial aggregations of census sectors, therefore obtaining demographic data for each UP is straightforward.

LBSN Data Sources

The location-based social networks Foursquare and Yelp were used as data sources. They were chosen because they have the following characteristics, as previously mentioned: (i) offer a public API allowing data gathering, (ii) contain a large number of users and places, and (iii) allow volunteered data contributions. Both services maintain datasets of publicly available services and businesses, which can be used as a source for IQVU indicators.

Data Gathering

Foursquare and Yelp APIs allow downloading data from points of interest that are located at a certain distance from a given reference point. In order to collect the full set of services for Belo Horizonte, an iterative procedure was implemented. First, a regular grid of points separated by 25 meters was generated, covering the bounding box of Belo Horizonte's city limits. The radius size of 25 meters was considered small enough to get all the available places for each social network, since larger radii might cause the APIs to return only a subset of the available places instead of the full list. Then, a geometric intersection between the grid and the city limits polygon was used to select the points inside the municipal territory. This operation resulted in a set of 530,044 distinct reference points. For each point, a call to the LBSN's API was made using the point's latitude, longitude and a radius of 25 meters as parameters. Since the points are separated also by 25 meters, calls overlap one another, making it unlikely to leave some point of interest behind.

Crawlers for each LBSN were implemented using Python and data were collected from October 2 to October 25, 2015. Crawlers had to pace themselves to comply to the API enforced limit on the number of daily searches. Yelp, for instance, allows 25,000 search requests per day, while Foursquare allows 5,000 requests per hour. The API returns JSON data that were initially stored as a single text file. Then the file was processed to extract, for each location, the ID, latitude, longitude and category. Finally, these data were stored in a PostgreSQL database with PostGIS extension. Redundant locations (those with same ID) were removed in the loading process, and the numeric latitude and longitude coordinates were transformed into a geographic point.

Data Analysis and Characterization

Data gathering produced a dataset containing 62,135 unique points of interest from Yelp and 91,816 from Foursquare. Figure 2 shows the spatial distribution of points (collected places) from Foursquare and Yelp across Belo Horizonte.

Foursquare and Yelp classify each place into one or more different categories. However, just part of them match the type of places used by IQVU. Yelp has 1,099 possible categories (Yelp, 2015) to classify a location, but only 488 were present in Belo Horizonte's dataset. Foursquare, on the other hand,

Figure 2. Point-density map of unique POIs

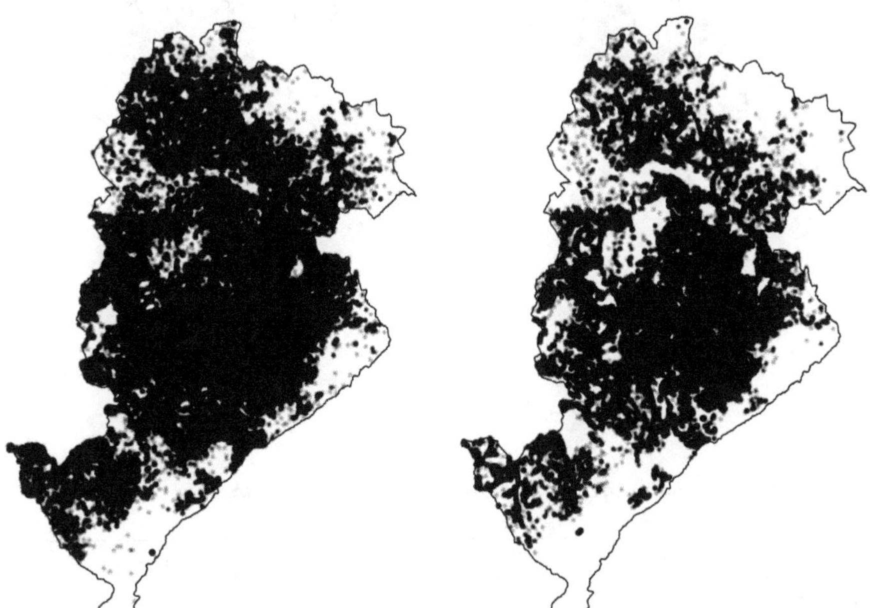

(a) Foursquare (b) Yelp

offers 398 possible categories (Foursquare, 2015), 379 of which appear in the dataset. Furthermore, many places collected from the LBSNs are not relevant to our study, since they are not used to calculate IQVU components. For example, the dataset has information on laundries and clothing stores, but these services are not considered in IQVU calculation.

Therefore, it was necessary to match the categories of locations in the datasets with those used by the IQVU. The approach used relied on relating each category offered by the LBSN with one of the IQVU indicators, using manual classification. Matches based on the interpretation on the meaning of the category were not considered. For example, many locations had the category "Alimentação" (Food), which has no exact match on IQVU indicators, but on a broader interpretation they could be classified as "food supply" services. After matching and selecting relevant categories, the Foursquare dataset remained with 6,827 locations left, and Yelp with 8,194. The cleaning process results are summarized in Table 1 and Figure 3 shows the spatial distribution of the resulting points on both datasets.

As expected, the distribution of places among categories and their location over space are not uniform. Figure 4 shows the distribution of locations collected for each UP. The horizontal axis represents the

Figure 3. Point-density map of unique POIs in relevant categories

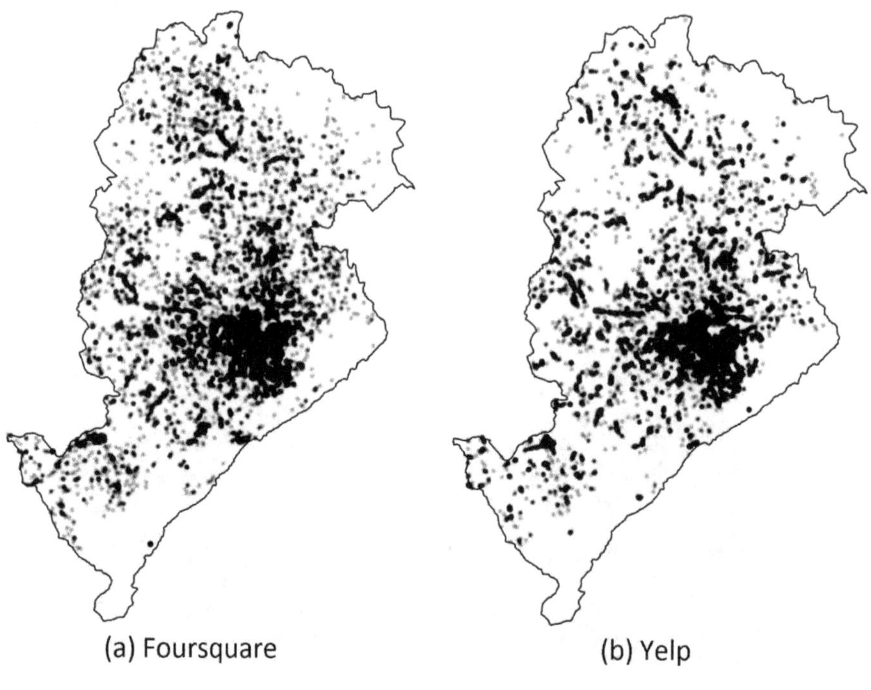

Table 1. Number of POIs at each step of the data source cleaning process

Operation on data	Foursquare	Yelp
Data Gathering (raw data)	286,227	354,278
Remove duplicate entries	91,816	62,135
Remove entries without compatible category	6,827	8,194

Figure 4. Distribution of locations collected for each UP

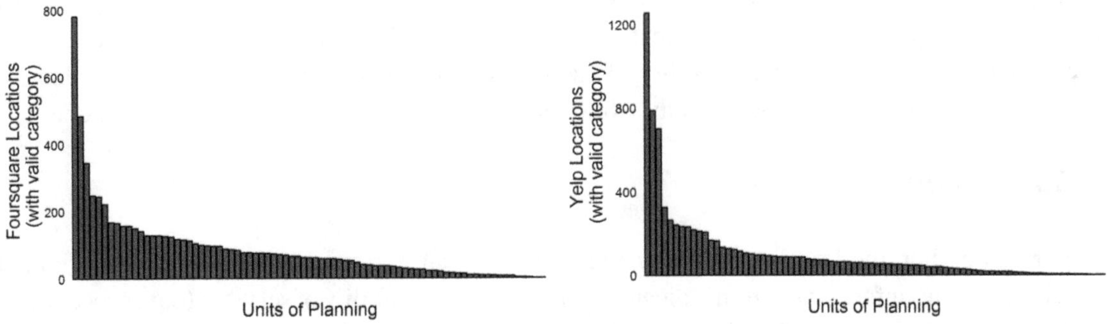

(a) Foursquare (b) Yelp

ordered list of UP by locations count (label omitted for better visualization). It can be seen that there are few UPs that contain most of the collected places, while others have very few. The distribution of collected places over the UPs was similar in both Foursquare and Yelp datasets. Figure 5 shows the spatial distribution of points among UPs. Darker color represents a higher locations count.

Figure 5. Spatial visualization of locations collected for each UP, Foursquare on left and Yelp on right

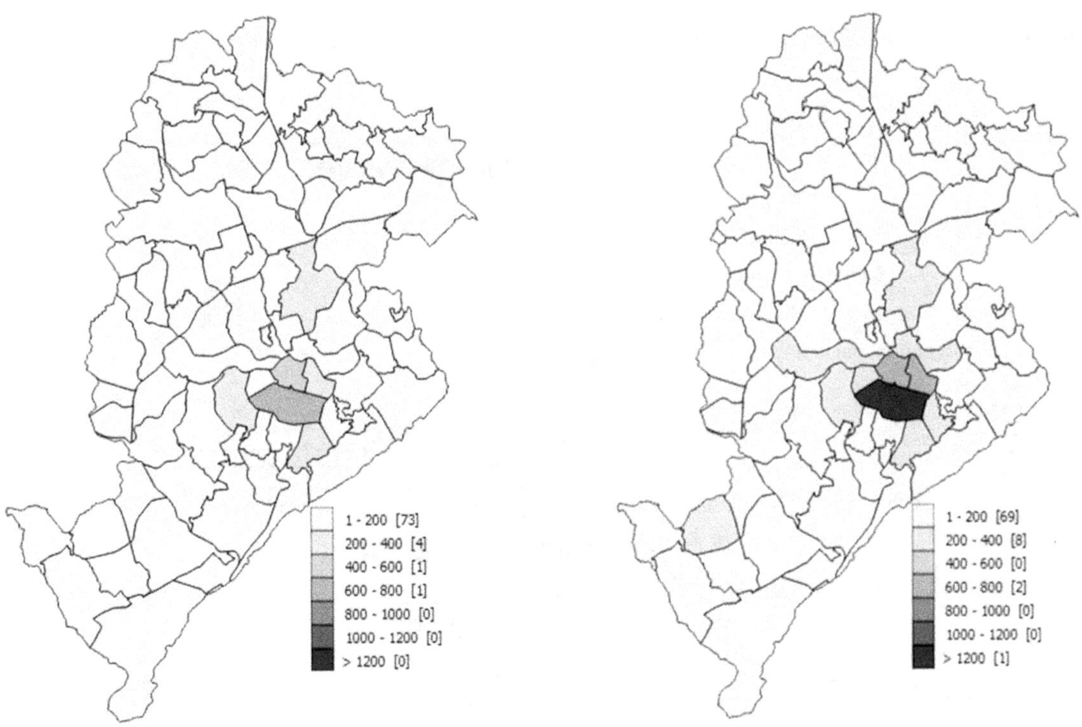

(a) Foursquare (b) Yelp

Some of the data used to calculate IOL and IQVU are not available in Foursquare and Yelp. For example, the number of registered students in schools and criminal occurrences. Table 2 presents the full list of variables, components and indicators of the IQVU, and identifies which data source is being used to calculate each one. The availability of LBSN data for calculating indicators varies: data from both LBSNs can be used for some of them, while for others only one can be used, or even none.

IOL Calculation

As described earlier, the Local Availability Index (IOL) is an important metric used to calculate the IQVU. It aggregates information from indicators that measure availability of services (e.g., grocery store), into IQVU variables (e.g., food supply), and the results are then weighted according to the importance of each IQVU variable.

To calculate the IOL variables, the number of places collected from the LBSNs inside each UP was counted for each indicator. This number is divided by the population of the UP to give the normalized indicator value. The indicator Green Area in the original IOL calculation is not based on location count, but uses the green surface (parks, woods, etc.), in square meters per thousand inhabitants. Naturally, this is not available from LBSNs. We decided to estimate it with the same approach used in other indicators, dividing the total count of places by population.

As expected, considering the differences in purpose and features between Foursquare and Yelp, the distribution of collected places over the indicators is not uniform, as presented in Figure 6. Some categories, such as grocery stores and drugstores, are more evenly matched than most.

Figure 6. Distribution of locations collected for IQVU indicators after category matching

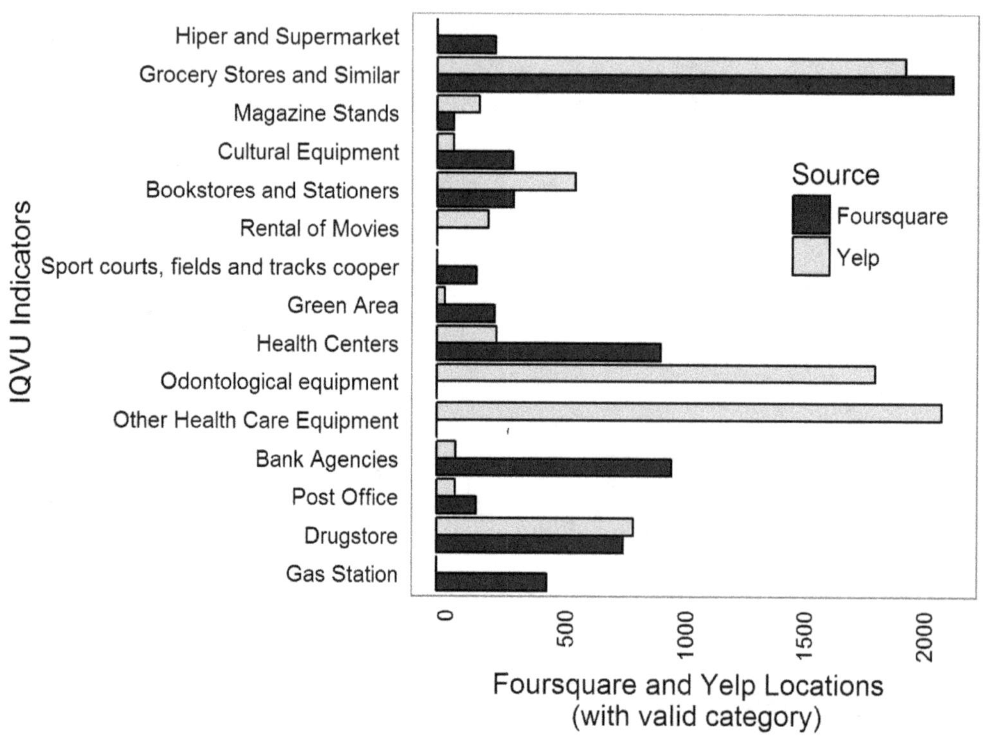

Since it was not possible to get LBSN data to calculate all indicators of IQVU, data from IQVU 2012 was used as a basis, in order to allow a comparative analysis. On IOL variables calculation, the indicators that lacked LBSN data by official data were replaced. We were able to calculate 15 of the 36 indicators using data from LBSNs (Foursquare or Yelp, or both), as shown in Table 2. Using LBSN data exclusively from Foursquare and Yelp, we were able to calculate the IOL for three of the ten IQVU variables. If combined with official data, six of the ten variables can be calculated.

RESULTS AND DISCUSSION

As presented in the Introduction section, this work aims to investigate if data from LBSN can be used to support measuring quality of urban life, more specifically as defined by the IQVU index methodology. The process to calculate IQVU comprises several steps, and this work achieved, as a result, the calculation of the IOL index of IQVU variables using data from LBSNs.

The first thing to do is to evaluate results from IOL indicators with data from Foursquare and Yelp. To do so, the obtained values were discretized in the same intervals used by the local government to present IQVU results. The values were divided into five intervals: [0,0.5), [0.5, 0.6), [0.6, 0.7), [0.7, 0.8), [0.8, 1], labeled "1","2","3","4" and "5", respectively. After classifying the results on the intervals, accuracy, precision and recall were calculated for IOL indicators, testing both datasets (Foursquare and Yelp) against official 2012 IOL results. Since this is a multiclass problem (five possible intervals) precision and recall for each interval were calculated using a "one-against-all-others" approach, and then the final value was calculated using the mean of all intervals.

Mixed results for accuracy were obtained, considering indicators and data sources as shown on Table 3. On Yelp some indicators performed quite well, such as Cultural Equipment (0.889), Green Area (0.886), Dental Service (0.886), Other Health Care Service (0.848) while others like Health Center (0.228) and Grocery Store and Similar (0.304) were not as good. On Foursquare the results were similar, but the indicators had a smaller accuracy than Yelp's. Foursquare data outperforms Yelp only on Bank Agency, with a value of 0.722 (0.506 for Yelp). There is a tie on Post Office and Health Center indicators as to accuracy, but Foursquare presents a slightly better precision on both and a better recall on the former.

In general, it was achieved good accuracy but poor results for precision and recall. To investigate this the values for each interval were analyzed to seek any abnormal behavior. As seen on Table 4 and Table 5, the precision and recall in intervals "1" and "5" were significantly better than intervals "2", "3" and "4". One possible cause is that intervals "1" and "5" have a wider range of values, so they become a bigger target to hit, increasing precision and recall. This polarization on the results of precision and recall may suggest that if a binary classification were to be used, for instance to identify deprived and well-off regions, good results could be achieved.

After the results on IOL indicators, we investigated how much the indicators calculated from LBSN data would impact IOL variables compared to official data. Four different datasets were used to evaluate this. The first two are the Foursquare and Yelp data that have been used so far. The third was created getting the IOL indicator data from Foursquare or Yelp with the best value for accuracy (Table 3). In case of a tie on accuracy, the precision and recall values, in this order, were used to untie. If the data were available from only one of the LBSN datasets, it was used. We called this dataset FUY (Foursquare–Union–Yelp). The fourth dataset is based on the intersection of Foursquare and Yelp. We used the same approach of FUY to choose the indicators, but only for those that are present on both datasets.

Table 2. IQVU variables, components, indicators, and their sources of data

Variable	Component	Indicator	Source
Food Supply	Food Supply equipment	- Hyper and supermarket	F
		- Grocery store and similar	F/Y
Culture	Commerce and cultural service	- Cultural equipment (e.g. theater)	F/Y
		- Bookstore and stationery	F/Y
		- Movie rental store	Y
		- Magazine stand	F/Y
Education	Childhood education	- % of students enrolled	O
	Elementary school	- % of students enrolled	O
		- Approval rating	O
	Secondary school	- % of students enrolled	O
		- Approval rating	O
Sport	Public space for recreation	- Sport court, field, jogging track	F
Housing	Quality of housing	- Adequate residential area per capita	O
		- Residential finishing	O
	Residential security	- Geological risk	O
Urban Infrastructure	Environmental healthiness	- Environmental healthiness	O
	Electrical power supply	- Electrical power supply	O
	Paving	- Possibility of access	O
	Collective transport	- Number of vehicles	O
		- Frequency of lines	O
Environment	Acoustic comfort	- Tranquility (lack of noise)	O
	Air quality	- Absence of collective polluters	O
	Green area	- Green area	F/Y
Health	Health care	- Health centers	F/Y
		- Other health care services	Y
		- Dental services	Y
	Health surveillance	- Absence of life years lost	O
Urban Service	Personal service	- Bank agency	F/Y
		- Gas station	F
		- Drugstore	F/Y
	Communication and IT service	- Post office	F/Y
		- Public space for digital inclusion	O
		- Public phone	O
Urban Security	Personal security	- Absence of crime against person	O
	Patrimonial security	- Absence of crime against patrimony	O
	Safety in traffic	- Absence of traffic accident	O

Source column: (F) for Foursquare, (Y) for Yelp and (O) for Official data source.
Indicator values were normalized between 0 and 1, and then aggregated on IOL variables for each UP by applying a simple average.

Table 3. Accuracy (acc.), Precision (prec.) and Recall (rec.) of indicators calculated from Foursquare and Yelp. The Precision and Recall values are the mean of the five classes considered

Indicators	Foursquare			Yelp		
	acc.	prec.	rec.	acc.	prec.	rec.
Hyper and supermarket	**0.468**	0.416	0.450	—	—	—
Grocery store and similar	0.127	0.112	0.156	**0.304**	0.200	0.204
Magazine stand	0.570	0.261	0.286	**0.582**	0.288	0.323
Cultural equipment	0.823	0.347	—	**0.899**	—	—
Bookstore and Stationery	0.532	—	0.197	**0.544**	0.196	0.221
Movie rental store	—	—		**0.544**	0.384	0.375
Sport court, field and jogging track	**0.481**	0.438	0.427	—	—	—
Green area	0.848	—	—	**0.886**	—	—
Health center	0.228	0.254	0.201	0.228	0.243	0.201
Dental service	—	—	—	**0.886**	0.497	0.469
Other health care service	—	—	—	**0.848**	0.488	0.403
Banking center	**0.722**	0.296	0.352	0.506	0.159	0.173
Post office	**0.734**	0.385	0.541	0.734	0.288	0.269
Drugstore	0.443	0.231	0.299	**0.608**	0.537	0.534
Gas station	**0.595**	0.403	0.398	—	—	—
Mean	0.547	0.314	0.331	0.631	0.328	0.317

Table 4. Precision for the five interval classes of IOL results calculated with data from Foursquare and Yelp

Indicators	Foursquare					Yelp				
	1	2	3	4	5	1	2	3	4	5
Hyper and supermarket	0.700	0.250	0.091	0.467	0.571	—	—	—	—	—
Grocery store and similar	0.136	0.000	0.167	0.067	0.190	0.143	0.000	0.125	0.182	0.548
Magazine Stand	0.867	0.000	0.000	0.143	0.294	0.796	0.000	0.143	0.000	0.500
Cultural Equipment	0.983	0.000	0.000	0.000	0.750	1.000	0.000	0.000	—	0.857
Bookstore and stationery	0.569	—	—	0.000	0.167	0.606	0.000	0.000	0.000	0.375
Movie Rental Store	—	—	—	—	—	0.821	0.000	0.250	0.250	0.600
Sport courts, fields and jogging tracks	0.528	0.143	0.417	0.571	0.529	—	—	—	—	—
Green Area	0.890	—	—	0.000	0.500	0.956	0.000	—	0.000	0.833
Health Centers	0.197	0.250	0.250	0.000	0.571	0.250	0.091	0.500	0.000	0.375
Dental services	—	—	—	—	—	0.984	0.000	0.000	0.500	1.000
Other Health care services	—	—	—	—	—	0.938	0.000	0.500	0.000	1.000
Banking Center	0.765	0.000	0.000	0.000	0.714	0.722	0.000	0.000	0.000	0.071
Post Office	0.885	0.167	0.000	0.250	0.625	0.839	0.000	0.000	0.000	0.600
Drugstore	0.556	0.000	0.100	0.000	0.500	0.714	0.444	0.455	0.571	0.500
Gas Station	0.829	0.286	0.000	0.364	0.538	—	—	—	—	—
Mean	**0.659**	0.110	0.102	0.155	**0.496**	**0.731**	0.045	0.179	0.137	**0.605**

Table 5. Recall for the five interval classes of IOL results calculated with data from Foursquare and Yelp

Indicators	Foursquare					Yelp				
	1	2	3	4	5	1	2	3	4	5
Hyper and supermarket	0.538	0.600	0.200	0.500	0.414	—	—	—	—	—
Grocery stores and Similar	0.375	0.000	0.250	0.063	0.093	0.250	0.000	0.125	0.250	0.395
Magazine stand	0.750	0.000	0.000	0.125	0.556	0.750	0.000	0.200	0.000	0.667
Cultural equipment	0.843	—	0.000	0.000	0.857	0.929	—	0.000	0.000	0.857
Bookstore and stationery	0.911	0.000	0.000	0.000	0.071	0.889	0.000	0.000	0.000	0.214
Movie rental store	—	—	—	—	—	0.767	0.000	0.182	0.273	0.652
Sport court, field and jogging track	0.528	0.143	0.500	0.400	0.563	—	—	—	—	—
Green area	0.970	0.000	0.000	—	0.286	0.970	0.000	0.000	—	0.714
Health center	0.571	0.167	0.077	0.000	0.190	0.619	0.167	0.077	0.000	0.143
Dental service	—	—	—	—	—	0.969	0.000	0.000	0.500	0.875
Other Health care service	—	—	—	—	—	0.968	0.000	0.333	0.000	0.714
Banking center	0.929	0.000	0.000	0.000	0.833	0.696	0.000	0.000	0.000	0.167
Post office	0.836	0.200	0.000	1.000	0.667	0.945	0.000	0.000	0.000	0.400
Drugstore	0.833	0.000	0.091	0.000	0.571	0.833	0.267	0.455	0.400	0.714
Gas station	0.791	0.200	0.000	0.364	0.636	—	—	—	—	—
Mean	**0.740**	0.119	0.093	0.223	**0.478**	**0.799**	0.039	0.114	0.129	**0.543**

We called this FIY (Foursquare–Intersects–Yelp). Table 6 shows the results of the four datasets by IOL variable. The variables without results did not use any LBSN data.

Yelp results had a better average accuracy, precision and recall than Foursquare. This was expected as the indicators calculated with Yelp data also performed better. This makes the different uses of Yelp and Foursquare more evident. Yelp has an approach to make people review and evaluate business places, so this reflects on a database of places that get more matches on categories that are used by IQVU. The FUY dataset had a better accuracy than Foursquare, but worse than Yelp alone. This was caused mainly by the sports variable, whose values came from Foursquare and had low accuracy (0.481). The FIY dataset presented the best accuracy (0.744), but this was expected as it is the one with less interference from LBSN data.

Aiming to visualize the spatial pattern of the results, maps were generated to compare the IOL values for the variables that were calculated for Foursquare and Yelp. Figures 7-11 show IOL results for the Units of Planning of Belo Horizonte for the variables Food Supply, Culture, Environment, Health and Urban Service, respectively. It is possible to observe that Foursquare and Yelp present a similar pattern and, as in the case of indicator results, accuracy in the "1" and "5" intervals is higher than in the intermediate ones.

Limitations

This work has three limitations that need to be observed. First, LBSN data can be biased. LBSNs and other online social networks data are tied to the profile of their users, generally young and technology-

Table 6. Accuracy (acc.), Precision (prec.) and Recall (rec.) of IOL variables calculated from Foursquare, Yelp, FUY and FIY

IOL Variables	Foursquare			Yelp			FUY			FIY		
	acc.	prec.	rec.	acc.	prec.	rec.	acc.	prec.	rec.	acc.	prec.	rec.
Food Supply	0.342	0.319	0.334	0.582	0.528	0.525	0.418	0.404	0.373	0.582	0.528	0.525
Culture	0.696	0.410	0.362	0.810	0.760	0.518	0.810	0.760	0.518	0.797	0.560	0.468
Education	—	—	—	—	—	—	—	—	—	—	—	—
Sport	0.481	0.438	0.427	—	—	—	0.481	0.438	0.427	—	—	—
Housing	—	—	—	—	—	—	—	—	—	—	—	—
Urban Infrastructure	—	—	—	—	—	—	—	—	—	—	—	—
Environment	0.734	0.530	0.524	0.886	0.872	0.825	0.886	0.872	0.825	0.886	0.872	0.825
Health	0.734	0.479	0.513	0.696	0.429	0.418	0.696	0.429	0.418	0.734	0.479	0.513
Urban Security	—	—	—	—	—	—	—	—	—	—	—	—
Urban Service	0.684	0.429	0.471	0.709	0.521	0.414	0.722	0.471	0.530	0.722	0.741	0.578
Mean	**0.612**	0.434	0.439	**0.737**	0.622	0.540	**0.669**	0.562	0.515	**0.744**	0.636	0.582

Figure 7. Spatial pattern of the results for the IOL variable Food Supply. (a) Official data, (b) Foursquare and (c) Yelp.

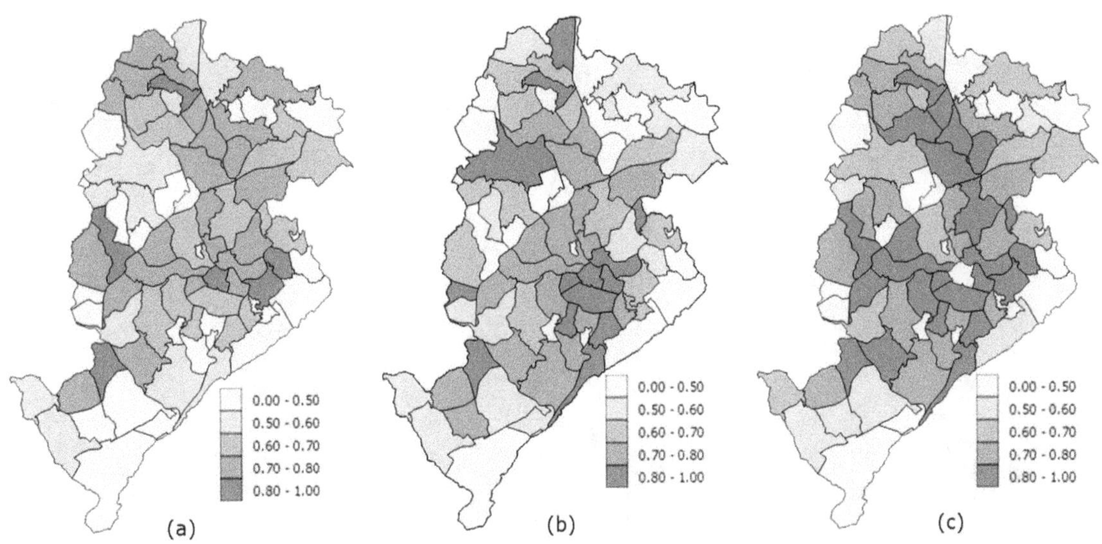

friendly (Quercia & Saez, 2014). This aspect can cause geographic coverage of those locations to be very good around neighborhoods where young population live, work or have fun (Rost, Barkhuus, Cramer, & Brown, 2013) and not so good on other areas. A region that does not have many locations on a LBSN does not necessarily represent the ground truth, as it can be a region in which smartphone usage is low. These biases decrease as the penetration rate of LBSN use increases. As smartphones and data plans get more affordable we expect to notice better quality of LBSN data in the near future.

Figure 8. Spatial pattern of the results for the IOL variable Culture. (a) Official data, (b) Foursquare and (c) Yelp.

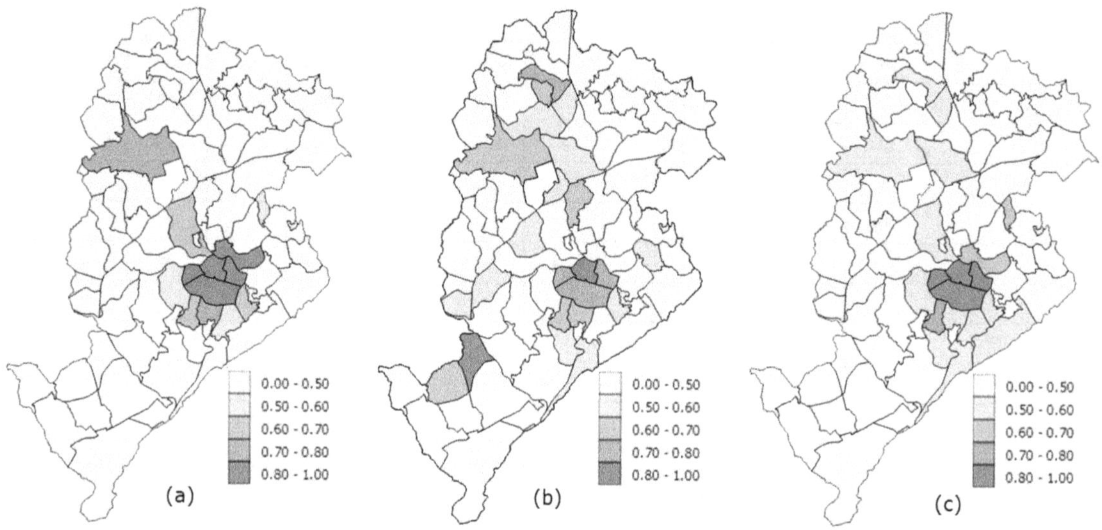

Figure 9. Spatial pattern of the results for the IOL variable Environment. (a) Official data, (b) Foursquare and (c) Yelp.

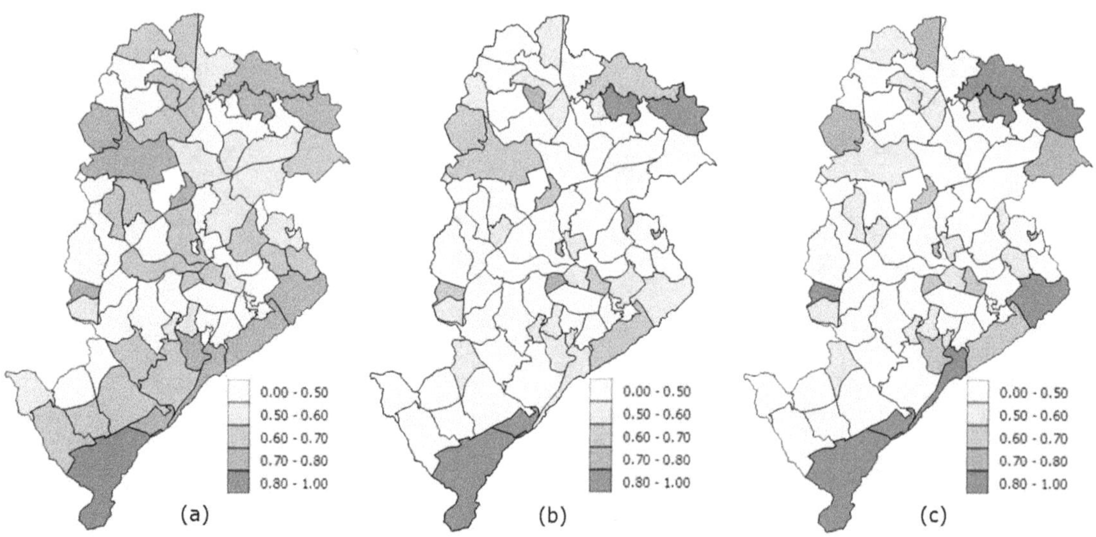

The second limitation regards the lack of some types of information from LBSN. As seen on Table 2, it was not possible to obtain all indicator values from the LBSNs used on this work, but this will not necessarily always be true. Currently, collecting the missing data required by this work depends on open data policies in place for government information-producing organizations. The authors believe that this can be fixed with upcoming open government data initiatives and legislation that supports information dissemination in machine-readable formats and services. PDF files containing images of tables and maps do not qualify as such.

Figure 10. Spatial pattern of the results for the IOL variable Health. (a) Official data, (b) Foursquare and (c) Yelp.

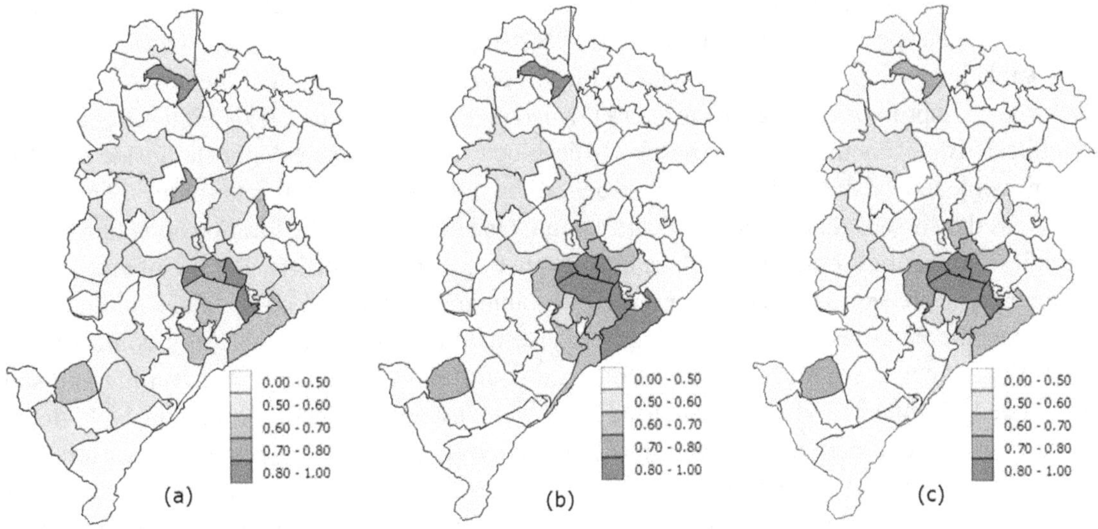

Figure 11. Spatial pattern of the results for the IOL variable Urban Services. (a) Official data, (b) Foursquare and (c) Yelp.

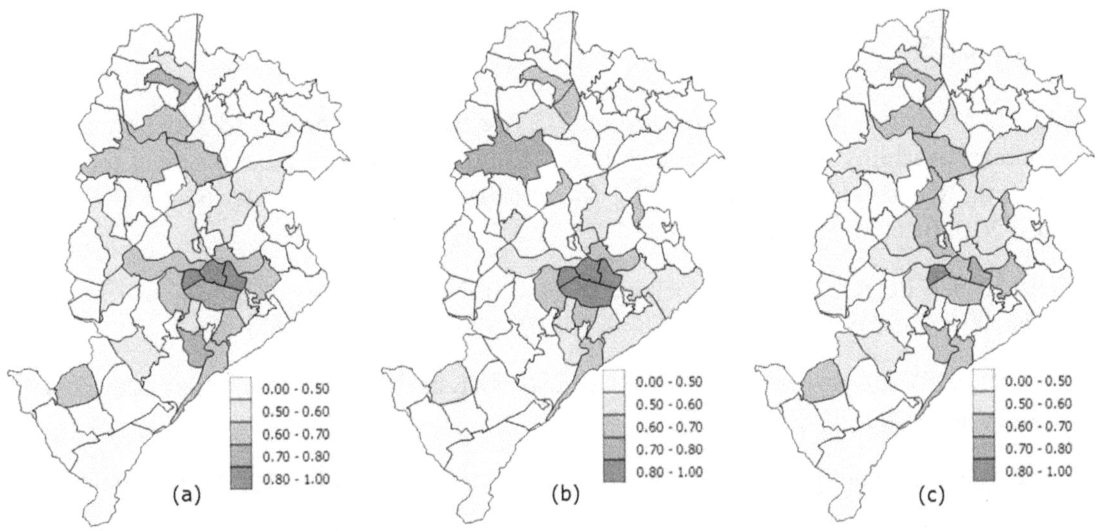

The third limitation, and a motivation for this work, is the lack of more frequently updated results of IQVU. All results for this work are based on data collected in the third quarter of 2015, but the latest IQVU results are from 2012. So we are comparing current LBSN data with official data that is three years old. Improving the results of this work, and possibly calculating a quality of urban life index with online and easily available social network and official government data may enable us to produce frequent results in the future.

FINAL REMARKS

This work shows the potential use of LBSN as data source to calculate a quality of life index for the city of Belo Horizonte. Besides the limitations of LBSN data, results encourage the expansion of research work towards improving the quality of data and methods for the calculation of the index.

One way to improve is to enhance the dataset with data from additional LBSNs and better integration of the collected data. For this purpose, we have to develop ways to eliminate duplicates inside each LBSN and the possible redundancies that may be introduced by integrating several data sources. This is not an easy task, as they do not share a unique identification. One can explore the location and other attributes to check if two locations represent the same business in two or more datasets. The use of ontologies and schema mapping are approaches used for spatial data integration that can be evaluated on future work.

Also, much data collected were not used because they did not have a category compatible with some IQVU indicator. We argue that some of them could affect the quality of urban life and should be considered like other services that are used daily by many people, such as restaurants. On the other hand, some indicators seem outdated (for instance, movie rentals and pay phones) to measure quality of urban life. We suggest that internet connection speed would be a better indicator of the availability of communication services, and required as infrastruçture for modern IT and online entertainment. The authors intend to use volunteered geographic information as a framework to survey popular perception of quality of urban life, and then dynamically measure and adjust the results according to city dynamics and citizen behavior.

REFERENCES

Badland, H., Whitzman, C., Lowe, M., Davern, M., Aye, L., Butterworth, I., & Giles-Corti, B. et al. (2014). Urban liveability: Emerging lessons from Australia for exploring the potential for indicators to measure the social determinants of health. *Social Science & Medicine*, *111*, 64–73. doi:10.1016/j.socscimed.2014.04.003 PMID:24762261

Blomquist, G. C., Berger, M. C., & Hoehn, J. P. (1988). New Estimates of Quality of Life in Urban Areas. *The American Economic Review*, *78*(1), 89–107. doi:10.2307/1814700

Foursquare. (2015). *Foursquare Categories*. Retrieved October 15, 2015, from https://developer.foursquare.com/categorytree

Foursquare. (2016). *About Us*. Retrieved July 14, 2016, from https://pt.foursquare.com/about

Gavrilidis, A. A., Ciocănea, C. M., Niţă, M. R., Onose, D. A., & Năstase, I. I. (2016). Urban Landscape Quality Index – Planning Tool for Evaluating Urban Landscapes and Improving the Quality of Life. *Procedia Environmental Sciences*, *32*, 155–167. doi:10.1016/j.proenv.2016.03.020

Hagerty, M. R., Cummins, R. A., Ferriss, A. L., Land, K., Michalos, A. C., Peterson, M., & Vogel, J. et al. (2001). Quality of Life Indexes for National Policy: Review and Agenda for Research. *Social Indicators Research*, *55*(1), 1–96. doi:10.1023/A:1010811312332

Helliwell, J. F. (2003). Hows life? Combining individual and national variables to explain subjective well-being. *Economic Modelling*, *20*(2), 331–360. doi:10.1016/S0264-9993(02)00057-3

IBGE. (2015). *Instituto Brasileiro de Geografia e Estatística*. Retrieved October 6, 2015, from http://www.ibge.gov.br/

Jacobs, J. (1961). *The Death and Life of Great American Cities*. Vintage Books.

Lemos, M. B., Esteves, O. de A., & Simões, R. F. (1995). Uma metodologia para a construção de um índice de qualidade de vida urbana. *Nova Economia*, *5*(2), 157–176. Retrieved from http://ideas.repec.org/a/nov/artigo/v5y1995i2p157-176.html

Marans, R. W. (2015). Quality of urban life & environmental sustainability studies: Future linkage opportunities. *Habitat International*, *45*(P1), 47–52. doi:10.1016/j.habitatint.2014.06.019

Nahas, M. I. P. (2002). *Bases teóricas, metodologia de elaboração e aplicabilidade de indicadores intra-urbanos na gestão municipal da qualidade de vida urbana em grandes cidades : o caso de Belo Horizonte*. Universidade Federal de São Carlos. Retrieved from http://www.bdtd.ufscar.br/htdocs/tedeSimplificado/tde_arquivos/2/TDE-2008-05-09T08:37:00Z-1824/Publico/1753.pdf

PBH. (2014). *Relatório geral sobre o cálculo do Índice de Qualidade de Vida Urbana de Belo Horizonte (IQVU-BH)*. Retrieved from http://portalpbh.pbh.gov.br/pbh/ecp/files.do?evento=download&urlArqPlc=relatorio_iqvu_rev-set14.pdf

PBH. (2015). *Portal PBH*. Retrieved October 15, 2015, from http://portalpbh.pbh.gov.br/pbh/

Pissourios, I. A. (2013). An interdisciplinary study on indicators: A comparative review of quality-of-life, macroeconomic, environmental, welfare and sustainability indicators. *Ecological Indicators*, *34*, 420–427. doi:10.1016/j.ecolind.2013.06.008

Pontes, T., Vasconcelos, M., Almeida, J., Kumaraguru, P., & Almeida, V. (2012). We know where you live. In *Proceedings of the 2012 ACM Conference on Ubiquitous Computing - UbiComp '12* (p. 898). New York: ACM Press. doi:10.1145/2370216.2370419

Quercia, D., & Saez, D. (2014). Mining Urban Deprivation from Foursquare: Implicit Crowdsourcing of City Land Use. *IEEE Pervasive Computing / IEEE Computer Society [and] IEEE Communications Society*, *13*(2), 30–36. doi:10.1109/MPRV.2014.31

Rezvani, M. R., & Mansourian, H. (2013). Developing Small Cities by Promoting Village to Town and its Effects on Quality of Life for the Local Residents. *Social Indicators Research*, *110*(1), 147–170. doi:10.1007/s11205-011-9921-7

Roback, J. (1982). Wages, Rents, and the Quality of Life. *Journal of Political Economy*, *90*(6), 1257–1278. doi:10.1086/261120

Rodrigues, M. C. P. (1991). O desenvolvimento social nas Regiões Brasileiras. *Ciência Hoje*, *13*(76), 39–45.

Rogerson, R. J. (1999). Quality of Life and City Competitiveness. *Urban Studies (Edinburgh, Scotland)*, *36*(5), 969–985. doi:10.1080/0042098993303

Rost, M., Barkhuus, L., Cramer, H., & Brown, B. (2013). Representation and communication. In *Proceedings of the 2013 conference on Computer supported cooperative work - CSCW '13* (p. 357). New York: ACM Press. doi:10.1145/2441776.2441817

Shelton, T., Poorthuis, A., & Zook, M. (2015). Social media and the city: Rethinking urban socio-spatial inequality using user-generated geographic information. *Landscape and Urban Planning*, *142*, 198–211. doi:10.1016/j.landurbplan.2015.02.020

Sung, H., Lee, S., & Cheon, S. (2015). Operationalizing Jane Jacobss Urban Design Theory: Empirical Verification from the Great City of Seoul, Korea. *Journal of Planning Education and Research*, *35*(2), 117–130. doi:10.1177/0739456X14568021

The World Bank. (2015). *Human Development Reports*. Retrieved from http://hdr.undp.org/en/content/human-development-index-hdi

UNDP. (2015). *Human Development Index (HDI)*. Retrieved from http://hdr.undp.org/en/content/human-development-index-hdi

Venerandi, A., Quattrone, G., Capra, L., Quercia, D., & Saez-Trumper, D. (2015). Measuring Urban Deprivation from User Generated Content. In *Proceedings of the 18th ACM Conference on Computer Supported Cooperative Work & Social Computing - CSCW '15* (pp. 254–264). New York: ACM Press. doi:10.1145/2675133.2675233

Węziak-Białowolska, D. (2016). Quality of life in cities – Empirical evidence in comparative European perspective. *Cities (London, England)*, *58*, 87–96. doi:10.1016/j.cities.2016.05.016

Yelp. (2015). *Yelp Categories*. Retrieved October 15, 2015, from https://www.yelp.com.br/developers/documentation/v2/all_category_list

Yelp. (2016). *About Us*. Retrieved August 15, 2016, from https://www.yelp.com/about

Yuan, J., Zheng, Y., & Xie, X. (2012). Discovering regions of different functions in a city using human mobility and POIs. In *Proceedings of the 18th ACM SIGKDD international conference on Knowledge discovery and data mining - KDD '12* (p. 186). New York: ACM Press. doi:10.1145/2339530.2339561

Zheng, Y. (2011). Location-based social networks: Users. In Y. Zheng & X. Zhou (Eds.), *Computing with Spatial Trajectories* (pp. 243–276). New York: Springer New York. doi:10.1007/978-1-4614-1629-6_8

KEY TERMS AND DEFINITIONS

API (Acronym for Application Programming Interface): A set of definitions of functions, protocols, tools and programming patterns used to access data or services provided by a software (application, library, platform, web-based).

Crawler: A software that systematically collects data from web resources.

Foursquare: A general purpose location-based social network in which users can share their location and search for places and friends.

IBGE (Acronym for *Instituto Brasileiro de Geografia e Estatística*, in Portuguese): Brazilian Institute of Geography and Statistics is a public organization responsible for collecting, management and providing data / information regarding several aspects such as demographic census.

IOL (Acronym for *Índice de Oferta Local*, in Portuguese):: It is a component of IQVU calculation methodology which measures the availability of services inside an administrative region (unit of planning).

IQVU (Acronym for *Índice de Qualidade de Vida Urbana*, in Portuguese): A multidimensional index created to measure urban inequalities, inside a city, regarding availability and accessibility of services.

Location Based Social Networks (LBSN): A social network where members share resources with / related to location information.

Python: A high-level general-purpose programming language widely used, among other aims, for gathering web data.

Unity of Planning:: A spatio-political subdivision of Belo Horizonte city in several administrative regions which are composed by the aggregation of census sectors.

Volunteered Geographical Information (VGI): It is a subtype of user-generated content services in which users create and share geographical contents.

Yelp: A business focused location-based social network where users can search nearby business places, access information and make reviews about them.

Chapter 11
Improving Accessibility Through VGI and Crowdsourcing

Igor Gomes Cruz
Federal University of Campina Grande, Brazil

Claudio Campelo
Federal University of Campina Grande, Brazil

ABSTRACT

Accessibility is an important element in the life of those who have certain limitations, such as the physically disabled and visually impaired people. However, one of the greatest challenges for this group is to find paths and areas adapted to their limitations while performing their daily activities, since not all environments they explore have these characteristics. Volunteered Geographic Information (VGI) and the crowdsourcing technique appear to be quite useful to develop solutions to overcome these challenges, since these techniques are naturally cheap as they rely on human sensors as the main agent of information delivery. In this chapter, we discuss how these techniques can help mitigate accessibility problems and present some existing research and applications in the field.

INTRODUCTION

According to the World Health Organization, "more than a billion people in the world today experience disability. These people generally have poorer health, lower education achievements, fewer economic opportunities and higher rates of poverty. This is largely due to the barriers they face in their everyday lives, rather than their disability. Disability is not only a public health issue, but also a human rights and development issue."

Urban mobility can be defined as the people's capacity to move around the city, to live and to interact with the space. This ability can be strongly affected by architectonic obstacles which represent a physical impediment to the exercise of citizenship by people with special needs, such as disabled elderly people. (Mirri, Prandi & Salomoni, 2016)

The term accessibility has various definitions and it can be understood as a medium to obtain barrier-free environments which affect the usability or the utilization of services associated to these areas, depending

DOI: 10.4018/978-1-5225-2446-5.ch011

on the context in which it is employed. Inside this group, we can highlight the environmental accessibility which refers to how accessible a place is to a group of people. This kind of accessibility describes the relation person-environment, in which the abilities of a person are opposed to the environmental properties for a given place. The environmental accessibility extends to a large group of users, which includes children, elders, people walking with baby carriages or with heavy luggage, among others. It basically has the goal of equality. (Laakso, 2014)

It is clear that human navigation is a quite complex activity, essential to people in environments which require movement. It is a process which involves the association of many locomotion components, discovery of paths/routes, vision (responsible for the capture of static and dynamic information of the environment) and the locomotion medium (on foot, cars, motorcycles, wheelchairs, among others). Thus, for someone with visual impairment or movement disability, the navigation in unknown environments becomes a hard task. This happens due to the specific necessities of each individual. Consequently, many disabled people deprive themselves from exploring unknown environments because of these problems and of the risk they can offer to their security and health. (Vanclooster, Weghe & Maeyer, 2016; Brock, Truillet, Oriola, Picard & Jouffrais, 2012; Coughlan & Shen, 2013; Schmitz, Becker, Blessing & Großmann, 2011; Santos, Forest, Dutra & Almeida, 2012; Neis, 2014)

Thus, environmental accessibility has a fundamental role for those who have some kind of special need, as these individuals have many difficulties when trying to move around public environments, since such places sometimes are not adequately adapted to users with some disability, whether it is physical or visual. People with disabilities should be in constant interaction with their sociocultural medium. Although the construction of accessible places is a goal in current society, urban areas, specifically the ones for pedestrians, are often inadequate for locomotion of the elderly and disabled, since these users have greater and more remarkable needs when it comes to accessibility. Therefore, these environments still represent one of the most real examples of the difficulty of locomotion felt by people with disabilities. Thus, geospatial data play an important role in offering help in choosing the best routes according to the population needs.

However, another barrier which hinders people's independent locomotion is the lack of information about the urban environment and its accessibility. The maintenance of up-to-date databases is considerably costly, thus the availability of information about urban accessibility has been traditionally limited for many urban centers. Consequently, traditional approaches based on databases (e.g., maintained by government authorities) have appeared to be not sufficiently effective in the field, thus lowering the chances of becoming a social inclusion medium (Mirri, Prandi & Salomoni, 2016).

In order to assist the users with special needs, a geospatial database must contain the maximum amount of information on the characteristics which affect the accessibility, since the pedestrians locomotion mediums are extremely large and suffer constant changes in their structure. Thus, crowdsourcing, social media and Volunteered Geographical Information (VGI) offer a great potential to the collection of data at pedestrian level. The OpenStreetMap – OSM (www.openstreetmap.org) is one of the most representative examples of dynamic maps. It is a voluntary project offering the users an online map, which grows with the help of non-official sensors, i.e., amateurs which contribute daily with information updates. (Qin, Aburizaiza, R. Rice, Paez & M. Rice, 2015; Laakso, 2014; Antoniou & Skopeliti, 2015; Basiouka, Potsiou & Bakogiannis, 2015)

The mapping of dynamic geographical phenomena is a hard task, due to the frequency of changes they are subject to. In urban areas, for instance, pedestrian walkways and the transportation infrastructure are among the ones which suffer the most with frequent changes, since they are affected by obstacles in sidewalks, construction detours, among others. For individuals with vision and mobility disabilities, changes in pedestrian walkways (even temporary) may cause various problems, due to the necessity of route change, inconvenience and the risks posed to their safety and health. Thereby, the updated mapping of these areas becomes something essential. Nevertheless, remote sensing and automatic characteristics extraction, in this scenario, are not a realistic approach, due to the cost and the requirement of frequent updates. Hence, a different approach is needed. Thereby, VGI, which is based in a new form of participative mapping and in the use of crowdsourcing, has been successfully applied to the mapping of such phenomena. (Brock et al., 2012; Wen, 2013)

It is generally accepted that maps offer significant help in activities like orientation, mobility and support to streets and buildings locations, as well as the estimate of the directions and distances between points. A great number of people with visual impairment use technology to perform tasks in their daily life. The increasing use of computers, smartphones and tablets can offer efficient forms of interaction, transforming these devices into valuable tools to offer users an intuitive interface, dedicated to improve their visualization and mobility. As a result, mobile platforms are rapidly evolving, aiming the development of supporting technologies. Recently, various smartphone apps were developed with the goal of helping disabled people. Such tools, although promising, still have proven incapable of offering some detailed information about the location of the user, such as precise orientation with respect to a pedestrian crossing, current status of signaling lights, barriers and surface inclination. (Coughlan & Shen, 2013; Koukourikos & Papadopoulos, 2015; Laakso, 2014; Prandi et al., 2014)

In the recent years, it could be noticed a constant effort to the development of applications capable of detecting such problems, making VGI a potential solution. Maps based on OSM arose with a variety of purposes, such as skiing, walking or public transportation. In addition, applications and projects like OpenRouteService and Open TripPlanner demonstrate VGI can represent a reliable source to the development of pedestrian routing and navigation software solutions applied to disabled users. (Laakso, 2014)

Moreover, we are witnessing the emergence of a new era of online geographic tools like Google Maps, Google Earth NASA World Wind, OpenStreetMap, among others, which have attracted an immense number of users. For example, in the five largest European economies, 50% of users access online maps through their personal computers, and 35% through mobile phones daily. However, not all people can have access to this type of information due to its graphical nature. Some groups of users, such as the visual impaired, are examples of it. (Calle-Jimenez & Luján-Mora, 2015)

This chapter discusses the importance of Volunteered Geographic Information applied to accessibility, demonstrating the efficacy of this phenomenon in the reduction of mobility problems, providing essential information in the context of physically and visually disabled people. In addition, the chapter presents a literature review, comparing and highlighting the strengths and weaknesses of existing research using VGI and crowdsourcing to mitigate accessibility issues for those user groups.

The next section introduces some relevant background concepts and discusses the motivations for development of VGI and crowdsourcing systems to physically disabled and visually impaired people. Then the following section describes how VGI and crowdsourcing phenomenon can help reduce the locomotion difficulties of disabled people. Then the next section highlights current researches and innovative accessible applications based on VGI and crowdsourcing to assist disabled people in their navigation activities. Finally, the last section concludes de chapter and points to future directions.

BACKGROUND AND MOTIVATION

This section presents some of the motivations for the development of solutions using crowdsourcing and VGI aiming to mitigate accessibility issues encountered by physical and visually impaired.

Physically Disabled People

According to World Report on Disability (2011), "Disability is part of the human condition. Almost everyone will be temporarily or permanently impaired at some point in life, and those who survive to old age will experience increasing difficulties in functioning. Most extended families have a disabled member, and many non-disabled people take responsibility for supporting and caring for their relatives and friends with disabilities" (p. 3).

According to World Health Organization (WHO), disability is a generic term, which encompasses activity limitation (difficulty in the execution of a task or action), participation restriction (problems caused to an individual when getting involved in daily life situations) and other disabilities in general. It can be represented as a problem in the body function or structure. (World Health Organization, n.d.b.).

Physical disability is characterized by complete or partial alterations which can affect one or more segments of human body, compromising its physical functions. People with such limitations have difficulties when performing their daily activities, like moving from one place to another; and therefore, many times, depend on other people for the execution of those. (Silva et al., 2015)

This limitation hinders the execution of simpler tasks. Frequently, the basic activity of moving becomes arduous, depending on other people's help to their execution. Wheelchair users, for instance, have a multitude of problems when trying to move through external environments. A simple deformation on the sidewalks surface can offer risks to the health of these individuals, resulting in accidents. However, construction companies do not frequently follow these rules and many environments are built in a way which does not offer the proper accessibility to the physically disabled people, preventing the utilization and locomotion through these places by this group of users.

Thus, the utilization of human sensors, in this context, can be extremely useful. Applications can, through the VGI techniques and crowdsourcing, offer information about the environments, presenting those who possess the proper alterations to accommodate the physically disabled people or relating the problems found in the attended places. So, as one identifies any type of problem in the environments used, one can insert information in the systems, about what type of inadequacy was found, like tight corridors, inexistence of accessible restrooms, inadequate surface elevations, among others. The utilization of the traditional mediums of acquisition of geographical information in this kind of situation is unfeasible due to the cost needed to update the data in a timely manner to the provision of correct information to the users.

Blind People

An individual characterized as blind has either the total loss of the visual capacity or only the loss of light perception. The one identified as having low or reduced vision, presents a reduction in the visual capacity, only being able to detect masses, colors, shapes, having limitations in long sight, and only having the possibility of identifying objects when they are close. One also presents difficulty to notice visual aspects such as: disproportional features in space, tridimensional representations, composed shapes, depth, movement, objects or materials over similar backgrounds, objects with low light and distinctive details in the forms and inside the figures. The individuals with this kind of disability, in spite of having great capacity of adaptation through other senses, still present some limitations like, for instance, the limitation of the communication capacity, considering that the process of communication consists of receiving, processing and transmitting written and or body information; and the capacity of locomotion, since, for example, individuals which are visually impaired use other senses to move in safety. (Brock et al., 2012; Coughlan & Shen, 2013; Favretto, Carvalho & Canini, 2008; Martinez-Sala, Losilla, Sánchez-Aarnoutse & García-Haro, 2015; Liao, Rakauskas & Rayankula, 2011)

Human beings gather information about the world through multiple sensorial channels (touch, audition and vision) and share them to enhance their comprehension of the world. Over many years, cartographers developed symbolic methods for encoding and communication in order to make spatial information available graphically. However, those strategies are being strongly studied with respect to their form and function, since, generally, the conclusion of these cartographic processes depends mainly on the vision as a medium of interaction. Nevertheless, some people, like the visually impaired, do not have the required visual acuity to the creation, use and comprehension of the traditional maps. Then, as an alternative for this group of users, several professionals of this area developed methods to adapt or substitute the presented information visually through other sensorial domains. (Rice et al., 2013)

The constant growth of the utilization of the technology and mobile applications increases the access to information available in the web, including geographical information. Nonetheless, for people with visual impairment, this interaction is hard when the maps are not accessible. Not all people have access to this kind of information through their devices due to the graphical nature of this geographical data. Visually impaired people, for instance, can have problems when trying to make use of this kind of application. (Calle-Jimenez & Luján-Mora, 2015)

Individuals with this type of disability, despite having great capacity of adaptation through other senses, still present some limitations, like the safe locomotion. Human navigation is a behavior which depends mainly on the vision, since it is through it the individuals capture a great amount of the information around them. Therefore, for a person with visual impairment, the locomotion becomes very hard. These individuals face great challenges when trying to move around external or internal environments due to the complexity of the environment around them and the presence of obstacles in their path (Koester, Lunt & Stiefelhagen, 2016).

Therefore, many of the visually impaired people desire to have the freedom and safety to explore the great variety of existent environments, just like the ones without this disability do. For this purpose, various types of studies are being developed to help the mobility of this group of people. Allied with the mobile technology, present in people's daily life, navigation apps based on maps and routing offer significant help in matters like orientation, mobility and location support, as well as the estimate of the directions and distances. Thus, a great number of people with visual impairment use these technologies to help their locomotion. However, this kind of people usually finds great obstacles to the access of spatial information, since the interaction with this type of application is performed mainly through the vision.

Accessibility Issues (Problems Related to Accessible Environments or People With Disabilities)

According to the United Nations (2007), "Accessibility is about giving equal access to everyone. Without being able to access the facilities and services found in the community, persons with disabilities will never be fully included. In most societies, however, there are innumerable obstacles and barriers that hinder persons with disabilities. These include such things as stairs, lack of information in accessible formats such as Braille and sign language, and community services provided in a form which persons with disabilities are not able to understand. Although some of the more costly accessibility provisions in the Convention can be implemented progressively, there are a number of low-cost, low-tech accessibility solutions that would have immediate benefits".

According to ABNT NBR 9050 (ABNT, 2015), accessibility is defined as the "possibility and reach, perception and understanding condition for utilization, with security and autonomy, of housing spaces, urban equipments, edifications, transportation, information and communication, including their systems and technologies, as well as other services and installations open to the public, of public or private collective use, in both urban and rural areas, by a person with disability or reduced mobility".

Therefore, accessibility is fundamental in all environments of collective usage. In order to be considered accessible, an environment must offer conditions to come and go with independence, comfort and safety whether the displacement is horizontal or vertical, with or without stairs, corridors, ramps and elevators, or any other medium, being properly signaled, if necessary, by plates, signs and any essential information. Thus, the presence of any kind of obstacle which can hamper this capacity is considered an architectonic barrier. (Silva et al., 2015)

Human mobility is related to various factors such as age, health, and/or social and economic status, and is directly related to the development of the cities, since all human actions involve movement around the space. Thus, locomotion difficulties can impact in a variety of areas, such as: economic, social, and even environmental. The equalization of the possibility of locomotion for various publics means making the social functions of the cities accessible to all. Nevertheless, urban locomotion is directed to assume that perfect mobility is something guaranteed, since motorization becomes more and more universal around the world. Thus, all exceptions to this rule tend to be treated as deviations, relating with social groups which are currently in disadvantage. The accessibility is, therefore, the way to make the urban space and city mobility socially just. In this sense, accessibility can support the sustainable urban development. (Abitante & Lima, 2014; Taylor & Józefowicz, 2012)

Accessibility is connected to both mobility factors and the physical environment, and must enable the displacement of the individual with plain safety. Thus, the possibility of locomotion is a fundamental aspect of the human being, whether disabled or not, directly related to sensations, perceptions, social history, social, economic, cultural and environmental contexts, among others. Consequently, accessibility offers access to all services, according to the necessities presented by people, at all levels of attention. Hence, the more individuals stay in environments which restrict their mobility and accessibility, the greater will be the difficulty for them to reach a plain and autonomous life. (Amaral et al., 2012; Taylor & Józefowicz, 2012)

One of the main challenges faced by people with disabilities is represented by the various problems of accessibility which can be found in the streets or in the development of their daily lives involving locomotion. Flooding, holes in the streets, objects inconveniently discarded in the streets, among others, are some of the questions which bring discomfort to all citizens, in addition to creating locomotion barriers to the disabled. The problem is not just the structure of the locomotion mediums, which affects where and how people move, but mainly the lack of mechanisms to determine the accessible areas of an environment. (Cardonha, Gallo & Avegliano, 2013; Hara, Le & Froehlich, 2013)

The US Department of Transportation and the US Access Board, for instance, describe some of the problems which hamper the accessibility of pedestrians, including: the lack of well-connected paths to destinations like schools, problems in the streets surfaces - blocked temporarily (by a vehicle, for instance) or permanently (light poles); hard crossing of tight streets and sidewalks with less than 60 inches (hampering the locomotion of wheelchair users) (Hara et al., 2013). Therefore, it can be noticed that accessibility is a fundamental factor in the life of the disabled people, once those need specificities in the execution of their activities.

Besides, the utilization of geographical information is currently becoming more and more important in society. Maps can help various people in their daily needs, such as children coming back from school, women avoiding unsafe areas at night, parents pushing baby carriages, among others. Nonetheless, the major part of this kind of information is inaccessible to the users with visual disability, once the interaction with these applications is performed mainly by vision, and usually made available through interactive maps. In addition, this group of users needs the knowledge of several characteristics of the environment in order to move safely and adequately, as sidewalk accessibility, points of interest, among others. However, this is not offered with efficiency by the current geographical data providers, due to the high costs and slowness in the collection of this type of information using only the traditional mediums. (Mirri et al., 2016; Kaklanis, Votis & Tzovaras, 2013a; Kaklanis, Votis & Tzovaras, 2013b; Prandi, Soave, Devigili, Amicis & Astyakopoulos, 2014)

HOW CAN VGI AND CROWDSOURCING MITIGATE ACCESSIBILITY BARRIERS?

One of the great difficulties to the production of accessible applications is the lack of details about the locomotion environment of the disabled public, such as the infrastructure of the pedestrian walkways, fixed and transitory obstacles, among others. These characteristics are important, since their presence can prevent or hamper the locomotion of this public in the place or even cause accidents, posing risks to the individual health.

If the information about areas used by pedestrians were reliable and complete, it would be possible to increase the number of pedestrians and the amount of physical activity performed by people in general. Thus, environments could become more accessible, offering the users the required information for the provision of accessible routes and about barriers and objects which hamper locomotion. (Laakso, 2014)

Therefore, crowdsourcing has recently received great attention, due to its capacity of providing with efficiency data and geographical synthesis. This technique has proven to be a new resource to make information more accessible, excluding precision gaps and gathering intellectual power of human beings. Due to its efficiency, it has obtained great interest from researchers from a variety of fields, starting from more specific to broader fields, even reaching the context of social media, demonstrating its importance and potential.

Along with that, during the last decade, the emergence of VGI offered an alternative to the provision of spatial data. The mechanisms used for acquisition and the nature of VGI can provide competitive advantages over official data, such as: the knowledge of the place, the update frequency, cost reduction to data acquisition and maintenance, temporal relevance, social and community engagement, among others. These benefits are intrinsically related to the participation of the public in the GIScience community and demonstrate crowdsourcing can serve as an essential technique in computational projects with geotechnological objectives, due to its association with the community, the final users and its real world application. Thus, these characteristics have drawn attention of both researchers and private sector institutions. This kind of information can enrich, complement or update datasets, products, or become the only source of such data. (Antoniou & Skopeliti, 2015; Rice et al., 2013)

Besides that, the advance and participation of technologies such as Global Positioning System (GPS), Internet, geospatial communication, mobile and wireless communication, automatic sound recognition and image recognition can offer alternatives to the development of solutions which can assist the sensorial capacity of disabled people, helping their displacement, security and health. However, even with the whole potential presented, these technologies are not well used in the field of accessibility, ignoring important factors like the special needs of each individual, environmental characteristics, among others (Mostafavi, 2015; Takagi et al., 2013)

Then, the union of all those characteristics makes the VGI approach a suitable alternative to be used in environments in which the frequency of changes in space is very high, which is the case with the environments used by pedestrians, generating great opportunities in the research field.

VGI and Crowdsourcing to Improve Accessibility

According to World Report on Disability (2011, p. XI), "Across the world, people with disabilities have poorer health outcomes, lower education achievements, less economic participation and higher rates of poverty than people without disabilities. This is partly because people with disabilities experience barriers in accessing services that many of us have long taken for granted, including health, education, employment, and transport as well as information. These difficulties are exacerbated in less advantaged communities".

Currently, much research is being developed to analyze the feasibility of the usage of human sensors and/or crowdsourcing to provide accessibility to visually and physically disabled people. In this field, therefore, it is worth mentioning some studies which have the objective to provide accessible maps with relevant geographical information for the disabled people and others which aim at providing accessible routes based on the users' limitations.

Some applications whose objective is to help the user in its navigation can be highlighted in the context of the physically disabled people. This can be done through the presentation of maps capable of exhibiting information directed specifically to this group. These can be locals and areas (restaurants, stores, boulevards, among others) which present adaptations to the perfect utilization by the physically disabled person, whether through restrooms, routes or any adaptation which enables the accessibility. Such information is obtained through the participation of the users in the system, informing the places which present the satisfactory adaptations and also the problems in this context.

Yet in this group, the creation of applications and research which work with the categorization and presentation of adapted routes using crowdsourcing has currently grown, for instance: studies on the collection, processing and presentation of the more viable routes for this public, considering their limitations. Thus, a map capable of generating accessible routes based on the specificities of the user and environment characteristics is commonly used.

With respect to the visually impaired people, a great amount of studies have emerged to facilitate the access and provision of information originating from maps through alternative mediums other than the visual channel (e.g., audio, vibration, among others). These individuals face daily a multitude of problems. The simple activity of crossing a busy street becomes something extremely hard due to their special needs. Therefore, the provision of geographical information capable of presenting which places have the required adaptations for the usage of this public (e.g., audible street signals or tactile paving) becomes essential.

Finally, as for the physically disabled people, some applications also emerged with the goal of generating the best routes (most accessible) to the user with some visual limitation. In this context, the main goal is to propose, considering the user's needs, the routes which have less obstacles, which can be: holes in the streets, steep slopes, trees, or any kind of change in the environment, whether it is fixed or transitory, capable of posing any risk to the health of the disabled person.

For the development of such applications, increasing efforts have been directed to use VGI and/or crowdsourcing as the origin of the information, whether for updating any internal information in a map of the application or to calculate accessible routes to the users with the described special needs, among other purposes.

ACCESSIBLE APPLICATIONS AND RESEARCHES

Any person, in order to develop some activities, necessarily has to get around from one place to another. But when this action is performed by the disabled people, the mobility is impaired due to their special needs. Locomotion for disabled individuals is a complex task, since they require tools adapted to their needs or some help of other people to facilitate their mobility. Moreover, if the environment has obstacles, this task becomes even more difficult, since these can block the passage of those with some disability or even offer them health risks. A wheelchair user, for example, requires unhindered passages that enable their navigation with the wheelchair. Therefore, steep slopes and narrow corridors can prevent the passage of these persons in certain environments. In the case of visually impaired people, the situation is also critical, because these individuals generally cannot get a detailed description of the environment around them, making it impossible for them to detect barriers, mainly transient, like potholes, slopes or any kind of obstacle in locomotion surfaces. Thus, these can easily suffer accidents.

Short-lived obstacles, such as obstructions on sidewalks, detours caused by construction and damaged surfaces are unpredictable and, therefore, considerably difficult to be detected in traditional mapping systems (Qin et al., 2015). The detection and rapid geographical location of transient obstacles can be of great help for the visually impaired (Rice et al., 2013). For that reason, alternative means should be developed to fill that gap. Some of the alternatives proposed by researchers aimed at solving this problem are described below.

Jayakody, Murray and Herrmann (2015) propose a model to help visually impaired people in indoor environments they are not yet familiar with, such as universities, hotels, houses and restaurants. Users usually use white canes, GPS navigation devices and guide dogs to assist them in these tasks. However, these instruments and animals do not provide sufficient information for the visually impaired to properly navigate in indoor spaces. The authors propose a mechanism for collecting useful information on locomotion areas and obstacles in internal environments (e.g., wash room, lift lobby, stair case, walls and doors), based on a mobile application that works with a set of sensors; and on information extracted from the web about Points of Interest (POI). Moreover, they propose a method of automatic labeling/tagging maps with semantic information on internal environments (e.g., room names, room description) to assist visually impaired navigation. This participatory methodology allows users to contribute with information on indoor environments in a collaborative way through their mobile devices. The use of cameras and internal sensors of the devices can increase the reliability of semantic information added to a crowdsourcing database.

Rice et al. (2013) propose a crowdsourcing system that consists of a base map and information on barriers and hazards environments. Such information may be represented in the form of events, such as those related to construction activities. The information is collected through crowdsourcing combined with Ambient Geographic Information (AGI) acquired from geosocial networks and by traditional means, that is, from government, state or municipal entities. Obtained data are geotagged using a method based on a gazetteer, filtered, labeled and evaluated in terms of quality. Finally, relevant information on obstacles hazardous sites is presented on a hybrid system.

Rice et al. (2013) argue that local construction activities, such as fencing across walkways, are usually mapped by authoritative entities. On the other hand, more transient obstacles, such as the temporary delivery of bulk construction materials onto a public right-of-way, are not mapped by these entities because of their temporary nature. Thus, through crowdsourcing, they gather information about those transient events and combine them with the authoritative reports. Although the authors propose a useful method of capturing obstacles which combine various techniques, no effort has been directed to acquire VGI information directly from the users. Furthermore, since the proposed system aims at supporting accessibility, it could be further improved by including a routing component to avoid obstacles and could provide an accessible interface to be used by their target audience.

In order to facilitate the identification of accessible environments for disabled people and wheelchair users, maps containing useful information for this public can be developed. One of the best and fastest ways to do this is to use the community knowledge. For instance, Wheelmap (http://wheelmap.org) is an online map based on VGI, available for web and mobile platforms that has the function of helping disabled people. This application can identify a variety of different environments such as gas stations, restaurants, banks, churches, medical centers, department stores, among others. Each environment is associated with a level of accessibility identified by anyone who wants to contribute. Available levels are "wheelchair-accessible", "limited accessibility", "not wheelchair-accessible" and "unknown". In addition, the user can search for locations based on different categories (e.g., tourism, shopping and food) and contribute by adding new places, editing place information (e.g., name, address) and by uploading photos of entrances. Users can easily add information about accessibility through an online application or a mobile application. It is an essential application for wheelchair users, since it facilitates the identification of places with affordable accommodation and avoids unpleasant experiences.

In the field of accessible maps, the research by Calle-Jiménez and Luján-Mora (2015) can be highlighted. The authors present an alternative solution to improve the accessibility of geographic maps using VGI and Scalable Vector Graphics Tiny (SVG Tiny - www.w3.org/TR/SVGTiny12/). The proposed application allows the users to interact with the system by adding accessibility information and other descriptions of elements that make up the map, such as the presence of ramps, escalators and parking lots available. Using this tool, users can select a map element and a window appears so that they can add accessibility information about that element or location using SVG Tiny code. Since SVG is a textual representation of a map, it helps the development of accessible maps through the use of text readers.

Kaklanis et al. (2013a, 2013c) present the "Open Touch / Sound Maps", a multimodal interactive map that replaces the visual information by alternative formats. The tool is a mobile application for Android operating system, which is aimed at improving accessibility of interactive maps for visually impaired users. Multimodal interaction is given by sound effects, Text-To-Speech (TTS) and vibration feedback, which allows access to OpenStreetMap data for the visually impaired and blind people. This application, unlike others previously described, does not implement the techniques of crowdsourcing or VGI directly. Rather, it relies on data from OSM, which is a comprehensive platform based on these techniques. Thus the intention of the authors is not the production of information from the collaboration of the community, but the availability of information to your target public (visually impaired people). This application enables the retrieval and the tactile exploration of any area of the map through a mobile device running Android, with the help of sound effects and messages generated by the TTS.

For the visually impaired, there is a need to identify routes that offer less risk to the user's health (routes that have fewer or less dangerous obstacles). As this user group has disabilities in an aspect of great importance to locomotion (the vision), they have to overcome this difficulty through the use of other senses. Thus, applications aimed at suggesting routes for the visually impaired must provide an essential feature: an accessible interface, which provides to individuals forms of usability which consider their needs. We now present some studies that aim at providing navigation features through accessible interfaces.

Zeng and Weber (2015) discuss the difficulty that the visually impaired have in navigating in an external environment and in finding access ports (entrances) to the unknown internal areas such as entrances of shopping malls, hotels and shops. One of the factors that cause this difficulty is the spatial complexity around the entrances of these locations (e.g., presence of trash cans, stairs and alternative corridors). From this, the authors propose the Coach system, a tool that unites crowdsourcing technology, location-based services and vibration effects and audio outputs in a mobile application. This research assumes that the currently available wayfinding systems by mobile devices cannot function properly near entrances because of GPS signal loss. Thus, the authors use information on environmental accessibility, like difference of textures in a way, and crowdsourcing information to help visually impaired people located in environments around the entrances and help them on the routes to these.

This research uses the crowdsourcing technique in two user groups. A group is formed by people who have normal vision and are volunteers who contribute information about environmental accessibility. This group is responsible for providing, through their mobile devices, structured information (presence of automatic doors, attached stairs or tactile paving) on a particular location. The other group is made up of visually impaired people who receive, on their mobile devices, the information provided by the first group to support navigation. This group is responsible for sharing their experiences providing unstructured information (text or audio recordings) about their experiences and the accessibility of a given environment. [Zeng& Weber, 2015]

Zeng and Weber (2015) offer a fairly complete system covering obstacle detection, accessible interface, generating routes and a component that the user guide to their destinations. In addition, the authors use the crowdsourcing technique both to get information from people who have no disability as for the experiences of the visually impaired. However, it is a very specific application intended only for providing a navigation system for detecting access to internal room doors. The way the crowdsourcing technique was used (accessibility information from user disabled and non-disabled) could be extended to navigations applications in outdoor and indoor environments.

Prandi et al. (2014) present the mPASS application (mobile Pervasive Sensing Social Accessibility). It is a system which aims to assist in the navigation of people with disabilities, by collecting data on urban accessibility and providing accessible paths and maps to citizens. The application uses experts, sensors and techniques of crowdsourcing and crowdsensing to obtain data. These techniques are responsible for detecting the accessible features of the environment and to adapt the routes to the limitations and preferences of each user. The authors introduce the term Accessibility Points of Interest (aPOI), which represents the presence or absence of a barrier. Thus, the users can share, through the mobile tool, their experiences in finding different aPOI and categorizing them according to the level of accessibility. From this, the system is able to generate routes based on the users profile and the categorization of aPOI provided by the users. Moreover, aiming at providing useful visual information, the mPASS can also capture multimedia information, such as pictures and videos. The mPass provide for disabled people a way to avoid environments unsuitable for their limitations, as steep and uneven surfaces. Through the application, users can prioritize paths that have ramps and accessible features. The system can avoid paths that have obstacles and risky crossings.

Table 1 compares the researches discussed above by identifying a set of desired features for an application aimed at improving accessibility for physically and visually impaired people. We believe that an ideal application should include all the features shown on the Table 1. Physically and visually impaired people suffer from the presence of obstacles in the environments they navigate, both indoor and outdoor environments. A feature to automatically detect barriers would be extremely useful, such as those based on movement patterns or by image analysis. However, unfortunately, envisioned solutions still have many drawbacks, such as those related to computational costs, financial problems or privacy. For example, movement pattern analysis depends on the continuous collection of the user's location (through the device's GPS), leading to excessive consumption of device's battery and to privacy concerns. Solutions based on computer vision to detect barriers from web pictures, for example, are not yet sufficiently accurate and are still costly. In this context, VGI-based techniques appear as suitable alternatives to identify barriers, even though this is not an automatic detection approach.

As users have different levels of limitation, there is a need for the development of a mobile client application equipped with a multimodal interaction interface. Moreover, the generation of routes avoiding obstacles would be an essential feature in these applications. As the automatic routing may contain

Table 1. Comparing the researches by a set of desired features

Research	Navigation Environment	Crowdsourcing	VGI	Obstacle Identification from VGI or Crowd-sourcing	Accessible Interface	Routing Based on crowd-sourcing or VGI	Guided Navigation
Jayakody et al. (2015)	Indoor	X	X	X			
Rice et al. (2013)	Indoor and Outdoor			X			
Wheelmap	Indoor and Outdoor			X			
Calle-Jiménez and Luján-Mora (2015).	Indoor and Outdoor	X		X	X		
Kaklanis et al. (2013a)	Outdoor	X	X		X		
Kaklanis et al. (2013c)	Outdoor	X	X		X		
Zeng and Weber (2015)	Outdoor				X		X
Prandi et al. (2016)	Outdoor	X		X	X	X	
Aplicação ideal	Indoor and Outdoor	X	X	X	X	X	X

errors, the VGI technique could help in this context, that is, the system could record the route of users, evaluate them as the presence of barriers and make them available in the system for use of the community. Audio and video messages about the environment could also be shared among users, aiming to provide information on a location in question for new users in that area. Finally, especially for visually impaired users, navigation functionality based on sound messages would be extremely useful. Users with this limitation need alternative ways to get located in relation to physical environment components. This can be easily implemented through the accessibility features available on mobile devices (e.g., Talkback present on the Android platform).

AN ACCESSIBLE PLATAFORM BASED ON VGI AND CROWDSOURCING

A research aimed at providing accessible outdoor navigation for visually impaired people is being conducted in our labs. We have developed a platform based on VGI and crowdsourcing to provide obstacle detection, recognition of users' environments, routing avoiding barriers and guided navigation. The solution comprises a mobile app equipped with an accessible interface which produces vibratory effects and sound messages.

The platform consists of four components, as shown in Figure 1. The map server retrieves information from the OSM service. The Talkback (accessibility service available on Android devices) provides the audio messages. The server is designed to receive, process and store information reported by the users. Finally, the client application (Mobile App) provides information to the end users and provide features to collect volunteered information.

Figure 1. The VGI-based platform

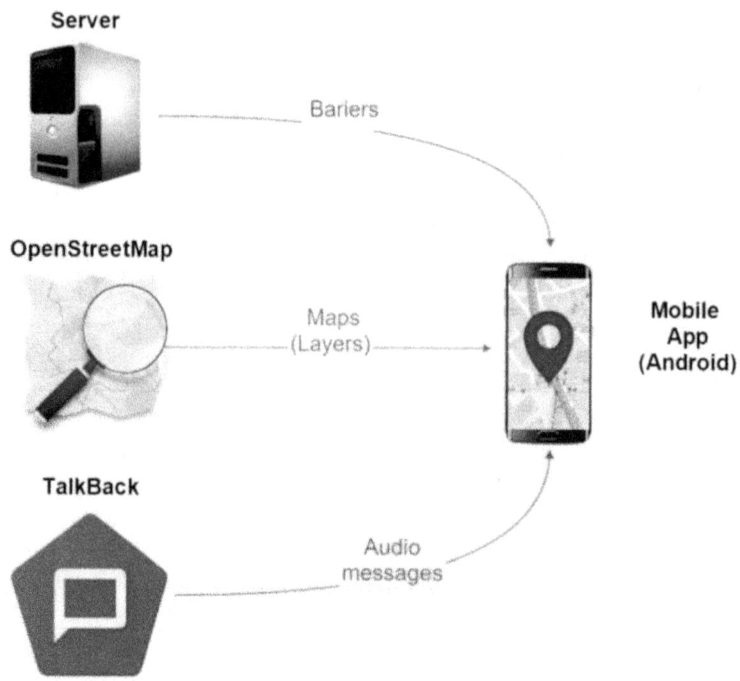

Figure 2 describes the process to report and visualize barriers in the platform. The numbers shown on the picture represent the order of actions within the process. First a user reports the barrier on the server using the mobile application, specifying its type (among several types available, e.g., public phones, elevation, tree) and its dangerous level. The server registers the barrier report in the database and performs validations on the data. Then barrier information is made available on the platform so that other users can view it on their own devices.

The accessible interface of the mobile app explores a variety of tactile gestures capable of running on Android devices was developed. The tool recognizes different touch commands (double-click, single click, drag fingers) and associates each of them a kind of action. For example, the double click command focuses the map at the selected location, facilitating the exploration and recognition of areas outside the initial focus; the movement of dragging fingers returns auditory messages containing the names of selected locations (e.g., streets, avenues). The output information or instructions for users are available through sound effects and audio messages.

In order to provide greater comfort for the visually impaired, when the application starts, the map is centered on the current position of the user. This is intended to facilitate navigation and identification of the individual's position in relation to other map elements due to their needs. Such elements include Points of Interest (POI), obtained from a set of external services. As the barriers represent a great danger for the visually impaired, the proposed solution provides participatory means of identifying barriers, both fixed and temporary. Beyond providing information about elements on the user environment, the tool is able to warn the users about barriers during their navigation.

Among the main challenges in providing such a VGI service is the development of algorithms to perform validations and analysis on the reported data. In our research, significant efforts have been

Figure 2. Collecting and using barriers information

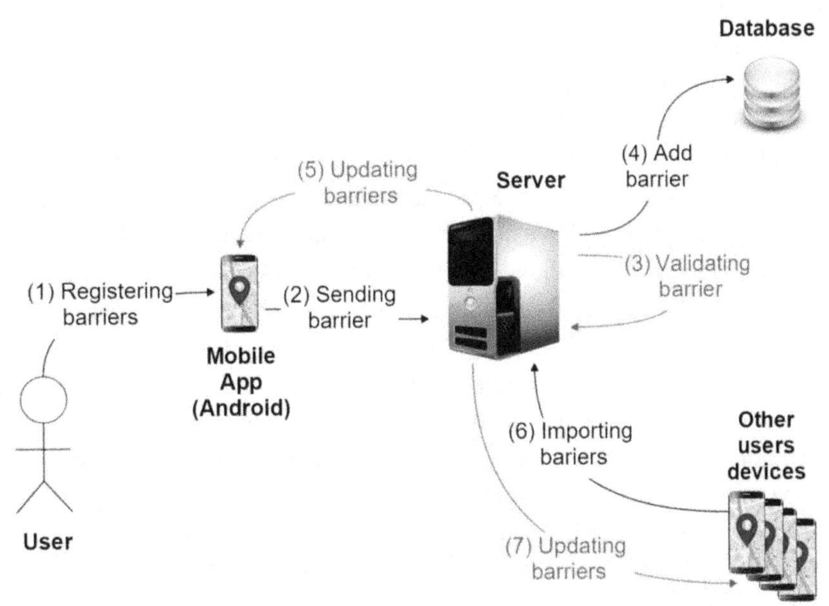

made in this direction. Several data quality issues must be considered in this task, since the location of a certain barrier differs considerably among different user reports. Moreover, the inaccuracy of information collected by users' GPS sensors adds significant challenges to the process, because the expected accuracy for this kind of information (i.e., barriers to be avoided by visually impaired) is much higher than in other location-based services.

CONCLUSION AND FUTURE DIRECTIONS

Physically and visually impaired people face serious locomotion difficulties in the various environments that are part of their daily activities. Often, small details such as trees, narrow hallways, small holes and slopes in surfaces can influence the locomotion, causing discomfort and offering danger to the health of these individuals. These people need effective solutions that can solve these problems, since mobility is essential to the development of the day-today tasks of any person. Therefore, it is clear that there is a great need for resources that can assist the navigation of this group of people; solutions that can indicate which environments are adapted to the needs of the individuals and the best paths to get to their destinations, considering the specificities of each disabled person.

However, one of the great difficulties in developing software to help disabled people in their navigation is to obtain detailed information about the environment, since many cities and countries do not have up-to-date databases that provide useful information on their physical space. Moreover, environments which have such information do not usually include certain details that are sometimes crucial for these people's navigation (e.g., trees, poles, holes and surface unevenness). These obstacles (fixed or transient) can strongly influence the movement of people with special needs, offering danger to their health and safety. Thus, the use of VGI and crowdsourcing techniques in applications for this purpose has been considerably efficient due to the volunteered methodology of data acquisition, low cost of the process and update speed.

Nevertheless, systems based on VGI or crowdsourcing have major challenges to be built, such as those related to the reliability of the data entered into the system, usually by thousands or millions of users who insert information several times a day. According to Coleman (2013), one of the main difficulties encountered in developing systems that have VGI as the basis is how to assess the credibility of the users and the quality of their contributions. However, methods of analyzing reputation, adapted for the kind of systems discussed in this chapter, have not yet been sufficiently developed.

Another problem still encountered in the development of VGI applications is the community participation (Chow, 2013; Qin et al., 2015). This is an essential feature in the application maintenance, since it depends on the users' motivation to keep inserting and updating data. Some application discussed in this chapter, such as those for help navigation avoiding barriers, face particular problems related to the users' motivation, since the number of reports for a barrier decreases considerably when the information is already available on the system. That is, the users will not report a barrier if the system is already aware of it. However, the system needs as much information (i.e., user reports) as possible to increase accuracy and to deal with temporary barriers. Thus, considerable efforts should be made to develop suitable methods of providing society awareness and user motivation in this kind of systems.

REFERENCES

Abitante, J. C., & Lima, J. P. (2014). A acessibilidade como fator para o desenvolvimento sustentável urbano: o caso de Campos do Jordão.*Proceedings of the IV Simpósio de Desenvolvimento, Tecnologias e Sociedade*.

Amaral, F. L. J. S., Holanda, C. M. A., Quirino, M. A. B., Nascimento, J. P. S., Neves, R. F., Ribeiro, K. S. Q. S., & Alves, S. B. (2012). Accessibility for people with disabilities or permanent mobility restrictions to the Unified Health System. *Ciencia & Saude Coletiva, 17*(7), 1833. doi:10.1590/S1413-81232012000700022 PMID:22872345

Antoniou, V., & Skopeliti, A. (2015). Measures and indicators of VGI quality: An overview. ISPRS Ann. In *ISPRS Annals of Photogrammetry* (Vol. 1, pp. 345–351). La Grande Motte, France: Remote Sensing and Spatial Information Sciences.

Associação Brasileira de Normas Técnicas (ABNT). (2015). *ABNT NBR 9050 – Accessibility to buildings, equipament and the urban environment* (3rd ed.). Rio de Janeiro, Brasil: ABNT.

Basiouka, S., Potsiou, C., & Bakogiannis, E. (2015). OpenStreetMap for cadastral purposes: An application using VGI for official processes in urban areas. *SurveyReview*, *47*(344), 333–341.

Brock, A., Truillet, P., Oriola, B., Picard, D., & Jouffrais, C. (2012). Design and user satisfaction of interactive maps for visually impaired people. In *International Conference on Computers for Handicapped Persons* (pp. 544-551). Springer Berlin Heidelberg. doi:10.1007/978-3-642-31534-3_80

Calle Jiménez, T., & Luján Mora, S. (2015). Using crowdsourcing to improve accessibility of geographic maps on mobile devices.*Proceedings of the Eighth International Conference on Advances in Computer-Human Interactions (ACHI 2015).*

Cardonha, C., Gallo, D., Avegliano, P., Herrmann, R., Koch, F., & Borger, S. (2013, May). A crowdsourcing platform for the construction of accessibility maps.*Proceedings of the 10th international cross-disciplinary conference on web accessibility*. doi:10.1145/2461121.2461129

Chow, T. E. (2013). "We Know Who You Are and We Know Where You Live": A Research Agenda for Web Demographics. In D. Sui, S. Elwood, & M. Goodchild (Eds.), *Crowdsourcing Geographic Knowledge: Volunteered Geographic Information (VGI) in Theory and Practice* (pp. 265–285). New York: Springer Dordrecht Heidelberg. doi:10.1007/978-94-007-4587-2_15

Coleman, D. J. (2013). Potential Contributions and Challenges of VGI for Conventional Topographic Base-Mapping Programs. In D. Sui, S. Elwood, & M. Goodchild (Eds.), *Crowdsourcing Geographic Knowledge: Volunteered Geographic Information (VGI) in Theory and Practice* (pp. 245–263). New York: Springer Dordrecht Heidelberg. doi:10.1007/978-94-007-4587-2_14

Coughlan, J. M. & Shen, H. (2013). Crosswatch: a system for providing guidance to visually impaired travellers at traffic intersections. *Journal of Assistive Technologies, 7*(2).

Csapó, Á., Wersényi, G., Nagy, H., & Stockman, T. (2015). A survey of assistive technologies and applications for blind users on mobile platforms: A review and foundation for research. *Journal on Multimodal User Interfaces*, *9*(4), 275–286. doi:10.1007/s12193-015-0182-7

Favretto, D. O., Carvalho, E. C. D., & Canini, S. R. M. D. S. (2008). Intervenções realizadas pelo enfermeiro para melhorar a comunicação com deficientes visuais. *Revista da Rede de Enfermagem do Nordeste*, *9*(3), 68–75.

Hara, K., Le, V., & Froehlich, J. (2013, April). Combining crowdsourcing and google streetview to identify street-level accessibility problems.*Proceedings of the SIGCHI conference on human factors in computing systems* (pp. 631-640). doi:10.1145/2470654.2470744

Intelligent Transportation Systems Institute, Center for Transportation Studies. (2011). *Development of Mobile Accessible Pedestrian Signals (MAPS) for Blind Pedestrians at Signalized Intersections*. Center for Transportation Studies.

Jayakody, J. A., Murray, I., & Herrmann, J. (2015, October). Analgorithm for labelling topological map store present point of interest for vision impaired navigation. In Proceedings of 2015 International Conferenceonthe Indoor Positioningand Indoor Navigation (IPIN), (pp. 1-8).

Kaklanis, N., Votis, K., & Tzovaras, D. (2013a, May). A mobile interactivemapsapplication for avisuallyimpairedaudience.*Proceedings of the 10th International Cross-DisciplinaryConferenceon Web Accessibility*.

Kaklanis, N., Votis, K., & Tzovaras, D. (2013b). Open Touch/Sound Maps: A system to convey street data through haptic and auditory feedback. *Computers & Geosciences*, *57*, 59–67. doi:10.1016/j.cageo.2013.03.005

Kaklanis, N., Votis, K., & Tzovaras, D. (2013c). Touching openstreetmap data in mobile context for the visually impaired. In *Proceedings of the 3rd Workshop on Mobile Accessibility-ACM SIGCHI Conference on Human Factors in Computing Systems* (pp. 68-74).

Koester, D., Lunt, B., & Stiefelhagen, R. (2016). Zebra crossing detection from aerial imagery across countries. In *Proceedings of the International Conference on Computers Helping People with Special Needs (ICCHP) part II* (pp. 27-34). doi:10.1007/978-3-319-41267-2_5

Koukourikos, P., & Papadopoulos, K. (2015). Development of cognitive maps by individuals with blindness using a multisensory application. In *Procedia Computer Science – Proceedings of the 6th International Conference on Software Development and Technologies for Enhancing Accessibility and Fighting Info-exclusion* (pp. 213-222). doi:10.1016/j.procs.2015.09.265

Laakso, M. (2014). *Improving accessibility for pedestrians with geographic information* (Doctoral dissertation). Available from Aalto University Library.

Lin, W. (2013). When Web 2.0 Meets Public Participation GIS (PPGIS): VGI and Spaces of Participatory Mapping in China. In D. Sui, S. Elwood, & M. Goodchild (Eds.), *Crowdsourcing Geographic Knowledge: Volunteered Geographic Information (VGI) in Theory and Practice* (pp. 83–104). New York: Springer Dordrecht Heidelberg. doi:10.1007/978-94-007-4587-2_6

Martinez-Sala, A. S., Losilla, F., Sánchez-Aarnoutse, J. C., & García-Haro, J. (2015). Design, implementation and evaluation of an indoor navigation system for visually impaired people. *Sensors (Basel, Switzerland)*, *15*(12), 32168–32187. doi:10.3390/s151229912 PMID:26703610

Mirri, S., Prandi, C., & Salomoni, P. (2016). Personalizing Pedestrian Accessible way-finding with mPASS. In 13th IEEE Annual Consumer Communications & Networking Conference (CCNC) (pp. 1119-1124). doi:10.1109/CCNC.2016.7444946

Mostafavi, M. A. (2015*).MobiliSIG: Development of a geospatial assistive technology for navigation of people with motor disabilities*. Paper presented at Spatial Knowledge and Information Conference. Banff, Alberta, Canada.

Neis, P. (2014). Measuring the reliability of wheel chair user route planning based on volunteered geographic information. *Transactions in GIS*, *19*(2), 188–201. doi:10.1111/tgis.12087

Prandi, C., Salomoni, P., & Mirri, S. (2014, January). mPASS: Integrating people sensing and crowdsourcing to map urban accessibility. In *Proceedings of the IEEE International Conference on Consumer Communications and Networking Conference* (pp. 10-13). doi:10.1109/CCNC.2014.6940491

Prandi, F., Soave, M., Devigili, F., Amicis, R., & Astyakopoulos, A. (2014) Collaboratively collected geodatato support routing service for disabled people.*Proceedings of the 11th International Symposiumon Location-Based Services (LBS)*.

Qin, H., Aburizaiza, A. O., Rice, R. M., Paez, F., & Rice, M. T. (2015). Obstacle characterization in a geocrowdsourced accessibility system. In *ISPRS Annals of Photogrammetry*. La Grande Motte, France: Remote Sensing and Spatial Information Sciences. doi:10.5194/isprsannals-II-3-W5-179-2015

Rice, M. T., Jacobson, R. D., Caldwell, D. R., McDermott, S. D., Paez, F. I., Aburizaiza, A. O., & Qin, H. et al. (2013). Crowdsourcing techniques for augmenting traditional accessibility maps with transitory obstacle information. *Cartography and Geographic Information Science*, *40*(3), 210–219. doi:10.1080/15230406.2013.799737

Santos, J. P., Forest, J. D., Dutra, G. A., & Almeida, M. (2012). *Uma avaliação da bengala eletrônica (tecnologia assistiva), para a melhoria de mobilidade dos deficientes visuais*. Paper presented at Simpósio de Exelência em Gestão e Tecnologia, Rio de Janeiro, Brasil.

Schmitz, B., Becker, S., Blessing, A., & Großmann, M. (2011, June). Acquisition and presentation of diverse spatial context data for blind navigation. In *Proceedings of the 2011 IEEE 12th International Conference on Mobile Data Management* (pp. 276-284). doi:10.1109/MDM.2011.66

Silva, D. C. N., Silva, T. M., Nogueira, M. S., Mendonça, R. M. C., Valente, P. H. F., Araújo, R. F., Brito, F. A. V. (2016). Acessibilidade de portadores de deficiência física ou mobilidade reduzida na Unidade Básica de Saúde Jonas Manoel Dias em São Luís de Montes Belos-GO. *Revista Eletrônica Faculdade Montes Belos*, *8*(3).

Takagi, H., Kosugi, A., Saito, S., & Teraguchi, M. (2013, May). Crowdsourcing platform for workplace accessibility.*Proceedings of the 10th International Cross-Disciplinary Conference on Web Accessibility*. doi:10.1145/2461121.2461135

Taylor, Z., & Józefowicz, I. (2012). Intra-urbandaily mobility of disabled people for recreational and leisure purposes. *Journal of Transport Geography*, *24*, 155–172. doi:10.1016/j.jtrangeo.2011.12.008

United Nations. (2007). *Accessibility: A guiding principle of the Convention*. Retrieved May 30, 2016, from http://www.un.org/esa/socdev/enable/disacc.htm

Vanclooster, A., Weghe, N. V., & Maeyer, P. (2016). Integrating indoor and outdoor spaces for pedestrian navigation guidance: A review. *Transactions in GIS*.

Wheelmap. (n.d.a). *What's Wheelmap all about?* Retrieved July 12, 2016, from https://community.wheelmap.org/en/about/

Wheelmap. (n.d.b) *Wheelmap*. Retrieved on June 17, 2016, from http://wheelmap.org/pt_BR/map#/?zoom=14

World Health Organization (WHO). (2011). *World report on disability 2011*. WHO.

World Health Organization. (n.d.b). *Disabilities*. Retrieved on June 17, 2016, from http://www.who.int/topics/disabilities/en/

World Health Organization. (n.d.a). *Disability and rehabilitation*. Retrieved May 30, 2016, from http://www.who.int/disabilities/en/

Zeng, L., & Weber, G. (2015, August). A Pilot Study of Collaborative Accessibility: How Blind People Find an Entrance. In *Proceedings of the 17th International Conference on Human-Computer Interaction with Mobile Devices and Services* (pp. 347-356). doi:10.1145/2785830.2785875

KEY TERMS AND DEFINITIONS

Accessibility: Ability to provide the inclusion of disabled people in the participation of any activity.

Barrier: Object that offers danger or discomfort to an individual and that is located in their path.

Disability: It is an intellectual, cognitive, sensory, mental or developmental impairment of an individual.

Location-Based Service: Software-level service that handles location-based data.

Mobility Impairment: Physical impairment that impairs an individual's mobility ability. Persons affected by mobility impairment often use assertive devices such as wheelchairs and artificial limbs to obtain mobility.

Multimodal Map: Map that enables interaction through other senses beyond vision.

Short-Lived Obstacles: Obstacles that disappear with the passage of time (for example, trash and stones).

Visual Impairment: Visual impairment that, even with the best optical correction, partially (between 40 and 60%) or totally compromises the vision.

Chapter 12
Gathering Road Safety Critical Information From Users

Ahmet Yıldız
Turk-Alman Üniversitesi, Turkey

ABSTRACT

Traffic roads are a core element of GIS and many volunteered systems like openstreetmaps have the goal to make road data publicly available. Road users collecting geographical information and sharing them according some rules are a great opportunity to make our roads a safer place. Traffic accidents are a major cause of death and with increase in urbanization and motorization the risk is expected to rise higher. Research regarding road safety is mostly reactive; sections of the road where a lot of accidents has already happened are investigated and possibly causes are identified and then improved. This means, that people have to die in order to make those road sections safe. The system described in this chapter is a proactive method and can be operated by the community or some responsible authority. The gathered data is also very useful for different research areas like social sciences or civil engineering.

INTRODUCTION

Traffic roads are a core element of GIS and many volunteered systems like openstreetmaps have the goal to make road data publicly available. Road users collecting geographical information about the road and sharing them according some rules are a great opportunity to make our roads a safer place.

Traffic accidents are still a major cause of death and with increase in urbanization and motorization the risk is expected to rise even higher. (World Health Organization 2015) Research regarding road safety is mostly reactive; sections of the road where a lot of accidents has already happened (black spots) are investigated and possibly causes are identified and then improved. This means, that people have to die in order to make those road sections safer.

Another approach is the proactive traffic conflict technique. Here a road section is observed for a few days and depending on the conflicts among road users, the safety of the section is evaluated. A conflict hereby is defined as a situation where an accident nearly happened but was evaded in the last second by the interaction of one of the drivers (breaking, dodging or similar). The most important quality of

DOI: 10.4018/978-1-5225-2446-5.ch012

this technique is that one can detect high risk sections on the road before an accident happens. But it is impossible to observe the whole road network by traffic safety experts.

There are some work utilizing image processing to use video footage of road cameras to automatically detect conflicts. But there are some limitations, like the need of good quality video cameras over the whole network. Also the algorithms behind the image processing have some pre-requirements, which cannot always be expected.

Volunteered Geographic Information presents a new opportunity for road safety research. Every conflict is observed by at least one user, namely the one involved. Most of the times there will be two users involved and several users witnessing the traffic conflict. Those are a valuable source of information. At this point the need for a system arises, where this information can be gathered.

There are several approaches to how this information flow can be managed. The first option would be an authority which operates the system and evaluates the information provided by the road users. The other option would be an open platform operated by the users themselves with public access to all provided data. Both approaches have their pro and cons.

In this chapter the primary outline of such a system that is required to make safety research possibly is discussed and a technical guideline for the implementation is drawn. Also the mentioned operating options are discussed in detail and a recommendation is given.

Another important point is that the system is easy to use. So, beside the technical needs of the system also some ideas about the user interaction are given. Also the motivation of participants has to be taken into account. The chapter ends with an outline of possible safety and research opportunities and a conclusive discussion.

BACKGROUND ON ROAD TRAFFIC SAFETY

Traffic is the movement of humans or goods, either by vehicle or on foot. Humans always strived for mobility, while the means of moving changed over time and advanced with the technological progress. The first roads would be the worn out paths, simply being created by frequent passage. With the invention of non-motorized carriages pulled by animals, more sophisticated roads were built to enable the wheels run smoothly. With growing vehicles which also became faster, safety on the road became an issue. Rules were set in order to prevent accidents and in case of an accident, to be able to determine the guilty ones.

Today, one distinguishes between traffic in the air, on the water and on the ground. While each have their safety issues, this chapter concentrates on traffic on the ground. With motorization of the vehicles and the increasing number of vehicles in traffic, the safety issue is more eminent than ever. According to the World Health Organization (2015, p. 2 ff), traffic accidents are among the top ranking causes of death, and in some countries even the number one cause of death.

There are three factors as part of the road traffic which have an influence on the safety. (Figure 1) First comes the human, whose need for mobility made the traffic in the first place. A human can have different roles in the traffic, like the driver of a vehicle, a passenger, a cyclist or a pedestrian. The biggest risk is the driver who often causes an accident. 90% of all accidents have in the chain of events that lead to that accident a human error or some law violation. (Hobbs, 1987, p.494) The human is a complex and unpredictable being and many things like his physical condition, his psychological attitude towards safety and rules, his state of mind have an influence on his errors. Making the roads safer with focus on the humans is therefore important, but not the topic of this chapter. Psychological work in this area concentrates on subjects like 'knowing the rules' and 'following the rules'. However, the author would like to take this opportunity and appeal to everyone reading this to be self-aware and try to make the roads safer by being a more conscious driver.

The second factor are the vehicles. Besides cases where the vehicle has some mechanical or electronical error, vehicles are mostly not the cause of accidents. But with advanced safety features, they can be very effective in preventing accidents or ease the outcome of them. The very simple seatbelt saved millions of lives worldwide. The seatbelt and airbags are examples for passive safety features. They don't prevent the accident from happening, but they safe lives. Assisting systems like ABS, EPS on the other hand are examples for active safety features. They can give the driver enough time to correct his error or correct the error directly and prevent an accident or help to decrease the speed and the power of the impact. It is possibly that in the future the roads will be much safer, when the vehicles can drive autonomously, thus taking the humans role as driver out of the equation. But this scenario is not guaranteed and in any case many years ahead, so it is still important to make the roads safer today in order to save lives.

The third entity is the road itself. While the road is only being used passively, its shape, its condition and its environment has a direct influence on the safety, affecting both the driver and the vehicle. Building a road according to some guidelines is easy, however, ensuring that it makes the road safe is not easy. Determining what makes a section of the road dangerous and correcting the issue is very hard. So in this chapter the author will try do outline a system which could help to make our roads safer by gathering information from the people who use them frequently and therefore know the risky parts and most likely also have an idea how to fix it.

PROACTIVE MEASURES ARE BETTER THAN REACTIVE CORRECTIONS

The safety of a road network or a road section is related to the amount of severe accidents. The number of fatalities and seriously injured people is used as an indicator to determine whether a road section needs to be primarily investigated or not. These are called reactive methods, since they need people to die in order to do something. This is obviously a bad thing. Proactive methods on the other hand try to determine risky spots and fix them before someone dies there. First, the most common reactive methods will be explained in some detail in order to better understand and value the proactive methods. Then, the main proactive methods are exposed.

Figure 1. Triangle of human – vehicle – road

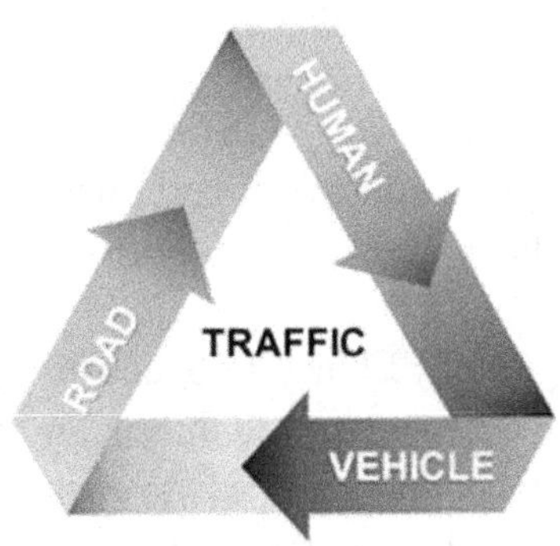

Black Spot Management and Network Safety Management

Both methods follow the same principle. First the road network is divided into sections. A natural divider would be junctions, but on higher road networks this would lead to very long sections, so often an upper limit for the section length is used. Some countries use a sliding window approach, where the accident pattern is being taken into account by dividing the sections. This approach is more complex, but gives better results. Sections in the Black Spot Management (BSM) are usually around 500 meters long, while in the Network Safety Management (NSM) sections up to 10 kilometres are possible. (Yıldız, 2009, p. 27)

Then, a statistical analysis is being performed using the count and the severity of the recorded traffic accidents. The BSM concentrates hereby on the accidents per section and decides whether the section is a black spot or not. The NSM differs in making a network wide statistical analysis. Bothe methods give a safety ranking of the sections.

While the theory of both methods is similar and simple, the implementation in real life is much more complicated. For the implementation of BSM, Sorenson and Elvik (2007) wrote a best practice guide. The easiest way to implement BSM is to use a not-model-based approach without considering road and traffic and using variables like the absolute number of accidents, the frequency (accidents per kilometre), the rate (accidents per vehicle kilometre) or combinations of those. A high number of false positives and false negatives is expected with such not-model-based approaches. Model-based approaches are better from a theoretical point of view, but they are very complex and one needs connected accident, road and traffic data.

NSM can be viewed as a derivative of BSM and is mostly used in France and Germany on motorways and two-laned rural roads (Sorenson & Elvik, 2007). Different to the BSM, in NSM one calculates an average or expected number of accidents and find network wide the sections with the biggest difference, thus identifying the sections with the biggest potential for improvement.

Road Safety Inspections

A Road Safety Inspection (RSI) is a regularly performed monitoring of existing roads with the goal of identifying traffic hazards related to the road environment characteristics and to propose measures to mitigate the detected hazards. As part of the EU Project RIPCORD-ISEREST, a survey was conducted to determine best practices for RSI with the participation of 14 countries (Moscari and Hollo, 2006, p28). Whereby, RSI is

- A preventive tool,
- Consisting of regular, systematic, on-site inspections of existing roads, covering the whole road network,
- Carried out by trained safety expert teams,
- Resulting in a formal report on detected hazards and safety issues,
- Requiring formal response by the relevant road authority.

A good practice guide for RSI is given by Elvik (2006) consisting of the seven points, including a list of important core elements that should be recognized during an inspection. Yıldız and Hauger (2010) introduced an approach to use the data gathered by RSIs to evaluate the safety of the road sections in a more precise and intuitive way without dividing the network into sections. An important contribution hereby is the definition of active and passive safety hazards, where an active hazard increases the risk of an accident (like rain grooves) and passive hazards have no influence on the accident risk, but on the outcome of an accident happening at that spot (like uncovered trees).

Road Safety Inspection is a good proactive method compared to BSM and NSM, but there are some limitations in the implementation. On one hand, it is very costly and therefore difficult to be performed over the whole road network on the suggested regular basis of two years. On the other hand, it has to be performed by specially trained experts and the results are limited with the knowledge and the perception of the inspector.

As a consequence, RSI becomes a similar tool like BSM in the countries implementing them. Since they can't inspect the whole network, they select sections with high accident numbers to perform RSI, thus turning a proactive method into another reactive method. A good overview of the practised RSI in Europe is given in their handbook by Nadler, Nadler and Strnad (2014). It states that a RSI can be initiated by:

- *When a concentration of accidents is observed, especially with diverse type structures*
- *When evidence, problems, or other information make an inspection prudent*
- *When there are safety deficiencies, hazard potential, or the same types of accidents on longer sections*

Traffic Conflict Technique

The Traffic Conflict Technique (TCT) is a very old approach, first introduced in 1967 by Perkins and Harris (1967) from the General Motors Company. The initial goal of the procedure was to observe to what extend cars from their company were involved in unsafe traffic situations. At the first Workshop on Traffic Conflict in 1977, the following definition was proposed and agreed upon (Amundsen and Hyden, 1997, p135):

A traffic conflict is an observable situation in which two or more road users approach each other in space and time to such an extent that there is a risk of collision if their movement remained unchanged.

Today TCT is essentially the more general definition of this formally stated procedure. Traffic conflicts, which can be described as "near-accidents" or "critical incidents" (Muhlrad, 1993, p.6) are observed by specially trained people who conduct the studies. Compared to RSI, these people don't need to be experts because they observe simple situations rather than the complex environment of the road. But the study has to have a duration of a few days to gather enough information, which, even so, in some cases is still not enough. Also one can only observe a very limited part of a road, so using this technique in its classical form is even less implementable than the Road Safety Inspections.

With the advance in computer technologies, some projects try to use captured video data to automatically detect conflicts. However, they also have some limitations such as the need of training the algorithm for every new camera position, which makes a universal use of such algorithms in near future improbable. This technology is mostly used to analyse the improvement on single spots after a safety measure was implemented. (Autey, Sayed and Zaki, 2012)

At this point, volunteered geographical information from road users can be the alternative to the improbable TCT. Since the goal of TCT is the observation of conflicts between two or more road participants, those participants are also observers of their conflict. And, possibly, a lot of other road participants will witness such conflicts. Motivating these people to share this information is the key point for the success of this proposed Social Traffic Conflict Technique (S-TCT). In the next sections the technical details of such a system will be addressed. Beforehand, however, some substantial issues of TCT and their counterpart in S-TCT should be discussed.

A point of argue in TCT was the definition and recognition of a conflict and the distinction between severe and non-severe conflicts. One method used mainly in Sweden was to use measured distances and speed information and use a pre-defined threshold for classification (Hyden, 1987). Other countries like Austria or Germany used a more semantical approach, where the observer would use expressions like "not so dangerous" or "collision averted by an inch" (Risser and Schützenhöfer, 1984). But studies showed that even classifications of the same observer can be inconsistent, which makes an overall comparison of observations from different people even more complicated (Chin and Quek, 1997).

Putting an untrained road participant in the role of the observer makes it necessary to treat the gathered data with some caution. Especially if the observer is involved in the conflict. In these cases, the observation may be biased either by neglecting the own errors and accusing the other participants or the road environment, or by blaming oneself and ignoring the other faults involved. Those questions will be partly considered in the technical outline of S-TCT but are subject for further research, in particular after such a system is running.

DATA MODEL AND SYSTEM REQUIREMENTS

The fundament for any such system is the availability of GPS coded road data. For many developed countries there are several commercial maps available, some with extended information about the environment. However, for this project we strongly recommend the usage of an open system where the information is gathered by the users, like openstreetmaps. The reason is not the cost factor, but the open nature of the data. Using, giving access, publishing results and so on are all problem points if the underlying data is not open.

Another point is the actuality of the data. If the road on some point has changed, the user can update the data faster than the commercial company which evaluates the cost/benefit factor of updating its data. Also for regions which are not covered by companies due to their financial unattractiveness, the users can build the road network. Recalling that the roads in underdeveloped countries and regions are much more dangerous than in developed countries, this argument gains more weight.

After the base road data is available, either open or commercial, the information volunteered by the users has to be stored. In this section some models for this data will be outlined as a draft, so any computer engineer can take it as a starting point for planning and developing a real system. Some words about system requirements are given in form of technology comparison.

Some systems exist where user volunteer geographical information related to the road or the traffic. An example would be Waze, where information about traffic load or gasoline prices is volunteered by the users. But such systems have a different data model, since they collect a small predefined set of data, which is only valid for a short period of time, and share this information with the other users in real-time.

Naming Convention and Data Model

It is important to establish a good naming convention to allow multiple people to work on such a project and make the system easy to understand. The first question to ask is, what are we collecting from the users. The core objective of the system is to gather reports of traffic conflicts. However, given the goal and structure of the system, it is possible that a road user recognizes a threat to the road safety (like an overgrown tree branch) and wants to report this also. For this purpose, everything that a user can report is called an 'event'.

In this extended meaning, an event can need different datatypes for different information. For example, a traffic conflict will happen on a single point, so a single GPS coordinate would be sufficient. But an event can cover a section of the road, so the data model needs to be able to handle also GPS routes as well as GPS points.

Another important point that can cause confusion is the severity of the reported incident. Two different severities are of interest. First; how close was the conflict, did it almost come to an accident or did the driver evade the situation easily? Second; how severe would have been the outcome of an accident if it had happened? If a car almost crashed into another car from behind at low speeds it would be a "light" accident with only material damage. If, on the other hand, a car almost crushed into a pedestrian or a bicycle frontal, it would be a severe accident with a high risk of fatality. In order to make an easy to understand distinction, the first severity will be called the accident risk, while the second severity will be called fatality risk.

Figure 2 shows a sample XML draft for the most important data. It is not a complete Scheme, but using this XML like representation enables one to give a quick overview of the data model and its relationships. XML is further recommended to be used by the system for data transfer because it is easily readable by machines as well as by humans and the data model can be adapted and extended very fast if the need occurs.

Figure 2. XML draft of data model


```
<?xml version="1.0"?>
- <event userid="1" dateteime="2016-05-29 16:34:47" id="0">
  - <route isPoint="true">
      <GPS_Coordinate timestamp="1464529085" numOfSat="7" long="29.101462"
        lat="41.143922"/>
    </route>
  - <participants>
      <vehicle id="1" type="car"/>
      <vehicle id="2" type="bicycle"/>
      <pedestrian id="1"/>
    </participants>
    <description>The driver of Vehicle#1 wanted to turn right, but concentrated on
      Pedestrian#1, who wanted to cross the street from the opposite direction. As
      a result, the driver of Vehicle#1 didn't see Vehicle#2 on the bicycle lane.
      Driver of Vehicle#2 had to drift of the bicycle lane and nearly fell of the
      bicycle in order to prevent an accident. </description>
    <accidentRisk>middle</accidentRisk>
    <fatalityRisk>low</fatalityRisk>
</event>
```

Together with a drawing of the incident on the map, showing position and direction of the participants, the information in Figure 2 would be the simplest traffic conflict report. More information and also the opinion of the reporter about cause and possible measures can be gathered.

System Requirements

In this section the technical requirements for the system to function are outlined. For every part, a brief introduction to the alternatives and a recommendation is given. Figure 3 shows an overview of the core elements of the system.

As mentioned before, the fundament of the system is a geographical database containing the road data for the area that will be covered with the system. The author recommend the usage of openstreetmaps.org as the source of this data, since it is openly available. Many databases are suitable for this task and give GIS functionality with Extension Packages or Add-Ons. Table 1 shows a comparison of basic functionality for the most used databases.

The author recommend the usage of PostGreSQL with the PostGIS extension. Besides the availability of more functions and good performance, the open source nature of both database and extension are in accordance with the nature of this project. Another benefit is that the data from openstreetmaps.org is available in two formats, namely XML and a specific binary format called PBF. For the PBF format, there exists a tool (osm2pgsql) that converts the data to PostGIS-enabled PostGreSQL databases.

Above the road database, a system database is needed for all data that is produced due to the functionality of the system. Besides administrative data (such as users and settings) the volunteered data from the users, namely the reports, will also be stored in this database. This database can run on the

Figure 3. Elements of the system

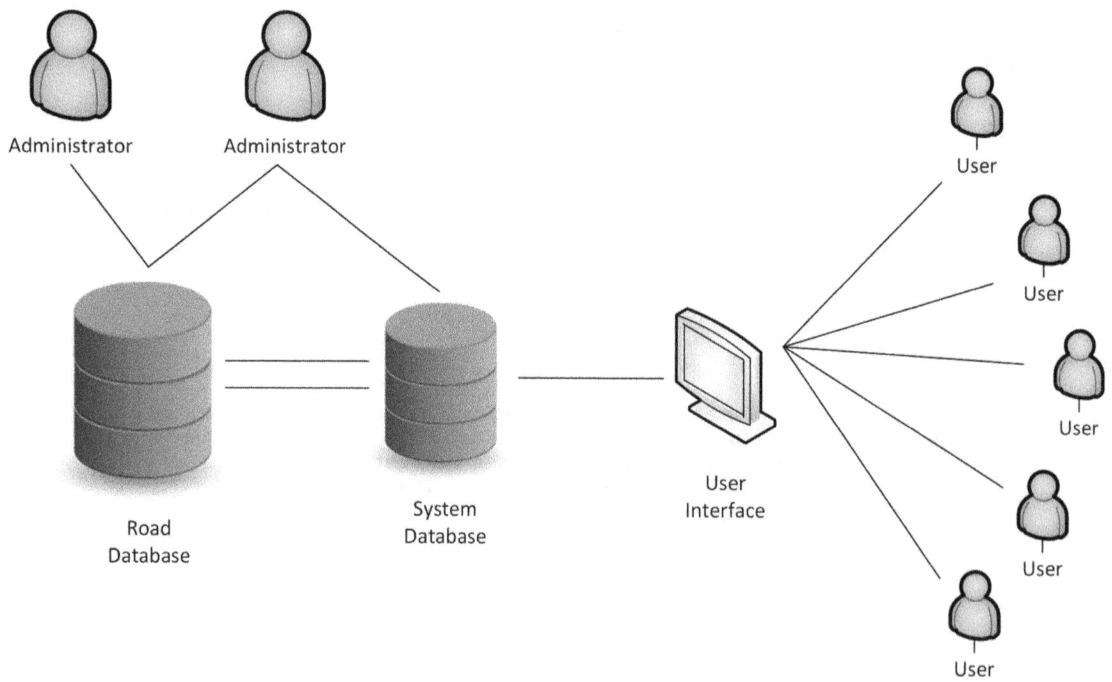

Table 1. Comparison of GIS functionality of common SQL Databases [1]

	PostGIS 1.5	Oracle	MS SQL Server 2008 R2	MySQL 6
Geometry Types	OGC basic 2D/2.5D, with curved support	OGC basic 2D/2.5D/3D, with curved support	OGC basic 2D/2.5D, no curved support	OGC basic 2D/2.5D/3D, no curved support
Output Functions	Binary, Text, SVG, GML, KML, GeoJson, WKT, WKB	WKB, WKT, GML, KML	Binary, Text, GML, TextZM	Binary, Text
Measurement	YES (Hausdorffdistance, spherical measurement)	Basic (Area, Length, Distance)	Basic (Area, Length, Distance)	Basic (Area, Length, Distance)
Linear Referencing	YES (referencing segments)	YES (manipulating segments)	Note included (open source add-on's available)	None
Spatial Aggregates	Very comprehensive	Basic spatial aggregates	Note included (open source add-on's available)	None

same instance as the road database, but it is recommended to use a separate instance in order to enable distributed systems giving a more flexible planning possibility and better performance.

The system database does not need to be the same as the road database since it is not mainly handling GIS operations. So, another open source database like MySQL could be used. However the author recommends still using also PostGreSQL for this database in order to better manage the system and ease the communication between road and system databases.

The most important part of the system is the user interface. Only if it is easy to use, enough people can be motivated to use it and volunteer the valuable information. The technology behind the interface becomes a minor point. Some ideas for the user interface will be given in the next section. At this point the author will outline some technical arguments.

The users must be able to fill a report both via computer and via smartphone. The first decision to take is whether to make separate mobile apps or to use a responsive design which can be used on mobile browsers. Developing a mobile app has the benefit of extended user interaction possibilities and a better performance. Both are important arguments since they increase the motivation of users to report an incident.

However, developing such an app that is easy to use is a non-trivial task and can be a big cost factor with very small rewards. An app has to be developed for the most common mobile operating systems. This will exclude users of some mobile phone vendors. Another cost factor is the continuation of the apps. Major updates of the operating system can cause the need for an update of the app.

Considering these arguments, the author proposes to concentrate on a modern, touch-friendly, responsive web design. This page can either be used on mobile browsers directly, or a simple wrapper app for some mobile operating systems can be developed easily. So most users would get the feeling of a standalone mobile app, while users of not supported vendors can still use the system.

The used programming language (PHP,.Net, JSP, Coldfusion etc.) for the web system is a minor secondary decision. It can be based on the skill of the project team and the availability of soft- and hardware resources. Important is the frontend for the users, which needs to be easily understandable and fast to use. Empowering HTML5 is strongly recommended since it gives multi-touch handling for browsers and make it possible to make fast and responsive interactive user interfaces.

OPERATING MODELS

The valuable data is volunteered by the users. They are the core element of the system. In this chapter the operating model is discussed. Who should operate the system, how should the user be involved, how is the trustworthiness of the information to be determined? Although there is no 'right' answer for this questions, some possibilities are elaborated and compared.

Community Operated Model

In this model, the system is not only used, but also operated by private persons. A non-commercial, non-governmental organization (NGO) which is run by the community, could operate such a system. Wikipedia can be named as a similar and very well-known system operated with this model. The content is created by users, while other users act as editors who examine and manage this content. The selection of the editors can either base on the contribution to the system and the community and/or on expertise and educational background.

Editors are needed mainly for two reasons. First, like any platform with user generated content, the system could be used to post opinions about sports, politics and other controversial subjects, thus flood the system and make the data unusable. For this kind of control, the editor doesn't need any special education or expertise. Second, real reports of traffic conflicts need to be classified correctly so any statistical analysis can give meaningful results. For this work, some knowledge about topics like traffic safety, road design would be helpful.

Another option would be to entrust the work of the editors also to the community itself. This would follow the idea of "wisdom of the crowd" (Giles, 2005)(Niederer & Van Dijck, 2010). The users could, for example, tag a report as spam and based on some weighing algorithms, after enough tags are given, the report would be deactivated. In a similar way, if a user thinks that a report is classified wrongly, she can suggest a correction and, accompanied by a discussion, the users would decide the final classification.

Public Authority Operated Model

The police department, the ministry of traffic or some other public authority responsible for some aspect of the road network and interested in road safety can officially operate such a system. The users would expect the authority to employ qualified staff to handle the above described work of editors and further analysis.

Moreover, the users would also expect and even demand reactions and actions to the reports. If a report isn't handled perceptibly, the motivation of the users would sink dramatically. If, for example, a junction is reported by different people for the same hazard, a counter measure by the authorities will be anticipated. The lacking of any official action would again decrease the motivation of the users.

Employing experts for the editorial work is not a bad thing, but two risks should be kept in mind. On the one hand, this people doing the same thing for a very long time can become careless. On the other hand, the education and opinion on some safety related matters could influence the classification of reports, making the data prejudiced.

Comparison of Operating Models

Two operating models where outlined above. Now, the models will be compared for some key subjects.

- **Cost Factors:** Operating such a system especially if it has a good acceptance and many users, can be very costly. The main cost factors are hosting and bandwidth for the web servers. Also other operational, administrative and advertising costs are unavoidable. For this subject, a public authority with an allocated budget has a clear advantage. The community operated system could work with donations by the users (like Wikipedia), ad revenue, public funding or any combination of those. But there is no guaranteed steady budget, so the cost is a big risk factor for the community model.
- **Publicity:** Any system where the content is created by the users need a big and active user base. If the system is operated by the community, every user can see himself as part of this community and work for the growth of it. A public authority on the other hand would have to advertise actively for the system to convince people to become users. This subject is a point for the community.
- **Motivation:** Having a wide user base is necessary, but not enough. The users must be kept motivated, the community needs to stay alive. The so called 1-9-90 rule most likely will be also valid for this system. It says that about 1 percent of the users create new content, 9 percent are active in editing or discussions and 90 percent are sole viewers. Research about community (Arthur, 2006) can help addressing the right topics to elevate motivation.
- Two driving motivations can be named for this system. First would be to do good by making the roads safer. For this, the reports need to have actual results. This depends on the attitude of the authorities and sometimes on the budget. If a junction needs to be completely rebuilt, it might take some years before the project and the budget for this is granted, but, in any case, the users should be fed with information about the process, like "project submitted", "project in revision" and so on. Small and inexpensive corrections on the other hand should be performed quickly. If for example many users report that a bad placed sign causes confusion and lead to dangerous situation, the local authority should replace the sign quickly. Making the correction is important, but it is also important to inform people that something was changed and that the user reports were the reason of this improvement.
- When the system is operated by a public authority, it might have some advantages for the communication with the local authorities responsible for the roads. But there is no guarantee since public authorities often have very strict regulations and limitations. So, more important then who operates the system is the question about the willingness for cooperation of all public authorities.
- **Technical Aspects:** The user interface is the most important aspect of the system. No matter how motivated the users are, if generating a report is a torture, if it takes too long or if one needs a training for the task, the number of participation will be very low. So the user interface needs to be intuitive and easy to use, and on the other hand guide the user to get the most important information in an accurate way.

Using simple but leading questions with predefined answer sets can gather more information than text areas. Text input should be kept to a minimum, at some points for additional comments on the made selection, and, at the end, to get the general opinion of the users.

This has two benefits. On the one hand, statistical analysis are much easier with predefined selectable answers than free text reports. On the other hand, discussions between users will be easier and more objective.

Another important part of the report is the sketch of the incident. It helps people to understand and reconstruct the situation. A simple user interface that can be used both on a personal computer with a mouse and on the touchscreen of a tablet or a mobile phone is the most challenging technical aspect of the project. As mentioned before, using state of the art web technologies can make the system independent of the operating system of the user.

Figure 4 shows a sample wireframe for the user interface to generate sketches. After the location of the incident is selected on the map, a plain drawing of the road is generated using the road data. The user can place and rotate the reported participants on the sketch and set the moving direction. Pre-defined objects like trees, bushes, traffic signs or placeholders for undefined objects can also be placed on the sketch. This interface needs to be multi-touch friendly to drag and rotate objects, but it also needs to be operable with a simple mouse.

Comparing the operating models regarding this technical aspect is difficult. If the community has a sufficient number of highly qualified software engineers, then a good system can be developed. But it is a time consuming and sometimes frustrating work, so this people need to have a lot of free time and energy. A public authority will most likely hire a company for the job. Here the tender regulations can limit the choice of the best developers, leading to a bad system.

Figure 4. Wireframe for a sample interface

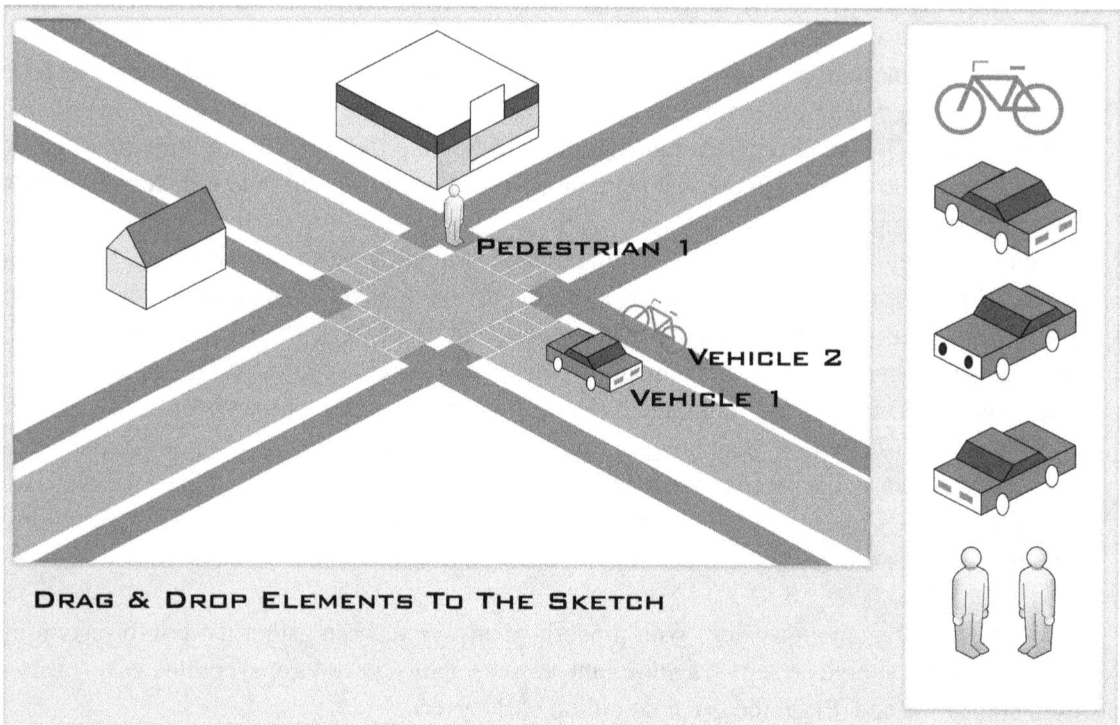

Table 2. Comparison of operating models

	Community Operated	Public Authority Operated
Cost Factor	Need donations, ad revenue or public funding	Advantage if allocated budget
Publicity	Users gather users	Needs advertisement and convincing
Motivation	Advantage of being a ommunity	Advantage of official correspondence
Technical Aspects	Potentially very good software developers in the community	Can assign professionals or companies, but regulations may be limiting

As can be seen in the overview in table 2, there is no clear winner. Both operation models have their strengths and their risks. The author therefore recommends a cooperation between a non-profit organisation and the public authorities. This will be elaborated during the discussion in the last chapter.

SAFETY AND RESEARCH OPPORTUNITIES

The primary goal of the described system is to make our roads safer, prevent accidents and thus safe life. This motivation should always be the number one priority. But during the operation of the system, a lot of data is gathered and stored. This data can and should be used by researchers. After safety opportunities, this section explains some research areas that can use this data and the benefits expected from this kind of research.

Safety Opportunities

Vision Zero is a philosophy officially accepted and funded by the Swedish parliament saying that it is ethically unacceptable that people are killed or seriously injured on the roads (Tingvall & Haworth, 2000). This is a goal worth working for, but no single measure or no single action plan can achieve such an assertive goal. This system can be but a part of many action steps towards this goal.

The system will have none to little contribution to the safety aspect of vehicles. The human aspect can be positively affected by allowing a better understanding of driver behaviour. The interaction between users and even only the knowledge about the existence of such a system can result in an awareness of the traffic participants.

The aim of the system, however, is to detect risky road segments so that they can be corrected before severe accidents happen. The effectiveness of the system hereby highly depends on the approach of the authorities. If they choose to ignore all reports and warnings of the road users, it would be a big missed opportunity.

Therefore, it is important that the authorities are at some point integrated into the system even if they don't join the operation of the system. The detected risks can be classified from two point of views. First, it is important how big the risk is. This depends on many variables like average speed, frequency. The second classification is the size of the needed measure. The nominal value for this classification will be most likely the cost factor. If, for example, only a tree needs to be trimmed this can be done very fast. But if a redesign of a whole junction is needed, this can take some time. In such cases, the authorities should publicly work on a timetable and take some easy measures like speed reduction or additional signs in the meanwhile.

Research in Social Sciences

Even though the system is about making roads safer, it has two connections to the human in its role as a social and psychological entity. The first role is the human as a driver, or, to be more general, a participant of the traffic environment. The design of a road and its environment has to take the psychology of the participants into account. A lot of research is done on how people react or interact with different infrastructures. Liu (2001) investigates how digital information systems affect the drivers. Al-Madani and Al-Janahi (2002) have taken a look the social characteristics of drivers comprehending traffic signs. In their book, Bonsall and Parry (1991) have analyzed the driver's compliance to route guidance advices.

As on can see by this very small outline, there are a lot of social questions and the introduced system can provide social scientists with the data they need without carrying out specific surveys. Most research in this area are performed with small groups of participants and sometimes simulators are used since a real world experiment is not feasible. But within the system a lot of valuable and required data will be publicly available for the researchers. This makes also the results more easily re-constructible by other researchers and would enhance the scientific quality.

The second role of the human is as a user of a social system. In the previous chapters some ideas were discussed on how to get people to participate and how to keep them motivated. Those questions are also important for psychological research. Since the data is publicly available, researchers can investigate the social behaviour of users.

Research in Architecture and Civil Engineering

Designing a road and its environment, building it, maintaining it is a living process. In which extent the safety regards are taken into account during each cycle of this process is very important. Safety should be always the first priority and main decision force, before costs, comfort, capacity or other concerns. But this is not always the case.

The data gathered in the system could help researchers to better analyse the safety of roads from the point of view of architects and civil engineers. Elvik, Vaa, Erke and Sorenson (2009) made a meta-analysis in their book. They made an extensive literature review and included qualified studies into their analysis. 20 measures for road design, 9 measures for road maintenance and 22 measures for traffic control are explained and inspected in detail (pages 145-542). This is a must-read handbook for everyone working related to road safety.

The book illustrates also how the data in the system could be used for research. In combination with already available data, like road geometry, and additional information from the users, many measures like roundabouts or road lightning can be evaluated regarding their safety effects.

CONCLUSION AND DISCUSSION

Road Traffic Safety is important. For everyone. But the roads are far from being save. The Vision Zero of zero deaths and severe injuries on roads is an assertive goal. Reactive methods are important to quickly fix very unsafe spots on the road network. However, proactive measures are needed to prevent severe accidents from happening before black spots even emerge. The system introduced in this chapter is a proactive method that can help coming closer to the goal of Vision Zero. It exploits a new source of valuable data, namely the road users themselves.

The traffic has three safety related aspects. The humans, the vehicles and the roads. The vehicles are an engineering section for themselves. This chapter concentrated on the humans as the endangered users who provide data to make the roads safer. Several technologies, which can be used to implement such a system, are briefly examined with the conclusion that the most important thing is a user-friendly front end, rather than high tech in the background.

Another important question is by whom the system will be operated. The possibilities of a public and a governmental system are explored. But it is hard to come to a conclusion or give a definite recommendation. However, a combination of both models, a cooperation between the people and the governmental authorities seems to be more promising. The government could host the system and bear the costs of operation. The community could, or should in return provide the needed software development work and management of the system. Every authority that is related to road safety and a possible executer of safety measures should have at least one commissioner assigned to the system, so that a direct and continuous communication with the community is possible.

Besides the direct effect of the system, namely detecting dangerous spots and finding measures to make them safer, the system also gathers a huge amount of data which can be used by scientist from different research areas. Such research can lead to improvements in the fundamental understanding of road safety and to safer roads from design to the built road.

The author wants to end the chapter with a plea for action. Regardless of the system, everyone has the possibility and the responsibility to make our roads a safer place.

Be awake, be aware, recognize what YOU could improve and do it.

REFERENCES

Al-Madani, H., & Al-Janahi, A. R. (2002). Assessment of drivers comprehension of traffic signs based on their traffic, personal and social characteristics. *Transportation Research Part F: Traffic Psychology and Behaviour*, *5*(1), 63–76. doi:10.1016/S1369-8478(02)00006-2

Amundsen, F., & Proc, C. H. (1977). *First workshop on traffic conflicts.* Institute of Transportation Economics. Oslo, Norway.

Arthur, C. (2006). What is the 1% rule? *The Guardian.*

Autey, J., Sayed, T., & Zaki, M. H. (2012). Safety evaluation of right-turn smart channels using automated traffic conflict analysis. *Accident; Analysis and Prevention*, *45*, 120–130. doi:10.1016/j.aap.2011.11.015 PMID:22269493

Bonsall, P., & Parry, T. (1991). *Using An Interactive Route-Choice Simulator To Investigate Drivers' compliance With Route Guidance Advice* (No. 1306). Academic Press.

Chin, H. C., & Quek, S. T. (1997). Measurement of traffic conflicts. *Safety Science*, *26*(3), 169–185. doi:10.1016/S0925-7535(97)00041-6

Elvik, R. (2006). *Road safety inspections: safety effects and best practice guidelines.* Transportøkonomisk institutt.

Elvik, R., Vaa, T., Erke, A., & Sorensen, M. (Eds.). (2009). *The handbook of road safety measures*. Emerald Group Publishing. doi:10.1108/9781848552517

Evans, L. (1991). *Traffic safety and the driver*. Science Serving Society.

Giles, J. (2005). Wisdom of the Crowd. *Nature*, *438*(7066), 281. doi:10.1038/438281a PMID:16292279

Hydén, C. (1987). The development of a method for traffic safety evaluation: The Swedish Traffic Conflicts Technique. *Bulletin Lund Institute of Technology, Department*, (70).

Liu, Y. C. (2001). Comparative study of the effects of auditory, visual and multimodality displays on drivers performance in advanced traveller information systems. *Ergonomics*, *44*(4), 425–442. doi:10.1080/00140130010011369 PMID:11291824

Moscari T. & Hollo P. (2006). *Common Understanding on Road Safety Inspections, RIPCORD ISEREST: Report of Workpackage 5*. BASt.

Muhlrad, N. (1993). Traffic conflict techniques and other forms of behavioural analysis: Application to safety diagnoses.*Proceedings of 6th ICTCT Workshop*, 48-64.

Nadler, B., Nadler, F., & Strnad, B. (2012). *Road Safety Inspection – Manual for Conducting RSI*. Vienna: Austrian Road Safety Fund.

Niederer, S., & Van Dijck, J. (2010). Wisdom of the crowd or technicity of content? Wikipedia as a sociotechnical system. *New Media & Society*, *12*(8), 1368–1387. doi:10.1177/1461444810365297

Perkins, S. R., & Harris, J. L. (1968). Traffic conflict characteristics-accident potential at intersections. *Highway Research Record*, 225.

Risser, R., & Schützenhöfer, A. (1984). Application of Traffic-Conflict Technique in Austria. In *International Calibration Study of Traffic Conflict Techniques* (pp. 141–151). Springer Berlin Heidelberg. doi:10.1007/978-3-642-82109-7_13

Sørensen, M. W., & Elvik, R. (2007). *Black Spot Management and Safety Analysis of Road Networks: Best Practice Guidelines and Implementation Steps*. Transportøkonomisk institutt.

Tingvall, C., & Haworth, N. (2000). Vision Zero: an ethical approach to safety and mobility. In *6th ITE International Conference Road Safety & Traffic Enforcement: Beyond* (Vol. 1999).

World Health Organization. (2015). *Global Status Report on Road Safety*. WHO.

Yildiz, A., & Hauger, G. (2010). An approach for a more precise road scoring. In *Proceedings of the Road Safety on Four Continents Conference* (Vol. 15, pp. 473-481).

ENDNOTE

1 Data gathered July 18, 2016, from http://www.bostongis.com

Chapter 13
Generation of a Data Model for Indoor Navigation Based on Volunteered Geospatial Information (VGI)

Rahim Ali Abbaspour
University of Tehran, Iran

Simin S. Mirvahabi
University of Tehran, Iran

ABSTRACT

Navigation has been an inseparable part of human life especially in modern days, when the structures of cities and their buildings' indoor environments have been more complex. More than 80% of routine life of a typical citizen is spent in indoor and the indoor environment are getting highly complex due to the increase in sizes of the buildings. An important factor to a successful indoor navigation is the precise suitable map for the inside of the buildings. Collection and generation of indoor geospatial data is very time consuming and costly for a building. Using the concept of volunteered geospatial information might be a suitable solution to deal with this problem. This chapter addresses the extraction of a data model for indoor navigation from VGI. An efficient methodology is proposed and evaluated to extract the navigation data model from OpenStreetMap automatically to use in indoor navigation applications.

INTRODUCTION

Navigation and route finding have been an inseparable part of human life especially in modern days, where the structures of cities and their buildings' indoor environments have been more complex. Navigation can be generally divided into two broad classes of indoor and outdoor according to the nature of the embedding space. While the navigation in outdoor is a most studied class from different perspectives such as positioning methods, application areas, and algorithms (Chatila, Devy, Lacroix, & Herrb, 1994; Langer, Rosenblatt, & Hebert, 1994; Levinson et al., 2011; Makri, Zlatanova, & Verbree, 2015;

DOI: 10.4018/978-1-5225-2446-5.ch013

Massidda, Bülthoff, & Stegagno, 2015; Urmson et al., 2008), the navigation and wayfinding in indoor environment is newer and more challenging.

More than 80% of the routine life of a typical citizen is spent in indoor environments. These are getting highly complex due to the increase in sizes of the buildings such as hospitals, airports, megamalls, and universities (Li & Lee, 2013). On the other hand, the progress in indoor positioning systems facilitates serving and using the location-based services (LBS) such as navigation and recommender systems. Therefore, the indoor navigation is one of the main applicable and useful services for citizens (Han, Jung, Lee, & Yoon, 2014; Möller, Kranz, Huitl, Diewald, & Roalter, 2012; Nakajima & Haruyama, 2013; Zheng et al., 2014). However, the indoor navigation has been a challenging research topic recently (Mortari, 2013; Karimi, 2015). The main challenges in this field are the lack of high precision in indoor navigation, possibility to free movements of agents, and difficulty to build a navigation network structure (Werner, 2014; Karimi, 2015). Moreover, collection and generation of indoor geospatial data are very time-consuming and costly. The main focus of the maps and geospatial data providers is on the production of formal outdoor maps. Using the concept of volunteered geographic information (VGI) might be a suitable solution to deal with this problem (Mortari, 2013). Using VGI greatly reduces time of data collecting and saves the costs related to it. On the other hand, accuracy of VGI data in most cases is acceptable for indoor navigation. Furthermore, important areas such as points of interest (POIs) and semantic information exist in many sources of VGI data such as OSM and Google Maps.

Most of the existing VGI projects such as Google Maps and Wikimapia have focused on the collection and enrichment of outdoor geospatial data. This functionality is similar to the elaboration of the traditional mapping agencies. Some of them even provide both 2D and 3D view of information. But, the number of projects which provide the users with functions to add or view the indoor environment is rare. Moreover, the usage of existing volunteered indoor datasets for navigation in the existing projects in the current form has some barriers. For example, the interior environments of buildings are not supported completely and the semantic information used for buildings is of heterogeneous formats in different countries due to the lack of standards for key values. Adding routines for supervised automatic entrance of indoor spatial data and assigning the existing or loaded semantic information to them in an acceptable standard format may be a feasible solution to this issue, which is also addressed in this paper.

An important factor to a successful indoor navigation is the precise suitable map for the inside of the buildings. The main characteristic of an indoor map, which separates it from a common outdoor map, is the difference in data model. A data model, which also is called location model, defines the structure of representation of location information (Werner, 2014). There is several buildings' indoor navigation software, which uses different data models. This heterogeneity results in impossibility or at least presents difficulty to share the data. The IndoorGML standard, which is proposed by Open Geospatial Consortium (OGC), is a useful format for indoor navigation applications (OGC, 2012). The main advantages of this standard are that it facilitates data exchange between different navigation software and supports the building modeling in 2D and 3D format.

One of the most popular projects in VGI is OpenStreetMap (OSM). The easy way of adding geospatial data to this project as well as access to the history of data are among the reasons which make this project more applicable rather than the other VGI-based projects. The OSM is generally used for outdoor wayfinding and navigation, but we present a methodology to add the indoor data model automatically to this site in order to enrich the project and facilitate the pervasive navigations.

This chapter addresses the extraction of a practical, standard data model for indoor navigation using VGI. The advantages and disadvantages of using each data model in indoor navigation based on geometric, topologic, and semantic aspects of models are discussed firstly and then, the most suitable one is introduced to be used. The next step, which is of importance in indoor navigation, is the extraction of desired data model from a VGI project practically. To this aim, the OpenStreetMap is used and the indoor data insertion is considered in this site. The automatic generation of indoor navigation data model from OSM data is carried out in a 9-steps algorithm using the IndoorGML standard. Therefore, the structure of this chapter is as follows. Related researches are explained in section 2, while in section 3, main data models for 3D modeling of interior of buildings are presented. Section 4 focuses on the standards for data models of indoor spaces and some important standards such as IFC, BIM, CityGML, and IndoorGML are explained. Because the VGI dataset used in this chapter is extracted from OpenStreetMap, section 5 introduces the data entry to this site. Finally, in section 6, automatically extraction of navigation data model from OpenStreetMap for using in a navigation application is discussed in detail according to the proposed methodology.

RELATED WORK

In recent years, owing to the advancements in information technology, navigation in indoor environment of complex buildings has attracted the attention of researchers and developers. There are several researches about navigation and wayfinding in indoor spaces in the literature. Cyberguide system was one of the early systems for indoor navigation. It was developed for tourist guidance both in indoor and outdoor environment (Abowd et al., 1997). The other navigation system, which consisted of a real-time wayfinding application capable to be installed in cellphones and PDAs, was iNAV (Kargl, Geßler, & Flerlage, 2007).

After introduction of the concept of crowdsourcing, the benefits of using such information leads some researchers to use them in navigation. One of the first systems for indoor navigation, which profit from crowdsourcing, was MARS (Höllerer, Feiner, Terauchi, Rashid, & Hallaway, 1999). This system was developed for navigation based on the personal information from interior of buildings in Columbia University. In 2012, an automatic approach is proposed to generate simulation-related data based on the IndoorOSM data to create multi-agent evacuation simulations (MATSim) (Goetz & Zipf, 2012). Moreover, the limitations of using IndoorOSM in this framework were discussed. Mortari used the CityGML data structure of buildings to extract connectivity graph of them for navigation purposes (Mortari, 2013). He proposed an automatic way to extract navigation graph from data in 3D structures of buildings using existing tags. Herrera et al. used IndoorOSM in the smartphones to provide a hybrid indoor localization approach (Herrera et al., 2014). In this system, the trajectories of users are recorded by the users mobile and a map of paths is generated, which was then converted to navigation graph. Moreover, a PDR is generated using the data in OSM to improve the navigation precision. Zaragozí et al. proposed a hybrid navigation system based on open data and published via standard protocols and web services (Zaragozí et al., 2015). The purpose of this research was increasing the access speed to navigation data in indoor environments in smart cities.

In these systems, the navigation data model is designed according to the desired application area. These models do not follow any standards. Moreover, there are not any automatic approaches to extract the navigation data model for indoor environment using VGI.

REPRESENTATION OF BUILDINGS INTERIOR

The initial and main challenge using the indoor environment in navigation is modeling of interior environment of buildings. The process of digital model generation of indoors for buildings should keep the information for key elements while ignore the redundant data. There are several models for modeling of interior of buildings, which are reviewed in this section briefly.

One of the popular models for the representation of buildings indoor spaces is location model (Werner, 2014). This model is introduced as a storage and retrieval method of spatial data. In location models, the ways to represent a model is diverse based on the different options for visualization. Among them, representation based on the dimension is most popular. For indoor navigation inside complex buildings, it is possible to use information in different dimensions, i.e. 2D, 2.5D, 2.75D, and 3D, according to the type of navigation (Figure 1). The 2D plan demonstrates a planar, thematic representation using only lines and polygons in the plane (Figure 1). The 2.5D model depicts 2D object embedded in 3D space, i.e. the structure is two-dimensional and the third dimension is represented by attributes (Figure 2) (Tse & Gold, 2004). The 2.75D model allows modeling of more advanced construction, e.g. tunnels and bridges, in which the z value is required. But it is not possible to construct 3D objects with separate volumes (Figure 3) (Tse & Gold, 2004). The 3D model is constructed from separate volumes connected together (Figure 4). Each volume is bounded by faces, edges, and points; thus, there are 3D volumes, 2D faces, 1D edges, and 0D points available in this model (Gold, 2006).

Raster and vector models are the other models, which are used the interiors of buildings, but using each of them has some shortcomings. Although vector maps have great flexibility, computing with them may become complicated. Raster maps can only be used practically in modeling 2D spaces. Moreover, because both of these models only represent geometry, they are not suitable for indoor navigation. For modeling the spatial relations between objects in geometric maps (e.g. is it possible to move from location i to location j?), environmental models are proposed. They can be divided into three categories: set-based, graph based, and hybrid (Figures 5-8) (Werner, 2014). Set-based model defines a finite set

Figure 1. 2D

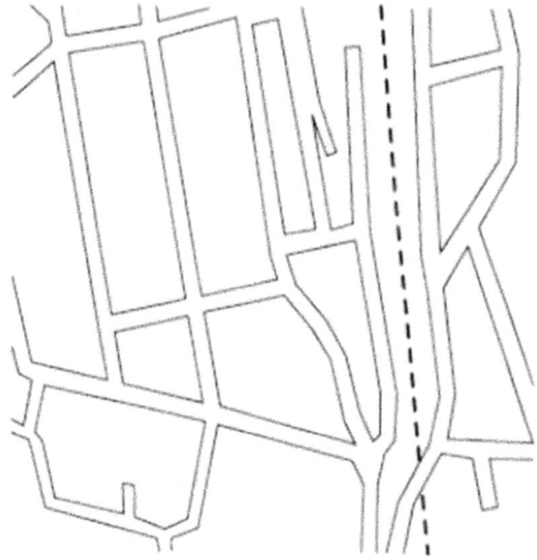

Figure 2. 2.5D

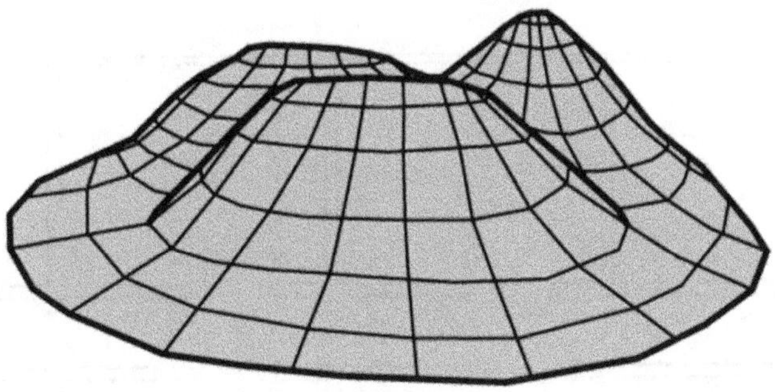

Figure 3. 3D

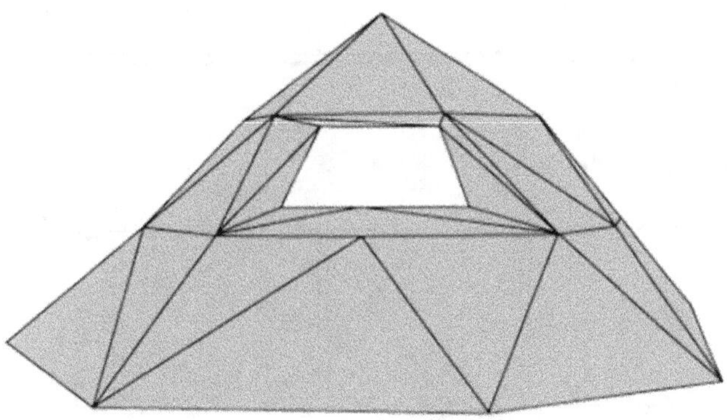

Figure 4. Representation of a building
(Boguslawski, 2011)

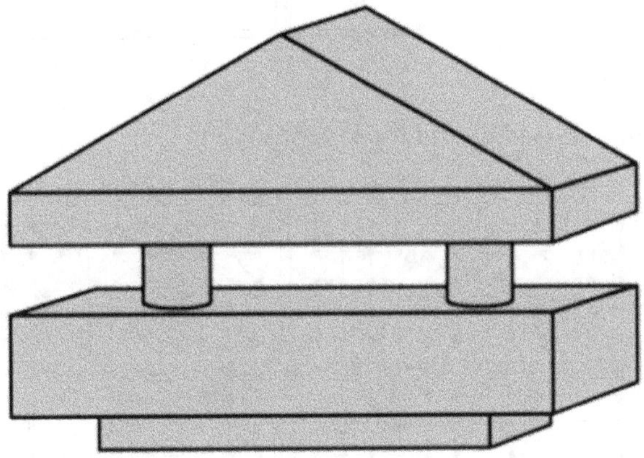

of atomic places and then complex places such as a floor are modeled as sets containing atomic places. This model cannot support distance calculation and finding the neighborhood of rooms. The graph-based environmental models cope with these issues using the graph theory concepts. This model is discussed in detail during the description of proposed methodology in order to support more applications; these two approaches are combined to hybrid environmental models.

Figure 5. Some of the environmental model: Set-based environmental model A partly completed corner graph

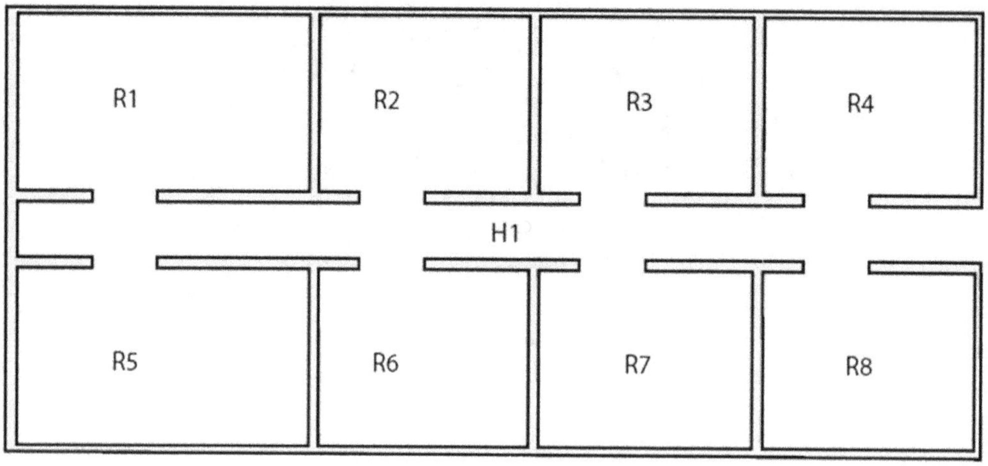

Figure 6. A partly completed corner graph

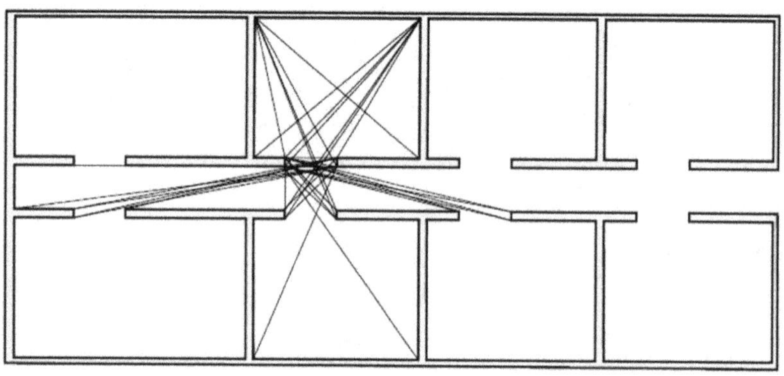

Figure 7. A regular grid graph based on the 8-corner system

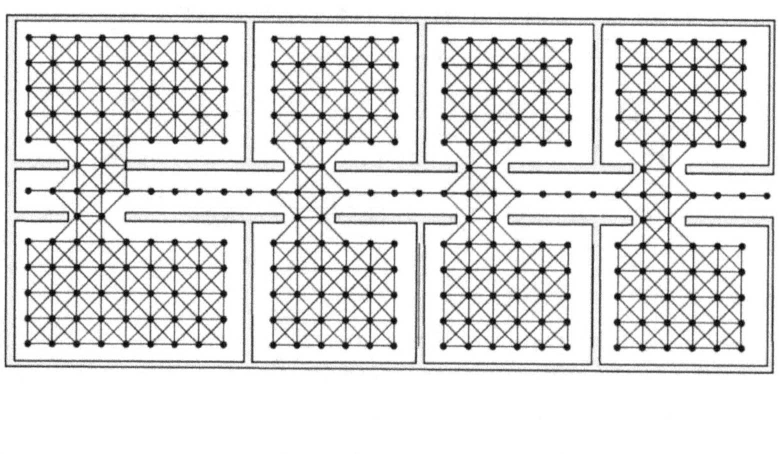

Figure 8. A graph with edges not contained in walkable space (Werner, 2014)

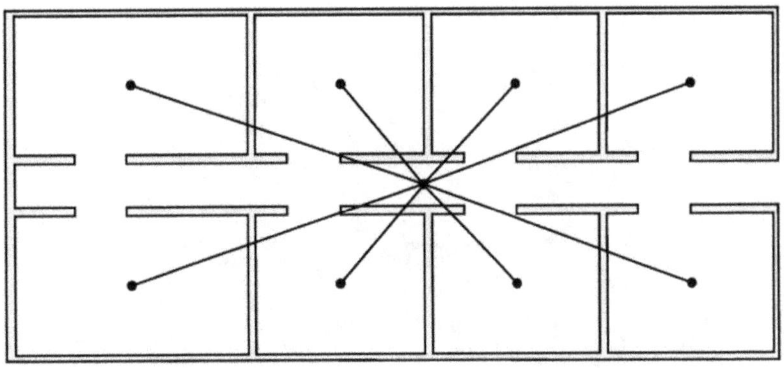

STANDARDS FOR DATA MODELS OF INDOOR SPACES

The modeling of interior parts of buildings has absorbed the attention of many researchers during the last decade (Azhar, 2011; Eastman, et al., 2011), which leads in several standards for explanation of indoor spaces especially the interior of the buildings. Each of these standards has its structure and implementation procedure, which is suitable for some specific applications. In this part, some of the well-known and popular standards are introduced. The main focus of this part is on two broadly used standards, namely CityGML and IndoorGML, and they are discussed in detail.

BIM

BIM stands for both building information modeling and building information model, but the former one is most cited and used in formal standards (Figure 9) (Azhar, 2011). BIM provides a comprehensive digital, virtual structure of the physical and semantic properties of a building and includes whole lifecycle steps of it from conceptual design to complete destruction. BIM, in fact, is not only a graphical model or tool for visualization of information about structures, but it provides a valuable database of physical and applicable properties of a building. This information is designed in structures which facilitate the transition between different applications. BIM covers light analysis, spatial relations, temporal information of construction and destruction as well as geometric information. The main benefits of this data structure are the high geometric precision in providing building elements, ease of data sharing, fast access to information, better quality in design and production, better control on costs, and generation of a complete, sound database for whole building elements. However, there are some problems using it, such as ownership of model, less control of data entry, and loss of complete software to work with (Eastman, et al., 2011).

IFC

Industry Foundation Classes (IFC) is introduced by the buildingSMART international institute and is a standard for data structure to support the construction purposes (Bazjanac, & Crawley, 1997). This format formalized different parts of a building in foundation classes to be used during the construction

Figure 9. Sample representation of a building using BIM

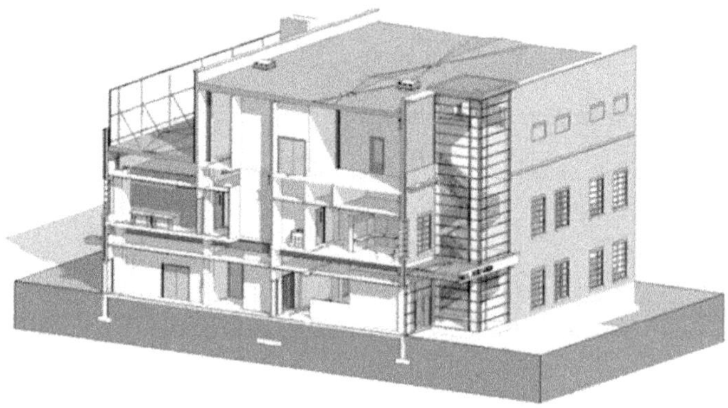

phase. IFC is the open data model for OpenBIM. In the other words, IFC is a standard to manipulate the BIM to narrow down it in order to organize. This standard provides BIM data model with an integrated nature which in turn facilitate the exchange between different software. IFC determine the schema and format for data, i.e. EXPRESS data specification language, which is defined in ISO 10303-11 standard and XML Schema definition language (XSD), which is defined in XML Schema W3C Recommendation. The definition of EXPRESS schema is the base and the XML schema is defined based on EXPRESS according to rules explained in ISO 10303-28 standard.

GML 3

Geography markup language (GML) is a data format to structure the geometry of geospatial data introduced by OGC. The purpose of this standard is separation of the contents from representation of features. This format is an XML-based, which increases the consistency with different software. There are several versions of GML, e.g. GML 1.0, GML 2.0, GML 3. Among them, GML 3 is used to represent 3D geospatial data because it supports the 3D geometry. Data models based on GML 3 can be defined in different coordinates systems such as polar, cylinder, spherical, and geodetic. The geometry in this data model supports point, line, polygon, surface, and solid (Figure 10). Moreover, it is possible to define the topology and semantic information. This data model can be shared in the web easily and geospatial web services such as WFS, WPS, and WCS use the GML for graphic exchange and representation. Because this data model uses XML-based syntax, it has the potential to simply be converted to other XML-based data models such as X3D.

CityGML

Recently, several 3D formats are proposed for buildings modeling. They are generally graphical and geometric, and pay less attention to the semantic aspects of geospatial information. These leads to decrease the level of semantic aspect of interoperability in geodata exchange. CityGML is an integrated

Figure 10. Geometry in GML3 (Cox et al., 2002)

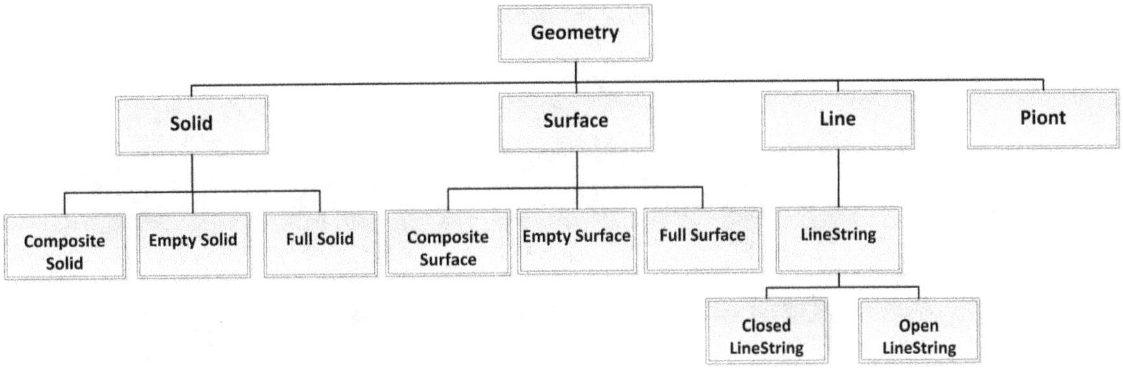

and structured data model and standard, introduced by OGC based on GML3 schema, for description of urban elements with focus on buildings. This standard is supported by most of the 3D modeling software.

The main body of this standard consists of two parts: geometric model and semantic model. The geomatic and topologic part of CityGML is supported by GML 3 using the Xlink concept. According to this concept, a feature is not defined several times, rather it is defined at first time met and is pointed using Xlink the next times. Because this model has 3D coordinated structure, unlike the image format, it is possible to scale and rotate the model easily. Moreover, having possibility to model the topologies besides the integrated structure facilitate using this standard in indoor navigation.

Semantic model, which is based on the attributes for different urban objects, presents a conceptual schema of 3D model besides the geometric model. It classifies the objects in several classes of Digital Terrain Models (DTM), sites (currently buildings; bridges and tunnels in the future), vegetation (areas, volumes, and solitary objects with vegetation classification), water bodies (volumes, surfaces), transportation facilities (both graph structures and 3D surface data), city furniture, generic city objects and attributes, and user-definable grouping.

The main focus of CityGML standard is on the representation of cities, structures, and buildings. Among them, the most useful class in CityGML, which also has more details, is building class. This class supports both geometric and semantic information. The geometry of a building is explained in a multiscale model with five well-defined consecutive Levels of Detail (LOD). Each level has more details from level 0 to level 4 (Gröger, Kolbe, Czerwinski, & Nagel, 2008). The LoD0 shows the 2.5D digital terrain model of a region. The next level, i.e. LoD1, represents buildings as the simple blocks with planar roofs, while the LoD2 supports the roofs with different shapes and geometries. In LoD3, it is possible to demonstrate the architectural model including walls, doors, windows, and more details of façade. In the last level, the focus is on the architectural interior plan including rooms, corridors, stairs, and furniture (Figure 11). Moreover, here are some intermediate levels to distinguish information in a more classes of details. For example, the LoD1.5 shows the LoD1 as well as the building floors.

Undoubtedly, one of the capabilities of CityGML is the building of 3D models for cities. Because this model supports semantic data, it could be used in several application areas such as urban planning, crisis management, architecture, tourism, 3D cadaster, 3D simulation, etc. This model can be easily build and edited by any text and XML-based editors. Furthermore, since the CityGML standard follows the GML3 schema, it is possible to show this model with GML-compatible software.

Figure 11. Different level of details for a building in CityGML standard
(Mortari, 2013)

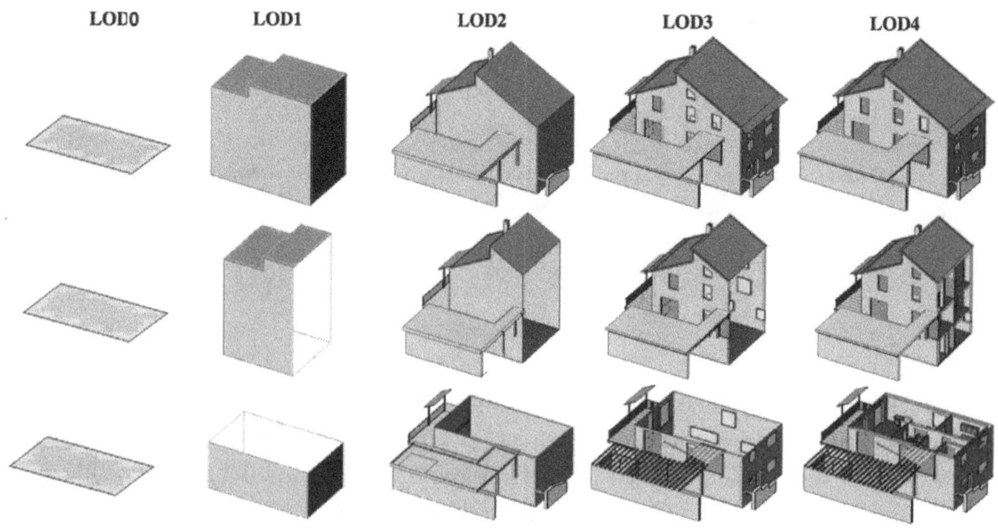

IndoorGML

The main focus of CityGML and IFC standards is on 3D geometry and semantic of buildings, but they lack necessary components required in indoor navigation applications. IndoorGML standard describes an appropriate data model based on XML schema for indoor spatial information (Figure 12), which focuses on modeling indoor spaces for navigation purposes. It is implemented as an application schema of the GML version 3.2.1. This standard consists of two basic components: a core data model (to describe topological connectivity and different contexts of indoor spaces) and a data model for navigation in indoor spaces (OGC, 2012). Indoor of buildings can be modeled as cellular spaces by non-overlapping cells. Each cell is defined as the smallest unit of a building such as rooms and corridors (Mortari, 2015) and has unique ID and common boundary. A cellular space S is defined as S = {c1, c2, c3... cn} where ci is the *i*th cell (OGC, 2012).

In IndoorGML structure, it is required to describe the topological relationships expressly. The Node Relation Graph (NRG) is used for building data to represent topological relationships, e.g. adjacency and connectivity among buildings (Lee, 2004). Furthermore, the Poincaré duality (Munkres, 1984) is used for representing topological relationships. This duality leads to simplification of spatial and topological relations between building elements. According to Poincaré duality, a k-dimensional object in an n-dimensional space changes into an n-k dimensional object in the dual space (Figures 14-16). For example, a cube in 3D space transform to a point in the dual space. Based on the Poincaré duality, it is possible to build NRG for the interior environment of a building. For this, each cell is assumed as a node and the connections between them are considered as the edges. In this standard, the nodes and the topologic relations are called state and transition, respectively.

Because the geometric representation of features in indoor space is not a major focus of IndoorGML, there are three ways for representing the geometry of buildings parts in IndoorGML data model (Figure 13) (OGC, 2012):

1. **External Reference**: An IndoorGML document only contains external links to objects defined in other data sets such as CityGML file.
2. **Geometry in IndoorGML**: Geometric representation of each building part may be included within an IndoorGML document. It is GM_Surface in 2D and GM_Solid in 3D spaces as defined in ISO 19107.
3. **No Geometry**: No geometric information is also included in IndoorGML document.

The UML diagram depicted in Figure 12 shows data model of a core module for IndoorGML. The model consists of two main classes: PrimalSpaceFeatures and MultiLayeredGraph. The Classes CellSpace and CellSpaceBoundary are two subclasses of class PrimalSpaceFeatures that demonstrate the geometry and semantic of cellular space according to the concept of geographic features (ISO 19109). To define one cell in Euclidean primal space, CellSpace class is represented as gml:Solid and gml:Surface in 3D and 2D space, respectively. CellSpaceBoundary is also used as gml:Surface and gml:Curve to describe 2D/3D boundary of each cell semantically.

Figure 12. IndoorGML core module data model
(OGC, 2012)

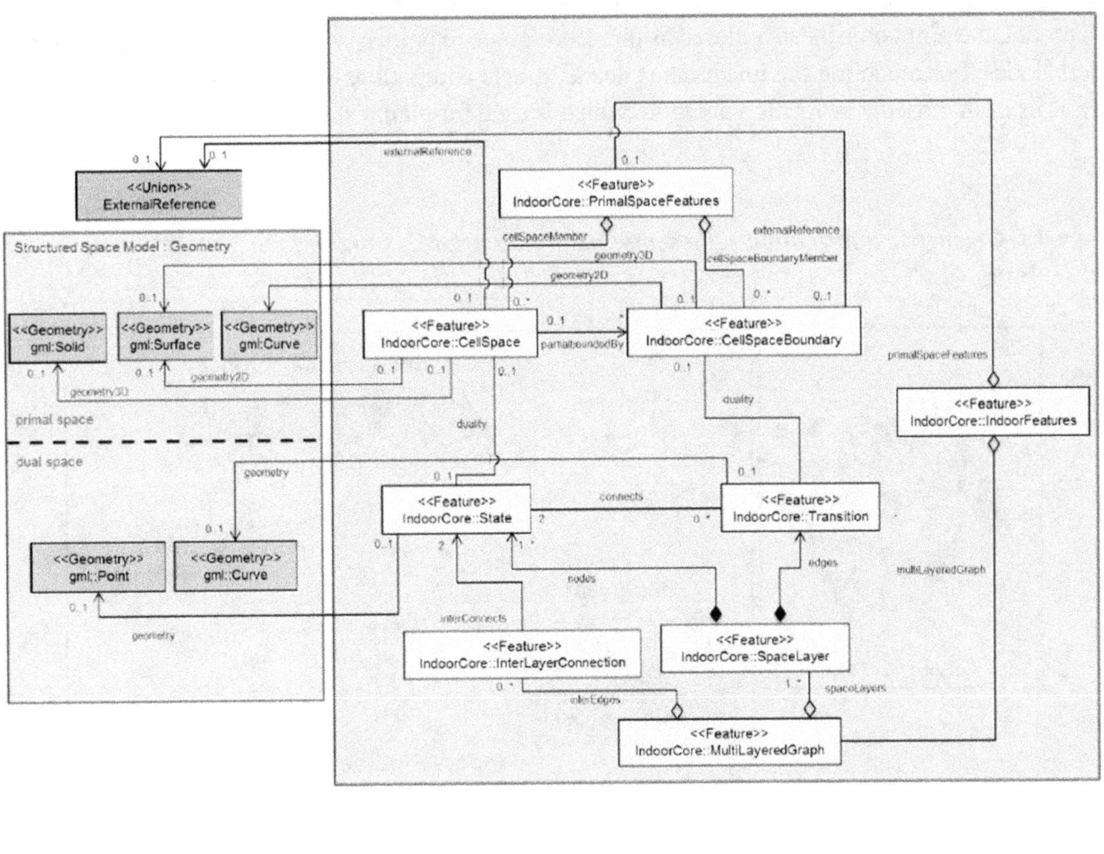

The class MultiLayeredGraph consists of several SpaceLayer classes. The multi-layered space model allows the coherent combination of different decompositions of space according to different semantics. Each *SpaceLayer* demonstrates the topological relations between cells (building parts) that are derived from the NRG graph. *States* and *Transitions* are also used as dual representations of cells in primal space.

OpenStreetMap

Data collections from interior space of buildings are both time-consuming and costly; hence, using the geospatial data collection capabilities in VGI-based databases is a suitable solution for coping with this issue. One of the well-known projects in VGI domain is OpenStreetMap (OSM). OSM was founded in 2004 and had more than 3,000,000 registered users until September, 2016. The main purpose of OSM is creation and storage of an on-line, free, and editable maps on the web (Goetz & Zipf, 2012; Lin, 2013). This project started by digitizing the pathways, but nowadays a broad range of spatial object classes such as buildings, land uses, and POIs are covered (Hashemi & Ali Abbaspour, 2015). The data entered into the OSM database is not uniformly distributed. Many details have been recorded in some areas, while the level of information sharing is low in the other areas. Figure 17 shows the information shown in the main page of OSM website.

Information is entered into the OSM database by users as the georeferenced points. These data are collected either based on the measurements by devices such as GPS or personal information and knowledge. Moreover, some satellite/aerial image providers' sites such as Bing facilitate the production of vector objects based on the embedding satellite images.

There are three levels of editors: iD, potlatch, and JOSM, for entering information by users according to their graphical skills. Figure 18 shows a part of OSM data for Tehran city in JOSM. Based on the OSM data model, the point objects are entered as the georeferenced points, which have latitude and longitude, called Nodes. For modeling the linear data such as streets and pathways, a collection of nodes, called a Way, is used. A closed way forms a polygon, which is used for planar shapes like building footprints and

Figure 13. Geometric information representation in IndoorGML standard
(OGC, 2012)

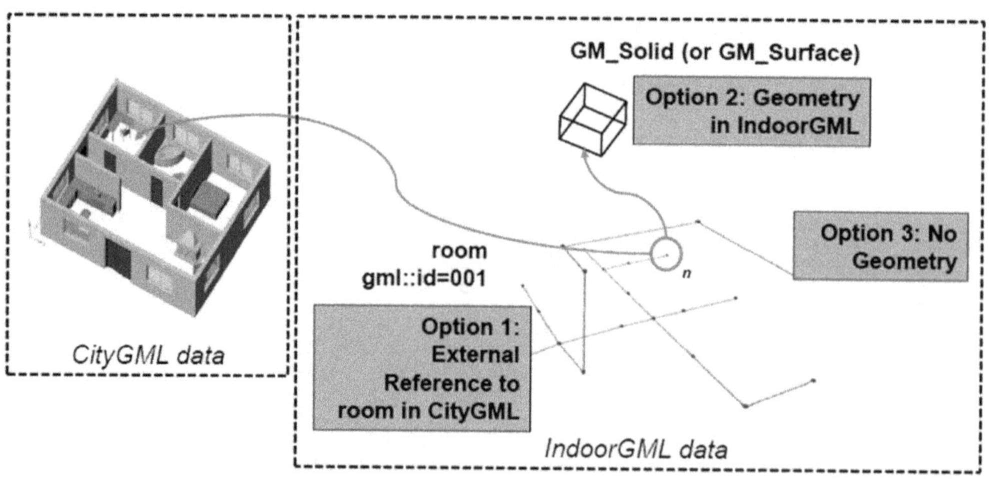

Figure 14. Poincaré duality in 2D space

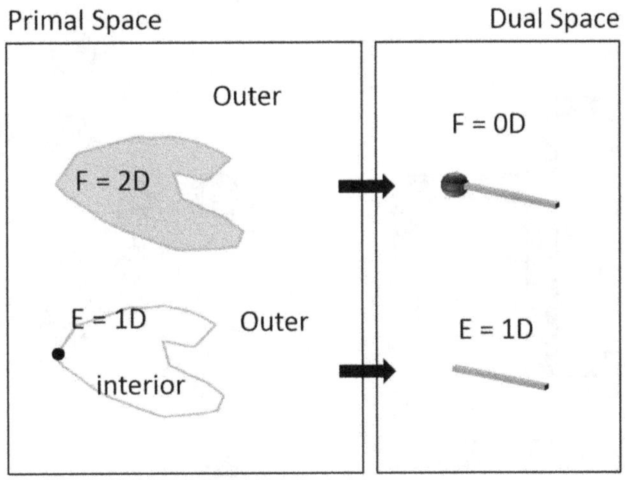

Figure 15. Poincaré duality in 3D space

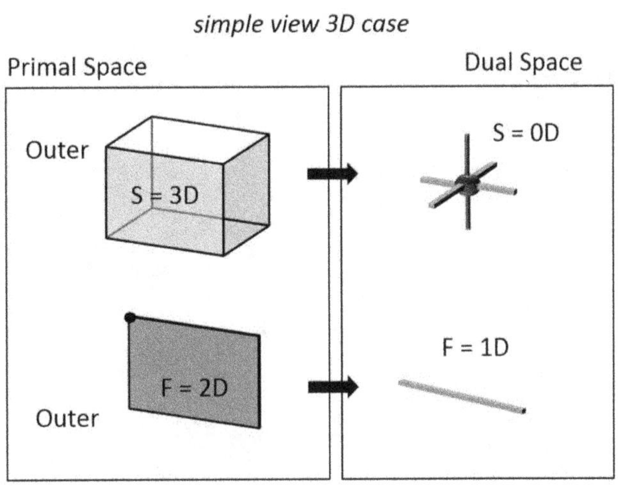

zones. A Relation, which is the combination of point and linear objects, is used for modeling complex objects like buildings with its interior parts (Goetz, 2013; Schilling et al., 2009).

In addition to the geometric information, there is possibility to add semantic information to objects. For this purpose, the concept of key-value pairs is used. There are no limitations for the users in creating these pairs. Data types are entered in this part according to the user's choice. Some optional instructions has been suggested on how to define the key-value pairs according to the types of objects (Goetz & Zipf, 2012). IndoorOSM introduced in 2011, allows users to capture and save data about indoor facilities such as hallways, doors, rooms, and stairs. The method used for data entry in IndoorOSM is the same as outdoor data entry used in OSM (Schilling et al., 2009).

Figure 16. NRG(left figure) for 2D space (right figure) (OGC, 2012)

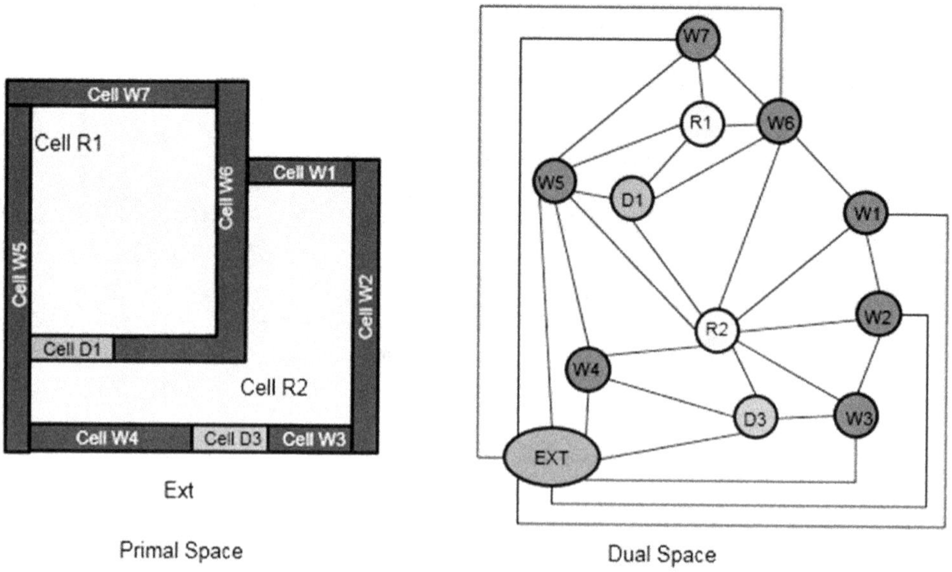

Figure 17. A view of OSM website

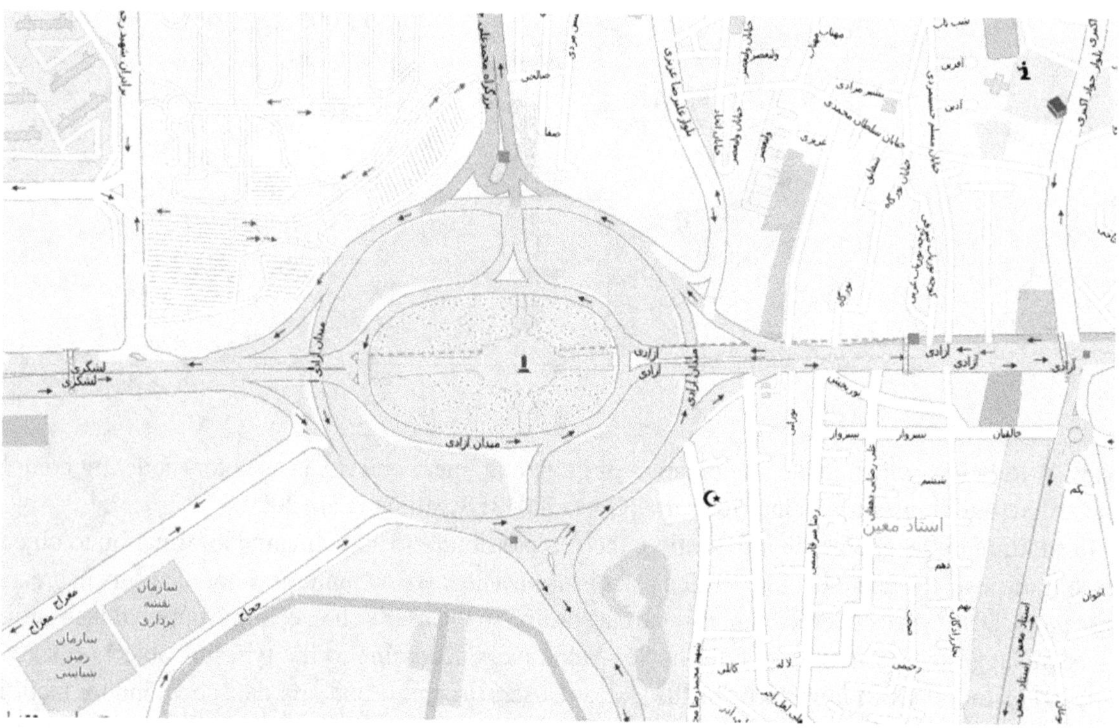

Figure 18. A view of OSM data for a part of Tehran city in JOSM editor

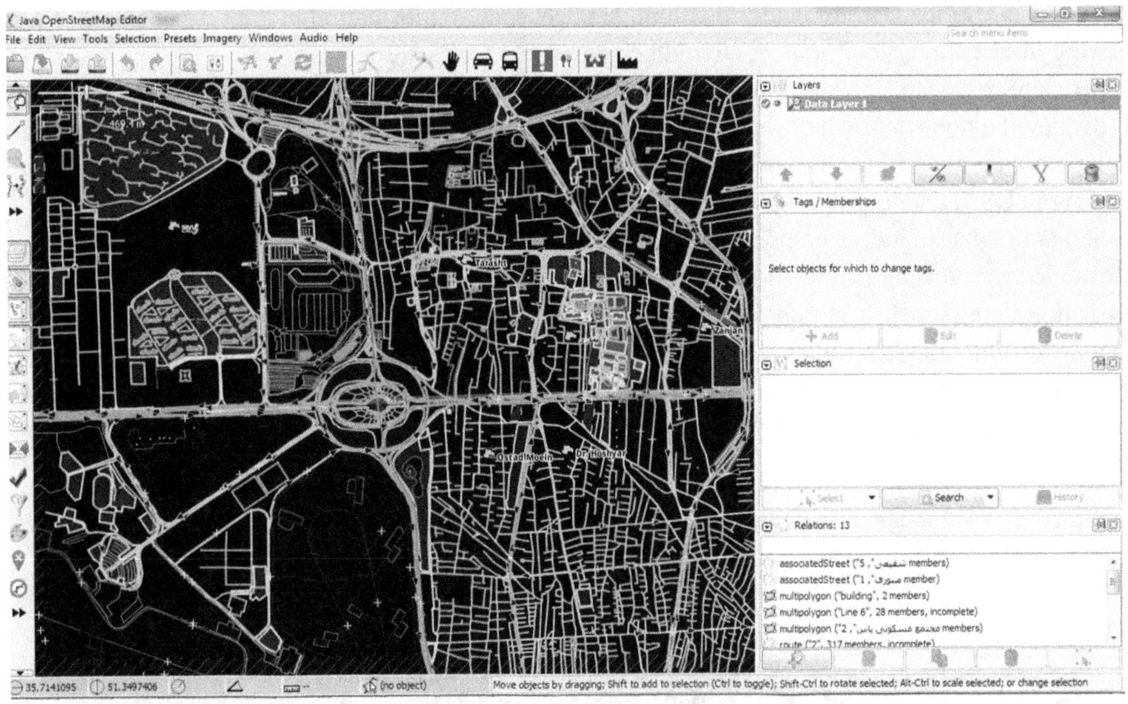

EXTRACTION OF NAVIGATION DATA MODEL FROM OpenStreetMap

In this section, the principles of construction of IndoorGML core data model based on volunteered geospatial information of OpenStreetMap will be described. In Figure 19, the overall steps of creation of IndoorGML data model based on OSM data are depicted briefly and symbolically. Each part will be explained in detail in the following subsections.

CREATION OF THE NAVIGATION DATA STRUCTURE IN OSM

At the first step, the information of intended building should be uploaded in OSM database. The use of the information that already exists in OSM database is also possible. For importing data to OSM, there

Figure 19. Methodology of creation IndoorGML data model
(Mirvahabi & Ali Abbaspour, 2015)

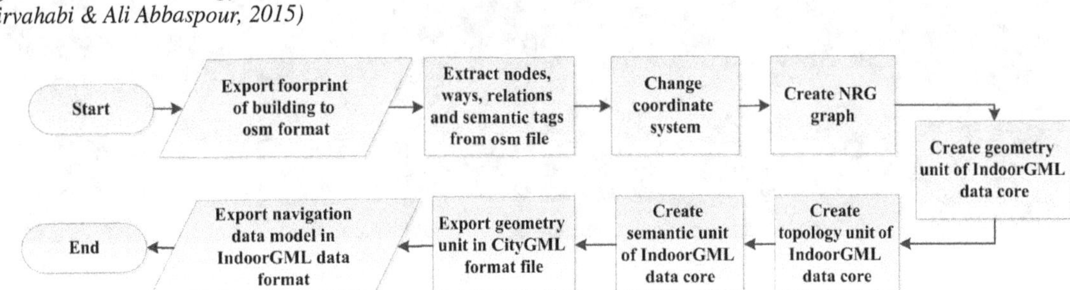

are no restrictions for buildings and interiors of buildings. Moreover, users can use online/offline editors, which are embedded for entering volunteered geospatial information in the OSM database. While entering information, floor number should be considered for each store.

In OpenStreetMap, polygons (Ways) are used for complex objects such as rooms, corridors, and halls. To display the doors and windows, points or nodes (Nodes) are utilized. A set of points and polygons (Relations) are assisted to display the floors of a building. Figure 20 shows a part of the interior of a building in OSM database, along with doors and windows on the JOSM editor.

Semantics data of indoor components are necessary to build IndoorGML navigation core data model. The advantage of using this method is the minimum semantic information required for creation of the navigation data model. It is only necessary to assign the number of floor to whole polygons of that floor and to enter the key-values for those nodes which are windows or doors (Figures 21 and 22). After generation of the output of building in the.osm format from OpenStreetMap, the process of generation the IndoorGML core data model will be started.

Figure 20. Indoor space with doors and windows in OSM

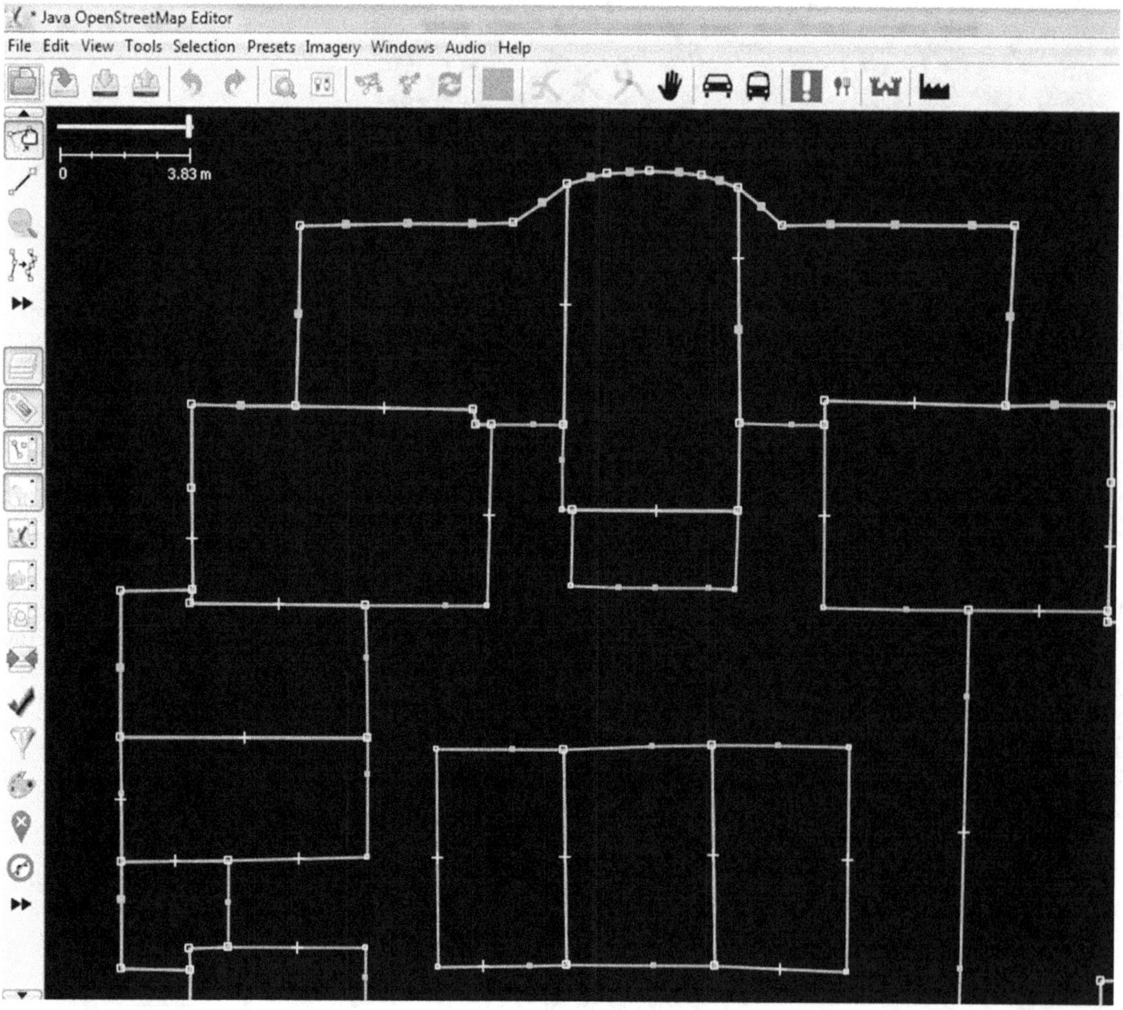

Figure 21. View of one room in building with semantic information

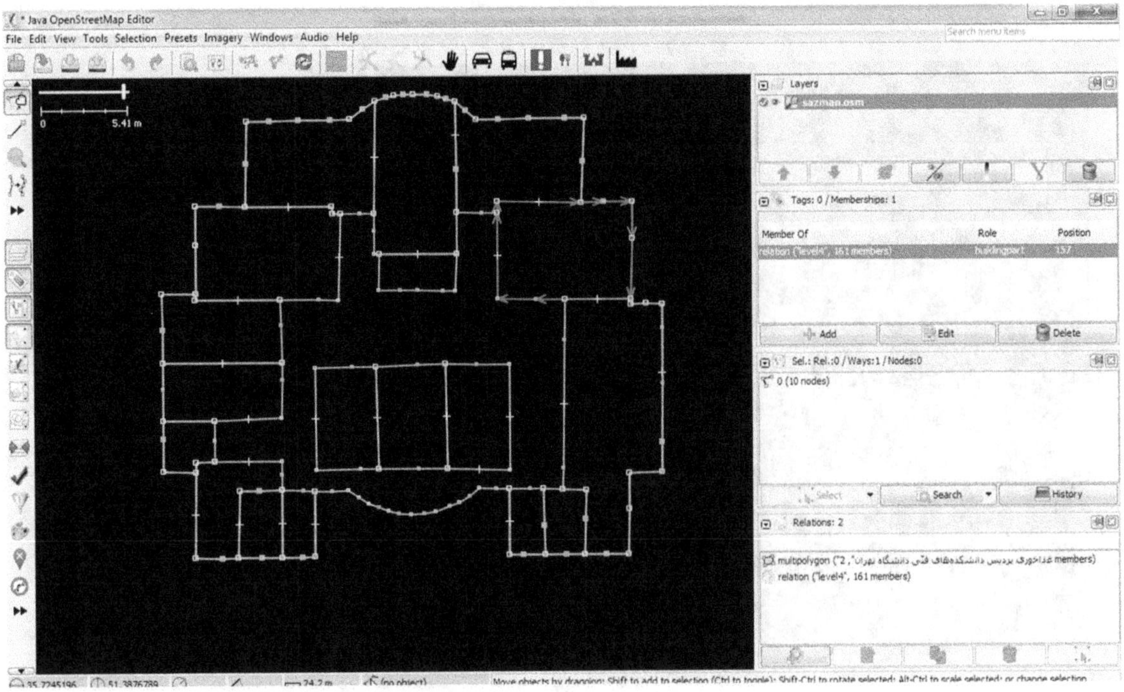

There is the possibility of adding other semantic information, but they are not essential. After entering suitable information for each floor in OSM database, the creation of output data phase is started. There are several approaches in OSM for exporting data. One of them is selection of the intended area and exports them in.osm format from OSM website. Furthermore, desktop editor, called JOSM, can be used for exporting data. Several formats exist for exporting data from OpenStreetMap, but the most common and most popular of them is.osm format. This data format saves geometric and semantic information in XML structure. In Figure 22, exported file of small part of a building in.osm format is displayed. In this part, nodes of the building and semantic information of them are defined.

Generation of the Core Data Model

In this section, the proposed method to create a core data model of IndoorGML is described in detail. This method is a programming tool in java language and has the ability to convert all information of the.osm data format to IndoorGML and 3D structure. In the first step, the.osm file related to the intended floor is loaded and read using the functions programmed. A projection from geographic coordinate system to Universal Transverse Mercator (UTM) projection system is also performed. Then, the structure of output file reviewed and essential data from it was extracted. As mentioned before, the output file has a geometric structure composed of three parts. Node, Way, and Relation sections are three geometry parts of the.osm files. Each of these parts has its semantic information. The information about the number of floor, checking a node whether it is a door or window, height of a window, the lengths and widths of

Figure 22. Representation of a node in OSM exported file

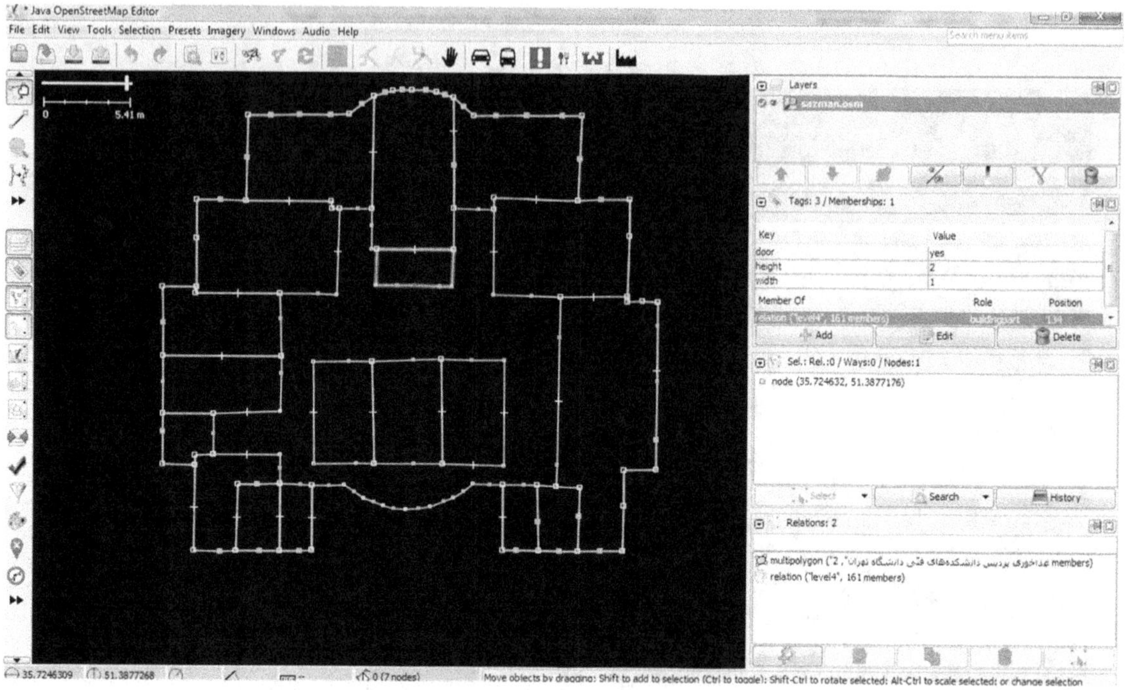

doors and windows is some of the semantic information extracted from output file based on the navigation model. In the proposed method, these three parts are separated and semantic information about each part is placed in the suitable data structure.

Geometry Unit of IndoorGML Core Data Model

After separation of the different parts of the file, relation of the intended floor of building, which is desired to create IndoorGML core data model from it, is extracted from all relations. Based on the type of existing data in database, each building has a relation, which includes the whole geometric and semantic information related to that building. Moreover, spatial information about all rooms, corridors, doors, and windows is separated from.osm location information. This information is placed in the specific categories in the right format to enter to the next stage of the process. Following the separation of geometric information, semantic information is extracted for each part.

Each relation is processes and the general information about a building such as building id, address, and usage is extracted. Then, this information is structured in the suitable format as the address and ID tags and placed in CityGML data model. (Mirvahabi & Ali Abbaspour, 2015)

When all the required information was extracted from OpenStreetMap output file, the construction of the geometric data model for navigation is started. To create geometry in IndoorGML data model, there are three main approaches: using of geometry in an external document and referencing to it, using conventional geometry introduced in IndoorGML structure, and using no geometry. To create the geometric unit of data model based on the first approach, geometric data should be changed to an appropriate 3D structure.

To create three dimensional structure, the desired floor is created using polygons related to the floor of the building in CityGML format and saved in the related tag. Then, the walls of that floor are built. Walls are divided in two groups: walls with holes (containing doors and/or windows) and simple walls (without doors and windows). For simple walls, they are created directly using the height of corresponding floor. But for the walls with holes, it is necessary to generate the doors and windows firstly. The heights of windows from the base ground height and the lengths and widths of windows as well as lengths and widths of doors are the necessary semantic information. The windows and doors are placed in the walls according to their coordinates. Figure 23 depicts a view of the semantic information of a wall with a door and a window.

When all of the interior and exterior walls as well as doors and windows of a desired floor are created in CityGML structure, the ceiling of the floor is formed according to OSM data using the related ceiling polygon. Figure 24 shows the map of selected floor (fourth floor) and Figures 25, 26, and 27 display three-dimensional model derived from presented algorithms to display the navigation data.

Information of created 3D model is stored in an appropriate data structure in a suitable file. Furthermore, information of this model placed in geometry unit of core IndoorGML data model. In the end, output file of IndoorGML navigation data model in XML structure will be generated. This file with geometry unit and navigation model of environment can be used for many analyses of applications. Figure 28 displays a small part of output file at the IndoorGML core data structure. Part (c) of Figure 28 shows a small part of geometry unit and part (a and b) of the figure shows a small part of topology unit of core data model.

Topology Unit of IndoorGML Core Data Model

The most important part of the core data model is building topology and proximity relations, because it determines navigation data model and the relationship between different areas in the building for navigation. As previously mentioned, to build the topology part of data model, the NRG graph should be created and for creation NRG graph, Poincare duality is used. The dimensions of both navigation space and the dual space are three. Thus, based on the principle of Poincare duality, all rooms and corridors and the entire three-dimensional volumes become point in dual space. Thus for each cell or room in three dimension, one point should be considered as representative. To achieve this, the geometric center (centroid) of each cell can be used as a representative of cell or room (Equations (1) and (2)):

$$C_x = \frac{1}{6A}\sum_{i=0}^{N-1}\left(x_i + x_{i+1}\right)\left(x_i y_{i+1} - x_{i+1} y_i\right) \tag{1}$$

$$C_y = \frac{1}{6A}\sum_{i=0}^{N-1}\left(y_i + y_{i+1}\right)\left(x_i y_{i+1} - x_{i+1} y_i\right) \tag{2}$$

where C_x and C_y are the coordinates of centroids of each cell or room and x and y are coordinates of the corners of a room or a cell. A is the area of a cell or room and calculated according to equation (3).

Figure 23. View of the semantic information of a wall with a door and a window


```
▼<osm version="0.6" upload="true" generator="JOSM">
  <bounds minlat="35.7233553" minlon="51.387198" maxlat="35.7251147" maxlon="51.3887" origin="CGImap 0.4.0 (13527 thorn-03.openstreetmap.org)"/>
 ▶<node id="-290" action="modify" visible="true" lat="35.72460476433" lon="51.38781826823">...</node>
 ▶<node id="-288" action="modify" visible="true" lat="35.72469743226" lon="51.38774417066">...</node>
 ▶<node id="-286" action="modify" visible="true" lat="35.72466424997" lon="51.38768756572">...</node>
 ▶<node id="-284" action="modify" visible="true" lat="35.72467324072" lon="51.38767819027">...</node>
 ▶<node id="-282" action="modify" visible="true" lat="35.72467355349" lon="51.38776157997">...</node>
 ▶<node id="-280" action="modify" visible="true" lat="35.72462682282" lon="51.3877987427">...</node>
 ▶<node id="-278" action="modify" visible="true" lat="35.72463195217" lon="51.38770556261">...</node>
 ▶<node id="-276" action="modify" visible="true" lat="35.72463197018" lon="51.38771761107">...</node>
 ▼<node id="-274" action="modify" visible="true" lat="35.72463178065" lon="51.3877350087">
    <tag k="door" v="yes"/>
    <tag k="height" v="2"/>
    <tag k="width" v="1"/>
  </node>
 ▼<node id="-272" action="modify" visible="true" lat="35.72459190529" lon="51.38775786906">
    <tag k="door" v="yes"/>
    <tag k="height" v="2"/>
    <tag k="width" v="1"/>
  </node>
 ▼<node id="-270" action="modify" visible="true" lat="35.72459155942" lon="51.38771688282">
    <tag k="door" v="yes"/>
    <tag k="height" v="2"/>
    <tag k="width" v="1"/>
  </node>
 ▼<node id="-268" action="modify" visible="true" lat="35.72459071246" lon="51.38767182507">
    <tag k="door" v="yes"/>
    <tag k="height" v="2"/>
    <tag k="width" v="1"/>
  </node>
 ▼<node id="-266" action="modify" visible="true" lat="35.72453275804" lon="51.38762397233">
    <tag k="door" v="yes"/>
    <tag k="height" v="2"/>
    <tag k="width" v="1"/>
  </node>
 ▼<node id="-264" action="modify" visible="true" lat="35.72452427786" lon="51.3876086207">
    <tag k="door" v="yes"/>
    <tag k="height" v="2"/>
    <tag k="width" v="1"/>
  </node>
 ▼<node id="-262" action="modify" visible="true" lat="35.72452409746" lon="51.38763832293">
    <tag k="door" v="yes"/>
    <tag k="height" v="2"/>
```


$$A = \frac{1}{2}\sum_{i=0}^{N-1}\left(x_i y_{i+1} - x_{i+1} y_i\right) \tag{3}$$

The coordinates related to each floor is extracted and the centroids are calculated based on the coordinates using equations (1) to (3) for all rooms and corridors in 3D space.

After choosing representative of each cell, NRG is created based on the available information using representative points. In this graph, cells are the nodes and the topology relations between cells are represented by edges. For modeling this graph and placing it in topology unit of IndoorGML core data model, two key elements are used: State and Transition. These two elements are the same points and edges which are used in NRG graph. To introduce the topology, <MultiLayeredGraph gml: id = “MG1”> tag with the specified ID is used. Each MultiLayeredGraph can be considered as several <spaceLayers gml: id = “SL1”>. Each spaceLayer is known with its ID and is considered as a relation that describes topology relationship. For example, it is possible to describe the topology relation of building in one spaceLayers and the domain of WiFi signal in another spaceLayers. Each spaceLayers consists of two main parts: node and edge. Figure 29 shows the node and edge units of output core data model that are the topology units of this data model.

In node part, information about each representative point of cells is placed. Each cell is converted to its dual and information of it is placed at one State in topology unit. Each State is placed in <stateMem-

Figure 24. View of a floor of a building

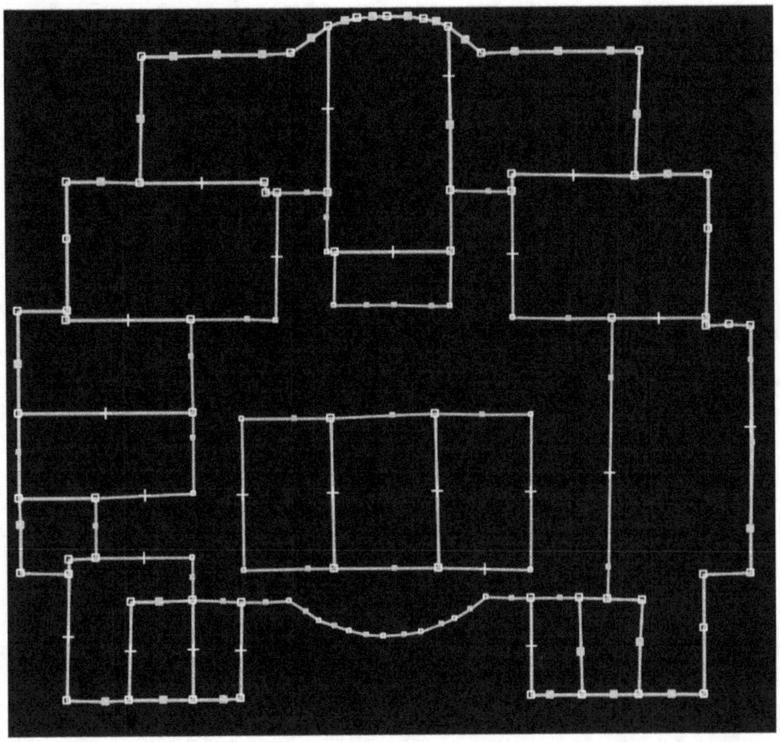

ber> and has a unique ID (R3) (Figure 30). Tag <duality xlink:href="#C13"/> of each State refers to ID of dual cell. Tags <connects xlink:href="#T45"/> shows the edges that are connected to this state with its ID. Finally, the coordinate of each State is placed in the position, i.e. x and y, tags.

Figure 25. CityGML geometry of the building

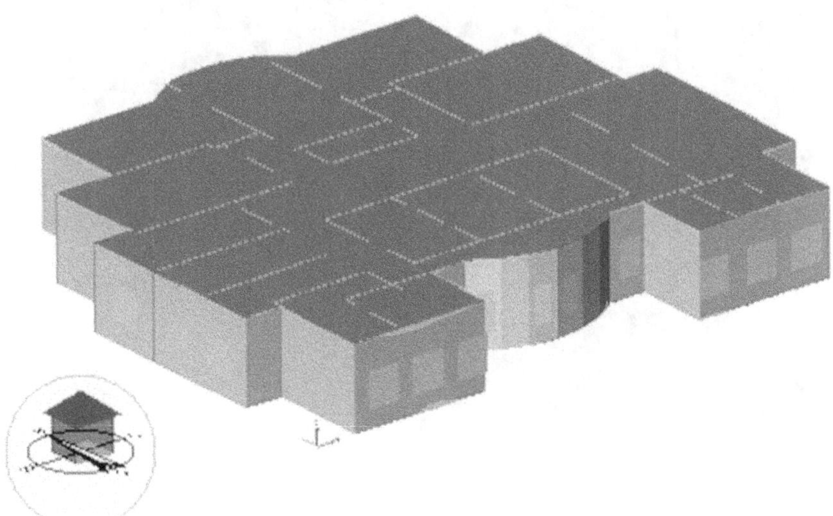

Figure 26. CityGML geometry of the building

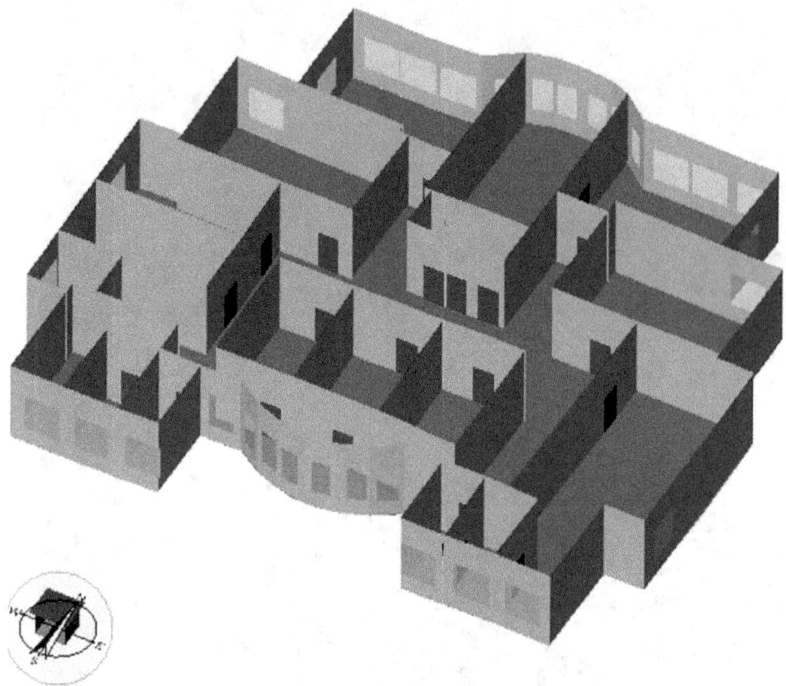

Figure 27. CityGML geometry of the building

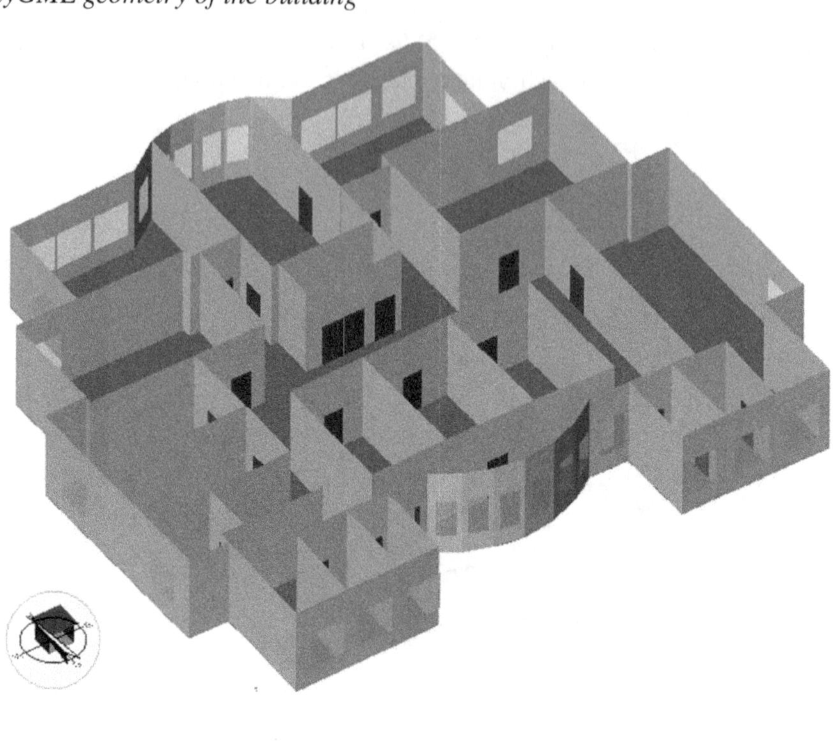

Figure 28a. Part of topology unit in IndoorGML data core model


```
- <stateMember>
  - <State gml:id="R9">
      <duality xlink:href="#C9"/>
      <connects xlink:href="#T19"/>
      <connects xlink:href="#T20"/>
      <connects xlink:href="#T21"/>
      <connects xlink:href="#T22"/>
    - <geometry>
      - <gml:Point gml:id="PR9">
          <gml:pos>535073.4994555828 3953473.3652144037</gml:pos>
        </gml:Point>
      </geometry>
    </State>
```


Figure 28b. Part of topology unit in IndoorGML data core model


```
- <transitionMember>
  - <Transition gml:id="T3">
      <weight>1</weight>
      <connects xlink:href="#Dyes"/>
      <connects xlink:href="#R16"/>
    - <geometry>
      - <gml:LineString gml:id="#TT3">
          <gml:pos>535052.7096039713 3953467.118050447</gml:pos>
          <gml:pos>535051.7257355151 3953471.7131331894</gml:pos>
        </gml:LineString>
      </geometry>
    </Transition>
  </transitionMember>
```


Figure 28c. Part of geometry unit in IndoorGML data core model


```
- <cellSpaceMember>
  - <cellSpace gml:id="C8">
      <gml:name>part=008</gml:name>
    - <Geometry2D>
      - <gml:surface gml:id="SF8">
        - <gml:exterior>
          - <gml:Shell>
            - <gml:curveMember>
              - <gml:LineString gml:id="LS8_1">
                  <gml:pos>535075.5588053722 3953460.297022492</gml:pos>
                  <gml:pos>535074.4480700651 3953460.279391854</gml:pos>
                </gml:LineString>
              </gml:curveMember>
            - <gml:curveMember>
              - <gml:LineString gml:id="LS8_2">
                  <gml:pos>535074.4480700651 3953460.279391854</gml:pos>
                  <gml:pos>535073.1235412486 3953460.2910623215</gml:pos>
                </gml:LineString>
              </gml:curveMember>
            - <gml:curveMember>
              - <gml:LineString gml:id="LS8_3">
                  <gml:pos>535073.1235412486 3953460.2910623215</gml:pos>
                  <gml:pos>535073.0717467434 3953461.9980561067</gml:pos>
                </gml:LineString>
              </gml:curveMember>
            - <gml:curveMember>
              - <gml:LineString gml:id="LS8_4">
                  <gml:pos>535073.0717467434 3953461.9980561067</gml:pos>
                  <gml:pos>535073.0054974192 3953464.181375631</gml:pos>
                </gml:LineString>
              </gml:curveMember>
            - <gml:curveMember>
              - <gml:LineString gml:id="LS8_5">
```

The next important section is the transitions that are inside the edge tag. Each transition is placed <TransitionMember> and has unique ID (T6) (Figure 31). It is possible to determine the weight for each transition in <weight> tag. <connects xlink:href="#R5"/> tag is used for displaying two state that are connected to intended transition. Finally, the geometry information of transition is paced in <geometry> tag.

When geometry and topology part is completed, semantic part of data model is completed in semantic unit of core data model. This part is created by semantic information of.osm file. Then, all of the file units are checked and output file is created in GML format. The data model was developed for using in navigation software and navigation/routing applications such as robots routing in the indoor spaces.

CONCLUSION

Due to the increase in the size of cities, navigation plays an important role in people's everyday lives. On the other hand, the extent of the buildings is growing, and this has been faced finding a proper route in indoor environments with challenges. That is why internal navigation is considered more and more.

Before addressing the indoor navigation, area within buildings should be modeled using suitable standard and data model. In this chapter, a number of data models and standards for building modeling are expressed. Then, based on this survey, IndoorGML standard format is used to create a graph-based navigation data model for indoors. Because the data collection for the interior of buildings is usually costly, using the volunteered geospatial data can play a vital role to overcome this problem. Moreover, most of the indoor wayfinding analysis can be carried out more precise and faster using 3D models of buildings.

Figure 29. Topology unit in IndoorGML core model


```
<?xml version="1.0"?>
<!--This file is the IndoorGML document-->
- <IndoorFeatures gml:id="IFs"
xmlns="http://www.opengis.net/indoorgml/1.0/core"
xsi:schemaLocation="http://www.opengis.net/indoorgml/1.0/core
indoorgmlcore.xsd" xmlns:xsi="http://www.w3.org/2001/XMLSchema-
instance" xmlns:xlink="http://www.w3.org/1999/xlink"
xmlns:gml="http://www.opengis.net/gml/3.2">
  - <PrimalSpaceFeatures>
    + <PrimalSpaceFeatures gml:id="PSF1">
    </PrimalSpaceFeatures>
  - <MultiLayeredGraph gml:id="MG1">
    - <spaceLayers gml:id="SL1">
      - <spaceLayerMember>
        - <spaceLayer gml:id="IS1">
          + <nodes gml:id="N1">
          + <edges gml:id="E1">
          </spaceLayer>
        </spaceLayerMember>
      </spaceLayers>
    </MultiLayeredGraph>
</IndoorFeatures>
```

Figure 30. View of one state in IndoorGML core model

```
- <stateMember>
  - <State gml:id="R3">
      <duality xlink:href="#C3"/>
      <connects xlink:href="#T45"/>
      <connects xlink:href="#T46"/>
      <connects xlink:href="#T47"/>
      <connects xlink:href="#T48"/>
    - <geometry>
      - <gml:Point gml:id="PR3">
          <gml:pos>535065.4077848047 3953485.4469054225</gml:pos>
        </gml:Point>
      </geometry>
    </State>
  </stateMember>
```

Figure 31. View of one Transition in IndoorGML core model

```
- <transitionMember>
  - <Transition gml:id="T6">
      <weight>1</weight>
      <connects xlink:href="#R1"/>
      <connects xlink:href="#R5'/>
    - <geometry>
      - <gml:LineString gml:id="#TT6">
          <gml:pos>535056.0801357726 3953479.3253566944</gml:pos>
          <gml:pos>535059.1150089665 3953475.500824427</gml:pos>
        </gml:LineString>
      </geometry>
    </Transition>
  </transitionMember>
```

The main idea of this chapter is to show a methodology to transform VGI-based interiors of buildings into a data model with 3D geometry structure suitable for indoor navigation. More specifically, the dataset in OSM was used to be converted to the IndoorGML-based navigation data model with 3D CityGML geometry. The novelty of the proposed method is automatic generation of navigation data model in IndoorGML format based on OSM data. The VGI is not a homogenous and this make automatic transformation face difficulties. To cope with this issue, a routine is introduced to add data to the OSM database in the proposed method.

The proposed method is a part of an integrated system for indoor navigations in buildings. The proposed method is useful for generation of navigation data model in one store in a building. But for 3D navigation, it is necessary to extend the method considering the relations between stores of a building in OSM.

REFERENCES

Abowd, G. D., Atkeson, C. G., Hong, J., Long, S., Kooper, R., & Pinkerton, M. (1997). Cyberguide: A mobile context-aware tour guide. *Wireless Networks*, *3*(5), 421–433. doi:10.1023/A:1019194325861

Azhar, S. (2011). Building information modeling (BIM): Trends, benefits, risks, and challenges for the AEC industry. *Leadership and Management in Engineering*, *11*(3), 241–252. doi:10.1061/(ASCE)LM.1943-5630.0000127

Bazjanac, V., & Crawley, D. B. (1997). *The implementation of industry foundation classes in simulation tools for the building industry*. Lawrence Berkeley National Laboratory.

Boguslawski, P. (2011). *Modelling and analysing 3d building interiors with the dual half-edge data structure*. University of Glamorgan.

Chatila, R., Devy, M., Lacroix, S., & Herrb, M. (1994). Perception system and functions for autonomous navigation in a natural environment.*Proceedings of the AIAA/NASA Conference on Intelligent Robots for Field, Factory, Service and Space*.

Cox, S., Cuthbert, A., Daisey, P., Davidson, J., Johnson, S., Keighan, E., Martell, R. (2002). *OpenGIS® Geography Markup Language (GML) Implementation Specification, version*. Academic Press.

Eastman, C., Eastman, C. M., Teicholz, P., Sacks, R., & Liston, K. (2011). *BIM handbook: A guide to building information modeling for owners, managers, designers, engineers and contractors*. John Wiley & Sons.

Goetz, M. (2013). Towards generating highly detailed 3D CityGML models from OpenStreetMap. *International Journal of Geographical Information Science*, *27*(5), 845–865. doi:10.1080/13658816.2012.721552

Goetz, M., & Zipf, A. (2012). Towards defining a framework for the automatic derivation of 3D CityGML models from volunteered geographic information. *International Journal of 3-D Information Modeling*, *1*(2), 1-16.

Gold, C. M. (2006). What is GIS and What is Not? *Transactions in GIS*, *10*(4), 505–519. doi:10.1111/j.1467-9671.2006.01009.x

Gröger, G., Kolbe, T. H., Czerwinski, A., & Nagel, C. (2008). *OpenGIS city geography markup language (CityGML) encoding standard, version 1.0. 0*. Academic Press.

Han, D., Jung, S., Lee, M., & Yoon, G. (2014). Building a practical Wi-Fi-based indoor navigation system. *IEEE Pervasive Computing / IEEE Computer Society [and] IEEE Communications Society*, *13*(2), 72–79. doi:10.1109/MPRV.2014.24

Hashemi, P., Ali Abbaspour, R., Jokar Arsanjani, J., Zipf, A., Mooney, P., & Helbich, M. (2015). *OpenStreetMap in GIScience, Experiences, Research, and Applications (LNCS)*. Springer Berlin Heidelberg.

Herrera, J. A., Plöger, P., Hinkenjann, A., Maiero, J., Flores, M., & Ramos, A. (2014). *Pedestrian indoor positioning using smartphone multi-sensing, radio beacons, user positions probability map and IndoorOSM floor plan representation*. Paper presented at the Indoor Positioning and Indoor Navigation (IPIN), 2014 International Conference on. doi:10.1109/IPIN.2014.7275538

Höllerer, T., Feiner, S., Terauchi, T., Rashid, G., & Hallaway, D. (1999). Exploring MARS: Developing indoor and outdoor user interfaces to a mobile augmented reality system. *Computers & Graphics*, *23*(6), 779–785. doi:10.1016/S0097-8493(99)00103-X

Kargl, F., Geßler, S., & Flerlage, F. (2007). *The iNAV indoor navigation system*. Paper presented at the International Symposium on Ubiquitious Computing Systems. doi:10.1007/978-3-540-76772-5_9

Karimi, H. A. (2015). *Indoor Wayfinding and Navigation*. CRC Press. doi:10.1201/b18220

Langer, D., Rosenblatt, J. K., & Hebert, M. (1994). A reactive system for off-road navigation. Robotics Institute, 441.

Lee, J. (2004). A spatial access-oriented implementation of a 3-D GIS topological data model for urban entities. *GeoInformatica*, *8*(3), 237–264. doi:10.1023/B:GEIN.0000034820.93914.d0

Levinson, J., Askeland, J., Becker, J., Dolson, J., Held, D., Kammel, S., & Pratt, V. (2011). *Towards fully autonomous driving: Systems and algorithms*. Paper presented at the Intelligent Vehicles Symposium (IV). doi:10.1109/IVS.2011.5940562

Li, K. J., & Lee, J. Y. (2013). Basic concepts of indoor spatial information candidate standard IndoorGML and its applications. *Journal of Korea Spatial Information Society*, *21*(3), 1–10. doi:10.12672/ksis.2013.21.3.001

Lin, W. (2013). *When Web 2.0 meets public participation GIS (PPGIS): VGI and spaces of participatory mapping in China. In Crowdsourcing geographic knowledge* (pp. 83–103). Springer.

Makri, A., Zlatanova, S., & Verbree, E. (2015). *An approach for indoor wayfinding replicating main principles of an outdoor navigation system for cyclists*. Paper presented at the International Archives of the Photogrammetry, Remote Sensing and Spatial Information Sciences, Indoor-Outdoor Seamless Modelling, Mapping and Navigation, Tokyo, Japan. doi:10.5194/isprsarchives-XL-4-W5-29-2015

Massidda, C., Bülthoff, H. H., & Stegagno, P. (2015). *Autonomous vegetation identification for outdoor aerial navigation*. Paper presented at the Intelligent Robots and Systems (IROS), 2015 IEEE/RSJ International Conference on. doi:10.1109/IROS.2015.7353806

Mirvahabi, S., & Ali Abbaspour, R. (2015). Automatic Extraction of IndoorGML Core Model from OpenStreetMap. *The International Archives of Photogrammetry, Remote Sensing and Spatial Information Sciences*, *40*(1), 459.

Möller, A., Kranz, M., Huitl, R., Diewald, S., & Roalter, L. (2012). *A mobile indoor navigation system interface adapted to vision-based localization*. Paper presented at the 11th international conference on mobile and ubiquitous multimedia. doi:10.1145/2406367.2406372

Mortari, F. (2013). *Automatic Extraction of Improved Geometric Network Model from CityGML for Indoor Navigation. (Master)*. TU Delft.

Munkres, J. R. (1984). *Elements of algebraic topology* (Vol. 2). Addison-Wesley Menlo Park.

Nakajima, M., & Haruyama, S. (2013). New indoor navigation system for visually impaired people using visible light communication. *EURASIP Journal on Wireless Communications and Networking*, (1): 1–10.

OGC. (2012). *IndoorGML Encoding Standard*. OGC.

Palacios, R. (2016). *deg2utm*. Retrieved from https://www.mathworks.com/matlabcentral/fileexchange/10915-deg2utm/content/deg2utm.m

Schilling, A., Over, M., Neubauer, S., Neis, P., Walenciak, G., & Zipf, A. (2009). Interoperable Location Based Services for 3D cities on the Web using user generated content from OpenStreetMap. *Urban and regional data management: UDMS annual*, 75-84.

Tse, R. O. C., & Gold, C. M. (2004). TIN meets CAD: Extending the TIN concept in GIS. *Future Generation Computer Systems*, *20*(7), 1171–1184. doi:10.1016/j.future.2003.11.007

Urmson, C., Anhalt, J., Bagnell, D., Baker, C., Bittner, R., Clark, M., & Geyer, C. (2008). Autonomous driving in urban environments: Boss and the urban challenge. *Journal of Field Robotics*, *25*(8), 425–466. doi:10.1002/rob.20255

Werner, M. (2014). *Building Modeling. In Indoor Location-Based Services: Prerequisites and Foundations* (pp. 101–131). Cham: Springer International Publishing.

Zaragozí, B., Gimýnez-Font, P., Navarro-Carriýn, J. T., Rico-Amorýs, A., Ramýn-Morte, A., & Belda-Antolý, A. (2015). *A proposal for design and implementation of an hybrid navigation system based on open data, augmented reality and big data: Applications for the smart cities*. Paper presented at the Geographical Information Systems Theory, Applications and Management (GISTAM), 2015 1st International Conference on.

Zheng, Y., Shen, G., Li, L., Zhao, C., Li, M., & Zhao, F. (2014). *Travi-navi: Self-deployable indoor navigation system*. Paper presented at the 20th annual international conference on Mobile computing and networking.

KEY TERMS AND DEFINITIONS

3D Modeling: The process of creation a 3D mathematical model of objects in software.

CityGML: It is an integrated and structured data model and standard, introduced by OGC based on GML 3 schema, for describing the urban elements with focus on buildings.

Data Model: A set of elements used to represent, describe, and organize a certain aspect of the real world entities in a computer environment.

IndoorGML: This standard describes an appropriate data model based on XML schema for indoor spatial information, which focuses on modeling indoor spaces for navigation purposes.

Indoor Positioning System (IPS): It is a system to locate objects (including people) inside of a building using some sensory information collected or sent by a mobile device.

Navigation: The process of monitoring the movement and guidance of any moving objects from one location to another.

OpenStreetMap (OSM): It is a volunteered project to create a free editable geodatabase and map of the world.

VGI (Volunteered Geospatial Information): A sort of user generated contents which consist of geospatial data.

Chapter 14
Adaptive Acquisition of VGI to Fill Out Gaps in Biological Observation Metadata

Daniel Cintra Cugler
Federal Institute of Triangulo Mineiro, Brazil

Claudia Bauzer Medeiros
University of Campinas, Brazil

ABSTRACT

Biological observation databases store data on species observations, being used in many kinds of research. Such observations are often queried and/or correlated primarily using metadata parameters (e.g., spatial queries on metadata concerning regions where observations were performed). However, metadata are often missing - either blank attributes, or lack of metadata records - which hampers the use of the observations databases. Filling these gaps is challenging because metadata requirements change as researchers acquire new knowledge about their problems. Related work is limited because it does not take this knowledge evolution into consideration. This chapter presents an approach to acquire missing metadata records, which fully supports dynamic on-the-fly evolution of metadata requirements. As proof of concept, we implemented a configurable software platform to collect data from "human sensors" and other sensors. Among its many dynamic characteristics, it allows deployment of context-sensitive forms to be filled by volunteers, according to a location and a research target.

INTRODUCTION AND MOTIVATION

Biological observation databases contain priceless information that can be used to provide knowledge for broad kinds of research, e.g., global warming, species behavior, food production, etc. Such information can be acquired in many distinct ways and from many distinct sources. Observation metadata (the data that describe observations) play a major role in scientific research involving biological observations; metadata are the primary means to computationally process, correlate, and access observational data.

DOI: 10.4018/978-1-5225-2446-5.ch014

As such, they are considered by scientists as "first-class citizens", being often indissociate from the observations themselves. Such metadata may differ according to the domain, but the most important fields are supported by all kinds of observations -- i.e., "what" (what was observed), "when", "where", "how" (methodology) and "who" (observer). We point out that, throughout this chapter, the main assumption is that research that uses biological observations is highly dependent on the "where" metadata attributes – i.e., though not always explicit, processing geolocation information is a key issue here. In fact, "where" has become one of the most important pieces of information in observations, since a growing spectrum of studies are sensitive to geospatial information. In many cases, observation data and metadata are intimately associated, so that the difference becomes fuzzy – see remarks in Section "*Background and Related Work*".

In the past, observations were made mainly by domain experts, who needed to spend from days to months traveling and collecting information in field trips. Such observations were usually made manually, and metadata were annotated in paper, in natural language, most of the time following no predefined data structure.

The way of collecting and persisting biological observations and metadata has changed drastically – and not only due to technological progress (e.g., use of GPS-enabled devices to automatically capture the "where"), but also because new paradigms have appeared for data capture methodologies. First, an increasing number of observations started to be provided not only by domain experts involved in projects, but also by non-expert volunteers. Such information, named VGI (Irwin, 1995; Goodchild, 2007) (Volunteered Geographic Information -- See Section *"Volunteered Geographic Information (VGI)"*), has introduced a new paradigm in the way of collecting data. Second, computational systems with predefined data structures are now commonly used to persist observational metadata. Third, different computational tools are being employed to acquire information, such as web-based forms and mobile applications. In spite of those changes, one feature remains the same -- there is often missing information in the metadata, which demands time- and cost-intensive curation processes.

Such missing information (here called *information gaps*) can bias research into wrong scientific conclusions or even hamper the use of the observations. There are countless reasons for gaps in biological observation metadata -- e.g., lack of equipment to measure environmental variables or lack of standards. The approach presented in this chapter generalizes and classifies such gaps into two groups: (a) incomplete information in the metadata records and (b) insufficient samples for some specific scenario.

- **Incomplete Information:** Here, not all record fields are filled out. For example, when recording animal vocalizations, identification of species is a hard task since many times the scientists do not see the animals whose vocalization they are recording. For this reason, metadata may, for instance, lack an animal scientific name, providing only partial taxonomic classification. Additional gaps may be caused, for instance, by lack of appropriate instruments at the observation site, or absence of adequate conditions to perform some measurement. This is a widely studied problem, for which the majority of solutions are based on data curation algorithms.
- **Insufficient Number of Samples:** Here, the problem is having an insufficient number of observation (and thus metadata) records that fit a scenario. This may occur, for instance, when a team of scientists wants to reuse data collected by another team, but the teams have distinct spatial sampling methodologies. This chapter is concerned with handling this second kind of gap, since, as we show, it is much more likely to profit from VGI as the other kind of gap. It must be pointed

out that, here, it is not just a matter of getting more observation records (i.e., this is not a quantity problem); rather, more samples must be collected to satisfy a combination of constraints on "where", "when", "how" and "what", with emphasis on the spatio-temporal factors.

However, the dynamics of VGI data collection and changes in the environment with time makes VGI-based techniques a challenging problem, as will be exemplified in the chapter. Filling out gaps in metadata requires adaptive techniques to collect the right information in the right moment and place. To the best of our knowledge, there is no work that supports such features in the selected target domain.

In order to overcome such challenges, this chapter discusses our context-sensitive approach to collect biological observation metadata using dynamic forms. Such forms are (re)configured on the fly by taking into account the metadata being acquired, in real time. The data acquisition forms are location-sensitive and are dynamically adapted according to scientists' requirements.

The rest of this chapter is organized as follows. Section *"Background and Related Work"* introduces basic concepts for the proposed approach and related work. Section *"Dynamic Data Acquisition Approach"* details the proposed approach. Section *"Prototype and Implementation Issues"* details the prototype, which collects VGI from mobile devices and sensors over the Arduino platform. The same section also presents a real case study using the prototype. Finally, Section *"Conclusions and Future Work"* presents the conclusions and future work.

BACKGROUND AND RELATED WORK

This book chapter is concerned with using VGI to collect biological observational data and the corresponding metadata. Thus, related work discussed in this section concerns these issues.

Biological Observation Metadata

Biological observations concern the recording of the occurrence of an organism or set of organisms detected at a given place and time according to some methodology. In other words, "an observation represents an assertion that a particular entity was observed and that the corresponding set of measurements were recorded (as part of the observation)" (Cornell, 2015). Nowadays, observations have been described by several kinds of data types, such as textual information, photos and sounds. Biological collections might concern broad kinds of species, or a specific category thereof. For example, the Cornell Lab of Ornithology (Cornell, 2015) contains sound recordings, photos and associated metadata restricted to birds (Bonney, 2007). The Fonoteca Neotropical Jacques Viellieard (FNJV) (FNJV, 2015) has sound recordings and associated metadata of all vertebrate groups (fishes, amphibians, reptiles, birds and mammals) and some groups of invertebrates (such as insects and arachnids). The plant database of the Kwantlen Polytechnic University (Kwantlen Polytechnic University, 2015) provides images of several species of plants and their associated metadata.

Biological observation metadata can be provided in two ways. Historically, observers went to the field, performed their observations and annotated the observational information in paper; eventually, observations were transcribed to spreadsheets. Here, the number of observations was low, since observations were mostly provided by a limited number of domain experts. Today, observational information is often acquired using mobile devices and sensors. The number of observations performed is much higher, and there is an increasing number of volunteers who contribute with their observations (Bonney, 2007).

Metadata must be curated, because there are several quality issues, including incomplete and incorrect information, which affect the quality of the associated observations. The acquisition of observational information from volunteers may affect the quality of the metadata provided, since volunteers are not always domain experts. There are approaches that overcome such problems, by focusing on quality aspects of the metadata, e.g., (Cugler, Medeiros, Shekhar, & Toledo, 2013; Lemos, 2013; J. G. Malaverri, Santanchè, & Medeiros, 2013; J. E. G. Malaverri, Mota, & Medeiros, 2013). Another problem is related to the fact that there is often missing information in such metadata. Distinct approaches have also been proposed for this problem, each of which focusing on different aspects, e.g., (Cugler et al., June 2012). Moreover, biological databases are always changing. As new observations are provided, gaps in the metadata are filled or new gaps arise. Filling out such gaps is an open research problem.

Different strategies - e.g., (Cugler, Medeiros, & Toledo, June 2012; Hartung et al., 2010) – have been proposed to fill out metadata gaps. Some approaches - e.g., (Cugler et al., June 2012) - acquire more metadata records by mining the appropriate information from web sources, often starting from spatial requirements. Others adopt VGI, which is the focus of this work, to get more records from a given region, predefined by the expert researchers. In (Hartung et al., 2010), for example, scientists create electronic forms to collect metadata from VGI through web-based and mobile applications. However, such electronic forms are static and the data acquisition strategy does not consider that metadata requirements may change as researchers acquire more knowledge about some question. Moreover, since these requirements vary with the goals of a research team, static forms lack the dynamicity required. As will be seen, our solution allows scientists to (re)configure forms on the fly as often as needed, which includes directing volunteers to move throughout a region – not only the form changes, but directives on where to collect data can be fed into the system.

Biological Observations: Metadata as First Class Objects

Research based on biological observations is commonly performed by linking the observations with their metadata. In all such cases, metadata is used not only to find, but also – and most importantly - to *correlate* observational data. Thus, VGI-based observation tasks require, above all, a great level of care in entering metadata.

Moreover, in environmental studies, many facts about plants and/or animal patterns can be explained based on metadata. For instance, in bioacoustics, amphibians usually produce less vocalizations when in the full moon phase because in light nights they are more visible to predators – here, the vocalization (i.e., sound recording) represents the *data* while the moon phase information represents the *metadata* associated to the data record. By the same token, in many observations, scientists treat data and metadata as one single set of "attributes". For instance, in many botanical field trips, the observation itself may be physical parts of a plant which are taken back to the laboratory or herbarium (leaves, flowers, fruit, branch, etc), while the metadata describes the plant. However, vegetation samples may be replaced by photos – and stored as attributes in the observation database together with the metadata.

Note that, in bioacoustics, the presence of a specific species in a given place may be detected by only listening to the animal vocalization (without visual contact). In ecological observations, sometimes the "object" being studied is an entire patch. This means that, although a biological observation commonly produces a database record in the pattern "<data, metadata>", there may have situations where only one of them is provided – e.g, (Cornell, 2015; McQuilton, et al., 2016).

Based on these remarks, this research assumes that if a new metadata record is inserted in the (metadata) database, it means that such metadata is necessarily associated to some observation. In particular, we concentrate our work on metadata aspects, given their relevance in ecological, biological and environmental studies. As mentioned in the introduction, spatial information is perhaps the single most important metadata field for this domain, together with species identification.

Volunteered Geographic Information (VGI)

Citizen science is the term often used to describe communities or networks of citizens who act as observers in some domain of science. Analogously, Volunteered Geographic Information (VGI) refers to data provided by citizens, in particular, including geographic information (Irwin, 1995; Goodchild, 2007; Shekhar et al, 2015).

The use of VGI has been shown to be successfully used by an increasing number of projects, however, acquiring information from volunteers is not a simple task. Besides the technical aspects involved, it also requires, among other things, defining strategies for recruiting and motivating people to perform observations (Rotman et al., 2012). Before collecting VGI, researchers must define which strategy they will adopt.

There are different projects and initiatives, in different domains of science, boosting the number of observations in their collections by using VGI. In the biology domain, for example, there is the Christmas Bird Count project (LeBaron et al., 2009), that is related to animal preservation and environmental studies. This project is an effort to perform a mid-winter census of bird populations in the North America. This kind of project considers, among others, information provided by citizen science and VGI to create geographic distribution range maps for several species. Another project (Beasley, Benson, Welch, Reid, & Mousseau, 2012) uses volunteers to record and map the 13-year periodical cicada in South Carolina/USA. Observations are entered through websites of the University of South Carolina (University of South Carolina, 2016) and the South Carolina Forestry Commission. Still in the domain of cicada observations, The Cicada project (Cicada Project, 2015) provided schematics to assemble sensors to measure soil temperature. This temperature is a key aspect on analyzing periodical cicadas. The temperature information is submitted through websites, in predefined forms. Another project is the Canada Nature Watch Programs (Whitelaw, Vaughan, Craig, & Atkinson, 2003), which receives observations from volunteers for several domains (e.g., plant phenology, worms, frogs, and ice). One possible way to send an observation is by printing a form, filling it out manually and sending it through mail to the project committee. Their purpose is to study the effects of climate change and other impacts on biodiversity. In spite of the emphasis on geospatial information (and thus the "G" in VGI), we point out that VGI for biological observations also requires gathering other kinds of metadata – i.e., no task can be defined only in terms of geospatial requirements.

There are also examples of VGI in non-biological domains. For instance, the Open Street Maps project (Haklay & Weber, 2008) provides editable maps of the world by using information from volunteers, including providing humanitarian help (Humantarian Open StreetMap Team, 2016). In the Geo-Referenced Field Photo Library project (Haklay & Weber, 2008) researchers created a mobile app in which volunteers use smartphones to take geo-referenced photos in the field in an effort to document observations of landscapes, agriculture, forests, natural disasters, and wildlife. Another example (Maisonneuve, Stevens,

Niessen, Hanappe, & Steels, 2009) uses citizens to measure people's personal exposure to noise in their everyday environment by using GPS-equipped mobile phones as noise sensors. The information is used to provide geographic distribution maps about noise-pollution. In (Longueville, Smith, & Luraschi, 2009), location-based social networks are used to acquire spatio-temporal data on forest fires.

The similarity among such projects is the fact that all of them are VGI (em)powered. Their differences come from the variables recorded and way they collect the data, which varies from simple printed forms sent by mail to tailored mobile software. They indicate the necessity of systems to support acquisition of VGI data in a wide range of domains.

Approaches to Collect VGI

In the past, before the advent of personal computers and smartphones, VGI was collected mainly through the use of printed forms, created by researchers. After volunteers filled out the forms, they were usually returned to the researchers via mail. The whole process of creating, deploying, and receiving the forms used to take several days or even months. Besides, depending on the number of forms received, it was hard to organize and process the information. Even nowadays, it is still possible to find projects that collect VGI using the same strategy, differentiated only by the fact that forms might be returned, in most cases, by both mail or email. An example is the MeadoWatch project at the Mount Rainier National Park (USA) in which the goal is to study climate change through observations of plants (MeadoWatch, 2015).

Two basic technologies are responsible for boosting the acquisition of VGI: geo-referencing and the Web 2.0 (Heipke, 2010). GPS devices and GPS-equipped smartphones easily enable geo-referencing and the Web 2.0 enables collecting information generated by volunteers, automatically persisting all information in databases with semantics. These technologies are the source of two popular ways of collecting VGI: (a) forms published in websites and (b) forms published in mobile applications.

In the first strategy (a), programming skills are required to create the websites containing the electronic forms. Besides requiring computing expertise, creating such websites is time consuming. Though there are tools that enable easy creation of electronic forms (such as Google docs), such tools are limited when providing ways of personalizing the forms. For instance, in the target domain of our work, the Nature's Notebook project (Posthumus & Crimmins, 2011) uses the strategy of web forms. Here, volunteers provide observations of plants and animals and all metadata is submitted in an electronic form embedded in the website. A hurdle in such a strategy is the fact that such forms are static, requiring time to be spent to change the web site if there is a need to ask volunteers other questions.

The second strategy (b) differs basically by the flexibility that mobile devices provide. Here, volunteers can go to the field and fill out the forms directly in the mobile devices. Another advantage here is that volunteers can use embedded sensors (e.g., GPS and camera) in the mobile devices in order to feed more data. However, in this strategy the forms are also static, requiring time to program and update the mobile applications to change the forms. One example of a project that uses such a strategy is the Creek Watch project (Kim, Robson, Zimmerman, Pierce, & Haber, 2011). It allows volunteers to report information about waterways, to support water management programs.

There are also approaches that support the creation of personalized questionnaires to collect VGI, such as the Open Data Kit (Hartung et al., 2010), wq (Sheppard, 2012), Sensr (Kim, Mankoff, & Paulos, 2013) and CitSci.org (Newman, Graham, Crall, & Laituri, 2011). However, none of them enables performing dynamic changes in the questionnaires by considering the gaps that may occur in the metadata as new VGI information is acquired or as researchers improve their knowledge about some problem. Moreover, there are no dynamic changes in the strategy of where to send volunteers. Our approach differs from other approaches by considering the necessity of such dynamism, as detailed in the Section *"Dynamic Data Acquisition Approach"*.

DYNAMIC DATA ACQUISITION APPROACH

The goal of this approach is to provide a strategy in which researchers take advantage of VGI to fill out gaps in biological observation metadata. As discussed in Section *"Introduction and Motivation"*, there are two kinds of gaps in biological observation metadata: empty fields in one or more records, or absence of records. The approach discussed here is concerned with this second kind of gap.

Our approach enables researchers to define what is the best strategy for their scenarios, since it facilitates improved collaboration between scientists and volunteers (Rotman et al., 2012). The underlying strategy considers that VGI will be used to provide additional records for situations in which the database already has records from previous projects. In other words, VGI is used to *add new records* to curated collections of observational metadata (as opposed to their use to create such a collection from scratch). The approach is context sensitive, i.e., researchers can define on the fly the VGI data that must be acquired. Its effectiveness is enhanced by also supporting the context-sensitive definition of spatial regions where information must be acquired.

Overview

Our approach is divided in three steps, as shown in Figure 1.

Figure 1. Approach for adaptive acquisition of VGI to fill out gaps in biological observation metadata

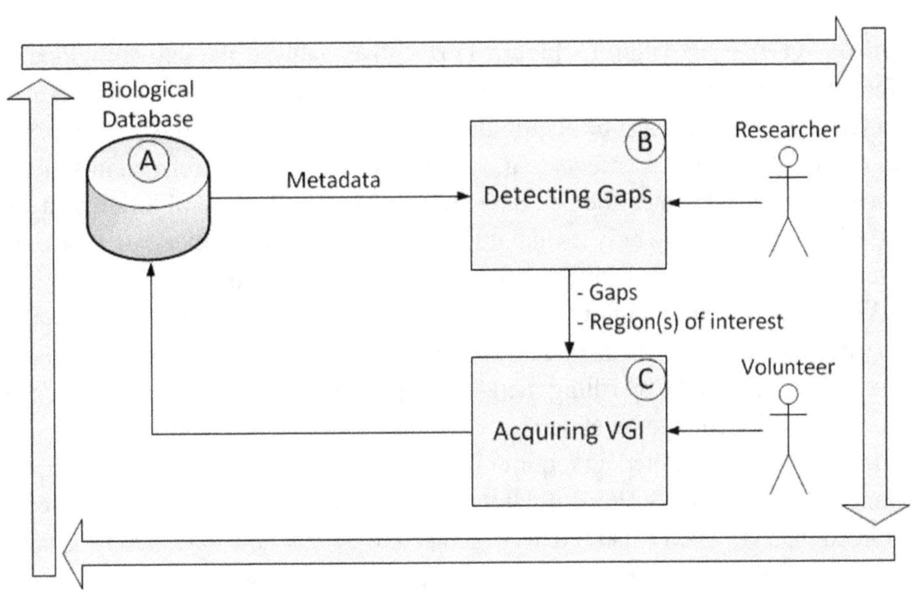

Step 1 - Detecting Gaps in the Metadata: Here, scientists will use a suite of algorithms for gap detection (item B in Figure 1) in a biological database. These algorithms detect gaps (i.e., missing records) in the metadata based on both scientist and research requirements. To do so, algorithms combine: (1) parameters provided by the scientists, (2) metadata records associated to the already collected biological data and (3) the collected biological data (i.e., in our case, sound recordings). The suite of algorithms detects gaps by considering the most important attributes in biological observation metadata, i.e., what (the species observed), where (location of the observation), when (date/time of the observation) and how (the methodology used). Figure 2 provides an overview of a gap detection algorithm, instantiated for a particular case study, illustrating how, given (1) parameters, (2) metadata, and (3) corresponding biological data, the algorithm returns a set of records that will allow the scientist to decide whether there are gaps or not. The algorithm works as follows.

First, the scientist defines the criteria that describe the scenario in which (s)he wants to check for gaps. This is illustrated, in Figure 2 by two parameters, named Parameters 1 and 2. The algorithm starts by retrieving from the metadata repository (2) the metadata records that match *Parameter 1* (here, species name, location, environmental conditions, and time frame). This is achieved via an SQL query against the metadata repository. This query retrieves a set of records that are sent as input to the next step, which will perform animal sound record analysis.

This step will then select, from the biological data repository (3), the sound records that correspond to the input metadata (i.e., sound recordings that match *Parameter1*). Next, these sound recordings are analyzed to check which recordings meet *Parameter2* (namely, decibel limits). We point out that metadata analysis is performed before the sound analysis in order to narrow the search space in the sound analysis step (which requires more computational power to process the inputs).

Finally, the algorithm provides a list of tuples in the format <metadata, sound recording> which match *Parameters 1 and 2*. Based on this result, the scientist will assess if there is a gap or not for the scenario of interest (e.g., for rare species, three observation records matching the criteria may be enough, whereas for commonly found species thirty records may not be enough).

Step 2 - Acquiring VGI: Here, (item C, Figure 1) scientists analyze the gaps provided by step 1, to determine what kinds of VGI should be acquired. As a result, researchers must define (a) a spatial area of interest where VGI must be acquired (*where*); (b) the focus of the VGI (i.e., the target species, *what*) and (c) extra requirements (e.g., observations must be performed in a given time of the day - *when* and/or *how*). Since the database fields may include several kinds of data types, VGI acquisition is not restricted to only textual information, comprising other data types, such as video, audio, etc. VGI metadata fields can be entered by humans and at the same time complemented by sensors. We point out that, though volunteers are asked to provide metadata, they may also provide the observations that go with the metadata (since the metadata describes details of the observations themselves). In other words, by filling metadata gaps, new observational data are acquired, which must be subsequently curated by experts.

Step 3 - Iterating the Process: Metadata requirements might change as researchers acquire new knowledge about their problems. Biological databases are always changing, with new records inserted in the collection, e.g., when experts detect the need to collect new data. This scenario guided the authors of this work to include a cycle in the approach, creating an interactive process in which

Figure 2. A gap detection algorithm. In this example the algorithm detects gaps considering the most important attributes in biological observation metadata (i.e., what, where, when and how)

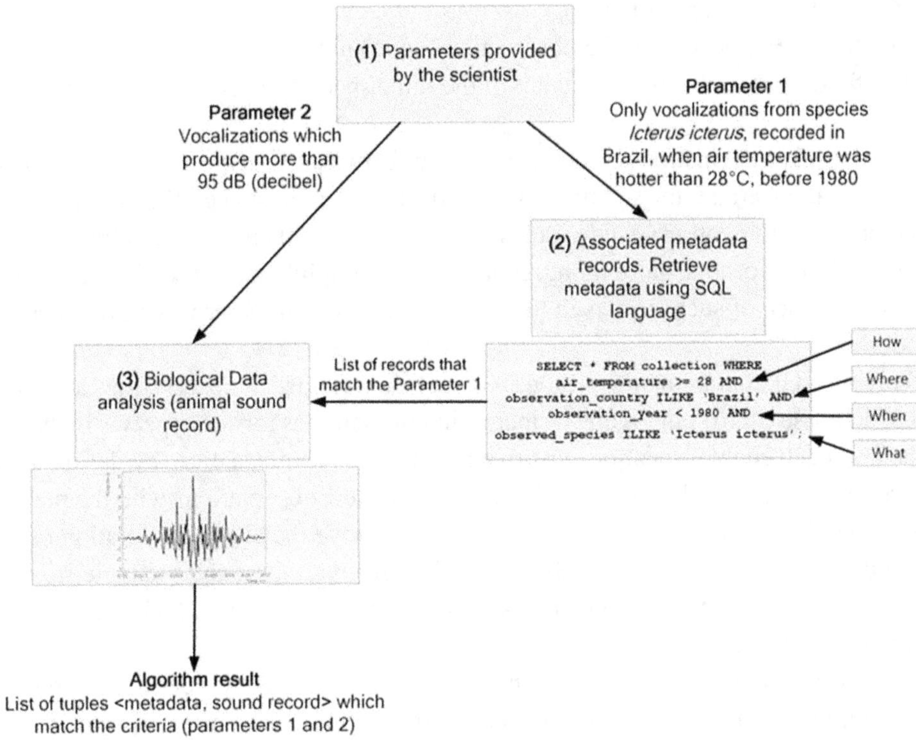

researchers can change *on the fly* any of the VGI collection directives (i.e., (a), (b) or (c) in step 2). This is the most important part of the contribution of this work, since other VGI approaches do not support on-the-fly gap specification.

Detection of Gaps

Gaps in Spatial Coverage: In order to illustrate the applicability of the approach, consider an example in which a researcher is studying the effect of a disease *D* that is infecting the species *Elachistocleis ovalis* (an amphibian species). To do so, (s)he needs to have at least *n* observations of such a species with spatial coverage *S* to perform experiments. In order to check if this condition is satisfied (i.e., check if there is a gap -- Step 1), the researcher firstly narrows the search space by selecting database records of interest. Afterwards, (s)he executes algorithms to detect spatial regions in which there is a low number of observations for the target species. If this spatial gap is detected, the researcher defines, for example, the following scenario to acquire VGI (Step 2):

- Target = Species *Elachistocleis ovalis*
- Region of interest = a circular area defined by the point(-30.118255, -50.422468) with a 10 kilometers radius
- Metadata field 1 = Time and date of the observation

- Metadata field 2 = Air temperature
- Metadata field 3 = Air humidity
- Metadata field 4 = Picture of the frog
- Metadata field 5 = Location of the observation (coordinates)
- Metadata field 6 = Number of individuals of the same species seen close to the individual observed

The scientist uses this information to configure a web form that is sent to volunteers to collect these data, for that spatial coverage, for sightings of the species of interest (i.e., the volunteers will provide this information when they observe individuals of this group). This notwithstanding, the main goal of the VGI task might be not to acquire more records of that amphibian, but to obtain more information concerning the presence of such species in the region of interest. In other words, the metadata become the observation.

After receiving VGI containing the pre-defined metadata fields, all observations are integrated with the biological database (item A in Figure 1), increasing the samples for such species in the target region. Note that the approach enables location-sensitive acquisition of information, by defining the region of interest where VGI must be collected. After integrating the new observations with the previous records, researchers can perform new data analyses. In order to improve the analyses, curation of VGI must be performed by the domain expert, since VGI data quality may vary broadly. Strategies to deal with data quality can be used, such as (Sousa, Cugler, Malaverri, & Medeiros, 2014; J. G. Malaverri et al., 2013; J. E. G. Malaverri et al., 2013).

Figure 3 shows the spatial coverage *S* that, in the example, refers to the circle defined by point(-30.118255, -50.422468) with radius of 10 kilometers. All VGI in the database (O_n) were expected to be performed within the region *S*. Blue points (from O_1 to O_{10}) are locations where observations were performed within region *S*. Opposite to the blue points, red points (from O_{11} to O_{17}) were observations performed outside the expected region *S* and the expert must decide if such observations are going to be inserted in the database or not.

Note that, in this example, just one set of database fields compose the form that guided the acquisition of VGI in the spatial coverage *S*. However, it could be also possible to define different forms to be deployed in different regions of interest simultaneously.

In this example, the gap detection algorithm detected a gap by checking a where field (the spatial coverage). However, the suite of algorithms for gap detection can also detect gaps by considering how, what and when fields. For completeness' sake, we briefly provide examples of *how* (Gaps in methodology), *what* (Gaps in species observed) and *when* (Gaps in temporal coverage) scenarios that typical algorithms must consider.

Gaps in Methodology: Methodology gaps can be divided into two scenarios.

Scenario 1: A researcher needs observations performed under a specific environmental condition. For example, all observations of species *s* were made in rainy days and the researcher needs samples in sunny days.

Scenario 2: A researcher needs animal sound recordings made with the equipment *E*, brand *B*. Depending on the equipment used, the quality of animal sound recordings might be affected.

Consider an example for *Scenario 1*, where the researcher acquired enough observations for species *Elachistocleis ovalis* using VGI. Also consider that after the acquisition of such observations the

Figure 3. Observations O_n provided by VGI. Blue points are observations provided within the expected spatial coverage S. Red points are out of the spatial coverage (Source map extracted from Google Maps (Google, 2016)

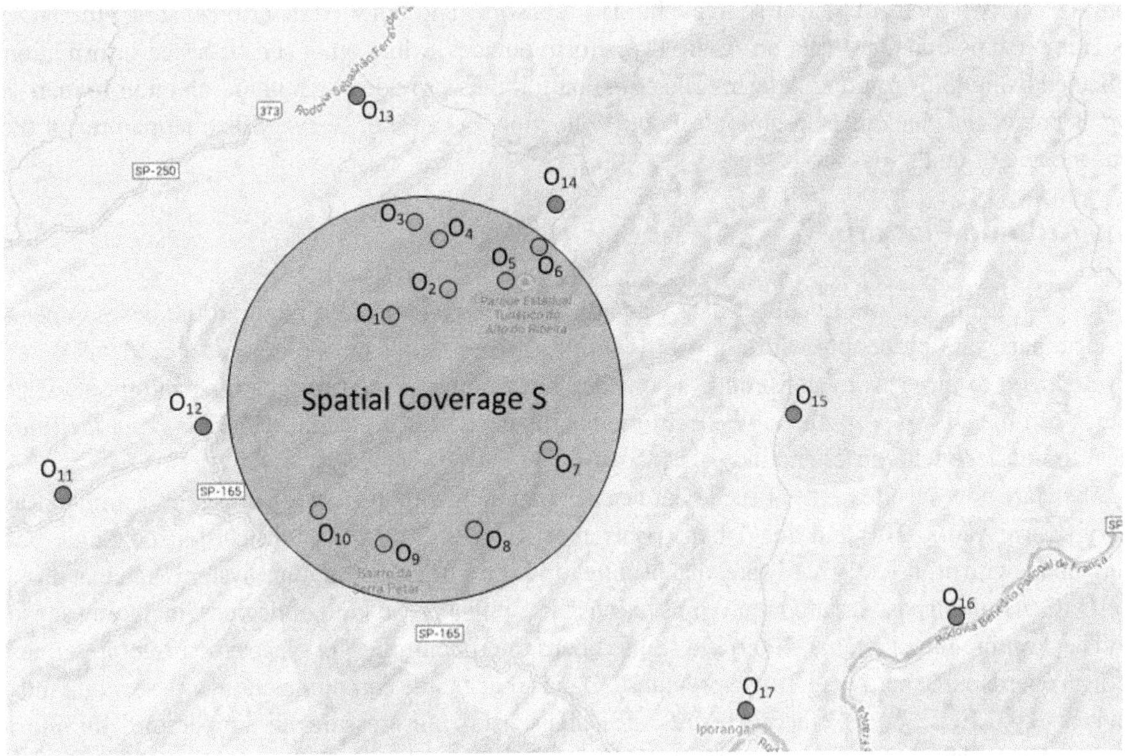

researcher detected another gap (using the suite of algorithms), by checking the absence of observations of species *Elachistocleis ovalis* in days with air humidity higher than 65%. The researcher could start over by defining the same database fields to compose the questionnaire to collect VGI. However, here (s)he would change a question's priority order by defining the first form questions as those that represent the most important gaps in the database (Step 3), e.g., defining the *air humidity* as the first form field, guiding volunteers towards focusing firstly on the most important information.

Gaps in Species Observed: Consider that the researcher needs observations of species *Aplastodiscus perviridis* and *Allobates marchesianus* (other kinds of amphibians). Here, a *what* gap might be absence or low number of observations of such species. The *what* gap can also take advantage of a *where* analysis, by contrasting *species* against *number of observations* and against *location of observations*, e.g., showing which species are missing in a sample within a region.

Gaps in Temporal Coverage: Consider that the researcher needs observations of amphibian species *Aplastodiscus perviridis* and *Allobates marchesianus* within a specific time frame. For example, there are no observations of such species between 1AM and 3AM. Here, similarly to the "*what*" example, it is also possible to take advantage of the "*where*" analysis by analysing "*when*" and "*where*" in conjunction.

For all such scenarios, the context sensitive approach enables researchers to change on the fly the list of data they want to acquire, by providing a never ending cycle that is adapted to their current requirements. Such requirements will be in turn affected by the metadata collection evolution.

PROTOTYPE AND IMPLEMENTATION ISSUES

We developed a prototype to enable adaptive acquisition of VGI, i.e., for item C of Figure 1. The VGI comes from two kinds of sensor sources: human "sensors" and other (standard) sensors. Human sensors are citizens that enter data on Android-platform devices, while other sensor-based information is collected using the Arduino platform. The combination data provided by human- and non-human sensors is one of the challenges in biological data collection, because of the very distinct natures of these sensors, and of quality issues.

The Arduino Platform

In the past, using sensors to collect data was not a simple task, since it required the development of specific hardware by people with high skills in electronics. This could represent a bottleneck in the development of projects, by consuming a considerable amount of the project's time and money. In the course of time, cheap programmable electronic boards have been created, accelerating and facilitating users to interface with different kinds of sensors.

There are now several kinds of such boards, each of which with distinct features. For example, Raspberry (Kurniawan, 2013) and Beaglebone (Barrett & Kridner, 2013)are credit-card-sized single-board computers, with proprietary hardware, that usually run Linux-based operational systems and that provide GPIO (General Purpose Input/Output) pins to interface with a broad kind of devices, including sensors.

The Arduino board (Banzi, 2009) is a single-board microcontroller for electronic prototyping, made with an open-source hardware. It also provides I/O pins that enable communication with several kinds of devices (e.g. LEDs, servos, ethernet boards, etc.) and sensors (air pressure, air temperature, luminosity, motion, RFID, etc.). Furthermore, Arduino is cheaper than single-board computers, and it is easier to use when interfacing with devices and sensors. Arduino can be programmable in C or C++ languages.

There are plenty of add-on hardware modules for Arduino, known as *shields*, which are expansion boards that plug into the Arduino pin-headers. Such shields are one of the features that make Arduino so attractive and broadly used. Shields can provide GPS, ethernet, LCD display, SD card interface, standardized connection pins, etc., letting users free of spending time on creating their own hardware. Another useful feature is that several kinds of sensors are provided using the Grove system. Grove is a ready-to-use tool set, similar to LEGO. It takes a building block approach to assemble electronics, simplifying their use. The Grove system is composed by a base shield and several modules with standardized connectors. This project is focused on providing an environment in which scientists do not need to have skills in programming and/or electronics. Thus, it is used Arduino in conjunction with Grove shields and Grove sensors, in order to simplify and enable the use of sensors to collect VGI. The prototype only requires that scientists assemble the hardware (similar to constructing a LEGO toy) and set up a XML file, as described on Subsection *"Example"*.

Defining on the Fly Which Data Fields Must be Acquired

The prototype is supported by a local web server that does not depend on third-party services. Therefore, it is ensured that the service will never be stopped by changes that might occur when using remote third-party services. Figure 4 shows an overview of the prototype, which enables collecting data from both humans and sensors. Collecting data from sensors and humans has different features, as follows.

Figure 4. Scientists remotely configure forms and sensors, defining which attributes to collect

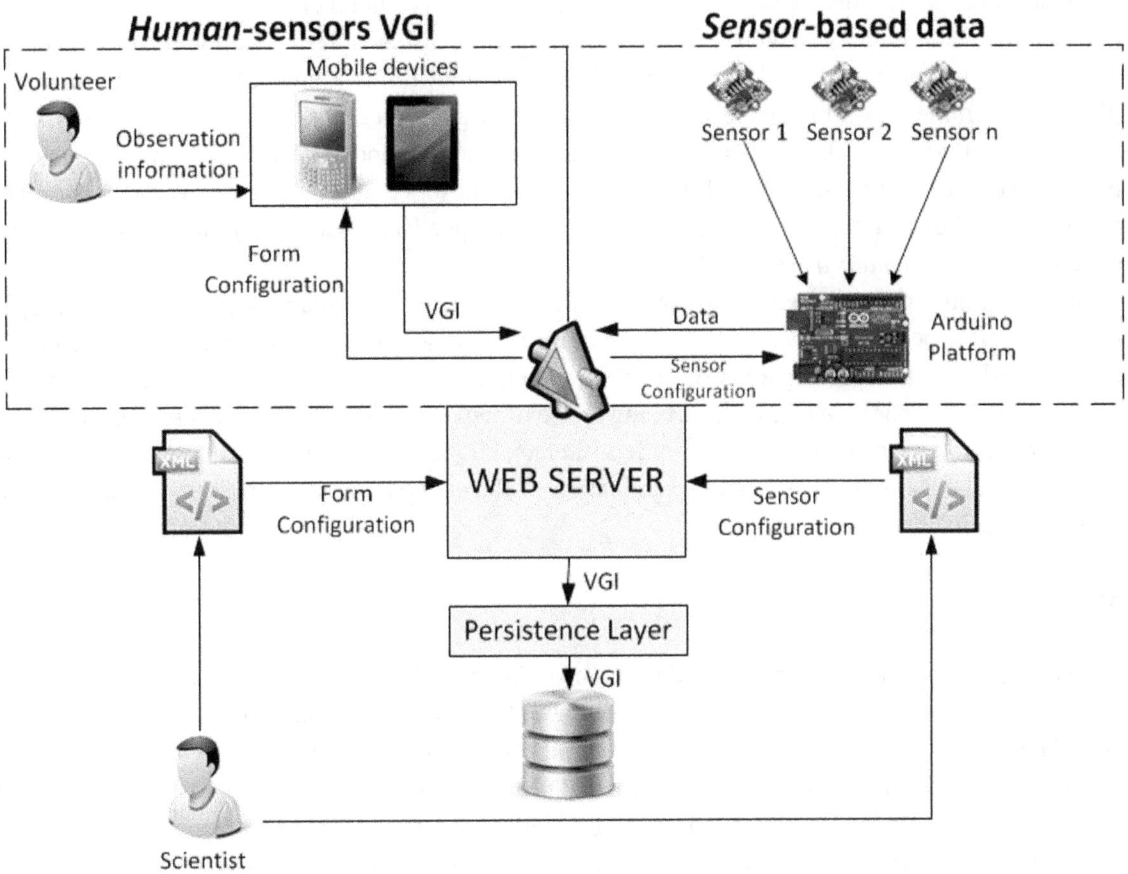

Humans as Sensors: Humans can act as sensors to provide VGI by observing things or phenomena. The prototype supports the whole process of observation, that includes (a) remotely configuring the form fields that will be presented to the volunteers, (b) publishing the form in mobile devices and (c) collecting and persisting the VGI. In (a), the form definitions must be set up through a XML file that is defined by experts. XML is used because it is a consensual standard format, and easier to integrate with other sources.

In such XML files, experts define the order and the fields that must compose the electronic form, e.g., location where the observation is performed, date and time of the observation, picture of what was observed and name of the species observed. Here XML files can be shared, enabling reuse of such configurations as templates. The form is made available by a web server through web services (developed in Java + Apache Tomcat server). Any mobile device using the proposed mobile application (developed in the Android platform) can connect into the web service, acquire data to configure the form, and create an electronic form dynamically. Afterwards, volunteers can record the observation, by filling out the form using mobile devices. Geographic coordinates and photos are taken directly from the GPS and camera embedded into the mobile device. All VGI is returned through the web service. Note that pictures are serialized before being submitted. Finally, the web server receives the VGI and persists all information into a CSV file. It has been chosen persisting VGI in CSV files because any software to manage spreadsheets can open CSV files. CSV files can also be easily imported into any database. Since spreadsheets are a major tool for scientists who deal with observations, this is a good choice.

Sensors as Data Sources for VGI: Similarly, the prototype also enables collecting data from sensors using the Arduino platform. To do so, we developed a framework that enables easy configuration of Arduino. The framework allows experts to (remotely) configure a set of sensors, such as GPS, air temperature, air humidity, gas and light sensors. Configuring the Arduino platform through the framework requires experts to create another XML file, defining the sensors that will be used. The whole process (from defining the information to acquire to collecting and persisting such data) follows the same procedure as the topic *humans as sensors* for VGI.

Since the prototype is configured through a XML file, whenever a researcher detects a different gap, the XML file can be updated to reflect the current research requirements. As mentioned subsequently, we intend to design and develop a visual interface to help experts (re)configure forms for humans, and directives for sensor deployment, without being aware of the underlying XML structure. Expert manual intervention, however, cannot be totally avoided – in fact, it is often a requirement imposed by scientists, so that they can ensure the appropriate specification of the observational constraints to meet a desired research scenario.

Example

To illustrate the use of the prototype, recall the scenario of Section *"Dynamic Data Acquisition Approach"*, in which a researcher is studying the effect of a disease *D* that is infecting the frog species *Elachistocleis ovalis*. The expert is interested in performing a spatial analysis to detect "*where*" gaps. Here, step 1 (gap detection) is performed by running an algorithm that is described in (Cugler et al., 2013). This algorithm is based on a geospatial approach that contrasts geographic distribution maps (regions in which species are expected to live) against the places in which the observations were made (as per location metadata). Figure 5 shows a gap detected by the proposed algorithm for records of the species *Elachistocleis ovalis*. The inputs are the 30,000 species observation records of the FNJV repository (FNJV, 2015).

The light gray polygons represent the regions (*R*) where the species is expected to live. Blue points are locations where observations were performed inside or at most 100Km far from the expected regions (*R*). The red point represents a location where the observation was performed farther than 100Km from the expected regions (*R*). The buffer (100Km) is adjustable and is set up by the domain expert. Note that in Figure 5 the red point represents an outlier (anomaly). In this example, such an outlier is considered a *spatial coverage gap*, since the red point evidences the presence of such a species in a region where there is just one observation. Therefore, here the gap is the low number of observations of the species *Elachistocleis ovalis* in a region, i.e., what (the species) and where (the location where there is a low number of observations).

After detecting the gap, the expert must firstly define the region where VGI must be acquired. Such a region can be specified in the prototype by defining a point coordinate and a radius. Afterwards, the researcher must define the data fields that will be presented to the volunteers as well as their respective data types. Listing 1 shows a XML configuration file where this information is defined. This XML file will configure forms that will be shown in smartphones and tablets. Note that Listing 1 defines the questions that must be deployed, in conjunction with their respective label, field name that will be associated with the data and the data type. It also defines the region where the questionnaire must be deployed. For

Figure 5. Output map generated by the gap detection algorithm for the Elachistocleis ovalis species (Amphibian). The anomaly (red point) led to the spatial gap detection

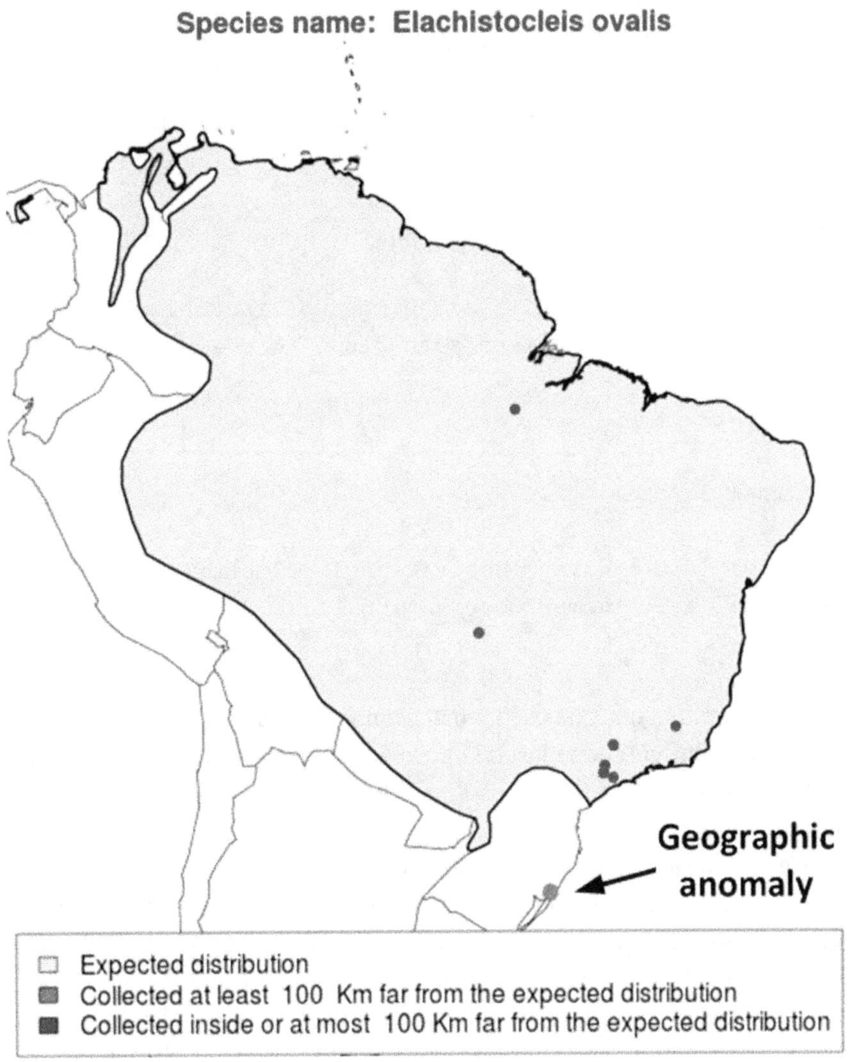

instance, line 3 defines the target species. Lines 13 to 18 request information on close-by individuals. Lines 4 to 8 define the region of interest, and so on.

Listing 1: XML Configuration File to Create the Form to Collect VGI

```
<?xml version="1.0" encoding="UTF-8"?>
<vgi source="mobile">
  <field type="target" field_name="species_name" label="Elachistocleis ova-
lis."/>
  <field type="spatial_coverage_definition">
    <lat>-30.118255</lat>
```

```
    <lon>-50.422468</lon>
    <radius>20</radius>
  </field>
  <field type="string" field_name="air_temperature" label="Inform the air tem-
perature."/>
  <field type="string" field_name="observer" label="Inform the air humidity."/>
  <field type="picture" field_name="species_picture" label ="Take a picture of
the species"/>
  <field type="location" field_name="location" label="Get the location of the
observation"/>
  <field type="radiobutton" field_name="number_of_individuals" label="How many
individuals of the same species were seen close to the species observed?">
    <item>1</item>
    <item>2</item>
    <item>3</item>
    <item>More than 3</item>
  </field>
  <field type="time" field_name="observation_local_time"/>
  <field type="date" field_name="observation_local_date"/>
</vgi>
```

The same kind of configuration must be used when acquiring data from sensors, except that additional information about sensor configuration must be informed. In the prototype, to collect data from sensors, it was used one Arduino board, one GPS sensor, one Ethernet shield, one Grove base shield and one humidity/temperature sensor.

The XML configuration file must be saved in the web server where the server application runs. Each volunteer who wants to provide observations by using the mobile platform must install the proposed mobile application prototype, open it and introduce the IP of the web server. After that, the dynamic form is created in the client side, based on the XML configuration file. Figure 6 shows a screenshot of the form created in the mobile device, based on the XML configuration file defined by the researcher. In the current version of the prototype, all observations are saved in a CSV file for future integration with the original database. Future versions of the prototype will integrate VGI data into the database automatically. The main barrier to automatic integration is data quality and curation.

The prototype is available on the internet (Cugler, 2016). Its current version enables defining one region at a time to collect VGI. Future versions will comprise several regions at the same time, including the definition of distinct lists of metadata attributes to be acquired in each region.

CONCLUSION AND FUTURE WORK

This chapter presented a context sensitive approach to collect VGI in order to fill out gaps in biological observation metadata, thereby improving the quality and extension of data available. Here, metadata are considered first-class data objects given their role in supporting environmental, ecological and biological studies. The proposed approach considers the dynamics of changing needs in research requirements, by

Figure 6. Screenshot of the Android application prototype. The form is created on the fly, based on the XML configuration file

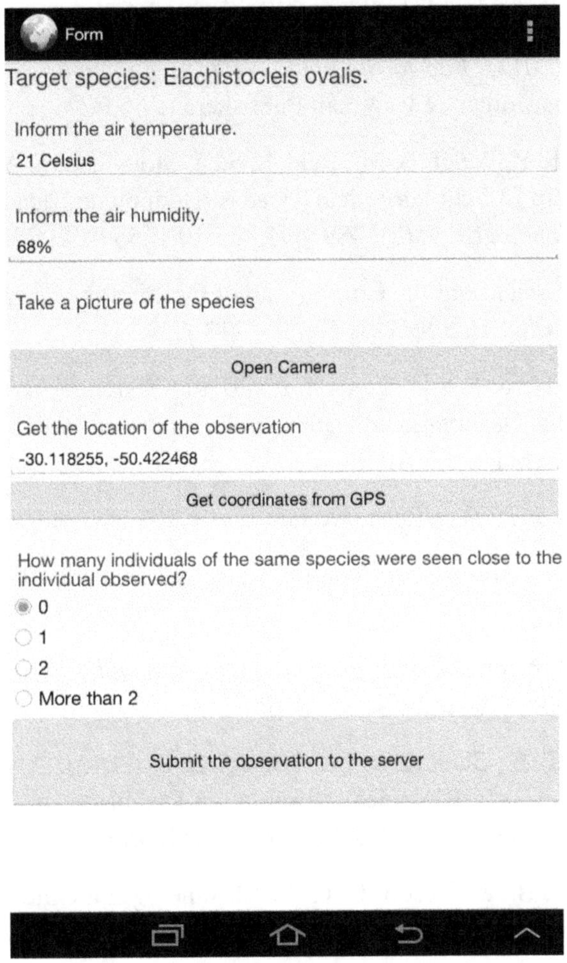

assuming that *gaps* may change as new observations are made or as researchers have new insights about some problem. The interactive location-sensitive approach enables researchers to define on-the-fly which information is required and the regions where information must be collected. Though the emphasis of this paper is on improving and curating metadata, the entire process also allows improving the number and quality of the observation records themselves – if new metadata records appear, they are associated with the observations themselves. However, given the important role of metadata curation in biological observation databases, we concentrated our effort in the metadata information rather than on the observations themselves.

As future work the authors envisage to focus on (a) improving the prototype by enabling the deployment of forms in regions with arbitrary shapes instead of regions defined by circular areas; (b) enabling the customization of deployment of forms in different regions at the same time; (c) automatically persisting the information acquired into the original database and (d) facilitating the creation of the XML configuration files, by creating a visual tool for selection of the database fields.

REFERENCES

Banzi, M. (2009). *Getting Started with Arduino*. O'Reilly Media, Inc.

Barrett, S., & Kridner, J. (2013). *Bad to the bone: Crafting electronic systems with beaglebone and beaglebone black* (Vol. 41). Morgan & Claypool Publishers.

Beasley, D. A. E., Benson, E. P., Welch, S. M., Reid, L. S., & Mousseau, T. A. (2012). The use of Citizen Scientists to Record and Map 13-Year Periodical Cicadas (Hemiptera: Cicadidae: Magicicada) in South Carolina. *The Florida Entomologist*, *95*(2), 489–491. doi:10.1653/024.095.0236

Bonney, R. (2007). Citizen science at the Cornell Lab of Ornithology. *Exemplary Science in Informal Education Settings: Standards-based Success Stories*, 213–229.

Bowers, S., Kudo, J., Cao, H., & Schildhauer, M. P. (2010). ObsDB: A System for Uniformly Storing and Querying Heterogeneous Observational Data. In *IEEE sixth international conference on e-science* (pp. 261–268). doi:10.1109/eScience.2010.24

Cicada Project. (2015). *Cicada Project*. Retrieved January 01, 2015 from http://project.wnyc.org/cicadas/

Cornell. (2015). *The Cornell Lab of Ornithology*. Retrieved January 01, 2015 from http://www.allaboutbirds.org

Cugler, D. C. (2016). *VGI Project. Dynamic form generator to acquire VGI*. Retrieved January 17, 2016 from http://www.ic.unicamp.br/~dcintra/files/VGI.zip

Cugler, D. C., Medeiros, C. B., Shekhar, S., & Toledo, L. F. (2013). A Geographical Approach for Metadata Quality Improvement in Biological Observation Databases. In *Proceedings of the 9th ieee international conference on e-science* (p. 212-220). doi:10.1109/eScience.2013.14

Cugler, D. C., Medeiros, C. B., & Toledo, L. F. (2012, June). An architecture for retrieval of animal sound recordings based on context variables. *Concurrency and Computation*, *25*(16), 2310–2326. doi:10.1002/cpe.2865

FNJV. (2015). *Online animal sound collection - Fonoteca Neotropical Jacques Vielliard*. Retrieved January 01, 2015 from http://proj.lis.ic.unicamp.br/fnjv

Goodchild, M. F. (2007). Citizens as sensors: The world of volunteered geography. *GeoJournal*, *69*(4), 211–221. doi:10.1007/s10708-007-9111-y

Google. (2016). *Google Maps*. Retrieved January 17, 2016 from http://maps.google.com

Haklay, M., & Weber, P. (2008). Openstreetmap: User-generated Street Maps. *Pervasive Computing, IEEE*, *7*(4), 12–18. doi:10.1109/MPRV.2008.80

Hartung, C., Lerer, A., Anokwa, Y., Tseng, C., Brunette, W., & Borriello, G. (2010). Open data kit: Tools to build information services for developing regions. In *Proceedings of the 4th acm/ieee international conference on information and communication technologies and development* (p. 18). doi:10.1145/2369220.2369236

Heipke, C. (2010). Crowdsourcing geospatial data. *ISPRS Journal of Photogrammetry and Remote Sensing*, *65*(6), 550–557. doi:10.1016/j.isprsjprs.2010.06.005

Humantarian Open StreetMap Team. (2016). *Humantarian Open StreetMap*. Retrieved February 13, 2016 from https://hotosm.org

Irwin, A. (1995). *Citizen science: A study of people, expertise, and sustainable development* (Vol. 136). Routledge Londres.

Kim, S., Mankoff, J., & Paulos, E. (2013). Sensr: Evaluating a Flexible Framework for Authoring Mobile Data-Collection Tools for Citizen Science. In *Proceedings of the 2013 conference on computer supported cooperative work* (pp. 1453–1462). doi:10.1145/2441776.2441940

Kim, S., Robson, C., Zimmerman, T., Pierce, J., & Haber, E. M. (2011). Creek Watch: Pairing Usefulness and Usability for Successful Citizen Science. In *Proceedings of the sigchi conference on human factors in computing systems* (pp. 2125–2134). doi:10.1145/1978942.1979251

Kurniawan, A. (2013). *Getting Started with Matlab Simulink and Raspberry Pi*. PE Press.

Kwantlen Polytechnic University. (2015). *Plant Database*. Retrieved January 01, 2015 from http://plantdatabase.kwantlen.ca/

LeBaron, G. S., Cannings, R. J., Niven, D. K., Butcher, G. S., Bancroft, G. T., Sykes, P. W. Jr, & Read, P. (2009). The 109th Christmas Bird Count. *American Birds*, *63*, 2–7.

Lemos, F. (2013). *Infrastructure and algorithms for information quality analysis and process discovery* (Doctoral dissertation). University of Versailles, France.

Longueville, B. D., Smith, R. S., & Luraschi, G. (2009). OMG, from here, I can see the flames!: a use case of mining location based social networks to acquire spatiotemporal data on forest fires. In *Proceedings of the 2009 international workshop on location based social networks* (pp. 73–80). doi:10.1145/1629890.1629907

Maisonneuve, N., Stevens, M., Niessen, M. E., Hanappe, P., & Steels, L. (2009). Citizen noise pollution monitoring. In *10th int. conference on digital government research* (pp. 96–103). Digital Government Society of North America.

Malaverri, J. E. G., Mota, M. S., & Medeiros, C. B. (2013). Estimating the Quality of Data Using Provenance: A Case Study in eScience. *19th Americas conference on information systems (AMCIS)*.

Malaverri, J. G., Santanchè, A., & Medeiros, C. B. (2013). *A Provenance-based Approach to Evaluate Data Quality in eScience. Int. J. Metadata, Semantics and Ontologies.*

McQuilton, P., Gonzalez-Beltran, A., Rocca-Serra, P., Thurston, M., Lister, A., Maguire, E., & Sansone, S. A. (2016). BioSharing: Curated and crowd-sourced metadata standards, databases and data policies in the life sciences. *Database*. PMID:27189610

MeadoWatch. (2015). *Citizen Science at the Mount Rainier National Park to Understand the Biological Impacts of Climate Change*. Retrieved January 01, 2015 from http://meadowatch.weebly

Newman, G., Graham, J., Crall, A., & Laituri, M. (2011). The Art and Science of MultiScale Citizen Science Support. *Ecological Informatics*, *6*(3), 217–227. doi:10.1016/j.ecoinf.2011.03.002

Posthumus, E., & Crimmins, T. (2011). Natures Notebook: A Tool for Education and Research. *Bulletin of the Ecological Society of America*, *92*(2), 185–187. doi:10.1890/0012-9623-92.2.185

Rotman, D., Preece, J., Hammock, J., Procita, K., Hansen, D., Parr, C., & Jacobs, D. (2012). Dynamic changes in motivation in collaborative citizen-science projects. In *Proceedings of the ACM 2012 conference on computer supported cooperative work* (pp. 217–226). doi:10.1145/2145204.2145238

Shekhar, S., Feiner, S. K., & Aref, W. G. (2015). Spatial computing. *Communications of the ACM*, *59*(1), 72–81. doi:10.1145/2756547

Sheppard, S. A. (2012). wq: A modular Framework for Collecting, Storing, and Utilizing Experiential VGI. In *Proceedings of the 1st acm sigspatial international workshop on crowdsourced and volunteered geographic information* (pp. 62–69). doi:10.1145/2442952.2442964

Sousa, R. B., Cugler, D. C., Malaverri, J. E. G., & Medeiros, C. B. (2014). A Provenance-based Approach to Manage Long Term Preservation of Scientific Data. In Data engineering workshops (ICDEW), ICDE (pp. 126–133). doi:10.1109/ICDEW.2014.6818316

University of South Carolina. (2016). *Cicada Project*. Retrieved January 17, 2016 from http://cricket.biol.sc.edu/cicada/contact.html

Whitelaw, G., Vaughan, H., Craig, B., & Atkinson, D. (2003). Establishing the Canadian community monitoring network. *Environmental Monitoring and Assessment*, *88*(1-3), 409–418. doi:10.1023/A:1025545813057 PMID:14570426

Xiao, X., Dorovskoy, P., Biradar, C., & Bridge, E. (2011). A Library of Georeferenced Photos from the Field. *Eos, Transactions, American Geophysical Union*, *92*(49), 453–454. doi:10.1029/2011EO490002

Chapter 15
Geospatial Digital Rights Management:
Challenge to Global Spatial Data Infrastructure

Titus M. Ng'ang'a
University of Nairobi, Kenya

Peter M. Wachira
University of Nairobi, Kenya

Tim J. L. Wango
Jomo Kenyatta University of Agriculture and Technology (JKWAT), Kenya

Joseph M. Ndung'u
University of Nairobi, Kenya

Margaret N. Ndungo
Technical University of Kenya, Kenya

ABSTRACT

This Chapter introduces the need for general Digital Rights Management (DRM) requirements. Further, it intertwines DRM with its spatial counterpart, Geospatial DRM (GeoDRM). However, unlike DRM, GeoDRM is far much complicated due to issues such as the development of Web Mapping technology among other issues. The Chapter discusses the ability of GeoDRM to mitigate transgression of Intellectual Property Rights (IPR). Highlighting economical and environmental wellbeing and other benefits of Spatial Data Infrastructure (SDI) geared towards global sustainable developments, the Chapter focuses on challenges of National Spatial Data Infrastructures (NSDIs) and Regional SDIs and the need to harmonize their standards for the upward mobility of global SDI (GSDI). Emphasizing the undisputed need for Local, Regional and Global Spatial Data Infrastructures (SDIs), in the presence of various Geo-communities and different GeoDRM models, the Chapter concludes that capacity building need to be urgently but carefully harnessed across all levels in order to develop cohesive GeoDRM policies.

DOI: 10.4018/978-1-5225-2446-5.ch015

INTRODUCTION

Intellectual Property

Intellectual Property (IP) is a broad concept that covers several types of legally recognized rights arising from some type of intellectual creativity, or otherwise related to ideas. In today's legal systems, IP typically includes at least copyrights, trademarks, patents, and trade secrets (Kinsella, 2001).

Open Geospatial Consortium (2006) observed that to create a marketplace, individuals who own something of value/resource must have some level of assurance that they will be able to obtain fair value for its use or purchase. However, in the digital world, most digital entities are not sold in the usual sense due to the nature of digital resources and commerce. When a user acquires an application, he actually acquires the right to use a copy of the application. Possession does not equate with ownership. A system of software and resource licensing has grown up that ensures that the user may legitimately act upon a resource if he has a corresponding license for that act, owner should maintain the resource such as fixing errors while still assuring a guaranteed level of functionality. Optionally, the user may be asked to pay the owner of the resource based upon agreed criteria, whether that is a one-time fee, a per-machine fee, a usage fee or some other mechanism stated in the legal contract or license between user and owner. The user agrees to protect the owner's rights based on the agreement. The owner agrees to maintain the resource and allow a reasonable access to the users for any fixes that may be required.

Geocommunities and Data Sharing

Over the recent years, geocommunities have emerged across the globe whereby members share data trough the creation of geoportals where members are expected to deposit their data. A lot of data have been collected by different organizations and individuals across the globe.

Despite the creation of geoportals, most of the geodata contains little emphasis on conditions under which data ought to be used. A visit to acquire data from a geodatabase manager can be shocking to realize that the only reason one may not acquire data is that the data-collector/owner never specified how it ought to be used yet a lot of funds may have been accrued in its collection. It then becomes inevitable to recollect the same data at additional cost. Therefore, the importance of geocommunities can not be underestimated especially for the noble aspect of sharing due to the often prohibitive high cost associated with spatial data collection.

Another disadvantage with most datasets available for sharing within geocommunities is that the data have low spatial resolution thus rendering it lesser useful in certain applications. However, such data is useful for learning purposes.

There exist some national and international organizations that avail spatial data without charges as long as the source of the data is properly referenced. The organizations require one to register as a user. User identification often includes names, organization and discipline of affiliation and the purpose of the data. The United States Geological Survey (USGS) archives remote sensing data that can be downloaded free of charge. The International Livestock Research Centre (ILRI) has a repository of geodata that has been of help to many institutions and individuals including students' research works. Environmental Systems Research Institute (ESRI) is an International organization that provides a platform whereby users can set up a public-facing website and share open data.

BACKGROUND

Digital Rights Management (DRM)

Digital Rights Management (DRM) stands for a set of technologies by publishers of digital content such as music, video and electronic texts to control the ways in which consumers are able to use information. DRM usually works by encrypting a digital object such as audio, video, or text file and providing some method for the user to decrypt it and use it only in ways specified by the publisher. The sender encrypts the digital object via a "key". The recipient or user decrypts it for use with a copy of the same "key". The information is protected from unauthorized users or uses who lack the "key". Vendors of online research databases often implement controls that prevent copying and pasting of text from electronic articles (Puckett, 2010). DRM therefore necessitates only the destined user to have access and block the unauthorized.

According to Flip Experts (2013), DRM refers to access control technologies that are used by hardware manufacturers, publishers, copyright holders and individuals to limit the use of digital content and devices. Therefore, DRM technologies control use of digital media by preventing copying or conversion to other formats by end users. Flip Experts states that the benefits of DRM includes protection of intellectual property; determination of who is accessing information; integration with other applications and formats; assuring direct access to recipients for latest version of a publication; among other benefits.

James (2003) explains the need for DRM, "Counterfeiting is a serious problem and the recording industry is becoming increasingly aggressive in pursuing counterfeiters through the courts. Legal remedies used against large-scale unlicensed CD pressing operations are usually too expensive and impractical to be used against millions of individuals, and therefore the music business is trying to control end-user redistribution by technological means instead. To this end, a sub-industry has emerged which promises control over the end users of music and other digital content by implementing DRM".

Examples of DRM users

British Library

DRM has been used by organizations such as the British Library in its secure electronic delivery service to permit worldwide access to substantial numbers of rare documents which, for legal reasons, were previously only available to authorized individuals actually visiting the Library's Document Centreat Boston in England (Azad *et al.*, 2010).

Microsoft

Microsoft has its own multimedia delivery system based around the Windows Media Player, and this has a system of inbuilt DRM(James, 2003). Inside Secure Website acknowledges thatMicrosoft DRM technologies are widely used today in Internet-based video and music services accessed with Windows Media Player (http://www.insidesecure.com).

Challenges of DRM

Despite the benefits that DRM offers, James (2003) notes that it takes only one frustrated teenager, for example, with a talent for cryptography to break the most expensive and complex DRM system. The reality is that there are thousands of people around the world who have that ability, and they have worked together to defeat nearly all of the significant DRM systems established so far.

Citing DRM challenges from Microsoft, James (2003) explains that nobody outside Microsoft was supposed to be able to understand the 'unique identifiers and one-time licences' that can't be shared by end users and which were done through encryption. However, in October 2001, an anonymous person(s) going by the name Beale Screamer posted a series of messages to a newsgroup by the name sci.crypt, revealing the secrets of version2 of the Microsoft DRM system and the source code for a program which could remove the DRM protection from Windows Media Audio files.

To offer a solution to challenges like those posed by the anonymous Beale Screamer,James (2003) recommends hardware based DRM as well. He explains the to make hardware-based DRM work, the CPU and motherboard manufacturer needs the cooperation of the company making the operating system. Microsoft's solution is to put DRM into the lowest level of Windows, where the user supposedly can't do much about it.

From DRM to GeoDRM

Digital geospatial datasets moving across computer networks can be easily copied, transformed, or incorporated into new value-added products and services. Geospatial data producers and owners are faced with the challenge of controlling the dissemination of their digital geospatial assets down-stream in the geospatial value chain. As technology pushes various industries toward the digital frontier, many type of content are becoming available solely in digital format, and geospatial data is no exception. Copyright-protected geospatial content used to be sold on paper sheets but is now available in digital format. As a result, such content can now be used by a variety of users and devices such as car navigation systems, corporate applications and business-related geospatial functions (Bishr *et al.*, 2006).

As geodata becomes more widely available in digital form over the internet, it becomes easier to distribute, share, copy and alter. While this is generally a good thing, many organizations involved in the production and trading of geodata need to be able to protect their Intellectual Property (IP) assets as they move through the digital distribution value chain. In the world of e-commerce, access to geodata is treated as a commodity to be priced, ordered, traded and licensed. Organizations want to specify, manage, control and track geodata distribution within safe, open and trusted environments. Direct monetary reward is often not the motivation or is only a minor motivation behind the desire for more rigorous control of IP assets (Vowles & McKee, 2005).

Despite the immeasurable benefits that geodata sharing has accrued, Intellectual Property Rights (IPR) are infringed when modified data loses its initial traits and degenerates yet the source of the data is referenced to the original data producer. This leads us to the moral question of who should take responsibility in terms of inaccuracies and consequences in case valued added information lacks the characteristics that the data producer intended. Procedures on geoinformation standards are rarely properly outlined especially in developing countries due to lack of legal and technical frameworks to guide the development of geoinformation policies.

Though geoinformation may be presented in picture formats such as JPEG rather than raw vectors and rasters, reusability of these formats and the inherent features their display can always be re-extracted through digitizing with possible compromise on spatial attributes. The conversion and reconversion of datasets from one format to another in innumerable formats may compromise on standards and the initial geodata producer can not therefore claim any attachment as such to information whose credibility has been compromised. To do so, would lead to owning any moral loss, at least, that may arise after value addition by other users. Fortunately with the employment of GeoDRM, value added information is licensed to the secondary producer and more licenses goes up the ladder as value-addition progresses thereby protecting every producer/enhancer from the other. This is not the case with DRM. Therefore, despite their common characteristics, we can comprehensively say that DRM and GeoDRM are not identical twins.

GeoDRM Versus DRM

There exist comparisons and contrasts between GeoDRM that cannot be overlooked. Interestingly, GeoDRM seems to harness the sharing of information rather that limits it unlike DRM. Cary (2006) observes that GeoDRM offers a way to digitally implement licensing agreements with third parties. Without such a capability, some organisations may not obtain licences because the risk of violating license terms is perceived as greater than the benefits of obtaining the licenced content. GeoDRM provides a clear audit trail for how information is derived thus providing a safe guard against unauthorized modification (Cary, 2006).

SPATIAL DATA INFRASTRUCTURE (SDI)

Spatial Data Infrastructure (SDI) is an initiative intended to create an environment in which all stakeholders can co-operate with each other and interact with technology, to better achieve their objectives at different political/administrative levels. Under current circumstances, governments of different nations and organizations within each nation and region should reach agreement on what fundamental datasets are required to meet their common interests, to what standards they should be collected and maintained, and what the priorities are for their collection. This is because it is necessary to integrate a number of datasets that may have been produced by different agencies within different nations for specific purposes, to their own specifications and priorities, and with little regard to the needs of other users. Many countries believe they can benefit both economically and environmentally from better management of their spatial information by taking a perspective that starts at a local level and proceeds through state, national and regional levels to a global level. This has resulted in the development of the Spatial Data Infrastructure (SDI) concept (Rajabifard & Williamson, 2001).

There exists various levels in which geodata of mutual importance can be shared. This Chapter mainly focuses on national, regional and global SDIs. Within National SDI (NSDI), different ministries, institutions and organizations within a country share geoinformation.

There exist different descriptions of NSDI. For example, the Australian and New Zealand Land Information Council (ANZLIC, 1998) defines NSDI as comprising four core components; an institutional framework, technical standards, fundamental datasets, and clearinghouse networks. The Federal

Geographic Data Committee (FGDC, 1997) defines the United States' NSDI as an umbrella of policies, standards, and procedures under which organizations and technologies interact to foster more efficient use, management, and production of geospatial data. However, these definitions do not alter the major objectives of what an NSDI stands for but moreover emphasizes the strength and benefits of SDI. This Chapter focuses on challenges of NSDIs and Regional SDIs and the need to harmonize their standards for the upward mobility of global SDI (GSDI).

Thematic SDIs

Few thematic SDIs exist across the globe at national and regional level. Most SDIs are multidisciplinary. There is need to develop various thematic SDIs at all levels as we gear towards the realization of a coherent GSDI. One of the most popular thematic SDI is the Marine SDI (MSDI).

MSDI is mostly found in countries that border oceans. Such countries share a lot in common and for that purpose, marine science is a highly researched field geared towards the internationally recognized marine industry. Marine flora and fauna, hydrography, ocean seasonality and fishing are vital aspects in the marine industry. Beyond marine science, even countries that have short coastlines and major rivers and canals need to develop and implement MSDI.

MSDI encompasses marine geographic and business information covering sea areas, inland navigable and non-navigable waters. This typically includes seabed topography, geology, marine infrastructure, administrative and legal boundaries, areas of conservation, marine habitats and oceanography (Hilario&Alexander, 2009). MCSDI provides information to decision makers and information products for environmental protection activities, research and development, private industry, military activities and public information (Pacheco, 2009).

Since coastlines are also used for other purposes other than marine sciences, such as port business and tourism, MSDI is in some countries is referred to as Marine and Ocean SDI (M/C SDI). Due to the growth of sea piracy, it has become increasingly important to further develop from National M/C SDIs to one integrated Global M/C SDI.

The main challenges of developing thematic SDIs such as a national M/C SDI is mainly due to the fact that it must lie within the frameworks specified by the guideline provided by the National SDI (NSDI). Therefore, from that dimension, a thematic SDI is a component of NSDI itself. Kenny, (2010) describes that MSDI inter reacts as a component framework within a National Spatial Data Infrastructure.

Human Expertise Hindrances to SDIs

The personnel resources for SDI in many countries are very limited since most of the GIS implementations being built up are understaffed. A pool of qualified staff has to be created if the projects are to become sustainable. What makes it difficult for developing countries is not only the number of specialists required, but also the working conditions offered. "Brain-drain" is a serious problem; the fact that skilled personnel are leaving their jobs too often, too soon. Human capacity development and long term career planning should be of prime concern to senior management. It includes the training, theoretical issues and practical hands-on capabilities to implement projects and programs, as well as the working conditions. Working conditions need to be considered not only with respect to salary, but even more importantly with respect to the work climate, motivation and professional perspectives (SDI Cookbook, 2004).

Financial Challenge to SDIs

Given the nature of GSDI, it would be vital for developed countries to allocate enough funds for capacity building and related endeavors towards the development and full realization and implementation of NSDIs across the board. To that effect, this would require continued support of strong GeoInformatics policies as a prerequisite towards a geospatially informed and enabled global society.

IPR and SDI

Chen (2007) explains the challenges facing data sharing emanating from increased concerns about national security and IPR by giving an example of the Unites States not releasing 30m SRTM for the whole world despite the geodata benefit for enumerable applications and the withdrawal of data due to increased IP restrictions. There exists loss of confidentiality through disclosing areas affected by danger during disasters where information on particular areas involved are shared among the general populace prior to informing the next of kin as protocols dictate.

Need and Challenges for GSDI

GSDI is the policies, organisational remits, data, technologies, standards, delivery mechanisms, financial and human resources necessary to ensure that those working at the global and regional scale are not impeded in meeting their objectives. GSDI is envisaged to support trans-national or global access to geographic information and it is seen by many as central to the response to the challenge of global sustainable development. It is an effective promotion of national and regional Spatial Data Infrastructures (SDI Cookbook, 2004). Actually, GSDI so envisaged would not be fully functional without the full development of all other SDIs at lower levels. In the 21st century, the necessity and benefits of a functional GSDI cannot be underestimated.

Due to climate change, scientists across the globe have continued with research studies on modeling, predicting and simulating future scenarios. Tremendous environmental changes are occurring faster than ever around the globe and often causing disasters. Disasters caused by these effects include El-Nino, hurricanes, flooding, drought, typhoons, and the list is endless. Thus, spatial data on environmental parameters are in great need today than ever. This has an impact on data producers about what to avail for SDIs to avoid losing organizational credibility after consecutive data use and reuse.

In the sense of geodata use and reuse, GeoDRM mandate is therefore not only about protecting use of geoinformation but credible use. Such credibility is perhaps a clear indication of why GSDI might remain only a wish in the near future. Perhaps due to the sensitivity of geodata and the fear of some of the mishaps of DRM, and generally the fear of the unknown, many people don't seem at ease about sharing their data. If this becomes a trend, the catastrophic effect would be that GSDI inevitably remains a concept primarily in the world of academia.

Many hurdles has also been on technical issues such as who actually owns data collected by government funded institutions or departments. Further, it has never been clear whether such data, having been funded through public kitty should be provided to the same public at an extra fee. Many developing countries are at different stages of SDI development. It would not be possible to develop a fully workable GSDI without functional local and regional SDIs.

From climate change, to environmental and economic sustainability, to terrorism, there are countless benefits that explain the needs for a functional GSDI for one global village. These needs have seen the coming up of interdisciplinary researches across the globe which translates into data sharing at the bottom line.

As droughts become intermittent, spatial information on agro-environmental factors needs to be shared. For example, the introduction of drought resistant and high yielding crops has seen geospatial analysts, agronomists and biostatisticians working together to determine the most favourable crop varieties. Such efforts and collaboration among different stakeholders regionally and globally are indications of globalizing the world into a digital village and geodata sharing is inevitable at a global scale.

SDI and Agro-Ecological Zoning

Agro-ecological zoning refers to the division of an area of land into land resource mapping units, having unique combination of landform, soil and climatic characteristics and/or land cover having a specific range of potentials and constraints for land use (FAO, 1996). Agro-ecological zones (AEZs) are geographical areas sharing similar climate characteristics with respect to their potential to support agricultural production (http://harvestchoice.org/topics/agroecological-domains).

Standardized delineation of agro-ecological zones across the whole globe will be facilitated by the development of SDIs at all levels since agro-ecological zoning parameters are independent of administrative boundaries. The necessity to harmonize Agro-ecological zoning models will have prolific benefits in that generalized AEZ parameters has resulted into large ecological zones that do not adequately determine the specific needs for the introduction of new crop varieties, among other requirements. Currently, there exist different criteria or models for defining agro-ecological zones. Further, the success with which these models meet their goals lies with the knowledge of the expert. FAO (1996) has provided guidelines for defining agro-ecological zones. The application of AEZ is limited by the lack of geospatial data (Quirez, 2000) whose solution is SDI.

GeoDRM and Spatial Data Infrastructure (SDI)

To illustrate how the GeoDRM spectrum can be implemented, geospatial digital watermarking can be used to track misuse of digital assets by identifying the sources of infringement. Thus, a combination of digital licensing and watermarking can assist geospatial data providers in an SDI with monitoring the usage of their assets for the correct purpose. Once a watermarked dataset is found in usage where it shouldn't have been used the providers can trace back the source of infringement and minimize the damage to their intellectual property (Bishr*et al.,* 2007).

Just as copyrights are vital, so are geodigital rights. This is not only considering the monetary value of spatial data but to protect the integrity of individuals or institution that acquired the primary data in case of its misuse. GeoDRMtherefore is a moral step towards preventing legal tussles as many countries move towards SDI. Beyond the issue of licensing, which has already been taken care by GeoDRM, there is need for geodata producers attaching a disclaimer as an extra measure of protecting themselves from unforeseeable problems.

GeoDRM Architectures

Various architectures of GeoDRM have been proposed and implemented. Generally from an architectural view, three major components can be identified; the content server, the license server, and the client (Duhl & Kevorkian S., 2001).

Rosenblatt *et al.* (2002) introduced a general high-level DRM reference architecture that is widely adopted in the DRM domain. It includes the most relevant components in DRM systems. That is, a publisher service that provides the original asset and facilities for delivery to the client. The licensing server is the administrative hub of the digital licensing server. It brokers negotiations between vendors and consumers and grants licenses. The licensing server generates the digital license, and the publisher attaches it to the asset being sent to the client. The client has a controller service which receives the user's request to exercise certain rights on the content. The controller acquires the user's identity information and obtains a license from the license server. It then retrieves the encryption keys from the license, decrypts the content, releases it to the rendering application and executes the license terms.

Trust

The challenges towards integrating various SDIs have also been hampered by security, privacy and integrity issues of geoinformation. There exist different geodata standards across all levels of SDIs. Continued efforts to reconcile these standards have always remained a short term measure. Due to lack or disharmony of technical, legal and institutional policies on the use of geoinformation, various organizations have teamed up to share data of similar needs based on mutual TRUST. This is due to the fact that as professional workshops, seminars and conferences continue to provide and suggest solutions towards an integrated geospatial data sharing within and across boundaries, the cost of acquiring own data is often prohibitive, time consuming and duplicative. As well, there exists no sound justification for an organization to either archive or discard geodata that surpasses its usefulness yet the same data can be valuable to someone else.

Open Geospatial Consortium (OGC) GeoDRM Model

The GeoDRM Reference Model offered by the Open GeoSpatial Consortium aims, among other objectives, to create a conceptual model for digital rights management of geospatial resources, providing a framework and reference for more detailed specification, provide a metadata model for the expression of rights that associates users to the acts that they can perform against a particular geospatial resource, and associated information, used in the enforcement and granting of those rights, assess the requirements that are placed on rights management systems for the enforcement of those rights (Korn*et al.*, 2007).

The OGC Geo Digital Rights Management Working Group denotes that the Reference Model defines the framework for web service mechanisms and rights languages to articulate, manage and protect the rights of all participants in the geographic information marketplace, including the owners of IP and the users who wish to use it (OGC, 2008).

Other GeoDRM Models

Recent GeoDRM models include the ISO 19153:2014 Reference Model for DRM. Its intention is to make sure that a larger market has access to geospatial resources through a mechanism that it understands and that is similar to and consistent with the ones already in use (http://www.iso.org/iso).

Inter-Operability

The competence and reliability of any software to thrive in the market, including GIS software, is its ability to accommodate different hardware and software standards and specifications. For example, ESRI's ArcGIS supports ISO Technical Committee 211 (ISO/TC 211) and Open OGC. The goal of GIS interoperability is to support users in finding, utilizing, and deploying data from various geographic information systems. In addition to interoperability, key success factors that enable GIS users to successfully work in heterogeneous environments include having a GIS infrastructure that fulfills security, reliability, scalability, extensibility and ease in usability among other requirements (Esri White Paper, 2013). Interoperability refers to mutual operability among software, hardware, data and different standards.

Integrating SDIs Inter-Operabilities

The successful development of a fully functional GSDI can never be simple given the predisposed challenges. In the process of developing a fully functional GSDI, various challenges, technical and non-technical arises depending on different regions. Bejar*et. al.* (2011) reports that in the process of establishing the horizontal interoperability among Spanish, French, Portuguese and Andorran geoportals, one of the technical hurdles to overcome was integrating the WMS-C services offered by the French geoportal in the Spanish SDI geoportal viewer, because these services are published following the GeoDRM reference model. The Spanish SDI geoportal viewer is a web application built on top of the OpenLayers open source viewer3, which supports both tiled and non-tiled map services. Tiled map services can be implemented following the WMS-C de facto standard, which provides a service interface that is a profile of the OGC Web Map Service standard. The report describes how the Spanish SDI geoportal viewer has implemented support for the GeoDRM reference model in order to access to the geoportal WMS-C services.

Rights Expression Language and GeoDRM

A Rights Expression Language (REL) is a language that provides syntax and vocabulary for the expression of agreed-upon rights and the conditions to which those rights are subject. To have a functional DRM system, it is necessary to have a standard REL throughout the entire framework. The choice of REL is integral to the design of a DRM system. The DRM framework must be able to distinguish among the types of data for which control is exercised, and the various types of control to be exercised (Wright, 2005).

Since it is too costly to develop a completely new REL, a better choice is to find an existing REL that is sufficiently flexible and allows the expression of interests unique to a specific geospatial community (Wright, 2005) or SDI. However, this is not an easy task. Having suffered several setbacks on several issues, Africa in general have joined hands and formed regional SDIs geared towards embracing GSDI.

Just as it has been in other cases such as France, Spain, Portugal and Andorra, Africa will have to solve interoperability issues amongst her nations' NSDIs. The challenges to harmonize the interoperability given the semantic challenges between and across the nations can not be underestimated. The worst fears that no one would like to be confirmed are the existence of a GSDI that can not communicate coherently within itself comparable to the biblical *house of Babylon.*

Africa is a continent with diversity of languages, some unrelated. For example, the Eastern Africa Kiswahili is different from the South African Africana and the horn of Africa's Amharic. Presumably, the more the lingual diversity the better since there exist many alternative words with similar meanings. In joining the GSDI as a block, the African regional SDI would therefore present an SDI with strong semantics.

Kuhn (2005) stresses that semantic interoperability is the only useful form of interoperability. In the real world, it is hard to imagine two agents interoperating without a shared understanding of the messages they exchange. Interoperability requires identifying the vocabulary or syntax of service interfaces and defining interoperability mathematically. Sematic interoperability is the technical analogy to human communication and cooperation. It serves as a technical goal justifying the formalization of semantics in GIS.

Drawing the arguments of Wright (2005) and Khun (2005), it is proper to conclude that Ethnophysiography is broadly an aspect of interoperability. Khun (2005) argues that languages in an information system need semantics, including programming, modeling and interface specification languages. By extension, languages used in an information system are the result of social agreements in information communities.

In Africa, United Nations Economic Commission for Africa (ECA) continues to support the member states in formulating geospatial policies and strategies and the development and implementation of NSDI. ECA is also leading the Global Geospatial Information Management (GGIM) initiative in Africa. GGIM reflects African issues and shapes its direction and dimension in addressing geospatial governance in Africa (http://www.uneca.org/acs-0/pages/geo-information)

GeoDRMChallenges in Developing Nations

The challenges facing GeoDRM are diverse and range from one region to another. As discussed by Kuhn (2005) and Wright (2005), the challenges facing geoDRM development are not just technical issues. The challenges range from social, cultural, literature, lingual, and the list is countless. SDI stakeholders especially in developing nations are yet to implement cohesive policies governing Freedom of Information (FOI). In many nations, FOI, which lays out rules and procedures for sharing spatial data is still underway. Once these policies have been successfully developed, the next step would be harmonization and standardization of the various geoDRMs. According to Cary (2006), the biggest technical challenge associated with geoDRM involves developing a shared model of what the concepts means to people and to business processes.

Government GeoDRMPolicies

In developing nations, the development of GeoDRM policy has been a great challenge especially due to the fact that the potential of geospatial technology has not been fully embraced both in the academia and in the industry. Geospatial technology is more or less decade's educational career in developing nations. However, upcoming issues such as global insecurity, climate change and disaster management have generated dire necessities in the transformation of diversified intellectual disciplines to combine efforts towards the benefits of Geospatial and Space Technology. The diversity of these disciplines as well as the retention of their various standards has made it inevitable towards the development of practical SDIs within and across geographical boundaries worldwide. Therefore, the idea and steps towards the development of various SDI ought not to be a shelved issue.

Standards of Geodata

The development of standards is not an easy task. It requires collaboration on global basis, consensus by many organizations, the approach of give and take, and is a repeatable process (Percivall, 2013). Content management and distribution platforms must be able to support multiple DRM technologies, insulating content owners from having to develop customized software applications to accommodate the underlying DRM technology or require specialized DRM expertise and construction of new distribution infrastructures (Vowles & McKee, 2005). In developing countries, there have been hindrances in developing standards for national geodata. This has hindered the development of regional SDI and eventually a global SDI.

Open Source

Rather than focusing on the rights of creators, open source software focuses on the rights of users: the right to have access to the software, the right to study and modify it, and the right to share it and distribute it further without any authorization. All the various open-source software licenses require that the source code be "open." The open-source model aims to ensure the widest possible distribution of the software by prohibiting restrictions on its distribution. Digital information products can be created, modified, and distributed at virtually no cost without diminishing their availability to others. Open-source software is geared toward follow-on innovation. Encouraging the widest possible distribution aims to provide the largest number of people with the opportunity to study, test, improve and extend the original creation, thereby generating the greatest number of improvements at the lowest possible cost. The open-source movement is important in its own right for its impact on the software market and as a laboratory for production through mass collaboration in a non market regime (Maxwell, 2006).Some examples of open source software are System for Automated Geoscientific Analyses (SAGA), Geo-graphic Resources Analysis Support System (GRASS), Integrated Land and Water Information System (ILWIS).

Openness of Open Standards

The European Committee for Interoperable Systems (ECIS) argues that a 'de facto' open standard is not necessarily an open standard. This creates the debate of how an open standard is actually open and if not fully open, what aspects of a particular open standard are closed? In simple terms, what percentage is an open standard open?

In trying to answer the above query, ECIS lays down the essential characteristics of an open standard. ECIS argues that the development and management process of the standard must be collaborative and democratic, and that the evolution and management process need to be transparent and open to all interested parties; Its approval must be by consensus among participants; implementations of the standard must interoperate; must be openly published (including specifications and documentation); must be available royalty-free or at minimal cost, with the only licensing restrictions being reciprocity and defensive suspension; and that the open standard must be platform-independent, vendor-neutral, and usable by an unrestricted number of competing implementations (http://www.ecis.eu/open-standards/)

OGC Standards

OGC is a non-profit, international voluntary consensus standards organization that is leading the development of standards for geospatial and location based services. It facilitates a consensus process in which government, private industry, NGOs, and academia collaborate to create open and extensible standards, encodings and architectural best practices for geospatial and other mainstream information technologies (Reed, 2012). Founded in 1994, it has a membership of over 480 organizations derived across industry, government and academia, and over 35 adopted OGC standards and several hundred software products implementing OGC standards (Trakas, 2014). OGC aims at making geographic information an integral part of the world's information infrastructure. Most of the OGC standards developed in recent years are standards for the Web services environment, and these standards are collectively referred to as OGC Web Services. (http://live.osgeo.org/en/standards/standards.html).

Notably, despite the universality in terms of OGC membership, Africa still lags far much behind followed closely by South America and Middle East in that order. Europe, North America and Asia pacific are much ahead. Given the benefits to reap from the continued development and implementation of open standards, stakeholders in every region needs to rise up given that the words that define an "open" open standard are no mean that collaboration, consensus, democratic, transparency, among other terms full of both virtue and vigor. This would be a further step towards the integration of a cohesive global village in its entire entirety. Notably, some universities offer courses on globalization in their academic programmes going by such names as degree in global studies, global management, global health, global politics, global economics, global business management, global governance, etc.

ISO/TC 211

The goal of ISO/TC 211 is to develop a family of international standards that will, support the understanding and usage of geographic information, increase the availability, access, integration, and sharing of geographic information, enable inter-operability of geospatially enabled computer systems, contribute to a unified approach to addressing global ecological and humanitarian problems, ease the establishment of geospatial infrastructures on local, regional and global level, and contribute to sustainable development(Granholm, 2012).

Benefits of Global Geodata Sharing

Chen (2007) outlines the benefits of sharing data globally. Sharing and utilizing geodata would reduce loss of life and property from natural and human-induced disasters; the understanding of environmental factors affecting human health and well-being; improvement of energy resources management; understanding, assessing, predicting, mitigating, and adapting to climate variability and change; Improvement of water resource management ; Improvement of weather information, forecasting and warning; Improvement and protection of ecosystems; Supporting sustainable agriculture and combating desertification; Understanding, monitoring and conserving biodiversity.

SOLUTIONS AND RECOMMENDATIONS

Irrespective of the challenges facing both DRM and GeoDRM, there is every need to gear forward to a hasty but workable GSDI. As proposed by James (2003), an integrated hardware/software (Geo)DRM would limit transgressions to IPR.

Since GeoDRM is safer than DRM in terms of licensing as it offers licensing agreements to third parties, it creates a buffer against irresponsibility to original owners of data and therefore it should not be a barrier towards the full realization of GSDI.

One of the worrying trends towards this great GSDI reality is the slow pace at which NSDIs and Regional SDIs are moving at as demonstrated by the Africa's tiny membership to the OGC. Mutual TRUST between organizations dealing with similar data needs to be promoted across the board.

Other than MSCDI and perhaps a few more, thematic SDIs needs to be developed and implemented at all levels of SDI in order to effectively tackle the growing international requirements such as climate change, global insecurity, global sustainable developments and international research collaborations. This has been hampered by poor/slow policy frameworks within some countries and regions.

FUTURE RESEARCH DIRECTIONS

Emerging challenges of DRM and GeoDRM needs to be studied more critically so that the challenges facing DRM do not infiltrate into GeoDRM systems. Besides addressing technical challenges facing SDIs, non-technical issues also need to be well studies.

CONCLUSION

Fear of the unknown, among other unconfirmed real or imagined fears, have always decelerated technology of all forms. Capacity building onGeoDRM policies needs to be adequately and steadfastly addressed in order to maintain genuine custody, use and reuse, of spatial data. Though there are innumerable and varying challenges and hindrances facing the full realization of GSDI, GeoDRM policies remain an issue that cannot be under-estimated. This is due to the fact that every GeoDRM models needs to be updated and harmonized across the globe in order to meet the ever evolving methods of information misuse. Due to the nature of GSDI, the challenges hindering the development and implementation of SDIs needs to be addressed both locally and globally. However, harmonizing GeoDRM, among other issues, between nations and regions is paramount towards this goal.

REFERENCES

ANZLIC (1998). *Discussion paper: Spatial Data Infrastructure for Australia and New Zealand*. ANZLIC.

Azad, M., Hasnat, A., Ahmed, S., & Alam, A. (2010). Article. International Journal of Computer Science and Network Security, 10(11).

Bejar, R., Latre, M. A., Valino, J., Zarazaga-Soria, F. J., & Muro-Medrano, P. R. (2011). *Implementing GeoDRM support in an OpenLayers based viewer: access to the French geoportal WMS-C services from the Spanish SDI*. Universidad de Zaragoza.

Bishr, M., Wytzisk, A., & Morales, J. (2007). *GeoDRM: Towards Digital Management of Intellectual Property Rights for Spatial Data Infrastructures, Institute for Geoinformatics (IFGI)*. University of Munster.

Cary, T. (2006, March). Geospatial Digital Rights management. *Geospatial Solutions*, *16*(3), 18.

Chen. (2007). *The GEO Data Sharing Challenge: Putting Principles into Practice*. Alliance for Earth Observations Briefing.

Cookbook, S. D. I. (2004). *Outreach and capacity building*. Retrieved from: http://gsdiassociation.org/images/publications/cookbooks/SDI_Cookbook_GSDI_2004_ver2.pdf

Duhl, J., & Kevorkian, S. (2001). *Understanding DRM Systems*. IDC White Paper.

ECIS. (n.d.). *Interoperability & Open Standards*. European Committee for Interoperable Systems. Retrieved from: http://www.ecis.eu/open-standards/

FAO. (1996). Guidelines: Agro-ecological zoning. Food and Agricultural Organisation, SoilsBulletin. Rome, Italy: FAO.

FGDC. (1997). Framework, introduction and guide, Book of Federal geographic data committee. Washington, DC: FGDC.

Flip Experts. (2013). *Digital rights management*. Retrieved from: http://searchsecurity.techtarget.com/magazineContent/Digital-Rights-Management-Microsoft-Windows-Rights-Management-Services

Granholm, L. (2012). *3D tie to mallinnuksenkansainvälinen standard ointi*. Academic Press.

Harvest Choice. (n.d.). *Agroecological Domains*. Retrieved from: http://harvestchoice.org/topics/agroecological-domains

Inside Secure. (n.d.). Retrieved from: http://www.insidesecure.com/Products/DRM-Content-Protection/Microsoft-DRM-based-Solutions

ISO. (2014). *ISO 19153:2014: Geospatial Digital Rights Management Reference Model*. Retrieved from: http://www.iso.org/iso/catalogue_detail.htm?csnumber=32571

James, D. (2003). *Digital Rights Management & Music, A barrier to creativity?* Academic Press.

Kenny, M. (2010). Marine Spatial data Infrastructure.*2nd IHO-HSSC meeting, HSSC2-INF3*.

Kinsella, N. S. (2001). Against Intellectual property. *The Journal of Libertarian Studies*, *15*(2).

Korn, N., Oppenheim, C., & Duncan, C. (2007). *IPR and Licensing issues in Derived Data*. Retrieved from: http://webarchive.nationalarchives.gov.uk/20140702233839/http:/www.jisc.ac.uk/media/documents/projects/iprinderiveddatareport.pdf

Kuhn, W. (2005). Geospatial semantics: Why, of what, and how? *Journal on Data Semantics, Special Issue on Semantic-based Geographical Information Systems*, *3534*, 1–24.

Lamotte, H. C., & Alexander, L. (2009). *Marine Spatial data infrastructure. Challenges and Opportunities: The Cuban experiences*. InternationlHydrographic Organization (IHO) Spatial Data infrastructure. *Seminar*.

Maxwell, E. (2006). *Open Standards, Open Source and Open Innovation, innovations*. MIT Press. Retrieved from: http://www.mitpressjournals.org/doi/pdf/10.1162/itgg.2006.1.3.119

OGC. (2008). *OGC web services initiative*. OGC.

OWS-6 Architecture Open Geospatial Consortium. (2006). *The OpenGIS Abstract Specification. Topic 18: Geospatial Digital Rights Management Reference Model (GeoDRM RM)*. Author.

Pacheco, M. B. (2009). *Public, private, and military applications flow from SDI*. Geospatial Foundation, InstitutoHidrografico, Portuguese Navy.

Percivall, G. (2013). *OpenSearch at OGC TC with CEOS23*. WGISS/OGC Meeting, OGC Technical Committee, Frascati, Italy.

Puckett, J. (2010). *Digital Rights Management as Information Access Barrier*. Georgia State University.

Quirez, R., Zorogastua, P., Baigorria, G., Barreda, C., Valdivia, R., Cruz, M., & Reinoso, J. (2000). *A dynamic definition of Agro-Ecological zones using modern information management tools*. Academic Press.

Rajabifard, A., & Williamson, I. P. (2001). Spatial Data Infrastructures: Concept, SDI Hierarchy and Future Directions. *Proceedings of GEOMATICS'80 Conference*, 10.

Reed, C. (2012). *Introduction to the OGC and Overview of Key Standards Focus areas*. ADSIC Presentation.

Rodríguez, F., Mas, S., Richard, D., Nataf, J. M., Furtado, D., Julião, R. P., … Alba, S. (2010). Horizontal Interoperability among National Geoportals of Spain, France, Portugal and Andorra. *INSPIRE Conference*.

Rosenblatt, B., Tripp, B., & Mooney, S. (2001). *Digital Rights Management: Business and Technology*. John Wiley and Sons.

Trakas, A. (2014). *OGC and Open Standards in action –Examples*. Workshop: Make the most of Geographic Information, *INSPIRE Conference*, Aalborg, Denmark.

UNECA. (n.d.). Retrieved from: http://www.uneca.org/acs-0/pages/geo-information

Vowles, G., & McKee, L. (2005). *Geospatial digital Rights (GeoDRM)*. Geoinformatics.

ESRI White Paper. (2013). *ESRI-Supported Open geospatial Consortium and ISO/TC211 standards*. ESRI.

Wright, D. J. (2005). *Geospatial Digital Rights management*. US Geological Survey. Retrieved from: https://pubs.usgs.gov/of/2005/1086/2005-1086.pdf

Compilation of References

Abdullah, S. A., & Nakagoshi, N. (2007). Forest fragmentation and its correlation to human land-use change in the state of Selangor, peninsular Malaysia. *Forest Ecology and Management, 241*(1), 39–48. doi:10.1016/j.foreco.2006.12.016

Abitante, J. C., & Lima, J. P. (2014). A acessibilidade como fator para o desenvolvimento sustentável urbano: o caso de Campos do Jordão.*Proceedings of the IV Simpósio de Desenvolvimento, Tecnologias e Sociedade*.

Abowd, G. D., Atkeson, C. G., Hong, J., Long, S., Kooper, R., & Pinkerton, M. (1997). Cyberguide: A mobile context-aware tour guide. *Wireless Networks, 3*(5), 421–433. doi:10.1023/A:1019194325861

Adler, B. T., Chatterjee, K., De Alfaro, L., Faella, M., Pye, I., & Raman, V. (2008). Assigning trust to wikipedia content. In *Proceedings of the 4th International Symposium on Wikis*. ACM.

Advertise with us - Gasbuddy Gas Prices. (n.d.). Retrieved July 3, 2016, from https://blog.gasbuddy.com/advertising.aspx

Aerts, R., & Honnay, O. (2011). Forest restoration, biodiversity and ecosystem functioning. *BMC Ecology, 11*(1), 29. doi:10.1186/1472-6785-11-29 PMID:22115365

AGORA. (2016). *AGORA - A Geospatial Open collaboRative Architecture for Building Resilience against Disasters and Extreme Events*. Retrieved from http://www.agora.icmc.usp.br/site/agora/

Agora. (2016). *Wikipedia, The Free Encyclopedia*. Retrieved from https://en.wikipedia.org/w/index.php?title=Agora&oldid=746994959

Agresti, A., & Finlay, B. (2009). Choice of sample size. In Statistical Methods for the Social Sciences (p. 126). Academic Press.

Albers, H. J., Ando, A. W., & Batz, M. (2008). Patterns of multi-agent land conservation: Crowding in/out, agglomeration, and policy. *Resource and Energy Economics, 30*(4), 492–508. doi:10.1016/j.reseneeco.2008.04.001

Albert, L., Rottensteiner, F., & Heipke, C. (2014). A two-layer conditional random field model for simultaneous classification of land cover and land use. *ISPRS International Archives of the Photogrammetry, Remote Sensing and Spatial Information Sciences, 1*, 17–24.

Albuquerque, J. P., Simon, E. J., Wahoff, J.-H., & Rolf, A. (2008). The Challenge of Transdisciplinarity in Information Systems Research. In *Information Systems Research Methods, Epistemology, and Applications* (Vol. 28, pp. 88–102). IGI Global. https://doi.org/10.4018/978-1-60566-040-0.ch005

Albuquerque, J. P., Eckle, M., Herfort, B., & Zipf, A. (2016). Crowdsourcing geographic information for disaster management and improving urban resilience: an overview of recent developments and lessons learned. In C. Capineri, M. Haklay, H. Huang, V. Antoniou, J. Kettunen, F. Ostermann, & R. Purves (Eds.), *European Handbook of Crowdsourced Geographic Information* (pp. 309–321). Ubiquity Press. doi:10.5334/bax.w

Albuquerque, J. P., Herfort, B., Brenning, A., & Zipf, A. (2015). A geographic approach for combining social media and authoritative data towards identifying useful information for disaster management. *International Journal of Geographical Information Science*, *29*(4), 667–689. doi:10.1080/13658816.2014.996567

Ali, A. L., & Schmid, F. (2014). Data Quality Assurance for Volunteered Geographic Information. In *Geographic. Information Science*, *8728*, 126–141.

Ali, A. L., Schmid, F., Al-Salman, R., & Kauppinen, T. (2014). Ambiguity and plausibility: managing classification quality in volunteered geographic information. In *Proceedings of the 22nd ACM SIGSPATIAL International Conference on Advances in Geographic Information Systems*. ACM. doi:10.1145/2666310.2666392

Ali, A. L., Schmid, F., Falomir, Z., & Freksa, C. (2015). Towards rule-guided classification for volunteered geographic information. *ISPRS Annals of Photogrammetry, Remote Sensing and Spatial Information Sciences*, *1*, 211–217.

Ali, A., & Ahmad, M. (2013). Geospatial data sharing in Pakistan: possibilities and problems?*Proceedings of Global Geospatial Conference 2013*.

Al-Madani, H., & Al-Janahi, A. R. (2002). Assessment of drivers comprehension of traffic signs based on their traffic, personal and social characteristics. *Transportation Research Part F: Traffic Psychology and Behaviour*, *5*(1), 63–76. doi:10.1016/S1369-8478(02)00006-2

Amaral, F. L. J. S., Holanda, C. M. A., Quirino, M. A. B., Nascimento, J. P. S., Neves, R. F., Ribeiro, K. S. Q. S., & Alves, S. B. (2012). Accessibility for people with disabilities or permanent mobility restrictions to the Unified Health System. *Ciencia & Saude Coletiva*, *17*(7), 1833. doi:10.1590/S1413-81232012000700022 PMID:22872345

Amelunxen, C. (2010). On the suitability of Volunteered Geographic Information for the purpose of geocoding. *GI_Forum*, 1–8.

Amundsen, F., & Proc, C. H. (1977). *First workshop on traffic conflicts*. Institute of Transportation Economics. Oslo, Norway.

Antoniou, V. (2011). *User Generated Spatial Content: An Analysis of the Phenomenon and its Challenges for Mapping Agencies*. Retrieved from http://discovery.ucl.ac.uk/1318053/

Antoniou, V., & Skopeliti, A. (2015). Measures and indicators of VGI quality: An overview. ISPRS Ann. In *ISPRS Annals of Photogrammetry* (Vol. 1, pp. 345–351). La Grande Motte, France: Remote Sensing and Spatial Information Sciences.

ANZLIC (1998). *Discussion paper: Spatial Data Infrastructure for Australia and New Zealand*. ANZLIC.

Arsanjani, J. J., Mooney, P., & Zipf, A. (2015). Quality assessment of the contributed land-use information from OpenStreetMap versus authoritative datasets. OpenStreetMap in GIScience, 37-58.

Arsanjani, J. J., Mooney, P., Helbich, M., & Zipf, A. (2015). An Exploration of Future Patterns of the Contributions to OpenStreetMap and Development of a Contribution Index. *Transactions in GIS*, *19*(6), 896–914. doi:10.1111/tgis.12139

Arthur, C. (2006). What is the 1% rule? *The Guardian*.

Assis, L. F. F. G., Behnck, L. P., Doering, D., de Freitas, E. P., Pereira, C. E., & Horita, F. E. A. … Albuquerque, J. P. de. (2016). Dynamic Sensor Management: Extending Sensor Web for Near Real-Time Mobile Sensor Integration in Dynamic Scenarios. In *Proceedings of the 30th International Conference on Advanced Information Networking and Applications (AINA 2016)* (pp. 303–310). IEEE. https://doi.org/10.1109/AINA.2016.100

Assis, L. F. F. G., Herfort, B., Steiger, E., Horita, F. E. A., & Albuquerque, J. P. (2015). Geographical prioritization of social network messages in near real-time using sensor data streams: an application to floods. In *Proc. of the XVI Brazilian Symposium on Geoinformatics (GEOINFO 2015)* (pp. 26–37). Campos do Jordão: MCTI/INPE.

Associação Brasileira de Normas Técnicas (ABNT). (2015). *ABNT NBR 9050 – Accessibility to buildings, equipament and the urban environment* (3rd ed.). Rio de Janeiro, Brasil: ABNT.

Autey, J., Sayed, T., & Zaki, M. H. (2012). Safety evaluation of right-turn smart channels using automated traffic conflict analysis. *Accident; Analysis and Prevention, 45*, 120–130. doi:10.1016/j.aap.2011.11.015 PMID:22269493

Azad, M., Hasnat, A., Ahmed, S., & Alam, A. (2010). Article. International Journal of Computer Science and Network Security, 10(11).

Azhar, S. (2011). Building information modeling (BIM): Trends, benefits, risks, and challenges for the AEC industry. *Leadership and Management in Engineering, 11*(3), 241–252. doi:10.1061/(ASCE)LM.1943-5630.0000127

Badland, H., Whitzman, C., Lowe, M., Davern, M., Aye, L., Butterworth, I., & Giles-Corti, B. et al. (2014). Urban liveability: Emerging lessons from Australia for exploring the potential for indicators to measure the social determinants of health. *Social Science & Medicine, 111*, 64–73. doi:10.1016/j.socscimed.2014.04.003 PMID:24762261

Baglatzi, A., Kokla, M., & Kavouras, M. (2012). Semantifying OpenStreetMap.*Proceedings of the 5th International Terra Cognita Workshop.*

Ballatore, A., & Zipf, A. (2015). A conceptual quality framework for volunteered geographic information. *12th International Conference on Spatial Information Theory, COSIT 2015.* Springer Verlag. doi:10.1007/978-3-319-23374-1_5

Ballatore, A., & Bertolotto, M. (2011). Semantically enriching VGI in support of implicit feedback analysis. In *Web and Wireless Geographical Information Systems* (pp. 78–93). Springer. doi:10.1007/978-3-642-19173-2_8

Ballatore, A., Bertolotto, M., & Wilson, D. C. (2013). Geographic knowledge extraction and semantic similarity in OpenStreetMap. *Knowledge and Information Systems, 37*(1), 61–81. doi:10.1007/s10115-012-0571-0

Ballatore, A., & Mooney, P. (2015). Conceptualising the geographic world: The dimensions of negotiation in crowdsourced cartography. *International Journal of Geographical Information Science, 29*(12), 2310–2327. doi:10.1080/13658816.2015.1076825

Banzi, M. (2009). *Getting Started with Arduino.* O'Reilly Media, Inc.

Baranoff, J., & Gonzales, R. I. (2015). Lantern: Empowering Refugees Through Community-Generated Guidance Using Near Field Communication. CHI' 15 Extended Abstracts, 7–12.

Barrett, A. M. (1976). Some reflections on those who taught me a personal view. *The Journal of Pharmacy and Pharmacology, 28*(4s), 338–341. doi:10.1111/j.2042-7158.1976.tb04176.x PMID:6736

Barrett, S., & Kridner, J. (2013). *Bad to the bone: Crafting electronic systems with beaglebone and beaglebone black* (Vol. 41). Morgan & Claypool Publishers.

Barron, C., Neis, P., & Zipf, A. (2014). A Comprehensive Framework for Intrinsic OpenStreetMap Quality Analysis. *Transactions in GIS, 18*(6), 877–895. doi:10.1111/tgis.12073

Bartoschek, T., & Keßler, C. (2013). VGI in education: From K-12 to graduate studies. In Crowdsourcing Geographic Knowledge (pp. 341-360). Springer Netherlands.

Basiouka, S., Potsiou, C., & Bakogiannis, E. (2015). OpenStreetMap for cadastral purposes: An application using VGI for official processes in urban areas. *SurveyReview, 47*(344), 333–341.

Basiri, A., Jackson, M., Amirian, P., Pourabdollah, A., Sester, M., Winstanley, A., ... Zhang, L. (2016). Quality assessment of OpenStreetMap data using trajectory mining. *Geo-Spatial Information Science, 19*(1), 56–68. 10.1080/10095020.2016.1151213

Basiri, A., Jackson, M., Amirian, P., Pourabdollah, A., Sester, M., Winstanley, A., & Zhang, L. et al. (2016). Quality assessment of OpenStreetMap data using trajectory mining. *Geo-spatial*[Taylor and Francis.]. *Information Science, 19*(1), 56–68.

Baskent, E. Z., & Jordan, G. A. (1995). Characterizing spatial structure of forest landscape. *Canadian Journal of Forest Research, 25*(11), 1830–1849. doi:10.1139/x95-198

Bazjanac, V., & Crawley, D. B. (1997). *The implementation of industry foundation classes in simulation tools for the building industry*. Lawrence Berkeley National Laboratory.

BC Marine Trails Network Association. (2016). *About the BC Marine Trails - BC Marine Trails Network: Kayak tours, adventure, kayaking maps*. Retrieved August 11, 2016, from https://www.bcmarinetrails.org/about-us/about-the-bc-marine-trails-network

Beasley, D. A. E., Benson, E. P., Welch, S. M., Reid, L. S., & Mousseau, T. A. (2012). The use of Citizen Scientists to Record and Map 13-Year Periodical Cicadas (Hemiptera: Cicadidae: Magicicada) in South Carolina. *The Florida Entomologist, 95*(2), 489–491. doi:10.1653/024.095.0236

Beischer, A., Cochrane, L., Corbett, J., Evans, M., Gill, M., & Millard, E. (2015). Mapping Experiences of Injustice: Developing a Crowdsourced Mapping Tool for Documenting Good Practices for Overcoming Social Exclusion. *Spatial Knowledge and Information Conference Proceedings*.

Bejar, R., Latre, M. A., Valino, J., Zarazaga-Soria, F. J., & Muro-Medrano, P. R. (2011). *Implementing GeoDRM support in an OpenLayers based viewer: access to the French geoportal WMS-C services from the Spanish SDI*. Universidad de Zaragoza.

Bergner, B., Zeile, P., & Papastefanou, G. (2011). Emotional Barrier-GIS – A new approach to integrate barrier-free planning in urban planning processes. In *Proceedings REAL CORP* (pp. 247–257). Retrieved from http://realcorp.at/archive/CORP2011_27.pdf

Biever, C. (2010). Twitter mood maps reveal emotional states of America. *New Scientist, 207*(2771), 14. doi:10.1016/S0262-4079(10)61833-7

Bishop, C. M. (2006). Pattern recognition. *Machine Learnng*.

Bishr, M., & Kuhn, W. (2007). Geospatial information bottom-up: A matter of trust and semantics. In *The European information society* (pp. 365–387). Springer. doi:10.1007/978-3-540-72385-1_22

Bishr, M., Wytzisk, A., & Morales, J. (2007). *GeoDRM: Towards Digital Management of Intellectual Property Rights for Spatial Data Infrastructures, Institute for Geoinformatics (IFGI)*. University of Munster.

Blake, A., Kohli, P., & Rother, C. (2011). *Markov random fields for vision and image processing*. MIT Press.

Blomquist, G. C., Berger, M. C., & Hoehn, J. P. (1988). New Estimates of Quality of Life in Urban Areas. *The American Economic Review, 78*(1), 89–107. doi:10.2307/1814700

Boguslawski, P. (2011). *Modelling and analysing 3d building interiors with the dual half-edge data structure*. University of Glamorgan.

Bollen, J., Mao, H., & Zeng, X. (2011). Twitter mood predicts the stock market. *Journal of Computational Science*, *2*(1), 1–8. doi:10.1016/j.jocs.2010.12.007

Bonney, R. (2007). Citizen science at the Cornell Lab of Ornithology. *Exemplary Science in Informal Education Settings: Standards-based Success Stories*, 213–229.

Bonsall, P., & Parry, T. (1991). *Using An Interactive Route-Choice Simulator To Investigate Drivers' compliance With Route Guidance Advice* (No. 1306). Academic Press.

Bordogna, G., Carrara, P., Criscuolo, L., Pepe, M., & Rampini, A. (2016). On predicting and improving the quality of Volunteer Geographic Information projects. *International Journal of Digital Earth*, *9*(2), 134–155. doi:10.1080/17538947.2014.976774

Bowers, S., Kudo, J., Cao, H., & Schildhauer, M. P. (2010). ObsDB: A System for Uniformly Storing and Querying Heterogeneous Observational Data. In *IEEE sixth international conference on e-science* (pp. 261–268). doi:10.1109/eScience.2010.24

Brabham, D. C. (2008). Crowdsourcing as a Model for Problem Solving. *Convergence (London)*, *14*(1), 75–90. doi:10.1177/1354856507084420

Brandeis, M. W., & Nyerges, T. (2015). Assessing Resistance to Volunteered Geographic Information Reporting with Local Government. *Transactions in GIS*, *20*(2), 203–220. doi:10.1111/tgis.12168

Bray, D. B., Ellis, E. A., Armijo-Canto, N., & Beck, C. T. (2004). The institutional drivers of sustainable landscapes: A case study of the 'Mayan Zonein Quintana Roo, Mexico. *Land Use Policy*, *21*(4), 333–346. doi:10.1016/j.landusepol.2003.11.001

Breiman, L. (1996). Bagging predictors. *Machine Learning*, *24*(2), 123–140. doi:10.1007/BF00058655

Breiman, L. (2001). Random forests. *Machine Learning*, *45*(1), 5–32. doi:10.1023/A:1010933404324

Brock, A., Truillet, P., Oriola, B., Picard, D., & Jouffrais, C. (2012). Design and user satisfaction of interactive maps for visually impaired people. In *International Conference on Computers for Handicapped Persons* (pp. 544-551). Springer Berlin Heidelberg. doi:10.1007/978-3-642-31534-3_80

Bröring, A., Jirka, S., & Foerster, T. (2010). The Sensor Bus – Integrating Geosensors and the Sensor Web. In *Proceedings of the 2nd Open Source GIS UK Conference (OSGIS)* (pp. 2–5). Nottingham, UK: OGC.

Brovelli, M. A., Minghini, M., & Molinari, M. (2016). An automated GRASS-based procedure to assess the accuracy of the OpenStreetMap Paris road network. In The International Archives of the Photogrammetry, Remote Sensing and Spatial Information Sciences (Vol. XLI-B7, pp. 919–925). ISPRS Archives. doi:10.5194/isprsarchives-XLI-B7-919-2016

Brovelli, M. A., Minghini, M., Molinari, M., & Zamboni, G. (2016). Positional accuracy assessment of the OpenStreetMap buildings layer through automatic homologous pairs detection: the method and a case study. In The International Archives of the Photogrammetry, Remote Sensing and Spatial Information Sciences (Vol. XLI-B2, pp. 615–620). ISPRS Archives. doi:10.5194/isprsarchives-XLI-B2-615-2016

Brovelli, M. A., Molinari, M. E., Hussein, E., Chen, J., & Li, R. (2015). The First Comprehensive Accuracy Assessment of GlobeLand30 at a National Level: Methodology and Results. *Remote Sensing*, *7*(4), 4191-4212. http://doi:10.3390/rs70404191

Brovelli, M. A., Minghini, M., Molinari, M., & Mooney, P. (2016). *Towards an Automated Comparison of OpenStreetMap with Authoritative Road Datasets. Transactions in GIS*. doi:10.1111/tgis.12182

Brown, G. (2006). Mapping landscape values and development preferences: A method for tourism and residential development planning. *International Journal of Tourism Research*, *8*(2), 101–113. doi:10.1002/jtr.562

Brown, G. (2012). Public Participation GIS (PPGIS) for Regional and Environmental Planning: Reflections on a Decade of Empirical Research. *Journal of Urban and Regional Information Systems Association*, *25*(2), 7–18.

Brown, G., Kelly, M., & Whitall, D. (2014). Which public? Sampling effects in public participation GIS (PPGIS) and volunteered geographic information (VGI) systems for public lands management. *Journal of Environmental Planning and Management*, *57*(2), 190–214. doi:10.1080/09640568.2012.741045

Brown, G., & Pullar, D. V. (2012). An evaluation of the use of points versus polygons in public participation geographic information systems using quasi-experimental design and Monte Carlo simulation. *International Journal of Geographical Information Science*, *26*(2), 231–246. doi:10.1080/13658816.2011.585139

Brown, G., & Reed, P. (2009). Public Participation GIS: A New Method for use in National Forest Planning. *Forest Science*, *55*(2), 166–182.

Brown, G., & Reed, P. (2012). Values Compatibility Analysis: Using Public Participation Geographic Information Systems (PPGIS) for Decision Support in National Forest Management. *Applied Spatial Analysis and Policy*, *5*(4), 317–332. doi:10.1007/s12061-011-9072-x

Brown, G., & Weber, D. (2013a). A Place-based Approach to Conservation Management using Public Participation GIS (PPGIS). *Journal of Environmental Planning and Management*, *56*(4), 455–473. doi:10.1080/09640568.2012.685628

Brown, G., & Weber, D. (2013b). Using Public Participation GIS (PPGIS) on the Geoweb to Monitor Tourism Development Preferences. *Journal of Sustainable Tourism*, *21*(2), 192–211. doi:10.1080/09669582.2012.693501

Brown, M., & Polk, K. (1996). Taking Fear of Crime Seriously: The Tasmanian Approach to Community Crime Prevention. *Crime and Delinquency*, *42*(3), 398–420. doi:10.1177/0011128796042003004

Brydon-Miller, M., Greenwood, D., & Maguire, P. (2003). Why Action Research? *Action Research*, *1*(1), 9–28. doi:10.1177/14767503030011002

Buehler, R., & Pucher, J. (2012). Cycling to work in 90 large American cities: New evidence on the role of bike paths and lanes. *Transportation*, *39*(2), 409–432. doi:10.1007/s11116-011-9355-8

Byrne, D., & Pickard, A. J. (2016). Neogeography and the democratization of GIS: A metasynthesis of qualitative research. *Information Communication and Society*, 1–18. doi:10.1080/1369118X.2015.1125936

CACL. (2006). *The employment and employability of Canadians with intellectual disabilities: Submission by the Canadian Association for Community Living to the House of Commons Standing Committee on Human Resources, Social Development and the Status of Persons with Disabilities*. Toronto: Canadian Association for Community Living.

Calle Jiménez, T., & Luján Mora, S. (2015). Using crowdsourcing to improve accessibility of geographic maps on mobile devices.*Proceedings of the Eighth International Conference on Advances in Computer-Human Interactions (ACHI 2015)*.

Camboim, S. P., Meza Bravo, J. V, & Sluter, C. R. (2015). An investigation into the completeness of, and the updates to, OpenStreetMap data in a heterogeneous area in Brazil. *ISPRS International Journal of Geo-Information, 4*(3), 1366–1388. 10.3390/ijgi4031366

Campagna, M., Floris, R., Massa, P., Girsheva, A., & Ivanov, K. (2015). The role of social media geographic information (SMGI) in spatial planning. In Planning Support Systems and Smart Cities (pp. 41-60). Springer International Publishing.

Camponovo, M. E., & Freundschuh, S. M. (2014). Assessing Uncertainty in VGI for Emergency Response. *Cartography and Geographic Information Science*, *41*(5), 440–455. doi:10.1080/15230406.2014.950332

Canada Import Status - OpenStreetMap Wiki. (2015). Retrieved July 3, 2016, from http://wiki.openstreetmap.org/wiki/Canada_Import_Status

Canavosio-Zuzelski, R., Agouris, P., & Doucette, P. (2013). *A Photogrammetric Approach for Assessing Positional Accuracy of OpenStreetMap© Roads. International Journal of Geo-Information, 2(2).*

Cardonha, C., Gallo, D., Avegliano, P., Herrmann, R., Koch, F., & Borger, S. (2013, May). A crowdsourcing platform for the construction of accessibility maps.*Proceedings of the 10th international cross-disciplinary conference on web accessibility*. doi:10.1145/2461121.2461129

Carls, N. (1947). *How to read aerial photographs for census work*. US Government Printing Office.

Cartier, M. (2016). Explaining Differences in the Canadian and American Response to the Syrian Refugee Crisis. *Virginia Policy Review*, 56-82.

Cary, T. (2006, March). Geospatial Digital Rights management. *Geospatial Solutions*, *16*(3), 18.

Castanhari, R. E. S., Rocha, R. dos S., Andrade, S. C., & Albuquerque, J. P. (2016). A Software Architecture to Integrate Sensor Data and Volunteered Geographic Information for Flood Risk Management. In *Proceedings of the 13th International Conference on Information Systems for Crisis Response and Management*. Rio de Janeiro, Brazil: ISCRAM.

CBC News. (2013). *CBC News - Google buys Waze mapping app*. Retrieved July 3, 2016, from http://www.cbc.ca/m/touch/business/story/1.1306058

Cha, M., Kwak, H., Rodriguez, P., Ahn, Y., & Moon, S. (2007). I tube, you tube, everybody tubes: analyzing the world's largest user generated content video system. *Proceedings of the 7th ACM SIGCOMM conference on Internet measurement*. Retrieved from http://portal.acm.org/citation.cfm?id=1298306.1298309

Chae, J., Thom, D., Jang, Y., Kim, S., Ertl, T., & Ebert, D. S. (2014). Public behavior response analysis in disaster events utilizing visual analytics of microblog data. *Computers & Graphics*, *38*(1), 51–60. doi:10.1016/j.cag.2013.10.008

Chainey, S., & Ratcliffe, J. (2013). *GIS and crime mapping*. John Wiley & Sons.

Chambers, R. (2006). Participatory mapping and geographic information systems: Whose map? Who is empowered and who disempowered? Who gains and who loses? *The Electronic Journal of Information Systems in Developing Countries*, *25*(2), 1–11.

Chander, G., Markham, B. L., & Helder, D. L. (2009). Summary of current radiometric calibration coefficients for Landsat MSS, TM, ETM+, and EO-1 ALI sensors. *Remote Sensing of Environment*, *113*(5), 893–903. doi:10.1016/j.rse.2009.01.007

Chatila, R., Devy, M., Lacroix, S., & Herrb, M. (1994). Perception system and functions for autonomous navigation in a natural environment.*Proceedings of the AIAA/NASA Conference on Intelligent Robots for Field, Factory, Service and Space*.

Chen. (2007). *The GEO Data Sharing Challenge: Putting Principles into Practice*. Alliance for Earth Observations Briefing.

Chen, J., Chen, J., Liao, A., Cao, X., Chen, L., Chen, X., & Mills, J. et al. (2015). Global land cover mapping at 30 m resolution: A POK-based operational approach. *ISPRS Journal of Photogrammetry and Remote Sensing*, *103*, 7–27. doi:10.1016/j.isprsjprs.2014.09.002

Chin, H. C., & Quek, S. T. (1997). Measurement of traffic conflicts. *Safety Science*, *26*(3), 169–185. doi:10.1016/S0925-7535(97)00041-6

Chow, T. E. (2013). "We Know Who You Are and We Know Where You Live": A Research Agenda for Web Demographics. In D. Sui, S. Elwood, & M. Goodchild (Eds.), *Crowdsourcing Geographic Knowledge: Volunteered Geographic Information (VGI) in Theory and Practice* (pp. 265–285). New York: Springer Dordrecht Heidelberg. doi:10.1007/978-94-007-4587-2_15

Cicada Project. (2015). *Cicada Project*. Retrieved January 01, 2015 from http://project.wnyc.org/cicadas/

Ciepłuch, B., Jacob, R., Mooney, P., & Winstanley, A. (2010). Comparison of the accuracy of OpenStreetMap for Ireland with Google Maps and Bing Maps. *Proceedings of the Ninth International Symposium on Spatial Accuracy Assessment in Natural Resources and Environmental Sciences*, 337. Retrieved from http://eprints.nuim.ie/2476/

Ciepluch, B., Jacob, R., Mooney, P., & Winstanley, A. (2010). Comparison of the accuracy of OpenStreetMap for Ireland with Google Maps and Bing Maps.*Proceedings of the Ninth International Symposium on Spatial Accuracy Assessment in Natural Resources and Environmental Sciences*, 337.

Ciepluch, B., Mooney, P., Jacob, R., & Winstanley, A. C. (2009). Using openstreetmap to deliver location-based environmental information in Ireland. *SIGSPATIAL Special*, *1*(3), 17–22. doi:10.1145/1645424.1645428

Ciepłuch, B., Mooney, P., & Winstanley, A. C. (2011). *Building generic quality indicators for OpenStreetMap. In 19th annual GIS Research UK*. GISRUK.

Cihlar, J. (2000). Land cover mapping of large areas from satellites: Status and research priorities. *International Journal of Remote Sensing*, *21*(6-7), 1093–1114. doi:10.1080/014311600210092

Cipełuch, B., Jacob, R., Mooney, P., & Winstanley, A. C. (2010). Comparison of the accuracy of OpenStreetMap for Ireland with Google Maps and Bing Maps. In *Proceedings of the Ninth International Symposium on Spatial Accuracy Assessment in Natural Resources and Environmental Sciences* (p. 337). CONF, University of Leicester.

Clemente, F., & Kleiman, M. B. (1977). Fear of Crime in the United States: A Multivariate Analysis. *Social Forces*, *56*(2), 519–531. doi:10.2307/2577738

Cochrane, L., Corbett, J., Evans, M., & Gill, M. (2016). Searching for Social Justice in Crowdsourced Mapping. *Cartography and Geographic Information Science*. doi:10.1080/15230406.2016.1212673

Cochrane, L., Corbett, J., & Keller, P. (2014). *Impact of Community-based and Participatory Mapping. Institute for Studies and Innovation in Community-University Engagement*. University of Victoria.

Coleman, D. J. (2013). Potential Contributions and Challenges of VGI for Conventional Topographic Base-Mapping Programs. In D. Sui, S. Elwood, & M. Goodchild (Eds.), Crowdsourcing Geographic Knowledge SE - 14 (pp. 245–263). CHAP, Springer Netherlands. doi:10.1007/978-94-007-4587-2_14

Coleman, D. J., Georgiadou, Y., & Labonte, J. (2009). Volunteered Geographic Information: The Nature and Motivation of Producer. *International Journal of Spatial Data Infrastructures Research*, *4*, 332–358.

Coleman, D. J., Georgiadou, Y., & Labonte, J. (2009). Volunteered geographic information: The nature and motivation of produsers. *International Journal of Spatial Data Infrastructures Research*, *4*(1), 332–358.

Connors, J. P., Lei, S., & Kelly, M. (2012). Citizen science in the age of neogeography: Utilizing volunteered geographic information for environmental monitoring. *Annals of the Association of American Geographers*, *102*(6), 1267–1289. doi:10.1080/00045608.2011.627058

Cookbook, S. D. I. (2004). *Outreach and capacity building*. Retrieved from: http://gsdiassociation.org/images/publications/cookbooks/SDI_Cookbook_GSDI_2004_ver2.pdf

Corbera, E., Estrada, M., May, P., Navarro, G., & Pacheco, P. (2011). Rights to land, forests and carbon in REDD+: Insights from Mexico, Brazil and Costa Rica. *Forests*, *2*(1), 301–342. doi:10.3390/f2010301

Corbett, J. (2003). *Empowering technologies? Introducing participatory geographic information and multimedia systems in two Indonesian communities* (Doctoral Dissertation). University of Victoria.

Corbett, J. M., & Keller, C. P. (2005). An analytical framework to examine empowerment associated with participatory geographic information systems (PGIS). *Cartographica: The International Journal for Geographic Information and Geovisualization*, *40*(4), 91–102. doi:10.3138/J590-6354-P38V-4269

Corbett, J. M., & Keller, C. P. M. (2004). Empowerment and participatory geographic information and multimedia systems: Observations from two communities in Indonesia. *Information Technologies and International Development*, *2*(2), 25–44. doi:10.1162/1544752044193425

Corbett, J., Cochrane, L., & Gill, M. (2016). Powering Up: Revisiting Participatory GIS and Empowerment. *The Cartographic Journal*, *53*(4), 335–340. doi:10.1080/00087041.2016.1209624

Corbett, J., & Mann, R. (2010). Tlowitsis re-imagined: The use of digital media to build nation and overcome disconnection in a displaced aboriginal community. *International Journal of Information Communication Technologies and Human Development*, *2*(3), 33–54. doi:10.4018/jicthd.2010070103

Corcoran, P., Mooney, P., & Bertolotto, M. (2013). Analysing the growth of OpenStreetMap networks. *Spatial. Statistics*, *3*, 21–32.

Corcoran, P., Mooney, P., & Bertolotto, M. (2013). Analysing the growth of OpenStreetMap networks. *Spatial Statistics*, *3*, 21–32. doi:10.1016/j.spasta.2013.01.002

Cornell. (2015). *The Cornell Lab of Ornithology*. Retrieved January 01, 2015 from http://www.allaboutbirds.org

Cornwall, A., & Jewkes, R. (1995). What is participatory research? *Social Science & Medicine*, *41*(12), 1667–1676. doi:10.1016/0277-9536(95)00127-S PMID:8746866

Cortes, C., & Vapnik, V. (1995). Support vector machine. *Machine Learning*, *20*(3), 273–297. doi:10.1007/BF00994018

Costanza, R., Fisher, B., Ali, S., Beer, C., Bond, L., Boumans, R., & Snapp, R. et al. (2007). Quality of life: An approach integrating opportunities, human needs, and subjective well-being. *Ecological Economics*, *61*(2), 267–276. doi:10.1016/j.ecolecon.2006.02.023

Coughlan, J. M. & Shen, H. (2013). Crosswatch: a system for providing guidance to visually impaired travellers at traffic intersections. *Journal of Assistive Technologies, 7*(2).

Cox, S., Cuthbert, A., Daisey, P., Davidson, J., Johnson, S., Keighan, E., Martell, R. (2002). *OpenGIS® Geography Markup Language (GML) Implementation Specification, version*. Academic Press.

Craglia, M., Ostermann, F., & Spinsanti, L. (2012). Digital Earth from vision to practice: Making sense of citizen-generated content. *International Journal of Digital Earth*, *5*(5), 398–416. doi:10.1080/17538947.2012.712273

Csapó, Á., Wersényi, G., Nagy, H., & Stockman, T. (2015). A survey of assistive technologies and applications for blind users on mobile platforms: A review and foundation for research. *Journal on Multimodal User Interfaces*, *9*(4), 275–286. doi:10.1007/s12193-015-0182-7

Cugler, D. C. (2016). *VGI Project. Dynamic form generator to acquire VGI.* Retrieved January 17, 2016 from http://www.ic.unicamp.br/~dcintra/files/VGI.zip

Cugler, D. C., Medeiros, C. B., Shekhar, S., & Toledo, L. F. (2013). A Geographical Approach for Metadata Quality Improvement in Biological Observation Databases. In *Proceedings of the 9th ieee international conference on e-science* (p. 212-220). doi:10.1109/eScience.2013.14

Cugler, D. C., Medeiros, C. B., & Toledo, L. F. (2012, June). An architecture for retrieval of animal sound recordings based on context variables. *Concurrency and Computation, 25*(16), 2310–2326. doi:10.1002/cpe.2865

Curry, M. (1997). The Digital Individual and the Private Realm. *Annals of the Association of American Geographers, 87*(4), 681–699. doi:10.1111/1467-8306.00073

D'Antonio, F., Fogliaroni, P., & Kauppinen, T. (2014). VGI edit history reveals data trustworthiness and user reputation. *17th AGILE Conference on Geographic Information Science.*

Daum'e, H., III. (2006). *Practical Structured Learning Techniques for Natural Language Processing* (PhD thesis). University of Southern California, Los Angeles, CA.

de Albuquerque, J. P., Fan, H., & Zipf, A. (2016). A conceptual model for quality assessment of VGI for the purpose of flood management. *19th AGILE Conference on Geographic Information Science.*

Degrossi, L. C., Albuquerque, J. P., Fava, M. C., & Mendiondo, E. M. (2014). Flood Citizen Observatory: a crowdsourcing-based approach for flood risk management in Brazil. In *Proc. of the 26th Int. Conf. on Software Engineering and Knowledge Engineering (SEKE 2014)* (pp. 1–6). Vancouver, Canada: Knowledge Systems Institute Graduate School.

Dietterich, T. G. (2002). Ensemble learning. The handbook of brain theory and neural networks, 2, 110-125.

DMTI Spatial Inc. (2015). *CanMap® Content Suite Support Manual.* Markham.

Dobson, M. (2013). VGI as a Compilation Tool for Navigation Map Databases. In D. Sui, S. Elwood, & M. Goodchild (Eds.), Crowdsourcing Geographic Knowledge SE - 17 (pp. 307–327). CHAP, Springer Netherlands. doi:10.1007/978-94-007-4587-2_17

Dolif, G., Engelbrecht, A., Jatobá, A., da Silva, A. J. D., Gomes, J. O., & Borges, M. R. S. ... de Carvalho, P. V. R. (2013). Resilience and brittleness in the ALERTA RIO system: A field study about the decision-making of forecasters. *Natural Hazards, 65*(3), 1831–1847. https://doi.org/10.1007/s11069-012-0448-2

Doran, B. J., & Burgess, M. B. (2011). *Putting fear of crime on the map: investigating perceptions of crime using geographic information systems.* Springer Science & Business Media.

Dorn, H., Törnros, T., & Zipf, A. (2015). Quality evaluation of VGI using authoritative data-a comparison with land use data in southern Germany. *ISPRS International Journal of Geo-Information, 4*(3), 1657–167110.3390/ijgi4031657

Duhl, J., & Kevorkian, S. (2001). *Understanding DRM Systems.* IDC White Paper.

Duncker, P. S., Raulund-Rasmussen, K., Gundersen, P., Katzensteiner, K., De Jong, J., Ravn, H. P., & Spiecker, H. (2012). How forest management affects ecosystem services, including timber production and economic return: Synergies and trade-offs. *Ecology and Society, 17*(4), 50. doi:10.5751/ES-05066-170450

Dunn, C. (2007). Participatory GIS – a peoples GIS? *Progress in Human Geography, 31*(5), 616–637. doi:10.1177/0309132507081493

Duží, B., & Jakubínský, J. (2013). Brownfield dilemmas in the transformation of post-communist cities: A case study of Ostrava, Czech Republic. *Journal of Studies and Research in Human Geography, 72*(2), 53–64. doi:10.5719/hgeo.2013.72.53

Eastman, C., Eastman, C. M., Teicholz, P., Sacks, R., & Liston, K. (2011). *BIM handbook: A guide to building information modeling for owners, managers, designers, engineers and contractors.* John Wiley & Sons.

ECIS. (n.d.). *Interoperability & Open Standards.* European Committee for Interoperable Systems. Retrieved from: http://www.ecis.eu/open-standards/

Eckle, M., & De Albuquerque, J. P. (2015). Quality assessment of remote mapping in OpenStreetMap for disaster management purposes. In L. A. Palen, T. Comes, M. Buscher, A. L. Hughes, & L. A. Palen (Eds.), *12th International Conference on Information Systems for Crisis Response and Management, ISCRAM 2015*. Retrieved from https://www.scopus.com/inward/record.uri?eid=2-s2.0-84947764602&partnerID=40&md5=772a654ed328051eca2e2591b6430aa9

EEA Corine Land Cover. (n.d.). Available online: http://www.eea.europa.eu/publications/COR0-landcover

EEA. (2011). *Mapping guide for a European Urban Atlas.* Available from: http://www.eea.europa.eu/data-and-maps/data/urban-atlas

El-Ashmawy, K. L. A. (2016). Testing the positional accuracy of OpenStreetMap data for mapping applications. *Geodesy and Cartography, 42*(1), 25–30. 10.3846/20296991.2016.1160493

Ellis, E., & Pontius, R. (2007). Land-use and land cover change. Encyclopedia of Earth.

Elvik, R. (2006). *Road safety inspections: safety effects and best practice guidelines*. Transportøkonomisk institutt.

Elvik, R., Vaa, T., Erke, A., & Sorensen, M. (Eds.). (2009). *The handbook of road safety measures*. Emerald Group Publishing. doi:10.1108/9781848552517

Elwood, S., Goodchild, M., & Sui, D. (2013). Prospects for VGI Research and the Emerging Fourth Paradigm. In D. Sui, S. Elwood, & M. Goodchild (Eds.), Crowdsourcing Geographic Knowledge SE - 20 (pp. 361–375). CHAP, Springer Netherlands. doi:10.1007/978-94-007-4587-2_20

Elwood, S. (2006). Critical issues in participatory GIS: Deconstructions, reconstructions, and new research directions. *Transactions in GIS, 10*(5), 693–708. doi:10.1111/j.1467-9671.2006.01023.x

Elwood, S. (2008). Volunteered geographic information: Future research directions motivated by critical, participatory, and feminist GIS. *GeoJournal, 72*(3-4), 173–183. doi:10.1007/s10708-008-9186-0

Elwood, S. (2008). Volunteered geographic information: Key questions, concepts and methods to guide emerging research and practice. *GeoJournal, 72*(3), 133–135. doi:10.1007/s10708-008-9187-z

Elwood, S. (2011). Geographic information science: Visualization, visual methods, and the geoweb. *Progress in Human Geography, 35*(3), 401–408. doi:10.1177/0309132510374250

Elwood, S., Goodchild, M. F., & Sui, D. Z. (2012). Researching volunteered geographic information: Spatial data, geographic research, and new social practice. *Annals of the Association of American Geographers, 102*(3), 571–590. doi:10.1080/00045608.2011.595657

Espírito-Santo, F. D., Gloor, M., Keller, M., Malhi, Y., Saatchi, S., Nelson, B., & Palace, M. et al. (2014). Size and frequency of natural forest disturbances and the Amazon forest carbon balance. *Nature Communications*, 5. PMID:24643258

ESRI White Paper. (2013). *ESRI-Supported Open geospatial Consortium and ISO/TC211 standards*. ESRI.

ESRI. (2011). *Spatial Literacy Program at the University of Redlands*. Accessed November 14 2015 from http://www.redlands.edu/prospective-students/6304.aspx

Estelles-Arolas, E., & Gonzalen-Ladron-de-Guevara, F. (2012). Towards an Integrated Crowdsourcing Definition. *Journal of Information Science, 38*(2), 189–200. doi:10.1177/0165551512437638

Estima, J., & Painho, M. (2015). Investigating the Potential of OpenStreetMap for Land Use/Land Cover Production: A Case Study for Continental Portugal. In J. Jokar Arsanjani, A. Zipf, P. Mooney, & M. Helbich (Eds.), OpenStreetMap in GIScience SE - 14 (pp. 273–293). CHAP, Springer International Publishing. doi:10.1007/978-3-319-14280-7_14

Estima, J., & Painho, M. (2013). Exploratory analysis of OpenStreetMap for land use classification. *Geocrowd, 13*, 39–46. doi:10.1145/2534732.2534734

Evans, L. (1991). *Traffic safety and the driver*. Science Serving Society.

Fan, H., Yang, B., Zipf, A., & Rousell, A. (2015). A polygon-based approach for matching OpenStreetMap road networks with regional transit authority data. *International Journal of Geographical Information Science, 30*, 1–17. 10.1080/13658816.2015.1100732

Fan, H., Zipf, A., & Fu, Q. (2014). Estimation of building types on openstreetmap based on urban morphology analysis. In *Connecting a Digital Europe Through Location and Place* (pp. 19–35). Springer. doi:10.1007/978-3-319-03611-3_2

Fan, H., Zipf, A., Fu, Q., & Neis, P. (2014). Quality assessment for building footprints data on OpenStreetMap. *International Journal of Geographical Information Science*, *0*(0), 1–20. doi:10.1080/13658816.2013.867495

FAO. (1996). Guidelines: Agro-ecological zoning. Food and Agricultural Organisation, SoilsBulletin. Rome, Italy: FAO.

Fava, M. C. (2015). *Modelo de Alerta Hidrológico com Base Participativa usando Sistema de Informações Voluntárias para Previsão de Enchentes* (Masters Thesis). São Carlos School of Engineering.

Fava, M. C., Abe, N., Restrepo-Estrada, C., Mendiondo, E. M., & Souza, V. C. B. (2015). Redução de incertezas na previsão de enchentes com uso de dados colaborativos. In *Proceedings of the XXI Simpósio Brasileiro de Recursos Hídricos* (pp. 1–8). Brasilia, DF: ABRH.

Favretto, D. O., Carvalho, E. C. D., & Canini, S. R. M. D. S. (2008). Intervenções realizadas pelo enfermeiro para melhorar a comunicação com deficientes visuais. *Revista da Rede de Enfermagem do Nordeste*, *9*(3), 68–75.

Feick, R., & Roche, S. (2013). Understanding the Value of VGI. In D. Sui, S. Elwood, & M. Goodchild (Eds.), Crowdsourcing Geographic Knowledge SE - 2 (pp. 15–29). CHAP, Springer Netherlands doi:10.1007/978-94-007-4587-2_2

FGDC. (1997). Framework, introduction and guide, Book of Federal geographic data committee. Washington, DC: FGDC.

Flanagin, A. J., & Metzger, M. J. (2008). The Credibility of Volunteered Geographic Information. *GeoJournal*, *72*(3-4), 137–148. doi:10.1007/s10708-008-9188-y

Flanagin, A. J., & Metzger, M. J. (2009). The credibility of volunteered geographic information. *GeoJournal*, *72*(3), 137–148.

Flip Experts. (2013). *Digital rights management*. Retrieved from: http://searchsecurity.techtarget.com/magazineContent/Digital-Rights-Management-Microsoft-Windows-Rights-Management-Services

Flores Guido, J. S., & Espejel Carvajal, I. (1994). *Tipos de vegetación de la Península de Yucatán* (No. Y/581.61 E8/3). Academic Press.

FNJV. (2015). *Online animal sound collection - Fonoteca Neotropical Jacques Vielliard*. Retrieved January 01, 2015 from http://proj.lis.ic.unicamp.br/fnjv

Fonte, C., Minghini, M., Antoniou, V., See, L., Brovelli, M. A., Patriarca, J., & Milčinski, G. (2016). An automated methodology for converting OSM data into a land use/cover map.*Proceedings of the 6th International Conference on Cartography & GIS.*

Foody, G. M., & Boyd, D. S. (2013). Using volunteered data in land cover map validation: Mapping West African forests. *IEEE Journal of Selected Topics in Applied Earth Observations and Remote Sensing*, *6*(3), 1305–1312. doi:10.1109/JSTARS.2013.2250257

Forghani, M., & Delavar, M. (2014). A Quality Study of the OpenStreetMap Dataset for Tehran. *ISPRS International Journal of Geo-Information*, *3*(2), 750–763. doi:10.3390/ijgi3020750

Forman, R. T. (2000). Estimate of the area affected ecologically by the road system in the United States. *Conservation Biology*, *14*(1), 31–35. doi:10.1046/j.1523-1739.2000.99299.x

Foursquare. (2015). *Foursquare Categories*. Retrieved October 15, 2015, from https://developer.foursquare.com/categorytree

Foursquare. (2016). *About Us*. Retrieved July 14, 2016, from https://pt.foursquare.com/about

Fram, C., Chistopoulou, K., & Ellul, C. (2015). Assessing the quality of OpenStreetMap building data and searching for a proxy variable to estimate OSM building data completeness.*Proceedings of the 23rd GIS Research UK (GISRUK) Conference*.

Friedl, M. A., Sulla-Menashe, D., Tan, B., Schneider, A., Ramankutty, N., Sibley, A., & Huang, X. (2010). MODIS Collection 5 global land cover: Algorithm refinements and characterization of new datasets. *Remote Sensing of Environment*, *114*(1), 168–182. doi:10.1016/j.rse.2009.08.016

Friedman, A., & Kohler, B. (2003). Bidimensional regression: Assessing the configural similarity and accuracy of cognitive maps and other two-dimensional data sets. *Psychological Methods*, *8*(4), 468–491. doi:10.1037/1082-989X.8.4.468 PMID:14664683

Fritz, S., McCallum, I., Schill, C., Perger, C., Grillmayer, R., Achard, F., & Obersteiner, M. et al. (2009). Geo-Wiki.Org: The use of crowdsourcing to improve global land cover. *Remote Sensing*, *1*(3), 345–354. doi:10.3390/rs1030345

Fritz, S., See, L., McCallum, I., Schill, C., Obersteiner, M., van der Velde, M., & Achard, F. et al. (2011). Highlighting continued uncertainty in global land cover maps for the user community. *Environmental Research Letters*, *6*(4), 44005. doi:10.1088/1748-9326/6/4/044005

Funke, S., Schirrmeister, R., & Storandt, S. (2015). Automatic Extrapolation of Missing Road Network Data in OpenStreetMap.*2nd International Workshop on Mining Urban Data, ICML.*

Furquim, G., Neto, F., Pessin, G., Ueyama, J., Albuquerque, J. P., & Clara, M., … Braun, T. (2014). Combining wireless sensor networks and machine learning for flash flood nowcasting. In *Proceedings of the IEEE 28th International Conference on Advanced Information Networking and Applications Workshops, IEEE WAINA 2014* (pp. 67–72). Victoria, Canada: IEEE. doi:10.1109/WAINA.2014.21

Gartner, G. (2012). *Putting Emotions in Maps–The Wayfinding Example*. Retrieved from http://www.mountaincartography.org/publications/papers/papers_taurewa_12/papers/mcw2012_sec3_ch08_p061-065_gartner.pdf

Gavrilidis, A. A., Ciocănea, C. M., Niţă, M. R., Onose, D. A., & Năstase, I. I. (2016). Urban Landscape Quality Index – Planning Tool for Evaluating Urban Landscapes and Improving the Quality of Life. *Procedia Environmental Sciences*, *32*, 155–167. doi:10.1016/j.proenv.2016.03.020

Geoghegan, J., Villar, S. C., Klepeis, P., Mendoza, P. M., Ogneva-Himmelberger, Y., Chowdhury, R. R., & Vance, C. et al. (2001). Modeling tropical deforestation in the southern Yucatan peninsula region: Comparing survey and satellite data. *Agriculture, Ecosystems & Environment*, *85*(1), 25–46. doi:10.1016/S0167-8809(01)00201-8

GeoNode. (2016). *GeoNode*. Retrieved from http://geonode.org/

Getis, A., & Ord, J. K. (1992). The analysis of spatial association by use of distance statistics. *Geographical Analysis, 24*(3), 189–206. doi:10.1111/j.1538-4632.1992.tb00261.x

Giannopoulos, G., Karagiannakis, N., Skoutas, D., & Athanasiou, S. (2015). Automatic recommendations of categories for geospatial entities. *LocalRec, 15*, 1405.

Gibbon, D. (1994). *The new production of knowledge: the dynamics of science and research in contemporary societies.* London, UK: SAGE Publications, Inc.

Giles, J. (2005). Wisdom of the Crowd. *Nature, 438*(7066), 281. doi:10.1038/438281a PMID:16292279

Girres, J. F., & Touya, G. (2010). Quality Assessment of the French OpenStreetMap Dataset. *Transactions in GIS, 14*(4), 435–459. doi:10.1111/j.1467-9671.2010.01203.x

Goetz, M., & Zipf, A. (2012). Towards defining a framework for the automatic derivation of 3D CityGML models from volunteered geographic information. *International Journal of 3-D Information Modeling, 1*(2), 1-16.

Goetz, M. (2013). Towards generating highly detailed 3D CityGML models from OpenStreetMap. *International Journal of Geographical Information Science, 27*(5), 845–865. doi:10.1080/13658816.2012.721552

Gold, C. M. (2006). What is GIS and What is Not? *Transactions in GIS, 10*(4), 505–519. doi:10.1111/j.1467-9671.2006.01009.x

Goodchild, M. F., & Hunter, G. J. (1997). A simple positional accuracy measure for linear features. *International Journal of Geographical Information Science, 11*(3), 299–306. doi:10.1080/136588197242419

Goodchild, M. (2007). Citizens as voluntary sensors: Spatial data infrastructure in the world of Web 2.0. *International Journal of Spatial Data Infrastructures Research, 2*, 24–32.

Goodchild, M. (2009). NeoGeography and the nature of geographic expertise. *Journal of Location Based Services, 3*(2), 82–96. doi:10.1080/17489720902950374

Goodchild, M. F. (2007a). Citizens as Sensors: Spatial data infrastructure in the world of web 2.0. *International Journal of Spatial Data Infrastructures Research, 2*, 24–32.

Goodchild, M. F. (2007b). Citizens and Sensors: The World of Volunteered Geography. *GeoJournal, 69*(4), 211–221. doi:10.1007/s10708-007-9111-y

Goodchild, M. F. (2008). Commentary: Whither VGI? *GeoJournal, 72*(3-4), 239–244. doi:10.1007/s10708-008-9190-4

Goodchild, M. F. (2009). What problem? Spatial autocorrelation and geographic information science. *Geographical Analysis, 41*(4), 411–417. doi:10.1111/j.1538-4632.2009.00769.x

Goodchild, M. F. (2013). The quality of big (geo)data. *Dialogues in Human Geography, 3*(3), 280–284. doi:10.1177/2043820613513392

Goodchild, M. F., & Glennon, J. A. (2010). Crowdsourcing geographic information for disaster response: A research frontier. *International Journal of Digital Earth, 3*(3), 231–241. doi:10.1080/17538941003759255

Goodchild, M. F., & Li, L. (2012). Assuring the quality of volunteered geographic information. *Spatial Statistics, 1*, 110–120.

Goodchild, M. F., & Li, L. (2012). Assuring the Quality of Volunteered Geographic Information. *Spatial Statistics, 1*, 110–120. doi:10.1016/j.spasta.2012.03.002

Google. (2015). Retrieved from http://www.google.com/mapmaker

Google. (2016). *Google Maps*. Retrieved January 17, 2016 from http://maps.google.com

Goranson, C., Thihalolipavan, S., & di Tada, N. (2013). VGI and public health: possibilities and pitfalls. In *Crowdsourcing Geographic Knowledge* (pp. 329–340). Springer Netherlands. doi:10.1007/978-94-007-4587-2_18

Gould, P. (1986). *Mental Maps*. Taylor & Francis.

Government of Canada. (2007a, March 31). *Sponsor a refugee*. Retrieved August 11, 2016, from http://www.cic.gc.ca/english/refugees/sponsor/index.asp

Government of Canada. (2015b, November). *Canada's response to the conflict in Syria*. Retrieved August 11, 2016, from http://www.international.gc.ca/development-developpement/humanitarian_response-situations_crises/syria-syrie.aspx?lang=eng

Government of Canada. (2016c, June 21). *Syria – by the numbers*. Retrieved August 11, 2016, from http://www.cic.gc.ca/english/refugees/welcome/commitment.asp

Government of the Czech Republic. (2016). *Report about the Roma minority in the Czech Republic in 2015*. Retrieved from http://www.vlada.cz/cz/ppov/zalezitosti-romske-komunity/dokumenty/zprava-o-stavu-romske-mensiny-za-rok-2015-146576/

Granholm, L. (2012). *3D tie to mallinnuksenkansainvälinen standard ointi*. Academic Press.

Graser, A., Straub, M., & Dragaschnig, M. (2014). Towards an open source analysis toolbox for street network comparison: Indicators, tools and results of a comparison of osm and the official Austrian reference graph. *Transactions in GIS*, *18*(4), 510–526. doi:10.1111/tgis.12061

Griffin, A. L., & Mcquoid, J. (2012). At the Intersection of Maps and Emotion : The Challenge of Spatially Representing Experience. *Kartographisch Nachrichten*, (6), 291–299.

Grimmett, G., & Stirzaker, D. (2001). *Probability and random processes*. Oxford University Press.

Grira, J., & Bédard, Y. (2009). Spatial data uncertainty in the VGI world: Going from consumer to producer. *Geomatica*, *64*(1), 61–71.

Grira, J., Bedard, Y., & Roch, S. (2009). Spatial data uncertainty in the VGI world. Going from consumer to producer. *Geomatica*, *46*(1), 61–71.

Gröchenig, S., Brunauer, R., & Rehrl, K. (2014). Estimating completeness of VGI datasets by analyzing community activity over time periods. In C. Granell, S. Schade, & J. Huerta (Eds.), *17th AGILE Conference on Geographic Information Science, AGILE 2014* (pp. 3–18). Kluwer Academic Publishers. doi:10.1007/978-3-319-03611-3_1

Gröger, G., Kolbe, T. H., Czerwinski, A., & Nagel, C. (2008). *OpenGIS city geography markup language (CityGML) encoding standard, version 1.0. 0*. Academic Press.

Guptill, S. C., & Morrison, J. L. (2013). *Elements of spatial data quality*. Elsevier.

Haddad, N. M., Brudvig, L. A., Clobert, J., Davies, K. F., Gonzalez, A., Holt, R. D., & Cook, W. M. et al. (2015). Habitat fragmentation and its lasting impact on Earths ecosystems. *Science Advances*, *1*(2), e1500052. doi:10.1126/sciadv.1500052 PMID:26601154

Hagenauer, J., & Helbich, M. (2012). Mining urban land-use patterns from volunteered geographic information by means of genetic algorithms and artificial neural networks. *International Journal of Geographical Information Science*, *26*(6), 963–982. doi:10.1080/13658816.2011.619501

Hagerty, M. R., Cummins, R. A., Ferriss, A. L., Land, K., Michalos, A. C., Peterson, M., & Vogel, J. et al. (2001). Quality of Life Indexes for National Policy: Review and Agenda for Research. *Social Indicators Research*, *55*(1), 1–96. doi:10.1023/A:1010811312332

Haklay, M. (2013). Citizen Science and Volunteered Geographic Information: Overview and Typology of Participation. In D. Sui, S. Elwood, & M. Goodchild (Eds.), Crowdsourcing Geographic Knowledge SE - 7 (pp. 105–122). CHAP, Springer Netherlands. doi:10.1007/978-94-007-4587-2_7

Haklay, M., Antoniou, V., Basiouka, S., Soden, R., & Mooney, P. (2014). Crowdsourced Geographic Information Use in Government. London: UCL-Geomatics.

Haklay, M. (2010). How good is volunteered geographical information? A comparative study of OpenStreetMap and ordnance survey datasets. *Environment and Planning. B, Planning & Design*, *37*(4), 682–703. doi:10.1068/b35097

Haklay, M. (2012). Citizen scince and volunteered geographic information: Overview and typology of participation. In D. Sui, S. Elwood, & M. Goodchild (Eds.), *Crowdsourcing geographic knowledge: Volunteered geographic information and knowledge* (pp. 105–122). New York: Springer.

Haklay, M., Basiouka, S., Antoniou, V., & Ather, A. (2010). How Many Volunteers Does it Take to Map an Area Well? The Validity of Linus Law to Volunteered Geographic Information. *The Cartographic Journal*, *47*(4), 315–322. doi:10.1179/000870410X12911304958827

Haklay, M., Singleton, A., & Parker, C. (2008). Web mapping 2.0: The neogeography of the GeoWeb. *Geography Compass*, *2*(6), 2011–2039. doi:10.1111/j.1749-8198.2008.00167.x

Haklay, M., & Weber, P. (2008). OpenStreet map: User-generated street maps. *IEEE Pervasive Computing / IEEE Computer Society [and] IEEE Communications Society*, *7*(4), 12–18. doi:10.1109/MPRV.2008.80

Hall, P. (2002). *Cities of tomorrow: an intellectual history of urban planning and design in the twentieth century*. Malden: Blackwell.

Han, D., Jung, S., Lee, M., & Yoon, G. (2014). Building a practical Wi-Fi-based indoor navigation system. *IEEE Pervasive Computing / IEEE Computer Society [and] IEEE Communications Society*, *13*(2), 72–79. doi:10.1109/MPRV.2014.24

Hara, K., Le, V., & Froehlich, J. (2013, April). Combining crowdsourcing and google streetview to identify street-level accessibility problems.*Proceedings of the SIGCHI conference on human factors in computing systems* (pp. 631-640). doi:10.1145/2470654.2470744

Harris, T., & Weiner, D. (1998). Empowerment, marginalization, and community integrated GIS. *Cartography and Geographic Information Systems*, *25*(2), 67–76. doi:10.1559/152304098782594580

Hartung, C., Lerer, A., Anokwa, Y., Tseng, C., Brunette, W., & Borriello, G. (2010). Open data kit: Tools to build information services for developing regions. In *Proceedings of the 4th acm/ieee international conference on information and communication technologies and development* (p. 18). doi:10.1145/2369220.2369236

HarvestChoice. (n.d.). *Agroecological Domains*. Retrieved from: http://harvestchoice.org/topics/agroecological-domains

Harvey, F. (2012). Code/space: Software and Everyday Life by Rob Kitchen and Martin Dodge. *International Journal of Geographical Information Science*, *26*(10), 1999–2001. doi:10.1080/13658816.2012.661059

Harvey, F. (2013). To volunteer or to contribute locational information? Towards truth in labeling for crowdsourced geographic information. In *Crowdsourcing Geographic Knowledge: Volunteered Geographic Information (VGI) in Theory and Practice* (Vol. 9789400745, pp. 31–42). Dordrecht: Springer Netherlands. doi:10.1007/978-94-007-4587-2_3

Harvey, F., Kwan, M. P., & Pavlovskaya, M. (2004). Introduction: Critical GIS. *Cartographica, 40*(4), 1–4. doi:10.3138/04L6-2314-6068-43V6

Hashemi, P., & Ali Abbaspour, R. (2015). Assessment of Logical Consistency in OpenStreetMap Based on the Spatial Similarity Concept. In J. Jokar Arsanjani, A. Zipf, P. Mooney, & M. Helbich (Eds.), OpenStreetMap in GIScience SE - 2 (pp. 19–36). CHAP, Springer International Publishing. doi:10.1007/978-3-319-14280-7_2

Hashemi, P., Ali Abbaspour, R., Jokar Arsanjani, J., Zipf, A., Mooney, P., & Helbich, M. (2015). *OpenStreetMap in GIScience, Experiences, Research, and Applications (LNCS)*. Springer Berlin Heidelberg.

Hashimoto, T., Kuboyama, T., Chakraborty, B., & Shirota, Y. (2012). Discovering topic transition about the east japan great earthquake in dynamic social media. In *Proceedings of the IEEE Global Humanitarian Technology Conference (GHTC 2012)* (pp. 259–264). IEEE. https://doi.org/10.1109/GHTC.2012.42

Hastie, T., Tibshirani, R., Friedman, J., & Franklin, J. (2005). The elements of statistical learning: Data mining, inference and prediction. *The Mathematical Intelligencer, 27*(2), 83–85. doi:10.1007/BF02985802

Hausdorf, F., & Aumann, J. R. (1914). *Grundzüge der mengenlehre*. Veit.

Hauthal, E., & Burghardt, D. (2016). Mapping Space-Related Emotions out of User-Generated Photo Metadata Considering Grammatical Issues. *The Cartographic Journal, 53*(1), 78–90. doi:10.1179/1743277414Y.0000000094

Hayakawa, T., Imi, Y., & Ito, T. (2012). Analysis of quality of data in OpenStreetMap. In *2012 IEEE 14th International Conference on Commerce and Enterprise Computing, CEC 2012* (pp. 131–134). doi:10.1109/CEC.2012.29

Hecht, R., Kunze, C., & Hahmann, S. (2013). Measuring Completeness of Building Footprints in OpenStreetMap over Space and Time. *International Journal of Geo-Information, 2*, 1066–1091. 10.3390/ijgi2041066

Heilman, G. E., Strittholt, J. R., Slosser, N. C., & Dellasala, D. A. (2002). Forest fragmentation of the conterminous United States: Assessing forest intactness through road density and spatial characteristics. *Bioscience, 5241*, 1–22.

Heipke, C. (2010). Crowdsourcing geospatial data. *ISPRS Journal of Photogrammetry and Remote Sensing, 65*(6), 550–557. doi:10.1016/j.isprsjprs.2010.06.005

Helbich, M., Amelunxen, C., & Neis, P. (2012). Comparative Spatial Analysis of Positional Accuracy of OpenStreetMap and Proprietary Geodata.*Proceedings of GI_Forum 2012: Geovisualization, Society and Learning*, 24–33. doi:10.1080/13658816

Helbich, M., Amelunxen, C., Neis, P., & Zipf, A. (2012). Comparative spatial analysis of positional accuracy of OpenStreetMap and proprietary geodata.*Proceedings of GI Forum*, 24–33.

Helliwell, J. F. (2003). Hows life? Combining individual and national variables to explain subjective well-being. *Economic Modelling, 20*(2), 331–360. doi:10.1016/S0264-9993(02)00057-3

Henn, A., Ro¨mer, C., Gro¨ger, G., & Plu¨mer, L. (2012). Automatic classification of building types in 3d city models. *GeoInformatica, 16*(2), 281–306. doi:10.1007/s10707-011-0131-x

Herrera, J. A., Plöger, P., Hinkenjann, A., Maiero, J., Flores, M., & Ramos, A. (2014). *Pedestrian indoor positioning using smartphone multi-sensing, radio beacons, user positions probability map and IndoorOSM floor plan representation.* Paper presented at the Indoor Positioning and Indoor Navigation (IPIN), 2014 International Conference on. doi:10.1109/IPIN.2014.7275538

Hirata, E., Giannotti, M. A., Larocca, A. P. C., & Quintanilha, J. A. (2015). Flooding and inundation collaborative mapping - use of the Crowdmap/Ushahidi platform in the city of Sao Paulo, Brazil. *Journal of Flood Risk Management*. https://doi.org/10.1111/jfr3.12181

Hochmair, H. H., & Zielstra, D. (2013). Development and Completeness of Points Of Interest in Free and Proprietary Data Sets: A Florida Case Study. *Gi_Forum 2013: Creating the Gisociety*, 39–48. 10.1553/giscience2013s39

Hochmair, H. H., Zielstra, D., & Neis, P. (2015). Assessing the Completeness of Bicycle Trail and Designated Lane Features in OpenStreetMap for the United States and Europe. *Transactions in GIS*, *19*(1), 63–81. doi:10.1111/tgis.12081

Hodgson, D., & Schroeder, R. (2002). Dilemmas of Counter-mapping Community Resources in Tanzania. *Development and Change*, *33*(1), 79–100. doi:10.1111/1467-7660.00241

Höllerer, T., Feiner, S., Terauchi, T., Rashid, G., & Hallaway, D. (1999). Exploring MARS: Developing indoor and outdoor user interfaces to a mobile augmented reality system. *Computers & Graphics*, *23*(6), 779–785. doi:10.1016/S0097-8493(99)00103-X

Homer, C. G., Dewitz, J. A., Yang, L., Jin, S., Danielson, P., Xian, G., & Megown, K. (2015). Completion of the 2011 National Land Cover Database for the conterminous United states-Representing a decade of land cover change information. *Photogrammetric Engineering and Remote Sensing*, *81*(5), 345–354.

Horita, F. E. A., Fava, M. C., Mendiondo, E. M., Rotava, J., Souz, V. C., Ueyama, J., & Albuquerque, J. P. (2014). AG-ORA-GeoDash: A Geosensor Dashboard for Real-time Flood Risk Monitoring. In *Proceeding of the 11th International Conference on Information Systems for Crisis Response and Management*. ISCRAM.

Horita, F. E. A., Link, D., Albuquerque, J. P., & Hellingrath, B. (2014). A Framework for the Integration of Volunteered Geographic Information into Humanitarian Logistics. In *Proceedings of the the 20th Americas Conference on Information Systems (AMCIS 2014)*. Savannah, GA: AIS.

Horita, F. E. A., Link, D., Albuquerque, J. P., & Hellingrath, B. (2016). oDMN: An Integrated Model to Connect Decision-Making Needs to Emerging Data Sources in Disaster Management. In *Proceedings of the 49th Hawaii International Conference on System Sciences (HICSS 2016)* (pp. 2882–2891). IEEE. https://doi.org/10.1109/HICSS.2016.361

Horita, F. E. A., Albuquerque, J. P., Degrossi, L. C., Mendiondo, E. M., & Ueyama, J. (2015). Development of a spatial decision support system for flood risk management in Brazil that combines volunteered geographic information with wireless sensor networks. *Computers & Geosciences*, *80*, 84–94. doi:10.1016/j.cageo.2015.04.001

Horita, F. E. A., Albuquerque, J. P., Marchezini, V., & Mendiondo, E. M. (2016). A qualitative analysis of the early warning process in disaster management. In *Proceedings of the 13th International Conference on Information Systems for Crisis Response and Management (ISCRAM)* (pp. 1–9).

Horita, F. E. A., Degrossi, L. C., Assis, L. F. F. G., Zipf, A., & Albuquerque, J. P. (2013). The use of Volunteered Geographic Information and Crowdsourcing in Disaster Management: a Systematic Literature Review. In *Proceedings of the Nineteenth Americas Conference on Information Systems* (pp. 1–10). Chicago, IL: AIS.

Horita, F. E., & De Albuquerque, J. P. (2013, May). An approach to support decision-making in disaster management based on volunteer geographic information (VGI) and spatial decision support systems (SDSS).*Proceedings of the 10th International Conference on Information Systems for Crisis Response and Management*, 1215.

Hosák, L., & Šrámek, R. (1980). *Místní jména na Moravě a ve Slezsku [Local names in Moravia and Silesia]* (Vol. 2). Brno: Academia. (In Czech)

Howe, J. (2006). The Rise of Crowdsourcing. *Wired*, *14*(6), 1–5.

Huang, H., Gartner, G., Klettner, S., & Schmidt, M. (2014). Considering Affective Responses towards Environments for Enhancing Location Based Services. ISPRS-International Archives of the Photogrammetry. *Remote Sensing and Spatial Information Sciences*, *1*, 93–96.

Huang, H., Kieler, B., & Sester, M. (2013). Urban building usage labeling by geometric and context analyses of the footprint data.*Proceeding of 26th international cartographic conference (ICC).*

Hughes, D., Ueyama, J., Mendiondo, E., Matthys, N., Horré, W., & Michiels, S. … Guan, S. U. (2011). A middleware platform to support river monitoring using wireless sensor networks. *Journal of the Brazilian Computer Society, 17*(2), 85–102. https://doi.org/10.1007/s13173-011-0029-3

Humantarian Open StreetMap Team. (2016). *Humantarian Open StreetMap*. Retrieved February 13, 2016 from https://hotosm.org

Hung, K. C., Kalantari, M., & Rajabifard, A. (2016). Methods for assessing the credibility of volunteered geographic information in flood response: A case study in Brisbane, Australia. *Applied Geography (Sevenoaks, England)*, *68*, 37–47. doi:10.1016/j.apgeog.2016.01.005

Hunter, G. J. (1999). New tools for handling spatial data quality: Moving from academic concepts to practical reality. *URISA Journal*, *11*(2), 25–34.

Hunter, G. J., Bregt, A. K., Heuvelink, G. B. M., De Bruin, S., & Virrantaus, K. (2009). Spatial Data Quality: Problems and Prospects. In G. Navratil (Ed.), *Research Trends in Geographic Information Science* (pp. 101–121). Berlin: Springer Berlin Heidelberg. doi:10.1007/978-3-540-88244-2_8

Hydén, C. (1987). The development of a method for traffic safety evaluation: The Swedish Traffic Conflicts Technique. *Bulletin Lund Institute of Technology, Department*, (70).

IBGE. (2015). *Instituto Brasileiro de Geografia e Estatística*. Retrieved October 6, 2015, from http://www.ibge.gov.br/

IDRC. (1996). *The Local Agenda 21 Planning Guide. The International Council for Local Environmental Initiatives (ICLEI), The International Development Research Centre (IDRC), The United Nations Environment Programme*. UNEP.

Imran, M., Castillo, C., Lucas, J., Meier, P., & Rogstadius, J. (2014). Coordinating human and machine intelligence to classify microblog communications in crises. In *Proceedings of the 11th International Conference on Information Systems for Crisis Response and Management (ISCRAM 2014)* (pp. 712–721). ISCRAM.

Index of /planet/full-history. (2016). Retrieved July 4, 2016, from http://planet.openstreetmap.org/planet/full-history/

Inside Secure. (n.d.). Retrieved from: http://www.insidesecure.com/Products/DRM-Content-Protection/Microsoft-DRM-based-Solutions

Intelligent Transportation Systems Institute, Center for Transportation Studies. (2011). *Development of Mobile Accessible Pedestrian Signals (MAPS) for Blind Pedestrians at Signalized Intersections*. Center for Transportation Studies.

IRC - OpenStreetMap Wiki. (2015). Retrieved July 4, 2016, from http://wiki.openstreetmap.org/wiki/IRC

Irwin, A. (1995). *Citizen science: A study of people, expertise, and sustainable development* (Vol. 136). Routledge Londres.

ISO, T. C. 211. (2002). *ISO 19113, Geographic Information–Quality Principles.* The International Organization for Standardisation.

ISO. (2014). *ISO 19153:2014: Geospatial Digital Rights Management Reference Model.* Retrieved from: http://www.iso.org/iso/catalogue_detail.htm?csnumber=32571

Jackson, S., Mullen, W., Agouris, P., Crooks, A., Croitoru, A., & Stefanidis, A. (2013). Assessing Completeness and Spatial Error of Features in Volunteered Geographic Information. *ISPRS International Journal of Geo-Information, 2*(2), 507–530. doi:10.3390/ijgi2020507

Jacobs, J. (1961). *The Death and Life of Great American Cities*. Vintage Books.

Jaeger, J. A., Schwarz-von Raumer, H. G., Esswein, H., Müller, M., & Schmidt-Lüttmann, M. (2007). Time series of landscape fragmentation caused by transportation infrastructure and urban development: A case study from Baden-Württemberg, Germany. *Ecology and Society, 12*(1), 22. doi:10.5751/ES-01983-120122

James, D. (2003). *Digital Rights Management & Music, A barrier to creativity?* Academic Press.

Jankowski, P., Czepkiewicz, M., Młodkowski, M., & Zwoliński, Z. (2015). Geo-questionnaire: A Method and Tool for Public Preference Elicitation in Land Use Planning. *Transactions in GIS*. 10.1111/tgis.12191

Javanmardi, S., Lopes, C., & Baldi, P. (2010). Modeling user reputation in wikis. *SStatistical Analysis and Data Mining, 3*(2), 126–139.

Jayakody, J. A., Murray, I., & Herrmann, J. (2015, October). Analgorithm for labelling topological map store present point of interest for vision impaired navigation. In Proceedings of 2015 International Conferenceonthe Indoor Positioningand Indoor Navigation (IPIN), (pp. 1-8).

Jha, A. K., Barenstein, J. D., Phelps, P. M., Pittet, D., & Sena, S. (2010). *Safer Homes, Stronger Communities*. The World Bank. https://doi.org/10.1596/978-0-8213-8045-1

Jiang, B. (2013). Volunteered Geographic Information and Computational Geography: New Perspectives. In D. Sui, S. Elwood, & M. Goodchild (Eds.), Crowdsourcing Geographic Knowledge SE - 8 (pp. 125–138). CHAP, Springer Netherlands. doi:10.1007/978-94-007-4587-2_8

Jíchová, J., & Temelová, J. (2012). Kriminalita a její percepce ve vnitřním městě: případová studie pražského Žižkova a Jarova [Criminality and its perception in the inner city: case studies of Žižkov and Jarov in Prague]. *Geografie, 3*(117), 329–348. Retrieved from http://geography.cz/sbornik/wp-content/uploads/downloads/2012/11/g03_2012_5jichova_temelova.pdf

Jilani, M. (2016). *Structured and Unstructured Machine Learning for Crowdsourced Spatial Data* (PhD Thesis). University College Dublin, Ireland.

Jilani, M., Corcoran, P., & Bertolotto, M. (2014). Automated highway tag assessment of openstreetmap road networks. In M. Schneider, M. Gertz, Y. Huang, J. Sankaranarayanan, & J. Krumm (Eds.), *22nd ACM SIGSPATIAL International Conference on Advances in Geographic Information Systems, ACM SIGSPATIAL GIS 2014* (*Vol. 04–07*–Nove, pp. 449–452). Association for Computing Machinery. doi:10.1145/2666310.2666476

Jilani, M., Corcoran, P., & Bertolotto, M. (2016). Probabilistic graphical modelling for semantic labelling of crowdsourced map data. Intelligent Systems Technologies and Applications, 213-224. doi:10.1007/978-3-319-23258-4_19

Jilani, M., Corcoran, P., & Bertolotto, M. (2013). Multi-granular street network representation towards quality assessment of OpenStreetMap data.*Proceedings of the Sixth ACM SIGSPATIAL International Workshop on Computational Transportation Science*, 19-24. doi:10.1145/2533828.2533833

Joachims, T. (1998). Text categorization with support vector machines: Learning with many relevant features. In *European conference on machine learning* (pp. 137-142). Springer Berlin Heidelberg. doi:10.1007/BFb0026683

John, G. H., & Langley, P. (1995). Estimating continuous distributions in bayesian classifiers. In *Proceedings of the Eleventh conference on Uncertainty in artificial intelligence*. Morgan Kaufmann Publishers Inc.

Johnson, B. A., & Iizuka, K. (2016). Integrating OpenStreetMap crowdsourced data and Landsat time-series imagery for rapid /land cover (LULC) mapping: Case study of the Laguna de Bay area of the Philippines. *Applied Geography (Sevenoaks, England), 67*, 140–149. doi:10.1016/j.apgeog.2015.12.006

Johnson, P. A., & Sieber, R. E. (2012). Motivations driving government adoption of the Geoweb. *GeoJournal, 77*(5), 667–680. doi:10.1007/s10708-011-9416-8

Jokar Arsanjani, J., & Vaz, E. (2015). An assessment of a collaborative mapping approach for exploring land use patterns for several European metropolises. *International Journal of Applied Earth Observation and Geoinformation, 35*(B), 329–337. 10.1016/j.jag.2014.09.009

Jokar Arsanjani, J., Helbich, M., Bakillah, M., & Loos, L. (2015). The emergence and evolution of OpenStreetMap: a cellular automata approach. *International Journal of Digital Earth, 8*(1), 74–88. 10.1080/17538947.2013.847125

Jokar Arsanjani, J., Mooney, P., Helbich, M., & Zipf, A. (2015). An Exploration of Future Patterns of the Contributions to OpenStreetMap and Development of a Contribution Index. *Transactions in GIS, 19*(6), 896–914. 10.1111/tgis.12139

Jokar Arsanjani, J., Mooney, P., Zipf, A., & Schauss, A. (2015). Quality Assessment of the Contributed Land Use Information from OpenStreetMap Versus Authoritative Datasets. In J. Jokar Arsanjani, A. Zipf, P. Mooney, & M. Helbich (Eds.), OpenStreetMap in GIScience SE - 3 (pp. 37–58). CHAP, Springer International Publishing. doi:10.1007/978-3-319-14280-7_3

Jokar Arsanjani, J., Zipf, A., Mooney, P., & Helbich, M. (2015). An Introduction to OpenStreetMap in Geographic Information Science: Experiences, Research, and Applications. In J. Jokar Arsanjani, A. Zipf, P. Mooney, & M. Helbich (Eds.), OpenStreetMap in GIScience SE - 1 (pp. 1–15). CHAP, Springer International Publishing. doi:10.1007/978-3-319-14280-7_1

Jokar Arsanjani, J., & Bakillah, M. (2014). Understanding the potential relationship between the socio-economic variables and contributions to OpenStreetMap. *International Journal of Digital Earth, 8*(10), 1–16. doi:10.1080/17538947.2014.951081

Jokar Arsanjani, J., Barron, C., Bakillah, M., & Helbich, M. (2013). Assessing the Quality of OpenStreetMap Contributors together with their Contributions.*16th AGILE Conference.*

Jokar Arsanjani, J., Helbich, M., & Bakillah, M. (2013). Exploiting volunteered geographic information to ease mapping of an urban landscape. *ISPRS-International Archives of the Photogrammetry. Remote Sensing and Spatial Information Sciences, 1*(1), 51–55.

Jokar Arsanjani, J., Helbich, M., Bakillah, M., Hagenauer, J., & Zipf, A. (2013). Toward mapping land-use patterns from volunteered geographic information. *International Journal of Geographical Information Science, 27*(12), 2264–2278. doi:10.1080/13658816.2013.800871

Jokar Arsanjani, J., & Vaz, E. (2015). An assessment of a collaborative mapping approach for exploring land use patterns for several European metropolises. *International Journal of Applied Earth Observation and Geoinformation, 35*, 329–337. doi:10.1016/j.jag.2014.09.009

Jokar Arsanjani, J., Zipf, A., Mooney, P., & Helbich, M. (2015). An introduction to OpenStreetMap in Geographic Information Science: Experiences, research, and applications. In J. Jokar Arsanjani, A. Zipf, P. Mooney, & M. Helbich (Eds.), *OpenStreetMap in GIScience* (pp. 1–18). Cham: Springer International Publishing. doi:10.1007/978-3-319-14280-7_1

Jokar Arsanjani, J., Zipf, A., Mooney, P., & Helbich, M. (Eds.). (2015). *OpenStreetMap in GIScience: Experiences, Research, and Applications. Lecture Notes in Geoinformation and Cartography*. Springer. doi:10.1007/978-3-319-14280-7

Kahila, M., & Kyttä, M. (2009). SoftGIS as a Bridge-Builder in Collaborative Urban Planning. In S. Geertman & J. Stillwell (Eds.), *Planning Support Systems Best Practice and New Methods* (Vol. 95, pp. 389–411). Dordrecht: Springer Netherlands; doi:10.1007/978-1-4020-8952-7_19

Kahila-Tani, M., Broberg, A., Kyttä, M., & Tyger, T. (2015). Let the Citizens Map—Public Participation GIS as a Planning Support System in the Helsinki Master Plan Process. *Planning Practice and Research*, 1–20. doi:10.1080/02697459.2015.1104203

Kaklanis, N., Votis, K., & Tzovaras, D. (2013a, May). A mobile interactivemapsapplication for avisuallyimpairedaudience.*Proceedings of the 10th International Cross-DisciplinaryConferenceon Web Accessibility.*

Kaklanis, N., Votis, K., & Tzovaras, D. (2013b). Open Touch/Sound Maps: A system to convey street data through haptic and auditory feedback. *Computers & Geosciences*, *57*, 59–67. doi:10.1016/j.cageo.2013.03.005

Kaklanis, N., Votis, K., & Tzovaras, D. (2013c). Touching openstreetmap data in mobile context for the visually impaired. In *Proceedings of the 3rd Workshop on Mobile Accessibility-ACM SIGCHI Conference on Human Factors in Computing Systems* (pp. 68-74).

Kalantari, M., & La, V. (2015). Assessing OpenStreetMap as an Open Property Map. In J. Jokar Arsanjani, A. Zipf, P. Mooney, & M. Helbich (Eds.), OpenStreetMap in GIScience SE - 13 (pp. 255–272). CHAP, Springer International Publishing. doi:10.1007/978-3-319-14280-7_13

Kargl, F., Geßler, S., & Flerlage, F. (2007). *The iNAV indoor navigation system.* Paper presented at the International Symposium on Ubiquitious Computing Systems. doi:10.1007/978-3-540-76772-5_9

Karimi, H. A. (2015). *Indoor Wayfinding and Navigation.* CRC Press. doi:10.1201/b18220

Kean, A. (2007). *The Cult of the Amateur: How Today's Internet is Killing Our Culture.* New York: Doubleday.

Kenny, M. (2010). Marine Spatial data Infrastructure.*2nd IHO-HSSC meeting, HSSC2-INF3.*

Kepler, C., & Groot, R. T. A. d. (2013). Trust as a proxy measure for the quality of Volunteered Geographic Information in the case of OpenStreetMap. Geographic Information Science at the Heart of Europe, 21–37.

Keßler, C., Theodore, R., & De Groot, A. (2013). Trust as a Proxy Measure for the Quality of Volunteered Geographic Information in the Case of OpenStreetMap. In *Geographic Information Science at the Heart of Europe* (pp. 21–37). Springer; doi:10.1007/978-3-319-00615-4_2

Keyes, L., & Winstanley, A. (2001). Using moment invariants for classifying shapes on large-scale maps. *Computers, Environment and Urban Systems*, *25*(1), 119–130. doi:10.1016/S0198-9715(00)00041-7

Keyes, L., Winstanley, A., & Hea, P. (2003). Comparing learning strategies for topographic object classification.*Geoscience and Remote Sensing Symposium, 2003. IGARSS'03. Proceedings. 2003 IEEE International*, 6, 3468– 3470. doi:10.1109/IGARSS.2003.1294824

Kim, S., Mankoff, J., & Paulos, E. (2013). Sensr: Evaluating a Flexible Framework for Authoring Mobile Data-Collection Tools for Citizen Science. In *Proceedings of the 2013 conference on computer supported cooperative work* (pp. 1453–1462). doi:10.1145/2441776.2441940

Kim, S., Robson, C., Zimmerman, T., Pierce, J., & Haber, E. M. (2011). Creek Watch: Pairing Usefulness and Usability for Successful Citizen Science. In *Proceedings of the sigchi conference on human factors in computing systems* (pp. 2125–2134). doi:10.1145/1978942.1979251

Kinsella, N. S. (2001). Against Intellectual property. *The Journal of Libertarian Studies*, *15*(2).

Klepeis, P., & Vance, C. (2003). Neoliberal policy and deforestation in southeastern Mexico: An assessment of the PROCAMPO program. *Economic Geography*, *79*(3), 221–240. doi:10.1111/j.1944-8287.2003.tb00210.x

Kloeckl, K., Senn, O., Di Lorenzo, G., & Ratti, C. (2011). Live singapore!-an urban platform for real-time data to program the city. In Computers in Urban Planning and Urban Management (Vol. 4, pp. 1–18). CUPUM. Retrieved from http://senseable.mit.edu/papers/pdf/20110708_Kloeckl_etal_LiveSingapore_ComputersUrban.pdf

Knowles, A. K., Westerveld, L., & Strom, L. (2015). Inductive Visualization: A Humanistic Alternative to GIS. *Geo-Humanities*, 1–33. 10.1080/2373566X.2015.1108831

Koester, D., Lunt, B., & Stiefelhagen, R. (2016). Zebra crossing detection from aerial imagery across countries. In *Proceedings of the International Conference on Computers Helping People with Special Needs (ICCHP) part II* (pp. 27-34). doi:10.1007/978-3-319-41267-2_5

Koller, D., & Friedman, N. (2009). *Probabilistic graphical models: principles and techniques*. MIT Press.

Kongthon, A., Haruechaiyasak, C., Pailai, J., & Kongyoung, S. (2012). The role of Twitter during a natural disaster: Case study of 2011 Thai Flood. In *2012 Proceedings of PICMET'12: Technology Management for Emerging Technologies* (pp. 2227–2232). Vancouver, Canada: PICMET.

Korn, N., Oppenheim, C., & Duncan, C. (2007). *IPR and Licensing issues in Derived Data*. Retrieved from: http://webarchive.nationalarchives.gov.uk/20140702233839/http:/www.jisc.ac.uk/media/documents/projects/iprinderived-datareport.pdf

Korpela, K. (2002). Children's environment.Handbook of Environmental Psychology, 363–373.

Kotsev, A., Pantisano, F., Schade, S., & Jirka, S. (2015). Architecture of a Service-Enabled Sensing Platform for the Environment. *Sensors (Basel, Switzerland)*, *15*(2), 4470–4495. doi:10.3390/s150204470 PMID:25688593

Koukoletsos, T., Haklay, M., & Ellul, C. (2012). Assessing Data Completeness of VGI through an Automated Matching Procedure for Linear Data. *Transactions in GIS*, *16*(4), 477–498. doi:10.1111/j.1467-9671.2012.01304.x

Koukourikos, P., & Papadopoulos, K. (2015). Development of cognitive maps by individuals with blindness using a multisensory application. In *Procedia Computer Science – Proceedings of the 6th International Conference on Software Development and Technologies for Enhancing Accessibility and Fighting Info-exclusion* (pp. 213-222). doi:10.1016/j.procs.2015.09.265

Kounadi, O. (2009). *Assessing the quality of openstreetmap data. Msc geographical information science*. University College of London Department of Civil, Environmental And Geomatic Engineering.

Kresse, W., & Fadaie, K. (2004). *ISO standards for geographic information*. Springer Science & Business Media. doi:10.1007/978-3-662-08039-9

Kschischang, F. R., Frey, B. J., & Loeliger, H.-A. (2001). Factor graphs and the sum-product algorithm. *Information Theory. IEEE Transactions on*, *47*(2), 498–519.

Kuhn, W. (2007). *Volunteered geographic information and GIScience*. NCGIA and Vespucci Workshop on VGI, Santa Barbara, CA.

Kuhn, W. (2005). Geospatial semantics: Why, of what, and how? *Journal on Data Semantics, Special Issue on Semantic-based Geographical Information Systems*, *3534*, 1–24.

Kumar, S., & Havey, T. (2013). Before and after disaster strikes: A relief supply chain decision support framework. *International Journal of Production Economics*, *145*(2), 613–629. doi:10.1016/j.ijpe.2013.05.016

Kurniawan, A. (2013). *Getting Started with Matlab Simulink and Raspberry Pi*. PE Press.

Kvasnička, R. (2010). *Description of socially excluded locations in Ostrava region*. Ostrava.

Kwantlen Polytechnic University. (2015). *Plant Database*. Retrieved January 01, 2015 from http://plantdatabase.kwantlen.ca/

Laakso, M. (2014). *Improving accessibility for pedestrians with geographic information* (Doctoral dissertation). Available from Aalto University Library.

Lafferty, J., McCallum, A., & Pereira, F. C. (2001). *Conditional random fields: Probabilistic models for segmenting and labeling sequence data*. Academic Press.

Lambin, E. F., Turner, B. L., Geist, H. J., Agbola, S. B., Angelsen, A., Bruce, J. W., & George, P. et al. (2001). The causes of land-use and change: Moving beyond the myths. *Global Environmental Change*, *11*(4), 261–269. doi:10.1016/S0959-3780(01)00007-3

Lamotte, H. C., & Alexander, L. (2009). *Marine Spatial data infrastructure. Challenges and Opportunities: The Cuban experiences*. InternationlHydrographic Organization (IHO) Spatial Data infrastructure. *Seminar*.

Langer, D., Rosenblatt, J. K., & Hebert, M. (1994). A reactive system for off-road navigation. Robotics Institute, 441.

LeBaron, G. S., Cannings, R. J., Niven, D. K., Butcher, G. S., Bancroft, G. T., Sykes, P. W. Jr, & Read, P. (2009). The 109th Christmas Bird Count. *American Birds*, *63*, 2–7.

Lee, J. (2004). A spatial access-oriented implementation of a 3-D GIS topological data model for urban entities. *GeoInformatica*, *8*(3), 237–264. doi:10.1023/B:GEIN.0000034820.93914.d0

Leeuw, J. D., Said, M., Ortegah, L., Nagda, S., Georgiadou, Y., & DeBlois, M. (2011). An assessment of the accuracy of volunteered road map production in Western Kenya. *Remote Sensing*, *3*(2), 247–256. doi:10.3390/rs3020247

Leitner, M. (2013). *Crime modeling and mapping using geospatial technologies* (Vol. 8). Springer Science & Business Media. doi:10.1007/978-94-007-4997-9

Lemos, F. (2013). *Infrastructure and algorithms for information quality analysis and process discovery* (Doctoral dissertation). University of Versailles, France.

Lemos, M. B., Esteves, O. de A., & Simões, R. F. (1995). Uma metodologia para a construção de um índice de qualidade de vida urbana. *Nova Economia*, *5*(2), 157–176. Retrieved from http://ideas.repec.org/a/nov/artigo/v5y1995i2p157-176.html

Leszczynski, A. (2012). Situating the geoweb in political economy. *Progress in Human Geography*, *36*(1), 72–89. doi:10.1177/0309132511411231

Levenshtein, V. I. (1966). Binary codes capable of correcting deletions, insertions and reversals. *Soviet Physics, Doklady*, *10*, 707.

Levinson, J., Askeland, J., Becker, J., Dolson, J., Held, D., Kammel, S., & Pratt, V. (2011). *Towards fully autonomous driving: Systems and algorithms*. Paper presented at the Intelligent Vehicles Symposium (IV). doi:10.1109/IVS.2011.5940562

Li, K. J., & Lee, J. Y. (2013). Basic concepts of indoor spatial information candidate standard IndoorGML and its applications. *Journal of Korea Spatial Information Society*, *21*(3), 1–10. doi:10.12672/ksis.2013.21.3.001

Lin, W. (2013). When Web 2.0 Meets Public Participation GIS (PPGIS): VGI and Spaces of Participatory Mapping in China. In D. Sui, S. Elwood, & M. Goodchild (Eds.), Crowdsourcing Geographic Knowledge SE - 6 (pp. 83–103). CHAP, Springer Netherlands. doi:10.1007/978-94-007-4587-2_6

Lin, W. (2013). *When Web 2.0 meets public participation GIS (PPGIS): VGI and spaces of participatory mapping in China. In Crowdsourcing geographic knowledge* (pp. 83–103). Springer.

Lin, W. (2013). When Web 2.0 Meets Public Participation GIS (PPGIS): VGI and Spaces of Participatory Mapping in China. In D. Sui, S. Elwood, & M. Goodchild (Eds.), *Crowdsourcing Geographic Knowledge: Volunteered Geographic Information (VGI) in Theory and Practice* (pp. 83–104). New York: Springer Dordrecht Heidelberg. doi:10.1007/978-94-007-4587-2_6

Liu, Y. C. (2001). Comparative study of the effects of auditory, visual and multimodality displays on drivers performance in advanced traveller information systems. *Ergonomics*, *44*(4), 425–442. doi:10.1080/00140130010011369 PMID:11291824

Loai Ali, A., Schmid, F., Falomir, Z., & Freksa, C. (2015). Towards rule-guided classification for Volunteered Geographic Information. In ISPRS Annals of the Photogrammetry, Remote Sensing and Spatial Information Sciences (Vol. II-3/W5). La Grande Motte, France: ISPRS Annals. doi:10.5194/isprsannals-II-3-W5-211-2015

Longueville, B. D., Smith, R. S., & Luraschi, G. (2009). OMG, from here, I can see the flames!: a use case of mining location based social networks to acquire spatiotemporal data on forest fires. In *Proceedings of the 2009 international workshop on location based social networks* (pp. 73–80). doi:10.1145/1629890.1629907

Lu¨scher, P., Weibel, R., & Burghardt, D. (2009). Integrating ontological modelling and bayesian inference for pattern classification in topographic vector data. *Computers, Environment and Urban Systems*, *33*(5), 363–374. doi:10.1016/j.compenvurbsys.2009.07.005

Ludwig, I., Voss, A., & Krause-traudes, M. (2011). A Comparison of the Street Networks of Navteq and OSM in Germany. In *Advancing Geoinformation Science for a Changing World* (Vol. 1, pp. 65–84). Berlin: Springer-Verlag; doi:10.1007/978-3-642-19789-5_4

MacEachren, A. M., Jaiswal, A., Robinson, A. C., Pezanowski, S., Savelyev, A., & Mitra, P., … Blanford, J. (2011). SensePlace2: GeoTwitter Analytics Support for Situational Awareness. In *Proceedings of the IEEE Conference on Visual Analytics Science and Technology* (pp. 181–190). Providence, RI: IEEE. doi:10.1109/VAST.2011.6102456

MacKerron, G., & Mourato, S. (2010). *Mappiness*. Retrieved September 12, 2015, from http://www.mappiness.org.uk/

Mailing lists - OpenStreetMap Wiki. (2016). Retrieved July 4, 2016, from http://wiki.openstreetmap.org/wiki/Mailing_lists

Maisonneuve, N., Stevens, M., Niessen, M. E., Hanappe, P., & Steels, L. (2009). Citizen noise pollution monitoring. In *10th int. conference on digital government research* (pp. 96–103). Digital Government Society of North America.

Makri, A., Zlatanova, S., & Verbree, E. (2015). *An approach for indoor wayfinding replicating main principles of an outdoor navigation system for cyclists.* Paper presented at the International Archives of the Photogrammetry, Remote Sensing and Spatial Information Sciences, Indoor-Outdoor Seamless Modelling, Mapping and Navigation, Tokyo, Japan. doi:10.5194/isprsarchives-XL-4-W5-29-2015

Malaverri, J. E. G., Mota, M. S., & Medeiros, C. B. (2013). Estimating the Quality of Data Using Provenance: A Case Study in eScience. *19th Americas conference on information systems (AMCIS).*

Malaverri, J. G., Santanchè, A., & Medeiros, C. B. (2013). *A Provenance-based Approach to Evaluate Data Quality in eScience. Int. J. Metadata, Semantics and Ontologies.*

Mantero, P., Moser, G., & Serpico, S. B. (2005). Partially supervised classification of remote sensing images through svm-based probability density estimation. *Geoscience and Remote Sensing. IEEE Transactions on*, *43*(3), 559–570.

Marans, R. W. (2015). Quality of urban life & environmental sustainability studies: Future linkage opportunities. *Habitat International*, *45*(P1), 47–52. doi:10.1016/j.habitatint.2014.06.019

Martinez-Sala, A. S., Losilla, F., Sánchez-Aarnoutse, J. C., & García-Haro, J. (2015). Design, implementation and evaluation of an indoor navigation system for visually impaired people. *Sensors (Basel, Switzerland), 15*(12), 32168–32187. doi:10.3390/s151229912 PMID:26703610

Martinho, N., & Fonte, C. C. (2015). *Assessing the applicability of OpenStreetMap data to validate Land Use/Land Cover Maps* (Working Paper No. 5). Coimbra, Portugal: Instituto de Engenharia de Sistemas e Computadores de Coimbra.

Mashhadi, A., Quattrone, G., & Capra, L. (2015). The Impact of Society on Volunteered Geographic Information: The Case of OpenStreetMap. In J. Jokar Arsanjani, A. Zipf, P. Mooney, & M. Helbich (Eds.), OpenStreetMap in GIScience SE - 7 (pp. 125–141). CHAP, Springer International Publishing. doi:10.1007/978-3-319-14280-7_7

Mashhadi, A., Quattrone, G., & Capra, L. (2013). Putting ubiquitous crowdsourcing into context. In *Proceedings of the 2013 conference on Computer supported cooperative work.* ACM.

Mashhadi, A., Quattrone, G., Capra, L., & Mooney, P. (2012). On the accuracy of urban crowd-sourcing for maintaining large-scale geospatial databases. In *Proceedings of the Eighth Annual International Symposium on Wikis and Open Collaboration.* ACM. doi:10.1145/2462932.2462952

Masli, M., Bouma, L., Owen, A., & Terveen, L. (2013, February). Geowiki+ route analysis= improved transportation planning. In *Proceedings of the 2013 Conference on Computer Supported Cooperative Work Companion* (pp. 213-218). ACM.

Masolo, C., Borgo, S., Gangemi, A., Guarino, N., Oltramari, A., & Schneider, L. (2003). Dolce: a descriptive ontology for linguistic and cognitive engineering. *WonderWeb Project, Deliverable D17, 2,* 1.

Massidda, C., Bülthoff, H. H., & Stegagno, P. (2015). *Autonomous vegetation identification for outdoor aerial navigation.* Paper presented at the Intelligent Robots and Systems (IROS), 2015 IEEE/RSJ International Conference on. doi:10.1109/IROS.2015.7353806

Maxwell, E. (2006). *Open Standards, Open Source and Open Innovation, innovations.* MIT Press. Retrieved from: http://www.mitpressjournals.org/doi/pdf/10.1162/itgg.2006.1.3.119

Mayaux, P., Eva, H., Gallego, J., Strahler, A. H., Herold, M., & Agrawal, S. … Roy, P. S. (2006). Validation of the global land cover 2000 map. *IEEE Transactions on Geoscience and Remote Sensing, 44*, 1728–1737. Retrieved from http://www.scopus.com/scopus/inward/record.url?eid=2-s2.0-33746330734&partnerID=40

McCallum, A., & Nigam, K. 1998, July. A comparison of event models for naive bayes text classification. In AAAI-98 workshop on learning for text categorization (Vol. 752, pp. 41-48).

McGarigal, K., Romme, W. H., Crist, M., & Roworth, E. (2001). Cumulative effects of roads and logging on landscape structure in the San Juan Mountains, Colorado (USA). *Landscape Ecology, 16*(4), 327–349. doi:10.1023/A:1011185409347

McGee, R., Edwards, D., Minkley, G., Pegus, C.-M., & Brock, K. (2015). *Making All Voices Count: Research and Evidence Strategy.* Brighton, UK: Institute of Development Studies.

McMaster, R. B. (1986). A statistical analysis of mathematical measures for linear simplification. *The American Cartographer, 13*(2), 103–116. doi:10.1559/152304086783900059

McQuilton, P., Gonzalez-Beltran, A., Rocca-Serra, P., Thurston, M., Lister, A., Maguire, E., & Sansone, S. A. (2016). BioSharing: Curated and crowd-sourced metadata standards, databases and data policies in the life sciences. *Database.* PMID:27189610

MeadoWatch. (2015). *Citizen Science at the Mount Rainier National Park to Understand the Biological Impacts of Climate Change.* Retrieved January 01, 2015 from http://meadowatch.weebly

Meier, J. C. H. (2015). *An Analysis of Quality for Volunteered Geographic Information*. JOUR.

Memon, G. R. (2007). Education in Pakistan: The key issues, problems and the new challenges. *Journal of Management and Social Sciences*, *3*(1), 47–55.

Mendes, H. C., & Mendiondo, E. M. (2007). Histórico da Expansão Urbana e Incidência de Inundações : O Caso da Bacia do Gregório, São Carlos - SP. *Revista Brasileira de Recursos Hídricos, 12*(1), 17–27. https://doi.org/10.21168/rbrh.v12n1.p17-27

Mihalcea, R., & Radev, D. (2011). *Graph-based natural language processing and information retrieval*. Cambridge University Press. doi:10.1017/CBO9780511976247

Mikola, J. (2015). *GeoJSON implementation for PHP*. Retrieved from https://github.com/jmikola/geojson

Miller, H. J., & Goodchild, M. F. (2015). Data-driven geography. *GeoJournal, 80*(4), 449–461. 10.1007/s10708-014-9602-6

Mirri, S., Prandi, C., & Salomoni, P. (2016). Personalizing Pedestrian Accessible way-finding with mPASS. In 13th IEEE Annual Consumer Communications & Networking Conference (CCNC) (pp. 1119-1124). doi:10.1109/CCNC.2016.7444946

Mirvahabi, S., & Ali Abbaspour, R. (2015). Automatic Extraction of IndoorGML Core Model from OpenStreetMap. *The International Archives of Photogrammetry, Remote Sensing and Spatial Information Sciences*, *40*(1), 459.

Mislove, A., Lehmann, S., Ahn, Y.-Y., Onnela, J.-P., & Rosenquist, J. N. (2010). *Pulse of the Nation: U.S. Mood Throughout the Day inferred from Twitter*. Retrieved September 12, 2015, from http://www.ccs.neu.edu/home/amislove/twittermood/?utm_campaign=Facebook+Page&utm_content=Pulse+of+the+Nation+US+Mood+Throughout+the+Day+inferred+from+Twitter&utm_medium=postitat&utm_source=facebook

Mitchell, T. M. (1997). *Machine learning*. McGraw-Hill.

Mnih, V. (2013). *Machine learning for aerial image labeling* (PhD thesis). University of Toronto.

Mody, R. N., Willis, K. S., & Kerstein, R. (2009). WiMo: location-based emotion tagging. In *Proceedings of the 8th international Conference on Mobile and Ubiquitous Multimedia* (p. 14).

Mohammadi, N., & Malek, M. (2015). Artificial intelligence-based solution to estimate the spatial accuracy of volunteered geographic data. *Journal of Spatial Science, 60*(1), 119–135. 10.1080/14498596.2014.927337

Möller, A., Kranz, M., Huitl, R., Diewald, S., & Roalter, L. (2012). *A mobile indoor navigation system interface adapted to vision-based localization.* Paper presented at the 11th international conference on mobile and ubiquitous multimedia. doi:10.1145/2406367.2406372

Mondzech, J., & Sester, M. (2011). Quality analysis of OpenStreetMap data based on application needs. *Cartographica: The International Journal for Geographic Information and Geovisualization*, *46*(2), 115–125. doi:10.3138/carto.46.2.115

Mooney, P., & Corcoran, P. (2012a). Characteristics of Heavily Edited Objects in OpenStreetMap. Future Internet, 4(4), 285–305. doi:10.3390/fi4010285

Mooney, P., & Corcoran, P. (2012b). The Annotation Process in OpenStreetMap. *Transactions in GIS, 16*(4), 561–579. Journal Article. 10.1111/j.1467-9671.2012.01306.x

Mooney, P., Corcoran, P., & Winstanley, A. C. (2010). A study of data representation of natural features in openstreetmap. Proceedings of GIScience, 150, 150-156.

Mooney, P., Corcoran, P., & Winstanley, A. C. (2010). Towards quality metrics for OpenStreetMap. *Proceedings of the 18th SIGSPATIAL International Conference on Advances in Geographic Information Systems GIS 10*, 514–517. doi:10.1145/1869790.1869875

Mooney, P., & Corcoran, P. (2012a). The annotation process in OpenStreetMap. *Transactions in GIS, 16*(4), 561–579. doi:10.1111/j.1467-9671.2012.01306.x

Mooney, P., & Corcoran, P. (2012b). Characteristics of heavily edited objects in OpenStreetMap. *Future Internet, 4*(1), 285–305. doi:10.3390/fi4010285

Mooney, P., & Corcoran, P. (2014). Analysis of Interaction and Co-editing Patterns amongst OpenStreetMap Contributors. *Transactions in GIS, 18*(5), 633–659. doi:10.1111/tgis.12051

Mooney, P., Corcoran, P., & Winstanley, A. C. (2010). Towards quality metrics for OpenStreetMap. In *Proceedings of the 18th SIGSPATIAL International Conference on Advances in Geographic Information Systems* (pp. 514-517). ACM.

Mooney, P., Corcoran, P., & Winstanley, A. C. (2010). Towards Quality Metrics for OpenStreetMap. In *Proceedings of the 18th SIGSPATIAL International Conference on Advances in Geographic Information Systems* (pp. 514–517). San Jose, CA: ACM.

Mooney, P., & Minghini, M. (in press). A review of OpenStreetMap data. In G. Foody, L. See, V. Antoniou, C. C. Fonte, S. Fritz, P. Mooney, & A.-M. Olteanu (Eds.), *Mapping and the Citizen Sensor*. London, UK: Ubiquity Press.

Moovit Company Overview. (2014). Retrieved July 3, 2016, from http://moovitapp.com/wp-content/uploads/2014/07/Moovit-At-A-Glance-USA-4-14.pdf

Moreira, R. de B., Degrossi, L. C., & Albuquerque, J. P. (2015). An experimental evaluation of a crowdsourcing-based approach for flood risk management. In *12th Workshop on Experimental Software Engineering* (pp. 393–403). Lima, Peru: SPC.

Mortari, F. (2013). *Automatic Extraction of Improved Geometric Network Model from CityGML for Indoor Navigation. (Master)*. TU Delft.

Moscari T. & Hollo P. (2006). *Common Understanding on Road Safety Inspections, RIPCORD ISEREST: Report of Workpackage 5*. BASt.

Mostafavi, M. A. (2015*).MobiliSIG: Development of a geospatial assistive technology for navigation of people with motor disabilities*. Paper presented at Spatial Knowledge and Information Conference. Banff, Alberta, Canada.

Muhlrad, N. (1993). Traffic conflict techniques and other forms of behavioural analysis: Application to safety diagnoses. *Proceedings of 6th ICTCT Workshop*, 48-64.

Mullen, W. F., Jackson, S. P., Croitoru, A., Crooks, A., Stefanidis, A., & Agouris, P. (2015). Assessing the impact of demographic characteristics on spatial error in volunteered geographic information features. *GeoJournal, 80*(4), 587–605. 10.1007/s10708-014-9564-8

Munich RE. (2013). *Münchener Rückversicherungs-Gesellschaft*. Geo Risiko Forschung, NatCatSERVICE.

Munkres, J. R. (1984). *Elements of algebraic topology* (Vol. 2). Addison-Wesley Menlo Park.

Murphy, K. P. (2012). *Machine learning: a probabilistic perspective*. MIT press.

Nadler, B., Nadler, F., & Strnad, B. (2012). *Road Safety Inspection – Manual for Conducting RSI*. Vienna: Austrian Road Safety Fund.

Nahas, M. I. P. (2002). *Bases teóricas, metodologia de elaboração e aplicabilidade de indicadores intra-urbanos na gestão municipal da qualidade de vida urbana em grandes cidades : o caso de Belo Horizonte*. Universidade Federal de São Carlos. Retrieved from http://www.bdtd.ufscar.br/htdocs/tedeSimplificado/tde_arquivos/2/TDE-2008-05-09T08:37:00Z-1824/Publico/1753.pdf

Nakajima, M., & Haruyama, S. (2013). New indoor navigation system for visually impaired people using visible light communication. *EURASIP Journal on Wireless Communications and Networking*, (1): 1–10.

Napolitano, M., & Mooney, P. (2012). Mvp osm: A tool to identify areas of high quality contributor activity in openstreetmap. *The Bulletin of the Society of Cartographers*, *45*(1), 10–18.

Natural Resources Canada. (2015). *National Road Network (NRN) - Full metadata*. Retrieved April 23, 2016, from http://geogratis.gc.ca/api/en/nrcan-rncan/ess-sst/c0d1f299-179c-47b2-bcd8-da1ba68a8032.html#full-screen-meta

Neis, P. (2014). Measuring the reliability of wheel chair user route planning based on volunteered geographic information. *Transactions in GIS*, *19*(2), 188–201. doi:10.1111/tgis.12087

Neis, P., Goetz, M., & Zipf, A. (2012). Towards automatic vandalism detection in OpenStreetMap. *ISPRS International Journal of Geo-Information*, *1*(3), 315–332. doi:10.3390/ijgi1030315

Neis, P., & Zielstra, D. (2014). Recent developments and future trends in volunteered geographic information research: The case of OpenStreetMap. *Future Internet*, *6*(1), 76–106. doi:10.3390/fi6010076

Neis, P., Zielstra, D., & Zipf, A. (2011). The Street Network Evolution of Crowdsourced Maps: OpenStreetMap in Germany 2007–2011. *Future Internet*, *4*(4), 1–21. doi:10.3390/fi4010001

Neis, P., & Zipf, A. (2012). Analyzing the Contributor Activity of a Volunteered Geographic Information Project — The Case of OpenStreetMap. *ISPRS International Journal of Geo-Information*, *1*(3), 146–165. doi:10.3390/ijgi1020146

Němečková, M. (2014). *Komparace vnímání bezpečnosti ve městech Most a Litvínov* [Comparison of safety perception in cities Most and Litvínov]. Masarykova Univerzita. Retrieved from http://is.muni.cz/th/363798/fss_m/dp_nemeckova.pdf

Newman, G., Graham, J., Crall, A., & Laituri, M. (2011). The Art and Science of MultiScale Citizen Science Support. *Ecological Informatics*, *6*(3), 217–227. doi:10.1016/j.ecoinf.2011.03.002

Niederer, S., & Van Dijck, J. (2010). Wisdom of the crowd or technicity of content? Wikipedia as a sociotechnical system. *New Media & Society*, *12*(8), 1368–1387. doi:10.1177/1461444810365297

Nold, C. (2009). *Emotional cartography: technologies of the self*. Retrieved from http://emotionalcartography.net/EmotionalCartography.pdf

Norris, F. H., Stevens, S. P., Pfefferbaum, B., Wyche, K. F., & Pfefferbaum, R. L. (2008). Community Resilience as a Metaphor, Theory, Set of Capacities, and Strategy for Disaster Readiness. *American Journal of Community Psychology*, *41*(1–2), 127–150. doi:10.1007/s10464-007-9156-6 PMID:18157631

Nowozin, S., & Lampert, C. H. (2011). Structured learning and prediction in computer vision. *Foundations and Trends R in Computer Graphics and Vision*, *6*(3–4), 185–365.

Oc, T., & Tiesdell, S. (1997). Safer City Centres: Reviving the Public Realm. *Sage (Atlanta, Ga.)*.

OGC. (2008). *OGC web services initiative*. OGC.

OGC. (2011). *Observations and Measurements*. Retrieved from http://www.opengeospatial.org/standards/om

OGC. (2012). *IndoorGML Encoding Standard*. OGC.

OGC. (2016). *Open Geospatial Consortium*. Retrieved from http://www.opengeospatial.org/

Okpala, D. C. I. (2001, October). Spatial Information: The basic tool for Sustainable human settlements development planning and management. In *International Conference on Spatial Information for Sustainable Development*.

Olteanu-Raimond, A.-M., Hart, G., Foody, G. M., Touya, G., Kellenberger, T., & Demetriou, D. (2015). The scale of VGI in map production: A perspective of European National Mapping Agencies. *Transactions in GIS*. doi:10.1111/tgis.12189

OMG. (2013). *Business Process Model and Notation (BPMN), Version 2.0 (formal/2013-12-09)*. Retrieved July 8, 2016, from http://www.omg.org/spec/BPMN/2.0/PDF/

OMG. (2014). *Decision Model and Notation (DMN) (dtc/2014-02-01)*. Retrieved from http://www.omg.org/spec/DMN/1.0/

OpenStreetMap in GIScience. (2015). *Experiences, Research, and Applications*. Springer.

OpenStreetMap Map Features. (2016). Retrieved from http://wiki.openstreetmap.org/wiki/Map_Features

OSGeo. (2016). *GeoNetwork opensource*. Retrieved from http://geonetwork-opensource.org/

OSMstats - Statistics of the free wiki world map. (2016). Retrieved July 3, 2016, from http://osmstats.neis-one.org/

OSMwiki. (2015). *OpenStreetMap Wiki Community*. Retrieved from http://wiki.openstreetmap.org/wiki

Ostrava-Poruba. (2016). *Demography of Ostrava-Poruba* [in Czech]. Retrieved from https://poruba.ostrava.cz/cs/o-porube/zakladni-udaje/demograficke-udaje

OWS-6 Architecture Open Geospatial Consortium. (2006). *The OpenGIS Abstract Specification. Topic 18: Geospatial Digital Rights Management Reference Model (GeoDRM RM)*. Author.

Pacheco, M. B. (2009). *Public, private, and military applications flow from SDI*. Geospatial Foundation, InstitutoHidrografico, Portuguese Navy.

Padarian, J., Minasny, B., & McBratney, A. B. (2015). Using Googles cloud-based platform for digital soil mapping. *Computers & Geosciences*, *83*, 80–88. doi:10.1016/j.cageo.2015.06.023

Palacios, R. (2016). *deg2utm*. Retrieved from https://www.mathworks.com/matlabcentral/fileexchange/10915-deg2utm/content/deg2utm.m

Pánek, J., Hrubeš, M., Kubásek, M., Valůch, J., & Zahumenská, V. (2014). GeoParticipace - jak používat prostorové nástroje v rozhodování o lokalitách ve kterých žijeme? [GeoParticipation - how to use spatial tool in decision making process about the locations we live in?]. Olomouc: Univerzita Palackého v Olomouci.

Pánek, J., & Pászto, V. (2017). Emotional Mapping in Local Neighbourhood Planning: Case Study of Příbram, Czech Republic. *International Journal of E-Planning Research*, *6*(1), 1–22. doi:10.4018/IJEPR.2017010101

Pánek, J., Pászto, V., & Marek, L. (2017). Mapping emotions: spatial distribution of safety perception in the city of Olomouc. In I. Ivan, A. Singleton, J. Horák, & T. Inspektor (Eds.), *Lecture Notes in Geoinformation and Cartography: The Rise of Big Spatial Data* (pp. 211–224). Ostrava, Czech Republic: Springer International Publishing; doi:10.1007/978-3-319-45123-7_16

Parr, D. A. (2015). *The production of volunteered geographic information: A study of OpenStreetMap in the United States*. Academic Press.

Patel, N. N., Angiuli, E., Gamba, P., Gaughan, A., Lisini, G., Stevens, F. R., & Trianni, G. et al. (2015). Multitemporal settlement and population mapping from Landsat using Google Earth Engine. *International Journal of Applied Earth Observation and Geoinformation*, *35*, 199–208. doi:10.1016/j.jag.2014.09.005

PBH. (2014). *Relatório geral sobre o cálculo do Índice de Qualidade de Vida Urbana de Belo Horizonte (IQVU-BH)*. Retrieved from http://portalpbh.pbh.gov.br/pbh/ecp/files.do?evento=download&urlArqPlc=relatorio_iqvu_rev-set14.pdf

PBH. (2015). *Portal PBH*. Retrieved October 15, 2015, from http://portalpbh.pbh.gov.br/pbh/

Pearce, M. W. (2008). Framing the days: Place and narrative in cartography. *Cartography and Geographic Information Science*, *35*(1), 17–32. doi:10.1559/152304008783475661

Peluso, N. L. (1995). Whose Woods are These? Counter-mapping Forest Territories in Kalimantan, Indonesia. *Antipode*, *27*(4), 383–406. doi:10.1111/j.1467-8330.1995.tb00286.x

Percivall, G. (2013). *OpenSearch at OGC TC with CEOS23*. WGISS/OGC Meeting, OGC Technical Committee, Frascati, Italy.

Perkins, S. R., & Harris, J. L. (1968). Traffic conflict characteristics-accident potential at intersections. *Highway Research Record*, 225.

Pickles, J. (1995). *Ground Truth: The Social Implications of Geographic Information Systems* (1st ed.). The Guilford Press.

Pissourios, I. A. (2013). An interdisciplinary study on indicators: A comparative review of quality-of-life, macroeconomic, environmental, welfare and sustainability indicators. *Ecological Indicators*, *34*, 420–427. doi:10.1016/j.ecolind.2013.06.008

Platt, J. (1998). *Sequential minimal optimization: A fast algorithm for training support vector machines*. Academic Press.

Pohl, D., Bouchachia, A., & Hellwagner, H. (2012). Supporting crisis management via sub-event detection in social networks. In *Proceedings of the Workshop on Enabling Technologies: Infrastructure for Collaborative Enterprises, WETICE* (pp. 373–378). Toulouse, France: IEEE. doi:10.1109/WETICE.2012.58

Pontes, T., Vasconcelos, M., Almeida, J., Kumaraguru, P., & Almeida, V. (2012). We know where you live. In *Proceedings of the 2012 ACM Conference on Ubiquitous Computing - UbiComp '12* (p. 898). New York: ACM Press. doi:10.1145/2370216.2370419

Poser, K., & Dransch, D. (2010). Volunteered geographic information for disaster management with application to rapid flood damage estimation. *Geomatica*, *64*(1), 89–98.

Posthumus, E., & Crimmins, T. (2011). Natures Notebook: A Tool for Education and Research. *Bulletin of the Ecological Society of America*, *92*(2), 185–187. doi:10.1890/0012-9623-92.2.185

Pourabdollah, A., Morley, J., Feldman, S., & Jackson, M. (2013). Towards an Authoritative OpenStreetMap: Conflating OSM and OS OpenData National Maps Road Network. *ISPRS International Journal of Geo-Information*, *2*(3), 704–728. doi:10.3390/ijgi2030704

Prandi, C., Salomoni, P., & Mirri, S. (2014, January). mPASS: Integrating people sensing and crowdsourcing to map urban accessibility. In *Proceedings of the IEEE International Conference on Consumer Communications and Networking Conference* (pp. 10-13). doi:10.1109/CCNC.2014.6940491

Prandi, F., Soave, M., Devigili, F., Amicis, R., & Astyakopoulos, A. (2014) Collaboratively collected geodatato support routing service for disabled people.*Proceedings of the 11th International Symposiumon Location-Based Services (LBS)*.

Puckett, J. (2010). *Digital Rights Management as Information Access Barrier*. Georgia State University.

Qin, H., Aburizaiza, A. O., Rice, R. M., Paez, F., & Rice, M. T. (2015). Obstacle characterization in a geocrowdsourced accessibility system. In *ISPRS Annals of Photogrammetry*. La Grande Motte, France: Remote Sensing and Spatial Information Sciences. doi:10.5194/isprsannals-II-3-W5-179-2015

Quality assurance - OpenStreetMap Wiki. (2016). Retrieved July 3, 2016, from http://wiki.openstreetmap.org/wiki/Quality_assurance

Quattrone, G., Mashhadi, A., Quercia, D., Smith-Clarke, C., & Capra, L. (2014). Modelling growth of urban crowd-sourced information. *Proceedings of the 7th ACM International Conference on Web Search and Data Mining - WSDM '14*, 563–572. doi:10.1145/2556195.2556244

Quercia, D., O'Hare, N. K., & Cramer, H. (2014). Aesthetic capital. In *Proceedings of the 17th ACM conference on Computer supported cooperative work & social computing - CSCW '14* (pp. 945–955). New York: ACM Press. doi:10.1145/2531602.2531613

Quercia, D., Schifanella, R., & Aiello, L. M. (2014). The shortest path to happiness. In *Proceedings of the 25th ACM conference on Hypertext and social media - HT '14* (pp. 116–125). New York: ACM Press. doi:10.1145/2631775.2631799

Quercia, D., & Saez, D. (2014). Mining Urban Deprivation from Foursquare: Implicit Crowdsourcing of City Land Use. *IEEE Pervasive Computing / IEEE Computer Society [and] IEEE Communications Society*, *13*(2), 30–36. doi:10.1109/MPRV.2014.31

Quinlan, J. R. (1996). Bagging, boosting, and c4. 5. AAAI/IAAI, 1, 725–730.

Quiquivix, L. (2014). Art of War, Art of Resistance: Palestinian Counter-cartography on Google Earth. *Annals of the Association of American Geographers*, *104*(3), 444–459. doi:10.1080/00045608.2014.892328

Quirez, R., Zorogastua, P., Baigorria, G., Barreda, C., Valdivia, R., Cruz, M., & Reinoso, J. (2000). *A dynamic definition of Agro-Ecological zones using modern information management tools*. Academic Press.

Rajabifard, A., & Williamson, I. P. (2001). Spatial Data Infrastructures: Concept, SDI Hierarchy and Future Directions. *Proceedings of GEOMATICS'80 Conference*, 10.

Raosoft. (2004). *Sample Size Calculator by Raosoft*. Retrieved February 12, 2016, from http://www.raosoft.com/samplesize.html

Raslan, R., Al-hagla, K., & Bakr, A. (2014). Integration of Emotional Behavioural Layer "EmoBeL" in City Planning. In REAL CORP 2014 (Vol. 8, pp. 309–317).

Raymond, E. S. (2001). *The Cathedral & the Bazaar: Musings on linux and open source by an accidental revolutionary*. O'Reilly Media, Inc.

Reed, C. (2012). *Introduction to the OGC and Overview of Key Standards Focus areas*. ADSIC Presentation.

Reeve, J. (2014). *Understanding motivation and emotion*. John Wiley & Sons.

Rezvani, M. R., & Mansourian, H. (2013). Developing Small Cities by Promoting Village to Town and its Effects on Quality of Life for the Local Residents. *Social Indicators Research*, *110*(1), 147–170. doi:10.1007/s11205-011-9921-7

Ribeiro, A., & Fonte, C. C. (2015). Methodology for assessing OpenStreetMap degree of coverage for purposes of land cover mapping. In *ISPRS Annals of the Photogrammetry, Remote Sensing and Spatial Information Sciences* (Vol. WG II/4). La Grande Motte, France: ISPRS Annals. Retrieved from http://www.isprs-ann-photogramm-remote-sens-spatial-inf-sci.net/II-3-W5/297/2015/isprsannals-II-3-W5-297-2015.pdf

Rice, M. T., Jacobson, R. D., Caldwell, D. R., McDermott, S. D., Paez, F. I., Aburizaiza, A. O., & Qin, H. et al. (2013). Crowdsourcing techniques for augmenting traditional accessibility maps with transitory obstacle information. *Cartography and Geographic Information Science*, *40*(3), 210–219. doi:10.1080/15230406.2013.799737

Risser, R., & Schützenhöfer, A. (1984). Application of Traffic-Conflict Technique in Austria. In *International Calibration Study of Traffic Conflict Techniques* (pp. 141–151). Springer Berlin Heidelberg. doi:10.1007/978-3-642-82109-7_13

Roback, J. (1982). Wages, Rents, and the Quality of Life. *Journal of Political Economy*, *90*(6), 1257–1278. doi:10.1086/261120

Rodrigues, M. C. P. (1991). O desenvolvimento social nas Regiões Brasileiras. *Ciência Hoje*, *13*(76), 39–45.

Rodríguez, F., Mas, S., Richard, D., Nataf, J. M., Furtado, D., Julião, R. P., ... Alba, S. (2010). Horizontal Interoperability among National Geoportals of Spain, France, Portugal and Andorra. *INSPIRE Conference*.

Rogerson, R. J. (1999). Quality of Life and City Competitiveness. *Urban Studies (Edinburgh, Scotland)*, *36*(5), 969–985. doi:10.1080/0042098993303

Rosenblatt, B., Tripp, B., & Mooney, S. (2001). *Digital Rights Management: Business and Technology*. John Wiley and Sons.

Rost, M., Barkhuus, L., Cramer, H., & Brown, B. (2013). Representation and communication. In *Proceedings of the 2013 conference on Computer supported cooperative work - CSCW '13* (p. 357). New York: ACM Press. doi:10.1145/2441776.2441817

Rotava, J., Mendiondo, E. M., & Souza, V. C. B. (2013). Simulação de instabilidade humana em inundações: primeiras considerações. In *Proceedings of the 20th Simpósio Brasileiro de Recursos Hídricos* (pp. 1–8).

Rotman, D., Preece, J., Hammock, J., Procita, K., Hansen, D., Parr, C., & Jacobs, D. (2012). Dynamic changes in motivation in collaborative citizen-science projects. In *Proceedings of the ACM 2012 conference on computer supported cooperative work* (pp. 217–226). doi:10.1145/2145204.2145238

Rusková, L. (2012). *Socially excluded localities in Ostrava*. Masaryk University Brno.

Russell, J. A. (1980). A circumplex model of affect. *Journal of Personality and Social Psychology*, *39*(6), 1161–1178. doi:10.1037/h0077714

Sakaki, T., Okazaki, M., & Matsuo, Y. (2010). Earthquake Shakes Twitter Users: Real-time Event Detection by Social Sensors. In *Proceedings of the 19th International Conference on World Wide Web* (pp. 851–860). New York, NY: ACM. doi:10.1145/1772690.1772777

Santos, J. P., Forest, J. D., Dutra, G. A., & Almeida, M. (2012). *Uma avaliação da bengala eletrônica (tecnologia assistiva), para a melhoria de mobilidade dos deficientes visuais*. Paper presented at Simpósio de Exelência em Gestão e Tecnologia, Rio de Janeiro, Brasil.

Santos, R. B. (2012). Crime analysis with crime mapping. *Sage (Atlanta, Ga.)*.

Scarborough, V. L., & Burnside, W. R. (2010). Complexity and sustainability: Perspectives from the ancient Maya and the modern Balinese. *American Antiquity*, *75*(2), 327–363. doi:10.7183/0002-7316.75.2.327

Scheider, S., Keßler, C., Ortmann, J., Devaraju, A., Trame, J., Kauppinen, T., & Kuhn, W. (2011). Semantic referencing of geosensor data and volunteered geographic information. In *Geospatial semantics and the semantic web* (pp. 27–59). Springer. doi:10.1007/978-1-4419-9446-2_2

Schilling, A., Over, M., Neubauer, S., Neis, P., Walenciak, G., & Zipf, A. (2009). Interoperable Location Based Services for 3D cities on the Web using user generated content from OpenStreetMap. *Urban and regional data management: UDMS annual*, 75-84.

Schmitz, B., Becker, S., Blessing, A., & Großmann, M. (2011, June). Acquisition and presentation of diverse spatial context data for blind navigation. In *Proceedings of the 2011 IEEE 12th International Conference on Mobile Data Management* (pp. 276-284). doi:10.1109/MDM.2011.66

Schnebele, E., Cervone, G., & Waters, N. (2014). Road assessment after flood events using non-authoritative data. *Natural Hazards and Earth System Sciences, 14*(4), 1007–1015. doi:10.5194/nhess-14-1007-2014

Schnebele, E., Oxendine, C., Cervone, G., Ferreira, C. M., & Waters, N. (2015). Using Non-authoritative Sources During Emergencies in Urban Areas. In M. Helbich, J. Jokar Arsanjani, & M. Leitner (Eds.), *Computational Approaches for Urban Environments* (pp. 337–361). Cham, Switzerland: Springer International Publishing. doi:10.1007/978-3-319-11469-9_14

See, L., Mooney, P., Foody, G., Bastin, L., Comber, A., & Estima, J. … Rutzinger, M. (2016). Crowdsourcing, Citizen Science or Volunteered Geographic Information? The Current State of Crowdsourced Geographic Information. *ISPRS International Journal of Geo-Information, 5*(5), 55. https://doi.org/10.3390/ijgi5050055

See, L., Comber, A., Salk, C., Fritz, S., van der Velde, M., Perger, C., & Obersteiner, M. et al. (2013). Comparing the quality of crowdsourced data contributed by expert and non-experts. *PLoS ONE, 8*(7), e69958. doi:10.1371/journal.pone.0069958 PMID:23936126

See, L., Fritz, S., Perger, C., Schill, C., McCallum, I., Schepaschenko, D., & Obersteiner, M. et al. (2015). Harnessing the power of volunteers, the internet and Google Earth to collect and validate global spatial information using Geo-Wiki. *Technological Forecasting and Social Change, 98*, 324–335. doi:10.1016/j.techfore.2015.03.002

See, L., Mooney, P., Foody, G., Bastin, L., Comber, A., Estima, J., & Liu, H. Y. et al. (2016). Crowdsourcing, citizen science or Volunteered Geographic Information? The current state of crowdsourced geographic information. *ISPRS International Journal of Geo-Information, 5*(5), 55. doi:10.3390/ijgi5050055

See, L., Mooney, P., Foody, G., Bastin, L., Comber, A., Estima, J., & Rutzinger, M. et al. (2016). Crowdsourcing, Citizen Science or Volunteered Geographic Information? The Current State of Crowdsourced Geographic Information, ISPRS Int. *J. Geo-Inf.*, *5*, 55.

See, L., Schepaschenko, D., Lesiv, M., McCallum, I., Fritz, S., Comber, A., & Siraj, M. A. et al. (2015). Building a hybrid land cover map with crowdsourcing and geographically weighted regression. *ISPRS Journal of Photogrammetry and Remote Sensing, 103*, 48–56. doi:10.1016/j.isprsjprs.2014.06.016

Sehra, S. S., Singh, J., & Rai, H. S. (2015). Quality assessment of crowdsourced data against custom recorded map data. *Indian Journal of Science and Technology, 8*(33). 10.17485/ijst/2015/v8i33/79884

Senaratne, H., Mobasheri, A., Loai Ali, A., Capineri, C., & Haklay, M. (2016). A review of volunteered geographic information quality assessment methods. *International Journal of Geographical Information Science*. DOI: 10.1080 13658816.2016.1189556

Sessar, K., & Sirotek, J. (2001). Společnost v období transformace a strach z kriminality. *Sociologicky Casopis, 37*(1), 7–22. Retrieved from http://www.jstor.org/stable/41131597?seq=1#page_scan_tab_contents

Sester, M. (2000). Knowledge acquisition for the automatic interpretation of spatial data. *International Journal of Geographical Information Science, 14*(1), 1–24. doi:10.1080/136588100240930

Shekhar, S., Feiner, S. K., & Aref, W. G. (2015). Spatial computing. *Communications of the ACM, 59*(1), 72–81. doi:10.1145/2756547

Shekhar, S., Yang, K., Gunturi, V. M., Manikonda, L., Olive, D., Zhou, X., & Lu, Q. (2012). Experiences with Evacuation Route Planning Algorithms. *International Journal of Geographical Information Science*, *26*(12), 2253–2265. doi:10.1080/13658816.2012.719624

Shelton, T., Poorthuis, A., & Zook, M. (2015). Social media and the city: Rethinking urban socio-spatial inequality using user-generated geographic information. *Landscape and Urban Planning*, *142*, 198–211. doi:10.1016/j.landurbplan.2015.02.020

Sheppard, S. A. (2012). wq: A modular Framework for Collecting, Storing, and Utilizing Experiential VGI. In *Proceedings of the 1st acm sigspatial international workshop on crowdsourced and volunteered geographic information* (pp. 62–69). doi:10.1145/2442952.2442964

Shumaker, B. P., & Sinnott, R. W. (1984). Astronomical computing: 1. Computing under the open sky. 2. Virtues of the haversine. *Sky and Telescope*, *68*, 158–159.

Siebritz, L., Sithole, G., & Zlatanova, S. (2012). Assessment of the homogeneity of volunteered geographic information in South Africa. In M. Shortis & M. Madden (Eds.), *22nd Congress of the International Society for Photogrammetry and Remote Sensing, ISPRS 2012* (Vol. 39, pp. 553–558). International Society for Photogrammetry and Remote Sensing. Retrieved from https://www.scopus.com/inward/record.uri?eid=2-s2.0-84901710768&partnerID=40&md5=e65dd4e6feb4732bcd9d94f1c39a4e19

Siebritz, L.-A. (2014). *Assessing the accuracy of openstreetmap data in south africa for the purpose of integrating it with authoritative data* (PhD thesis). University of Cape Town.

Silva, D. C. N., Silva, T. M., Nogueira, M. S., Mendonça, R. M. C., Valente, P. H. F., Araújo, R. F., Brito, F. A. V. (2016). Acessibilidade de portadores de deficiência física ou mobilidade reduzida na Unidade Básica de Saúde Jonas Manoel Dias em São Luís de Montes Belos-GO. *Revista Eletrônica Faculdade Montes Belos*, *8*(3).

Simard, M., Pinto, N., Fisher, J. B., & Baccini, A. (2011). Mapping forest canopy height globally with spaceborne lidar. *Journal of Geophysical Research: Biogeosciences*, *116*(G4), G04021. doi:10.1029/2011JG001708

Smith, L., Liang, Q., James, P., & Lin, W. (2015). Assessing the utility of social media as a data source for flood risk management using a real-time modelling framework. *Journal of Flood Risk Management*. https://doi.org/10.1111/jfr3.12154

Smith, M., Bondi, L., & Davidson, J. (2012). *Emotional geographies*. Ashgate Publishing, Ltd.

Soden, R., & Palen, L. (2014). From crowdsourced mapping to community mapping: The post-earthquake work of OpenStreetMap Haiti. In C. Rossitto, L. Ciolfi, D. Martin, & B. Conein (Eds.), *COOP 2014 - Proceedings of the 11th International Conference on the Design of Cooperative Systems, 27-30 May 2014, Nice (France)* (pp. 311–326). Springer International Publishing. Retrieved from http://link.springer.com/chapter/10.1007/978-3-319-06498-7_19

Sørensen, M. W., & Elvik, R. (2007). *Black Spot Management and Safety Analysis of Road Networks: Best Practice Guidelines and Implementation Steps*. Transportøkonomisk institutt.

Sousa, R. B., Cugler, D. C., Malaverri, J. E. G., & Medeiros, C. B. (2014). A Provenance-based Approach to Manage Long Term Preservation of Scientific Data. In Data engineering workshops (ICDEW), ICDE (pp. 126–133). doi:10.1109/ICDEW.2014.6818316

Spinsanti, L., & Ostermann, F. (2013). Automated geographic context analysis for volunteered information. *Applied Geography (Sevenoaks, England)*, *43*, 36–44. doi:10.1016/j.apgeog.2013.05.005

Stasíková, L. (2011). Relevantnost výskumu strachu z kriminality v urbánnej geografii. *Geograficky Casopis*, *63*(4), 325–343.

State Administration of Land Surveying and Cadastre. (2014). *Fundamental Base of Geographic Data of the Czech Republic*. Retrieved September 21, 2015, from http://geoportal.cuzk.cz/(S(z2w1esrjen1dfactx0vyepfg))/Default.aspx?mode=TextMeta&text=dSady_zabaged&side=zabaged&menu=24

Statistics Canada. (2006). *Participation and Activity Limitation Survey 2006: Analytical Report*. Retrieved August 14, 2016, from http://www.statcan.gc.ca/pub/89-628-x/89-628-x2007002-eng.htm

Statistics Canada. (2012). Focus on Geography Series, 2011 Census. Statistics Canada Catalogue no. 98-310-XWE2011004. Ottawa: Author.

Steiger, E., Albuquerque, J. P., & Zipf, A. (2015). An Advanced Systematic Literature Review on Spatiotemporal Analyses of Twitter Data. *Transactions in GIS*, *19*(6), 809–834. doi:10.1111/tgis.12132

Stein, K., Kremer, D., & Schlieder, C. (2015). Spatial Collaboration Networks of OpenStreetMap. In J. Jokar Arsanjani, A. Zipf, P. Mooney, & M. Helbich (Eds.), OpenStreetMap in GIScience SE - 9 (pp. 167–186). CHAP, Springer International Publishing. doi:10.1007/978-3-319-14280-7_9

Sui, D., & DeLyser, D. (2012). Crossing the Qualitative-Quantitative Chasm I: Hybrid Geographies, the Spatial Turn, and Volunteered Geographic Information (VGI). *Progress in Human Geography*, *36*(1), 111–124. doi:10.1177/0309132510392164

Sui, D., Elwood, S., & Goodchild, M. (2012). *Crowdsourcing geographic knowledge: Volunteered geographic information and knowledge*. New York: Springer.

Sung, H., Lee, S., & Cheon, S. (2015). Operationalizing Jane Jacobss Urban Design Theory: Empirical Verification from the Great City of Seoul, Korea. *Journal of Planning Education and Research*, *35*(2), 117–130. doi:10.1177/0739456X14568021

Surowiecki, J. (2004). *The Wisdom of Crowds: Why the Many are Smarter than the Few and How Collective Wisdom Shapes Business, Economies, Societies, and Nations*. New York: Doubleday.

Taboada, M., Brooke, J., Tofiloski, M., Voll, K., & Stede, M. (2011). Lexicon-based methods for sentiment analysis. *Computational Linguistics*, *37*(2), 267–307. doi:10.1162/COLI_a_00049

Takagi, H., Kosugi, A., Saito, S., & Teraguchi, M. (2013, May). Crowdsourcing platform for workplace accessibility. *Proceedings of the 10th International Cross-Disciplinary Conference on Web Accessibility*. doi:10.1145/2461121.2461135

Taylor, Z., & Józefowicz, I. (2012). Intra-urbandaily mobility of disabled people for recreational and leisure purposes. *Journal of Transport Geography*, *24*, 155–172. doi:10.1016/j.jtrangeo.2011.12.008

Tenney, M. (2014). *Quality Evaluations on Canadian OpenStreetMap Data*. Retrieved from http://rose.geog.mcgill.ca/ski/system/files/fm/2014/tenney.pdf

The World Bank. (2015). *Human Development Reports*. Retrieved from http://hdr.undp.org/en/content/human-development-index-hdi

Tingvall, C., & Haworth, N. (2000). Vision Zero: an ethical approach to safety and mobility. In *6th ITE International Conference Road Safety & Traffic Enforcement: Beyond* (Vol. 1999).

Tobler, W. R. (1970). A computer movie simulating urban growth in the Detroit region. *Economic Geography*, *46*, 234–240. doi:10.2307/143141

Tobler, W. R. (1994). Bidimensional regression. *Geographical Analysis*, *26*(3), 187–212. doi:10.1111/j.1538-4632.1994.tb00320.x

Tombone. (2015). Retrieved from http://www.computervisionblog.com/2015/04/deep-learning-vsprobabilistic.html

Torheim, T., Malinen, E., Kvaal, K., Lyng, H., Indahl, U. G., Andersen, E. K., & Futsaether, C. M. (2014). Classification of dynamic contrast enhanced MR images of cervical cancers using texture analysis and support vector machines. *IEEE Transactions on Medical Imaging*, *33*(8), 1648–1656. doi:10.1109/TMI.2014.2321024 PMID:24802069

Touya, G., & Reimer, A. (2015). Inferring the Scale of OpenStreetMap Features. In J. Jokar Arsanjani, A. Zipf, P. Mooney, & M. Helbich (Eds.), OpenStreetMap in GIScience SE - 5 (pp. 81–99). CHAP, Springer International Publishing. doi:10.1007/978-3-319-14280-7_5

Touya, G., & Brando-Escobar, C. (2013). Detecting Level-of-Detail Inconsistencies in Volunteered Geographic Information Data Sets. *Cartographica: The International Journal for Geographic Information and Geovisualization*, *48*(2), 134–143. doi:10.3138/carto.48.2.1836

Trakas, A. (2014). *OGC and Open Standards in action –Examples*. Workshop: Make the most of Geographic Information, *INSPIRE Conference*, Aalborg, Denmark.

Tse, R. O. C., & Gold, C. M. (2004). TIN meets CAD: Extending the TIN concept in GIS. *Future Generation Computer Systems*, *20*(7), 1171–1184. doi:10.1016/j.future.2003.11.007

Tsochantaridis, I., Hofmann, T., Joachims, T., & Altun, Y. (2004). Support vector machine learning for interdependent and structured output spaces. In *Proceedings of the twenty-first international conference on Machine learning*. ACM. doi:10.1145/1015330.1015341

Turing, A. M. (1950). Computing machinery and intelligence. *Mind*, *59*(236), 433–460. doi:10.1093/mind/LIX.236.433

Turner, B. L., Villar, S. C., Foster, D., Geoghegan, J., Keys, E., Klepeis, P., & Plotkin, A. B. et al. (2001). Deforestation in the southern Yucatán peninsular region: An integrative approach. *Forest Ecology and Management*, *154*(3), 353–370. doi:10.1016/S0378-1127(01)00508-4

Turner, M. G. (1989). Landscape ecology: The effect of pattern on process. *Annual Review of Ecology and Systematics*, *20*(1), 171–197. doi:10.1146/annurev.es.20.110189.001131

Tu, Y., Li, Q., & Liu, R. (2009). A geospatial information portal for emergency management of natural disasters. In *Proceedings of the IEEE International Geoscience and Remote Sensing Symposium (IGARSS 2009)* (Vol. 2, pp. II-404–II-407). Cape Town, South Africa: IEEE. doi:10.1109/IGARSS.2009.5418100

UNDP. (2015). *Human Development Index (HDI)*. Retrieved from http://hdr.undp.org/en/content/human-development-index-hdi

UNECA. (n.d.). Retrieved from: http://www.uneca.org/acs-0/pages/geo-information

United Nations. (2007). *Accessibility: A guiding principle of the Convention*. Retrieved May 30, 2016, from http://www.un.org/esa/socdev/enable/disacc.htm

University of South Carolina. (2016). *Cicada Project*. Retrieved January 17, 2016 from http://cricket.biol.sc.edu/cicada/contact.html

Urmson, C., Anhalt, J., Bagnell, D., Baker, C., Bittner, R., Clark, M., & Geyer, C. (2008). Autonomous driving in urban environments: Boss and the urban challenge. *Journal of Field Robotics*, *25*(8), 425–466. doi:10.1002/rob.20255

Ushahidi. (2016). *Ushahidi*. Retrieved from https://www.ushahidi.com/

Van Exel, M., Dias, E., & Fruijtier, S. (2010). The impact of crowdsourcing on spatial data quality indicators. *Proceedings of GiScience*, *2011*, 1–4.

Van Exel, M., Dias, E., & Fruijtier, S. (2011). *The impact of crowdsourcing on spatial data quality indicators*. GiScience.

Van Oort, P. (2006). *Spatial data quality: from description to application. Publications on Geodesy 60*. DISS, WUR Wageningen UR. Retrieved from http://library.wur.nl/WebQuery/wdab/1788022

van Oort, P. A. (2006). *Spatial data quality: from description to application* (Doctoral dissertation). Publications on Geodesy 60, NCG.

van Winden, K., Biljecki, F., & van der Spek, S. (2016). Automatic Update of Road Attributes by Mining GPS Tracks. *Transactions in GIS*, *20*(5), 664–683. doi:10.1111/tgis.12186

Vanclooster, A., Weghe, N. V., & Maeyer, P. (2016). Integrating indoor and outdoor spaces for pedestrian navigation guidance: A review. *Transactions in GIS*.

Vandecasteele, A., & Devillers, R. (2015). Improving Volunteered Geographic Information Quality Using a Tag Recommender System: The Case of OpenStreetMap. In J. Jokar Arsanjani, A. Zipf, P. Mooney, & M. Helbich (Eds.), OpenStreetMap in GIScience SE - 4 (pp. 59–80). CHAP, Springer International Publishing. doi:10.1007/978-3-319-14280-7_4

Vasiliu, A. (2006). *Dashboards and scorecards: Linking management reporting to execution*. Retrieved from http://dssresources.com/papers/features/vasiliu/vasiliu04302006.html

Vauglin, F. (1997). *Modèles statistiques des imprécisions géométriques des objets géographiques linéaires*. DISS.

Vaz, E., & Jokar Arsanjani, J. (2015). Crowdsourced mapping of land use in urban dense environments: An assessment of Toronto. *Canadian Geographer*, *59*(2), 246–255. 10.1111/cag.12170

Venerandi, A., Quattrone, G., Capra, L., Quercia, D., & Saez-Trumper, D. (2015). Measuring Urban Deprivation from User Generated Content. In *Proceedings of the 18th ACM Conference on Computer Supported Cooperative Work & Social Computing - CSCW '15* (pp. 254–264). New York: ACM Press. doi:10.1145/2675133.2675233

Vieweg, S., Castillo, C., & Imran, M. (2014). Integrating Social Media Communications into the Rapid Assessment of Sudden Onset Disasters. In *Proceedings of the 6th International Conference on Social Informatics* (pp. 444–461). Barcelona, Spain: Springer International Publishing. doi:10.1007/978-3-319-13734-6_32

Vowles, G., & McKee, L. (2005). *Geospatial digital Rights (GeoDRM)*. Geoinformatics.

Wainwright, J., & Bryan, J. (2009). Cartography, Territory, Property: Postcolonial Reflections on Indigenous Counter-mapping in Nicaragua and Belize. *Cultural Geographies*, *16*(2), 153–178. doi:10.1177/1474474008101515

Walter, V., & Luo, F. (2011). Automatic interpretation of digital maps. *ISPRS Journal of Photogrammetry and Remote Sensing*, *66*(4), 519–528. doi:10.1016/j.isprsjprs.2011.02.010

Wang, M., Li, Q., Hu, Q., & Zhou, M. (2013). Quality analysis of open street map data. In E. Guilbert, B. Wu, & J. Shi (Eds.), *8th International Symposium on Spatial Data Quality* (2W1, Vol. 40, pp. 155–158). International Society for Photogrammetry and Remote Sensing. Retrieved from https://www.scopus.com/inward/record.uri?eid=2-s2.0-84929074534&partnerID=40&md5=cc6cb4934bc6f9cdeea8a05303363311

Werder, S., Kieler, B., & Sester, M. (2010). Semi-automatic interpretation of buildings and settlement areas in user-generated spatial data. In *Proceedings of the 18th SIGSPATIAL international conference on advances in geographic information systems*, (pp. 330–339). ACM. doi:10.1145/1869790.1869836

Werner, M. (2014). *Building Modeling. In Indoor Location-Based Services: Prerequisites and Foundations* (pp. 101–131). Cham: Springer International Publishing.

Węziak-Białowolska, D. (2016). Quality of life in cities – Empirical evidence in comparative European perspective. *Cities (London, England), 58*, 87–96. doi:10.1016/j.cities.2016.05.016

Wheelmap. (n.d.a). *What's Wheelmap all about?* Retrieved July 12, 2016, from https://community.wheelmap.org/en/about/

Wheelmap. (n.d.b) *Wheelmap*. Retrieved on June 17, 2016, from http://wheelmap.org/pt_BR/map#/?zoom=14

White, D. A., & Hood, C. S. (2004). Vegetation patterns and environmental gradients in tropical dry forests of the northern Yucatan Peninsula. *Journal of Vegetation Science, 15*(2), 151–160. doi:10.1111/j.1654-1103.2004.tb02250.x

Whitelaw, G., Vaughan, H., Craig, B., & Atkinson, D. (2003). Establishing the Canadian community monitoring network. *Environmental Monitoring and Assessment, 88*(1-3), 409–418. doi:10.1023/A:1025545813057 PMID:14570426

Wickham, J. D., Riitters, K. H., Wade, T. G., & Vogt, P. (2010). A national assessment of green infrastructure and change for the conterminous United States using morphological image processing. *Landscape and Urban Planning, 94*(3), 186–195. doi:10.1016/j.landurbplan.2009.10.003

Wilson, M. W. (2011). Training the eye: Formation of the geocoding subject. *Social & Cultural Geography, 12*(4), 357–376. doi:10.1080/14649365.2010.521856

World Bank. (2014). *Strategic Framework for Mainstreaming Citizen Engagement in World Bank Group Operations: Engaging with Citizens for Improved Results*. Washington, DC: World Bank Group.

World Bank. (2016). *Evaluating Digital Citizen Engagement: A Practical Guide*. Washington, DC: World Bank Group.

World Health Organization (WHO). (2011). *World report on disability 2011*. WHO.

World Health Organization. (2015). *Global Status Report on Road Safety*. WHO.

World Health Organization. (n.d.a). *Disability and rehabilitation*. Retrieved May 30, 2016, from http://www.who.int/disabilities/en/

World Health Organization. (n.d.b). *Disabilities*. Retrieved on June 17, 2016, from http://www.who.int/topics/disabilities/en/

Wright, D. J. (2005). *Geospatial Digital Rights management*. US Geological Survey. Retrieved from: https://pubs.usgs.gov/of/2005/1086/2005-1086.pdf

Xiao, X., Dorovskoy, P., Biradar, C., & Bridge, E. (2011). A Library of Georeferenced Photos from the Field. *Eos, Transactions, American Geophysical Union, 92*(49), 453–454. doi:10.1029/2011EO490002

Xu, Y., Maitland, C., & Tomaszewski, B. (2015). Promoting Participatory Community Building in Refugee Camps with Mapping Technology.*Proceedings of the Eighth International Conference on Information and Communication Technologies and Development - ICTD '15*. doi:10.1145/2737856.2737883

Yang, A., Fan, H., & Jing, N. (2016). *Amateur or professional: Assessing the expertise of major contributors in openstreetmap based on contributing behaviors. ISPRS International Journal of Geo-Information, 5(2)*. doi:10.3390/ijgi5020021

Yelp. (2015). *Yelp Categories*. Retrieved October 15, 2015, from https://www.yelp.com.br/developers/documentation/v2/all_category_list

Yelp. (2016). *About Us*. Retrieved August 15, 2016, from https://www.yelp.com/about

Yildiz, A., & Hauger, G. (2010). An approach for a more precise road scoring. In *Proceedings of the Road Safety on Four Continents Conference* (Vol. 15, pp. 473-481).

Yuan, J., Zheng, Y., & Xie, X. (2012). Discovering regions of different functions in a city using human mobility and POIs. In *Proceedings of the 18th ACM SIGKDD international conference on Knowledge discovery and data mining - KDD '12* (p. 186). New York: ACM Press. doi:10.1145/2339530.2339561

Yu, L., & Gong, P. (2012). Google Earth as a virtual globe tool for Earth science applications at the global scale: Progress and perspectives. *International Journal of Remote Sensing*, *33*(12), 3966–3986. doi:10.1080/01431161.2011.636081

Zaragozí, B., Gimýnez-Font, P., Navarro-Carriýn, J. T., Rico-Amorýs, A., Ramýn-Morte, A., & Belda-Antolý, A. (2015). *A proposal for design and implementation of an hybrid navigation system based on open data, augmented reality and big data: Applications for the smart cities.* Paper presented at the Geographical Information Systems Theory, Applications and Management (GISTAM), 2015 1st International Conference on.

Zeng, L., & Weber, G. (2015, August). A Pilot Study of Collaborative Accessibility: How Blind People Find an Entrance. In *Proceedings of the 17th International Conference on Human-Computer Interaction with Mobile Devices and Services* (pp. 347-356). doi:10.1145/2785830.2785875

Zhang, X., & Ai, T. (2015). How to model roads in openstreetmap? A method for evaluating the fitness-for-use of the network for navigation. In Y. Leung & F. Harvey (Eds.), *16th International Symposium on Spatial Data Handling, SDH 2014* (*Vol. 19*, pp. 143–162). Springer Berlin. doi:10.1007/978-3-319-19950-4_9

Zhao, Y. J., Zhou, X. G., Li, G. Q., & Xing, H. F. (2016). *A Spatio-Temporal VGI Model Considering Trust-Related Information. ISPRS International Journal of Geo-Information, 5(2), 19.* doi:10.3390/ijgi5020010

Zheng, Y., Shen, G., Li, L., Zhao, C., Li, M., & Zhao, F. (2014). *Travi-navi: Self-deployable indoor navigation system.* Paper presented at the 20th annual international conference on Mobile computing and networking.

Zheng, S., & Zheng, J. (2014). Assessing the completeness and positional accuracy of OpenStreetMap in China. In T. Bandrova, M. Konecny, & S. Zlatanova (Eds.), *Thematic Cartography for the Society* (pp. 171–189). Springer International Publishing. doi:10.1007/978-3-319-08180-9_14

Zheng, Y. (2011). Location-based social networks: Users. In Y. Zheng & X. Zhou (Eds.), *Computing with Spatial Trajectories* (pp. 243–276). New York: Springer New York. doi:10.1007/978-1-4614-1629-6_8

Zhou, P., Huang, W., & Jang, J. (2014). Validation analysis of OpenStreetMap data in some areas of China. In J. Jiang & H. Zhang (Eds.), *ISPRS Technical Commission IV Symposium 2014* (4th ed.; *Vol. 40*, pp. 383–391). International Society for Photogrammetry and Remote Sensing. doi:10.5194/isprsarchives-XL-4-383-2014

Zielstra, D., & Zipf, A. (2010). A comparative study of proprietary geodata and volunteered geographic information for germany. *13th AGILE international conference on geographic information science.*

Zielstra, D., & Zipf, A. (2010). *A Comparative Study of Proprietary Geodata and Volunteered Geographic Information for Germany.* Paper presented at the 13th AGILE International Conference on Geographic Information Science, Guimarães, Portugal.

Zielstra, D., & Zipf, A. (2010). *A comparative study of proprietary geodata and volunteered geographic information for Germany.* Presented at the 13th AGILE International Conference on Geographic Information Science 2010, Guimarães, Portugal. Retrieved from http://www.agile-online.org/Conference_Paper/CDs/agile_2010/ShortPapers_PDF/142_DOC.pdf

Zielstra, D., Hochmair, H. H., & Neis, P. (2013). Assessing the effect of data imports on the completeness of openstreetmap - A United States case study. *Transactions in GIS*, *17*(3), 315–334. doi:10.1111/tgis.12037

Zielstra, D., & Zipf, A. (2010). A comparative study of proprietary geodata and volunteered geographic information for Germany. *13th AGILE International Conference on Geographic Information Science 2010 Guimarães, Portugal, 1*, 1–15.

Zook, M., Graham, M., Shelton, T., & Gorman, S. (2010). Volunteered Geographic Information and Crowdsourcing Disaster Relief: A Case Study of the Haitian Earthquake. *World Medical & Health Policy, 2*(2), 6–32. https://doi.org/10.2202/1948-4682.1069

Zook, M., Graham, M., Shelton, T., & Gorman, S. (2010). Volunteered geographic information and crowdsourcing disaster relief: A case study of the Haitian earthquake. *World Medical & Health Policy*, *2*(2), 7–33. doi:10.2202/1948-4682.1069

About the Contributors

Claudio Campelo is a Lecturer in the Systems and Computing Department at the Federal University of Campina Grande (UFCG/Brazil). He obtained his Ph.D. from the School of Computing at the University of Leeds (United Kingdom) and holds M.Sc. and B.Sc. degrees from the Systems and Computing Department at UFCG/Brazil. He is currently the department head of the undergraduate program in computer science. His current research interests include topics in data modelling, knowledge representation, information retrieval, handling of conceptual vagueness, and user-aware/context-aware systems. He is particularly interested in spatial and spatio-temporal information: modelling geospatial semantics; event/process modelling; trajectory modelling and prediction; Geographic Information Retrieval (GIR); and Volunteered Geographic Information (VGI).

Michela Bertolotto is an Associate Professor at the School of Computer Science and a member of the Earth Institute at University College Dublin.

Padraig Corcoran is a lecturer in the School of Computer Science and Informatics at Cardiff University. He holds a BSc and PhD in Computer Science from Maynooth University. Prior to joining Cardiff University he completed a European Marie Curie International Outgoing Fellowship where the outgoing phase of this fellowship was spent as a visiting researcher at Massachusetts Institute of Technology.

* * *

Munir Ahmad is a Ph.D. Scholar in Computer Sciences with particular interests in Spatial Data Infrastructure (SDI), Volunteered Geographic Information (VGI) and Spatial Data Quality Assessment and Enhancement. He is working in Survey of Pakistan since 1999 and contributing his services in the discipline of Geospatial Data Production and Management using proprietary as well as open source tools. In 2004 he got his Master Degree in Computer Sciences (MCS). He earned PGD in Land Survey in 2007 from Survey Training Institute, Islamabad-Pakistan and in 2013 he got Master of Science Degree in Computer Sciences (MSCS) from AIOU, Islamabad-Pakistan. Presently he is working on geospatial metadata profiling and standards in the context of NSDI for Pakistan. He is adding geospatial knowledge as practitioner, educator, developer, and researcher. For more info see https://pk.linkedin.com/in/munirahmad-spatial.

Asmat Ali is a Ph.D. student at PMAS-Arid Agriculture University Rawalpindi with particular interest in Spatial Data Infrastructure (SDI) and Policy Development. He is working in Survey of Pakistan as

Deputy Director. Presently, he is working on SDI-Pakistan. He holds master's degree in Geo-information Management in addition to professional master's degree in Geoinformatics from Faculty of Geo-Information Science and Earth Observation (ITC) University of Twente, The Netherlands. He has PGD in Land Survey from Survey Training Institute Survey of Pakistan, Islamabad, and diploma in Cadastre from National Land Survey of Sweden. He has been identified by SDI-Asia and Pacific as focal point for SDI in Pakistan. Over 20 research papers and articles of him on SDI, GIS, Remote Sensing, E-governance, as well as Land Administration have been published in various international conferences and renowned journals. He is contributing towards geospatial knowledge and technologies as practitioner, educator, researcher and author. For more info see https://pk.linkedin.com/in/asmatali.

Rahim Ali Abbaspour received the B.Sc. degree in surveying engineering from the University of Tehran, Tehran, Iran, in 2000, and the M.Sc. and Ph.D. degrees in GIS from the University of Tehran, Tehran, Iran, in 2002 and 2009, respectively. In 2010, he joined the School of Surveying and Geospatial Engineering, as an assistant professor. His current research interests include geospatial information science, artificial intelligence, and operational research and spatial optimization, Spatial Decision Support System (SDSS), spatio-temporal data mining and knowledge discovery, and volunteered Geographic Information (VGI). Dr. Ali Abbaspour is a Life Member of the Iran Construction Engineering Organization (IRCEO); the National Academy of Science, Iran; and Iran Society of Engineering Education (ISEE). He was the Scientific Chair of the 1st ISPRS International Conference on Geospatial Information Research (GIResearch 2014) held in Tehran.

Sidgley Camargo de Andrade received his Bachelor's in Computer Science from the College of Applied Sciences of Cascavel (FACIAP), Brazil, in 2006, and the Master's in Computer Science from the State University of Maringá (UEM), Brazil, in 2012. Currently, he is PhD student in Computer Science at the Institute of Mathematical and Computing Sciences (ICMC) of the University of São Paulo (USP). Since 2013, he has been a lecturer of Computer Science at the Federal Technological University of Paraná (UTFPR). Research interests include information fusion, geographic information system, and collaborative information system.

Vyron Antoniou is an Army Officer serving in the Hellenic Military Geographical Service. He has studied Surveying and Rural Engineering at National Technical University (NTUA) of Athens, Greece. He holds an MSc in Geoinformation from NTUA and a PhD from UCL, UK in Geomatics. His research focused on the analysis of VGI and its challenges for mapping agencies. His research interests are in GIS, spatial databases, spatial analysis, Web mapping applications, VGI, spatial data quality, OGC specifications, ISO standards, XML technologies and Web 2.0 technologies. He is Lecturer at the Hellenic Military Academy and a guest Lecturer at the National Technical University of Athens.

Maria Brovelli, degree with honors in Physics, PhD in Geodesy. She is Professor of GIS at Politecnico di Milano. From 2001 to 2011 she was the Head of the Geomatics Laboratory of the Politecnico di Milano and from 2006 to 2011 she lectured also GIS at ETH of Zurich. Since 2011 she is Vice Rector for the Como Campus of Politecnico di Milano. She is co-chair of ISPRS WG IV/5 "Web and Cloud Based Geospatial Services and Applications", Associate Editor of Applied Geomatics Journal (Springer); Member of the Scientific Committee of the Italian Photogrammetric and Topography Society (SIFET); Charter Member of OSGeo; Member of the Advisory Board of the ICA-OSGeo Labs Network. She organized (in Como) the first WebMGS Conference in 2010 and the FOSS4G Europe in 2015. In 2015, she was awarded the Sol Katz Prize by OSGeo.

Lívia Castro Degrossi is a PhD student in Computer Science at University of São Paulo, Brazil. She carried out a research internship at Heidelberg University, Germany, in 2016. She received a Master's Degree in Computer Science by University of São Paulo, Brazil, in 2014 and Bachelor's Degree in Computer Science by São Paulo State University, Brazil, in 2011. Research interests include collaborative information systems, volunteered geographic information and information quality.

Jon Corbett is an Associate Professor at UBC Okanagan, the director of the SpICE Lab (Spatial Information for Community Engagement) and participatory cartographer. All aspects of his research include a core community element; this means that the research is of tangible benefit for the communities with whom he works and that those communities feel a strong sense of ownership over the research process.

Clodoveu Davis Jr. is a civil engineer from Universidade Federal de Minas Gerais (1985), Master's in Computer Science from Universidade Federal de Minas Gerais (1992) and PhD in Computer Science from Universidade Federal de Minas Gerais (2000). He is a professor at the Computer Science Department of UFMG, teaching graduate and undergraduate courses. Main research interest subjects: spatial databases, spatial data infrastructures, geographic information systems (GIS), geographic data modeling, geocoding and urban GIS applications.

João Porto de Albuquerque is an Associate Professor at the Centre for Interdisciplinary Methodologies in the University of Warwick, Coventry, United Kingdom. He received his PhD in Computer Science from the State University of Campinas, Campinas, Brazil in 2006. He is also a core member of the Warwick Institute for the Science of Cities and an associate member of the Human-Centred Computing Group of the Department of Computer Science. His research centres on the study of collaborative and geospatial relationships in sociotechnical information systems and his current research combines quantitative, qualitative and spatial methods to investigate relationships between collaboratively-produced information (e.g. volunteered ge¬ographic information from social media, collaborative maps, and citizen sensing) and other official data sources (e.g. environmental sensors and socioeconomic data). He is particularly interested in developing innovative methods to improve our understanding about urban contexts and supporting decision-making with a view to make cities more resilient against disasters and extreme events triggered by natural hazards.

Cidália C. Fonte has a PhD in Geomatic Engineering and is an Assistant Professor at the Department of Mathematics - Faculty of Sciences and Technology - University of Coimbra, Portugal, and a researcher and member of the board of directors of the Institute for Systems Engineering and Computers at Coimbra. Her main research interests are quality assessment of geographic information and uncertainty modelling, with applications in the areas of remote sensing, GIS and collection and use of Volunteered Geographic Information. She currently chairs Working Group 4 of the EU COST Action TD1202 "Mapping and the Citizen Sensor".

Flávio Eduardo Aoki Horita is a PhD student in the Institute of Mathematics and Computing Sciences (ICMC) in the University of São Paulo (USP), São Carlos, Brazil. He received his MSc in

Computer Science from State University of Londrina, Londrina, Brazil in 2013. He spent one year as a visiting PhD student at European Research Center for Information Systems (ERCIS) in the University of Münster, Germany. His research project relates to the area of decision-making in disaster management. Particularly, he focuses on the modeling of decision-makers' tasks and decisions, the development of spatial decision support systems, and the integration of heterogeneous data sources (static and mobile sensors data and volunteered geographic information).

Musfira Jilani is a researcher at the Insight Centre for Data Analytics at National University of Ireland Galway.

Werner Leyh is a Postdoctoral Researcher in the Institute of Mathematics and Computing Sciences (ICMC) in the University of São Paulo (USP), São Carlos, Brazil. He received his PhD in Robotics at the Technische Universitat München (TUM). He conducted a large number of consultancies for UN agencies (UNESCO, UNDP, WHO, OPAS) and the Brazilian Federal Government in the field of spatial data infrastructure, public health management and environmental monitoring. He also participated in various R&D- projects funded by the European Union.

Tiago França Melo de Lima is an Assistant Professor at the Department of Computer and Systems (DECSI) of the Federal University of Ouro Preto (UFOP/Brazil), where he heads the Laboratory of Engineering and Development of Systems (www.leds.ufop.br). He is a Ph.D student in Computer Science (UFMG/Brazil), and holds a M.Sc. degree in Analysis and Modeling of Environmental Systems (UFMG) and a B.Sc. in Computer Science (UFOP).

Jacek Malczewski is Professor of Geography at Western University, London, Ontario, Canada. He holds degrees in geography from Jagiellonian University, Krakow, Poland (M.Sc.) and from the Institute of Geography and Spatial Organization, Polish Academy of Sciences, Warsaw (Ph.D.). His research interests include geographic information sciences, multi-criteria decision analysis, and spatial decision support. He authored or coauthored over hundred refereed papers, book chapters, and books. His research has been funded by the GEOIDE (Geomatics for Informed Decisions) Network of the Canadian Network of Centres of Excellence, the Natural Science and Engineering Research Council of Canada, and the Social Science and Humanities Research Council of Canada.

Gavin McArdle is a lecturer in the UCD School of Computer Science. Dr McArdle received a PhD from University College Dublin in 2008. Dr McArdle is also a Research Associate with the Programmable City Project at Maynooth University. His research focus includes location-based services, user profiling, geovisual analysis, smart city technology and urban dynamics. Gavin has an extensive number of peer-reviewed publications and received several grants from National and European Funding agencies to support collaboration with researchers in academia and industry.

Marco Minghini obtained a BSc degree with honors in Environmental Engineering at Politecnico di Milano (2008) with a thesis on the ortho-rectification of satellite images using LiDAR DTM. In 2010 he obtained an MSc degree with honors in Environmental and Geomatic Engineering at Politecnico di Milano with a thesis on the georeferencing and WebGIS publishing of historical maps. In 2014 he obtained his PhD with merit in Environmental and Infrastructures Engineering with a thesis on the use of open source software, including mobile data collection and virtual globes for VGI and citizen science applications. Since 2014, he is a temporary research fellow at the GEOlab of the Como Campus of Politecnico di Milano. Driven by the use of open source software, his current research focuses on geoWeb services and geoWeb applications, citizen science and VGI. He is an active member of COST Action TD1202 "Mapping and the Citizen Sensor", a Charter Member of OSGeo, an OpenStreetMap contributor and a proud volunteer of the Humanitarian OpenStreetMap Team (HOT).

Simin S. Mirvahabi received the B.Sc. degree in surveying engineering from the University of Tehran, Tehran, Iran, in 2013, and the M.Sc. degrees in GIS from the University of Tehran, Tehran, Iran, in 2016. She is now the PhD student at University of Tehran. His current research interests include geospatial information science, volunteered Geographic Information (VGI), and 3D Modeling.

Margaret N. Ndungo works at Technical University of Kenya, located in Nairobi, Kenya. Areas of specialization: Hydrology and GIS.

Joseph M. Ndung'u works at the Department of Geospatial and Space Technology, University of Nairobi, Kenya. Areas of specialization: Geospatial Science and Remote Sensing.

Titus M. Ng'ang'a works at the Department of Geospatial and Space Technology, University of Nairobi, Kenya. Areas of specialization: GIS and Remote Sensing.

Jiří Pánek holds B.Sc. and M.Sc. in Geography and GIS from the Department of Geoinformatics and PhD from the Department of Development Studies, both Palacky University in Olomouc, Czech Republic. His research is focused on GIS in development cooperation and humanitarian aid, with main focus on Participatory GIS (PGIS/ PPGIS). He has experienced mapping in Kenya and South Africa, studied GIS in India and currently he is developing participatory platform for collection perceptions about urban environment.

Joaquim Patriarca has a Master Degree in Geographic Information Technologies and is a NonPhD Researcher at the Institute for Systems Engineering and Computers at Coimbra, Portugal. His main research interests are: construction and implementation of algorithms to solve spatial problems; GIS software testing; technology adoption at organizations; network analysis; risk assessment.

Camilo Restrepo-Estrada is a PhD student in Hydraulic and Sanitation Engineering at the School of Engineering of São Carlos of the University of São Paulo, São Carlos, Brazil and Visiting Professor at the University of Antioquia, Medellín, Colombia. He received his MSc in Environmental Engineering from the University of Antioquia, Medellín, Colombia in 2011.

Roberto Rocha is a PhD student at the Institute of Mathematical and Computing Sciences of the University of São Paulo, Brazil. He received his MSc in Information Systems from University of São Paulo, Brazil in 2012. He is fellow of the Marie Skłodowska-Curie Actions and spent a research stay at Department for Information Systems and Supply Chain Management of the University of Münster, Germany in 2016. His current research focusses on Collaborative Systems and Volunteered Geographic Information.

Linda See is a Research Scholar in the Ecosystems Services and Management Program at the International Institute for Applied Systems Analysis in Laxenburg Austria. She was awarded a PhD in spatial applications of fuzzy logic from University of Leeds in 1999, where she was a Senior Lecturer in Computational Geography and GIS for more than a decade. She spent one year at the Max Planck Institute for Atmospheric Sciences in Germany followed by four years at the Food and Agriculture Organization in Rome, working in agrometeorology and early warning for food security. She is an active member of the Geo-Wiki team and has worked on a number of crowdsourcing aspects including quality issues, community building, and a branch of Geo-Wiki concerned with validation of urban land cover.

Rodrigo Smarzaro received his B.Sc. in computer Science from Federal University of Viçosa (1999), M.Sc. in Computer Science from Federal University of Minas Gerais (2002) and is a PhD student in Computer Science at Federal University of Minas Gerais. He works as professor at Federal University of Viçosa with main research interests on spatial databases and urban computing.

Ali Tahir received his BSc (Hons) degree in Computing and Information Systems in 2004 from Oxford Brookes University and subsequently a Masters degree in Geographical Information Science (GIS) from University of Nottingham in 2006. Ali completed his PhD in Geoinformatics from the School of Computer Science and Informatics at University College Dublin, Ireland. Recently, he has completed his post-doctorate at the Geoenvironmental Research Centre, School of Engineering, Cardiff University. Currently he is an Assistant Professor at IGIS-NUST.

Peter M. Wachira works at the School of Biological Science, University of Nairobi, Kenya. Area of specialization: Biodiversity.

Tim J. L Wango works at the Department of Geomatic Engineering and Geospatial Information Systems, Jomo Kenyatta University of Agriculture and Technology, Kenya. Area of specialization: Conservation GIS.

Di Yang is a Ph.D. student in Geography department of University of Florida, USA. She received the Master in Environmental Engineering from Texas A &M University – Kingsville (2013). The general theme of her work is to gain a quantitative understanding of the relationships between humans, landscape patterns and ecological processes. Current projects include mapping forest management unit, road ecology by applying Volunteered Geographic Information (VGI) at the macrosystems scale.

Ahmet Yildiz completed his MSc in Computer Science in 2004 at the Technical University Vienna. He worked in an EU-Project about Road Safety and completed his PhD at the Technical University Vienna about using computer technologies for road safety research.

Hongyu Zhang a master's student in geography at Western University, London, Ontario, Canada. He received his bachelor's degree in Geomatics from the University of Waterloo in 2015. His research interests include geographic information science and user-generated geospatial content.

Index

T

U

V

W

Y

Premier Reference Source
Solutions for High-Touch Communications in a High-Tech World

Premier Reference Source
Advanced Research on Biologically Inspired Cognitive Architectures

Premier Reference Source
Workforce Development Theory and Practice in the Mental Health Sector

Premier Reference Source
Resource Management and Efficiency in Cloud Computing Environments

www.ingramcontent.com/pod-product-compliance
Lightning Source LLC
LaVergne TN
LVHW060411300926
849162LV00006B/1396